# Genetic Structure and Local Adaptation in Natural Insect Populations

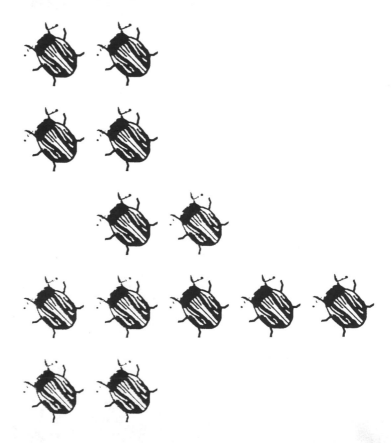

# Join Us on the Internet

WWW:      http://www.thomson.com
EMAIL:    findit@kiosk.thomson.com

*thomson.com* is the on-line portal for the products, services and resources available from International Thomson Publishing (ITP).

This Internet kiosk gives users immediate access to more than 34 ITP publishers and over 20,000 products. Through *thomson.com* Internet users can search catalogs, examine subject-specific resource centers and subscribe to electronic discussion lists. You can purchase ITP products from your local bookseller, or directly through *thomson.com*.

---

A service of I(T)P®

# Genetic Structure and Local Adaptation in Natural Insect Populations

## Effects of Ecology, Life History, and Behavior

Edited by

**Susan Mopper**

Department of Biology
University of Southwestern Louisiana
Lafayette, LA

**Sharon Y. Strauss**

Section of Evolution and Ecology
University of California, Davis
Davis, CA

 Springer-Science+Business Media, B.V.

Cover design: Susan Mopper, Karl Hasenstein, Curtis Tow Graphics

Copyright © 1998 by Springer Science+Business Media Dordrecht
Originally published by Chapman & Hall in 1998

1 2 3 4 5 6 7 8 9 10 XXX 01 00 99 98

**Library of Congress Cataloging-in-Publication Data**

Genetic structure and local adaptation in natural insect populations :
  effects of ecology, life history, and behavior / [compiled by] Susan
  Mopper and Sharon Y. Strauss.
      p.   cm.
    Includes bibliographical references and index.
    ISBN 978-1-4757-0904-9     ISBN 978-1-4757-0902-5 (eBook)
    DOI 10.1007/978-1-4757-0902-5
    1. Phytophagous insects--Genetics.  2. Insect populations.
  3. Phytophagous insects--Adaptation.  I. Mopper, Susan.
  II. Strauss, Sharon Y.
  QL493.G45   1997
  595.7'135--DC21                                     97-7322
                                                      CIP

**British Library Cataloguing in Publication Data available**

To order this or any other  Springer-Science+Business Media, B.V.
book, please contact **International Thomson Publishing,**
**7625 Empire Drive, Florence, KY 41042.**  Phone: (606) 525-6600 or 1-800-842-3636.
Fax: (606) 525-7778. e-mail: order@chaphall.com.

# Contents

# Preface

This volume marks the first compilation of studies investigating genetic structure in natural phytophagous insect populations, with a special focus on local adaptation. For some time, insect populations were considered large, panmictic assemblages, but recent experimental studies and genetic analyses indicate that many are subdivided into semi-isolated demes. Genetic differentiation within populations can reflect local adaptation at small spatial scales, and may ultimately influence rates of speciation, and community biodiversity. For example, generalist insect species are typically not globally polyphagous, but rather a collection of local subpopulations that exploit a limited number of host plant species. Although host-plant use may be extremely broad across the entire species distribution, at specific locations 'host races' may be locally adapted to a reduced subset of plant resources. Similarly, at smaller spatial scales insect biotypes may specialize on particular host-plant genotypes. Such population subdivision may ultimately promote sympatric speciation through local isolation of gene pools. The focus of this volume is to understand the underlying mechanisms and evolutionary implications of population genetic heterogeneity from intrademic to host-race spatial scales. The chapters address how genetic traits, life-history patterns, and natural selection produce adaptive and stochastic genetic structure.

The contributed chapters are a diverse yet interrelated assemblage of experimental and theoretical approaches to evolutionary ecology. Most chapters review the relevant literature and also include original work not previously published. Part I presents tests of local adaptation in natural insect populations and discusses some of the strengths and weaknesses of the experimental methods employed. Part II addresses the mechanisms that produce adaptive genetic structure ranging, from specific genetic traits to ecological agents of selection. Part III describes how behavioral and life-history patterns produce, adaptive and nonadaptive genetic structure within and between populations. Part IV combines empirical and theoretical methods to investigate the mechanisms producing genetic structure and adaptation along a continuum from metapopulations, to host races, to species. Insects are ideal models for understanding the central principles of evolutionary ecology. Many natural populations are structured by a shifting balance between adaptive and stochastic forces. Our goal is to integrate this dynamic evolutionary ecology into a larger theoretical framework that reveals how complex evolutionary processes operate at larger spatial and temporal scales.

# Acknowledgments

First and foremost, we thank the authors whose work has contributed so much to a growing and dynamic field. This volume would have been impossible without the efforts of Keli Landau, Karl Hasenstein, Peter Van Zandt, and Lance Gorham, who labored long but not thankless hours. S.Y.S. thanks Mark Schwartz for his patience in letting her work late nights on many occasions. We are indebted to the following reviewers for their comments on the chapter manuscripts: Robert Bourgeois, Donna Devlin, Anita Evans, Sarah Faragher, Caitlin Gabor, Cary Guffey, Charles Fox, John Hatle, Robert Jaeger, Keli Landau, Paul Leberg, Allen Moore, Thomas McGinnis, Timothy Mousseau, H. B. Shaffer, Julie Smith, Christine Spencer, Art Weis, and the contributing authors who each reviewed several chapters. Special thanks to Jenny Thibodeaux for her help throughout the production of this book. S.M.'s contribution was supported by a 1996 University of Southwestern Louisiana (USL) Summer Research Fellowship, the Lafayette Parish Medical Society Endowed Professorship (1993–1999), and the following research grants: NSF BSR90-07144 (1990–1993), EPSCoR NSF/LEQSF-ADP-02 (1992-1996), LEQSF-RD-A-37 (1994–1996), and NSF (1996–1999) DEB-9632302. S.Y.S. was supported by NSF HRD-91-03471.

# Contributors

Donald N. Alstad
Department of Ecology, Evolution
and Behavior
100 Ecology Building
University of Minnesota
St. Paul, MN 55108 USA

May R. Berenbaum
Department of Entomology
320 Morrill Hall
University of Illinois
Urbana, IL 61801 USA

Stewart H. Berlocher
Department of Entomology
320 Morrill Hall
University of Illinois
Urbana, IL 61801 USA

William J. Boecklen
Department of Biology
Box 30001/Dept. 3AF
New Mexico State University
Las Cruces, NM 88003 USA

Neil S. Cobb
Department of Biological Sciences
Northern Arizona University
Flagstaff, AZ 86011 USA

James T. Costa
Museum of Comparative Zoology
Harvard University
Cambridge, MA 02138 USA

Current address:
Department of Biology
Western Carolina University
Cullowhee, NC 28723

Timothy P. Craig
Department of Life Sciences
P. O. Box 37100
Arizona State University West
Phoenix, AZ 85069 USA

Robert F. Denno
Deptartment of Entomology
University of Maryland
College Park, MD 20742 USA

Dieter Ebert
Zoologishes Institut
Universitat Basel
Rheinsprung 9
CH-4051 Basel Switzerland

Jeffrey L. Feder
Department of Biological Sciences
University of Notre Dame
Notre Dame, IN 46556 USA

Sylvain Gandon
Laboratoire d'Ecologie
CNRS-URA 258
Universite Pierre et Marie Curie
7eme etage, 7, quai Saint Bernard
case 237, F-75252
Paris Cedex 05 France

Peter W. Goff
Department of Biology
Vanderbilt University
Nashville, TN 37235 USA

Lawrence M. Hanks
Deptartment of Entomology
320 Morrill Hall
University of Illinois
Urbana, IL 61801 USA

John D. Horner
Department of Biology
P.O. Box 32916
Texas Christian University
Ft. Worth, TX 76129 USA

Joanne K. Itami
Department of Life Sciences
P. O. Box 37100
Arizona State University West
Phoenix, AZ 85069 USA.

Richard Karban
Department of Entomology
University of California
Davis, CA 95616 USA

David E. McCauley
Department of Biology
Vanderbilt University
Nashville, TN 37235 USA

Yannis Michalakis
Laboratoire d'Ecologie
CNRS-URA 258
Universite Pierre et Marie Curie
7eme etage, 7, quai Saint Bernard
case 237, F-75252
Paris Cedex 05 France

Susan Mopper
Department of Biology
University of Southwestern Louisiana
Lafayette, LA 70504 USA

Isabelle Olivieri
Lab Genetique et Environment
Insitut des Sciences de l'Evolution
Universite Montpellier II
CNRS-UMR 5554
34095 Montpellier CEDEX 05 France

Susan B. Opp
Department of Biology
California State University
    at Hayward
Hayward, CA 94542 USA

Merrill A. Peterson
Department of Entomology
University of Maryland
College Park, MD 20742 USA

Anthony M. Rossi
Department of Biology
University of South Florida
Tampa, FL 33620 USA

MaryCarol Rossiter
Institute of Ecology
University of Georgia
Athens, GA 30602 USA

Michael C. Singer
Department of Zoology
University of Texas
Austin, TX 78712 USA

Peter Stiling
Department of Biology
University of South Florida
Tampa, FL 33620 USA

Sharon Y. Strauss
Section of Evolution and Ecology
2320 Storer Hall
University of California
Davis, CA 95616 USA

Chris D. Thomas
Department of Biology
University of Leeds
Leeds, LS2 9JT UK

Thomas G. Whitham
Department of Biological Sciences
Northern Arizona University
Flagstaff, AZ 86011 USA

Arthur R. Zangerl
Department of Entomology
320 Morrill Hall
University of Illinois
Urbana, IL 61801 USA

# Genetic Structure and Local Adaptation in Natural Insect Populations

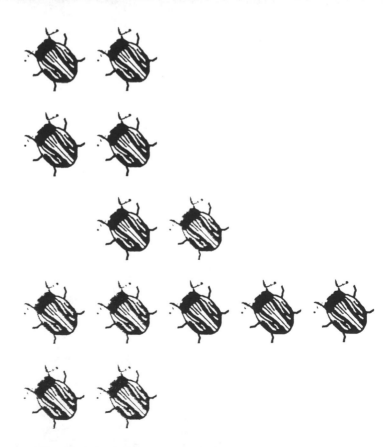

# PART I

# Local Adaptation: Empirical Evidence from Case Studies

# 1

# Population Structure and the Conundrum of Local Adaptation

*Don Alstad*
Department of Ecology, Evolution, and Behavior
University of Minnesota, St. Paul, MN

## 1.1 Introduction

The black pineleaf scale insect (*Nuculaspis californica* Coleman) is a parasite of western yellow pine (*Pinus ponderosa* Lawson) and 11 other conifer species (Ferris 1938; Furniss and Carolin 1977). The insects are short-lived relative to their host trees, largely sedentary, and achieve persistent, damaging infestations in areas where airborne dust or the drift of orchard insecticide compromises biological control agents. The abundance of black pineleaf scale varies within an infested stand and correlates with the age and size of the trees; larger, older pines harbor more scales than smaller, younger ones. In the same paper that laid out this basic biology, George Edmunds (1973, p. 765) was first to suggest that "scale populations apparently become adapted to specific host individuals, and population densities can become high only with genetic fitness of the population to the host species and individual."

I began graduate work under Edmunds's tutelage just as the 1973 paper was published, shared his interest in local adaptation, and worked sporadically with this system for nearly twenty years, both in collaboration with George, and after his retirement. In this chapter, I provide a retrospective of that work, beginning with background on the scales' natural history and basic hypotheses pertaining to population structure; summarizing three areas of empirical research, including transfer experiments, analyses of scale population biology, and allozyme studies of population genetic structure; and closing with my interpretation of these results and their implications.

## 1.2 Background

At field sites in the northwestern United States, black pineleaf scale insects complete one generation per year. Beginning in early July, after new pine needles become fully elongated, individual females lay up to 70 eggs over a three-week period. Eggs hatch quickly, and teneral larvae harden beneath the maternal scale cover before walking to feeding positions on new needles. Young scales are vulnerable to desiccation, and most settle on the same twig that bore their mothers. Nevertheless, larvae of both sexes may be passively dispersed by the wind during this "crawler" stage, and colonies are founded on previously uninfested host trees by windblown females (Edmunds 1973; Edmunds and Alstad 1981; Alstad and Edmunds 1983b).

After inserting their mouthparts through a stomate into mesophyll cells, larvae secrete a protective scale cover, which is firmly attached to the pine needle, and molt to a legless second instar. Females never move again after this initial larval settlement (Baranyovits 1953; Beardsley and Gonzalez 1975; Miller and Kosztarab 1979). Scales feed through fall and pass a winter diapause *in situ* before reaching the third instar in early spring. Males are sedentary until mid-May, when they metamorphose and crawl from beneath scale covers to mate with females on the same and nearby pine needles (Alstad et al. 1980). The winged males may also fly between trees, providing a second vehicle for gene flow beyond the natal host.

Like the olive scale (*Parlatoria oleae* Culvee) on California citrus crops (Murdoch et al. 1984) and the pine needle scale (*Chionaspis pinifoliae* Fitch) that infest numerous conifer species (Luck 1973; Luck and Dahlsten 1974, 1975), the black pineleaf scale has important natural enemies, and damaging scale outbreaks occur where these biological control agents are compromised or absent. *Coccinelid* predation affects the dynamics of *Nuculaspis*, but the principal determinant of its abundance is the *aphelinid* parasitoid *Coccobius varicornis* (Howard) (J. Wooley pers. comm.; Hayat 1983). Our major study sites are all in areas affected by dust from unpaved roads, or insecticide drift from orchards, both of which reduce the wasps' effectiveness and allow variations in scale insect density.

## 1.3 Hypotheses

Tree-to-tree differentiation and genetic structure in a scale insect population could have several potential causes. Natural selection might increase the local frequency of insect traits adapted to individual trees (a local adaptation hypothesis). Local adaptation hypotheses are further divisible according to the nature of the host-plant characteristics that mediate the selective process. At one extreme, insects might become adapted to intrinsic host traits that are under strong genotypic control. Alternatively, if environmental influences are relatively constant from generation to generation, insects might adapt to extrinsic features of the phenotype that may be entirely of environmental origin, or the product of interactions between the tree's genotype and environmental circumstances; for example, scales might become adapted to trees that are in especially good or bad condition with respect to nutrient availability, water relations, or the stress imposed by pathogens or herbivores such as scale insects themselves. These mechanistic alternatives distinguish two local adaptation hypotheses that I will call the intrinsic and extrinsic Local Adaptation Hypotheses. The essential difference between them is the spatial distribution of plant traits that cause the adaptive response. The intrinsic hypothesis assumes that all foliage with the same genotype (i.e., most of the trees with the apical dominance of conifers) provides a homogeneous selection regime, whereas the extrinsic hypothesis does not.

As an alternative to both local adaptation hypotheses, it is also possible that the sedentary habit of scale insects might allow variations to arise by chance (a drift hypothesis). This implies that the limited gene flow resulting from movements of larvae and winged males is insufficient to homogenize spatial variations in allelic frequency that arise as sampling effects. The limited number of insects on individual branches, their extreme polygyny (see below), and their haplodiploid inheritance will all foster sampling effects; thus, the basic natural history of black pineleaf scale suggests that drift may be important.

## 1.4 Transfer Experiments

We began empirical work on the pine–scale interaction with an adaptationist perspective. We were especially drawn to the intrinsic hypothesis, because trees that produce recombinant progeny might be less likely to pass preadapted pests to their offspring, and parasites such as the black pineleaf scale might therefore drive the obligate sexuality of pines (Williams 1975; Maynard Smith 1978; Linhart et al. 1979, 1981; Hamilton 1982; Rice 1983; Herre 1985; Lively 1987; Seger and Hamilton 1988; Michod and Levin 1988; Frank 1993). In this context, the logical first step was a series of transfer experiments, moving scale insects within and between individual host trees and quantifying their survival.

Our first such experiment involved the transfer of insects from 10 infested trees near Spokane, Washington, to 10 uninfested trees, with three replicates in each of 100 combinations. The resulting analysis of variance (ANOVA) showed significant differences in insect survivorship attributable to (1) the donor trees that served as sources of transferred insects, (2) the receptor trees (to which insects were moved), and (3) donor–receptor interaction. We interpreted these effects as evidence of local adaptation to intrinsic traits of individual pines (Edmunds and Alstad 1978). This experiment and its weaknesses became fairly well known (cf. Unruh and Luck 1987; Hairston 1989). In particular, we were concerned about moving insects that had caused persistent damage in the Spokane area, and hence chose receptor trees 30 km away in Deer Park, where we knew that weather would gradually snuff out the aftermath of our experiment. A more restricted test of the hypothesis that insects are adapted to individual host trees is a reciprocal transfer, where the same trees serve both as the source and destination of manipulated insects, eliminating locality as a source of performance variation.

Following the work presented in the 1978 paper, we developed techniques for removing resident scales from individual branches intended to receive experimental transfers, improved our procedure for manipulating the samples, strengthened

*Table 1.1* Average Survival of Scale Insects among Replicate Transfers in 25 Pairwise Combinations.

| DONOR | RECEPTOR | | | | |
|---|---|---|---|---|---|
| | 615 | 634 | 651 | 652 | 653 |
| 615 | **0.506** | 0.329 | 0.548 | 0.304 | 0.856 |
| 634 | 0.637 | **0.742** | 0.493 | 0.568 | 0.624 |
| 651 | 0.444 | 0.845 | **0.290** | 0.455 | 0.269 |
| 652 | 0.532 | 0.642 | 0.266 | **0.416** | 0.301 |
| 653 | 0.374 | 0.677 | 0.333 | 0.424 | **0.455** |

Values represent the ratio of survivors to initial colonists. Within-tree survival values are on the diagonal.

both quantitative detail and the range of response parameters, and employed these improved methods in three fully reciprocal transfer experiments at three different field sites. The example I present here is a 1988–1989 experiment performed at Dryden, Washington, within a 5-hectare area adjacent to orchards in the Wenatchee River Valley. We moved insects within and between five host trees in all combinations. Scale-infested pines were chosen with wide variation in physical condition, as evidenced by needle elongation and retention, to maximize the probability of detecting local adaptation associated with host vigor. The experiment was designed for a two-way analysis of variance, and there were four replicate transfers in each donor–receptor combination. Sixteen samples were lost to tip moths, scarabaeid grazing, and other causes. Transfer twigs were moved in mid-July of 1988, 11,400 initial colonists were counted in place on marked needles in early August, and surviving scales were harvested for analysis in May 1989. The matrix of average survival values for different donor–receptor combinations (Table 1.1) shows no indication of significant differences in the survival of insects that were moved within (on the diagonal) and between (off of the diagonal) trees.

ANOVA of the ratio of surviving insects to initial colonists (Table 1.2) demonstrated a marginally significant receptor effect (some trees were more difficult to colonize than others), but neither the donor treatment nor the donor–receptor interaction contributed significantly to survival patterns. A similar analysis of surviving sex ratios (rationale for use of the late-instar sex ratio as an experimental response parameter is given in the next section) yielded the same result with more statistical confidence; the receptor treatment contributed significant variance with $p < 0.02$. Again, neither the donor treatment nor the donor–receptor interaction was significant. In a stepwise analysis, the strongest predictor of insect survival on individual pine needles was the density of initial colonists; scales suffered higher mortality on needles that received a high-density inoculation.

The survival data in this fully reciprocal design were different from those we reported for the 1978 experiment and inconsistent with the hypothesis that scales

*Table 1.2*  ANOVA Table for the Reciprocal Transfer Experiment, Calculated on the Ratio of Survivors to Initial Colonists after Angular Transformation.

| Source | DF | SS | MS | F | P |
|---|---|---|---|---|---|
| Donor | 4 | 0.425 | 0.106 | 1.34 | 0.264 |
| Receptor | 4 | 0.775 | 0.194 | 2.45 | 0.055 |
| Donor–Receptor | 16 | 1.686 | 0.105 | 1.33 | 0.208 |
| Error | 59 | 4.666 | 0.079 | | |
| | | | | | |
| Total | 83 | 7.552 | | | |

The treatments are Donor-Tree Identity and Receptor-Tree Identity.

were adapted to intrinsic genotypic characteristics of individual trees. The fact that this experiment has gone unpublished to date reflects my struggle to make sense of the two results. In addition to genotype, there were undoubtedly many extrinsic factors that varied between the trees at Spokane and Deer Park, and any between-tree transfer design that is executed without a common-garden arrangement controlling environmental variables (including the one reported by Edmunds and Alstad 1978) will confound intrinsic and extrinsic attributes of the host plants.

## 1.5  Scale Insect Population Biology

If insects become locally adapted, then survival and reproductive success ought to index their relative adaptation. To explore this possibility, we began keeping annual records of scale density and sex ratios on individually numbered trees from study sites near Spokane and Dryden, Washington, and The Dalles, Oregon. We chose trees with a range of insect densities, presuming that this sample would reflect differing levels of local adaptation.

These census data showed wide variation in the sex ratio measured just before male eclosion (when they fly and can no longer be counted). Tree-to-tree variation in the proportion of males ranged from 1% to 6% in 1984 (Fig. 1.1) and 1% to 30% in other years (Fig. 1.2). In addition, the proportion of males on a tree was correlated with density; trees with many scale insects had a higher proportion of males (Alstad and Edmunds 1983a).

Sex ratios might vary locally through differential dispersal of the sexes, differences in primary allocation to sons and daughters, or differential survival. To test the dispersal hypothesis, we counted early-instar sex ratios just after larval settlement. These data were comparable to the primary ratios in Figure 1.1, showing little difference in the dispersal of males and females (Alstad and Edmunds 1989). None of the many optimality models of primary sex allocation predict a sex ratio lower than 25% sons in a multiple-foundress system (Hamilton 1967; Bulmer and Taylor 1980; Taylor and Bulmer 1980; Werren 1980; Wilson and Colwell

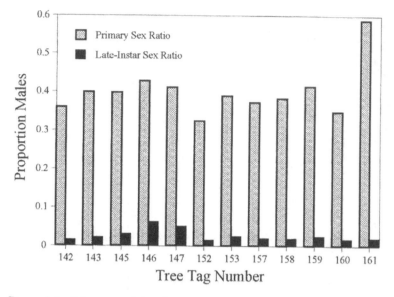

*Figure 1.1* Primary- and late-instar sex ratios. Primary ratios showed a higher male frequency than late-instar ratios from the same generation. Samples were collected at Spokane in July 1983 and May 1984. Redrawn after Alstad and Edmunds 1989.

1981; Charnov 1982; Nunney 1985a, 1985b), so we presumed that much of the variation we were observing was due to survival differences. After learning to sex first-instar larvae (by their sensillae and setal pattern; Stoetzel and Davidson 1974), we confirmed that primary sex allocation is modestly female biased, varying little from tree to tree and from year to year, and ranging among trees from 35% to 45% sons (Alstad and Edmunds 1989). Thus, most of the tree-to-tree sex-ratio variation we observed late in the insects' life cycle was attributable to differential survival of males and females in the interval between settlement and mating. Differences in mortality following treatment with the insecticide Malathion also caused the surviving sex ratio to become increasingly female biased (Edmunds and Alstad 1985). We interpret these within-generation changes in the sex ratio as evidence of selection, revealed by a method comparable to the "cohort analyses" of Endler (1986).

Scale insects in the family *Diaspididae* are haplodiploid. Development of both sexes is initiated by obligate fertilization, but the paternal chromosomes of males then become dysfunctional, so that sons are haploid and hemizygous at all loci. Daughters express both maternal and paternal components of their diploid genotype (Bennett and Brown 1958; Brown 1958, 1965; Brown and McKenzie 1962; Nur 1967, 1971; Bull 1983). Because the insects are completely sedentary in the

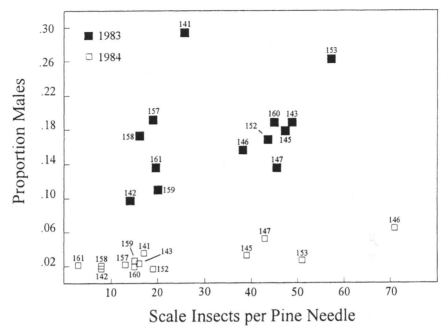

*Figure 1.2* Late-instar sex ratios with scale density during the 1983 and 1984 seasons. Numbers identify individual trees for comparison. Redrawn after Alstad and Edmunds 1989.

interval between larval settlement and mating, variations in the survival of males and females can be associated with both the local selective regime and genetic attributes of the scales (sex and ploidy). In particular, any polymorphic locus with a dominant, locally adaptive allele will express the adaptive phenotype at frequency $p$ in males and $2pq + p^2$ in females. With maximum genetic variance at $p = q = 0.5$, half again as many females as males are expected to express that adaptive character. This female advantage will fall and then disappear as selection raises the frequency of locally adaptive traits to fixation. This simple genetic model suggests that the late-instar sex ratio (which is a correlate of insect density) is also a correlate of genetic variance and local adaptation (Alstad and Edmunds 1983a, 1989). It predicts that relative male survival and the late sex ratio will rise as genetic variance falls under selection, and local adaptation of the insect deme increases.

We tested the hypothesis that late-instar sex ratios vary inversely with genetic variance by measuring late sex ratios on the adjacent and isolated sides of paired trees standing in close proximity. We chose tree pairs whose limbs touched and interlaced on one side, but were separated on the other side by at least 5 meters from the foliage of any other tree (Fig. 1.3). We reasoned that the limited movement of scale insects would cause more gene flow in the contact zone where foliage

| Tree | A | B | C | D | Tree |
|------|------|--------|------|--------|------|
| 183 | .225 | > .054 | .091 | < .112 | 184 |
| 185 | .040 | > .030 | .035 | < .108 | 186 |
| 187 | .171 | > .117 | .129 | < .190 | 188 |
| 189 | .073 | < .116 | .092 | < .113 | 190 |
| 199 | .123 | > .054 | .097 | < .118 | 200 |
| 26 | .135 | > .048 | .086 | > .074 | 27 |
| 29 | .161 | > .076 | .113 | < .146 | 38 |
| 30 | .169 | > .076 | .077 | < .126 | 31 |
| 32 | .106 | < .127 | .056 | < .112 | 33 |
| 35 | .094 | > .058 | .058 | < .069 | 34 |
| 37 | .196 | > .153 | .052 | < .068 | 36 |

*Figure 1.3* Late-instar sex ratios from near- and farsides of adjacent tree pairs. Each row represents four samples from a single pair of pines. Columns *A* and *D* show sex ratios from the farsides, whereas *B* and *C* are from the nearsides. Cases violating the prediction are highlighted. Redrawn after Alstad and Edmunds 1983b © Academic Press, Inc.

of two different trees touched than on their isolated sides. Gene flow should lead to the introgression of maladaptive alleles from the adjacent tree, increase genetic variance, and reduce the survival of haploid male scales relative to diploid females. Nineteen of 22 late-instar sex-ratio comparisons showed a greater female bias on the adjacent side relative to the isolated side of the same tree. Scale insect densities were also lower near the contact zone. We interpreted these data as evidence of an outbreeding depression in the contact zone between adjacent demes (Alstad and Edmunds 1983a, 1987). In retrospect, two possible mechanisms could produce this result. The outbreeding interpretation assumes that selection is homogeneous across the tree, and insects vary as a result of nearside introgression of maladaptive alleles. It is also possible that insects are panmictic across the tree, and selection imposed by differences in the pine foliage varies from one side to the other. There is variation across the foliage of an individual pine in carbon–nitrogen ratios, terpenoid composition and concentration, and the number and extent of resin ducts in needles (Johnson, Young, and Alstad unpublished data); it is possible that these traits are consistently distributed with respect to our nearside, farside dichotomy, biasing the sex ratio without causing differentiation to persist beyond the current generation (Jaenike 1981).

Just as scale survival and density vary from tree to tree, they also vary from year to year. The relative survival of males was correlated with total scale density on individual trees, as seen for two years of different average density in Figure 1.2. If densities and the surviving sex ratio reflect intrinsic local adaptation, one would expect succeeding annual sex ratios observed on individual trees to be autocorrelated. Although 8 of 10 trends were positive, only 1 of 10 between-year comparisons of successive sex ratios on individual trees showed a significant rank correlation (1979–1981, $n = 11$, $r_s = 0.683$, $p < 0.05$), and one more approached significance (1979–1980, $n = 11$, $r_s = 0.524$, $0.05 < p < 0.10$; Alstad and Edmunds 1989). Year-to-year reversals in the rank order of sex ratios and densities on individual trees in Figure 1.2 illustrate this point. The weakness of year-to-year density correlations suggests that selection pressures change seasonally. Tree genotypes remain constant over time, so the selection process driving these patterns of insect survival is likely to be an extrinsic, rather than an intrinsic, attribute of the host tree. These data also suggest that extrinsic factors are "noisy," causing substantial mortality (and selection), but varying over such short time periods that adaptive responses may correspond only to long-term averages.

Long-term observations also provide anecdotal information about the importance of weather in the interaction between scales and pine. There has been a damaging population density of scale insects adjacent to apple and pear orchards at Dryden since the mid-1950s. When I began work there in 1985, many trees were heavily infested, and almost all of the pines carried some scales. The summer of 1989 was drier than usual (Fig. 1.4), and precipitation that fall and winter (when most of the annual water budget accumulates) was about 25% of its 10-year average. In the summer of 1990, scale insect densities increased dramatically, and by 1991, all of the trees at that study site were dead. A similar anecdote is available from our field site at The Dalles. When I first began work there in 1979, I feared that I would soon have to abandon the site because scales would overwhelm and kill all the pines. Seven years later, scale densities had fallen to the point where I had difficulty finding sufficiently infested trees for my experiments, and almost all of the pines were in much better condition. In the ensuing

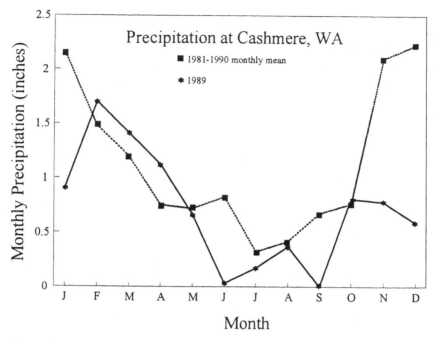

*Figure 1.4* Precipitation at a gauging station 10 km from the study site at Dryden, Washington, where a 1989 drought led to increased scale insect density and widespread pine mortality. Data provided by Ellis Darley (unpublished data).

decade, scale densities have recovered. The nearest weather records are much farther from this location, so I cannot cite rainfall data such as those in Figure 1.4, but both of these examples demonstrate that extrinsic factors have a strong influence on the fitness of scale insects and their interaction with pines.

### 1.6 Allozyme Studies of Population Genetic Structure

The drift hypothesis suggests that scale insect populations are sufficiently viscous (i.e., the insects move so little) that local differentiation can develop by chance. To test this, we chose 13 infested pines from three plots on adjacent city blocks at the Spokane study site, and collected five twig samples from each tree for a hierarchical analysis that partitioned the insects' genetic variance among plots, among trees within plots, and among twigs within individual trees within plots. All 13 trees grew within a radius of 100 meters. After investigating 30 enzyme systems, we found three with sufficient genetic variation to be useful: acid phosphatase (*Acp*, EC 3.1.3.2), 6-phosphoglucose isomerase (*Pgi*, EC 5.3.1.9), and NADP-dependent malate dehydrogenase or "malic enzyme" (*Me-1*, EC 1.1.1.40). Because the average persistence of neutral alleles is proportional to effective population size, it is not surprising that periodic founder effects, haplodiploidy, and extreme polygyny would reduce the genetic variance of black pineleaf scale insects relative to most diploid, sexual species (Crow and Kimura 1970).

We used allozyme data from 2,350 diploid female scale insects to estimate two of Wright's $F$ statistics. The fixation index, $F_{ST}$ (Wright 1951, 1965, 1978; Weir and Cockerham 1984; Weir 1990), gives a standardized genetic variance between subpopulations, normalized against the limit to that variance at the observed allelic frequencies.

$$F_{ST} = \frac{\sigma^2}{\bar{p}\bar{q}}$$

It can also be interpreted as a measure of the heterozygote deficiency associated with subdivision of a population into drifting, genetically isolated demes,

$$F_{ST} = \frac{H_T - H_S}{H_T},$$

where $H_T$ is the expected heterozygosity for the entire population, calculated as $2\bar{p}\bar{q}$ using global allelic frequency estimates, and $H_S$ is the expected heterozygosity of demic subunits calculated as a weighted average that incorporates corrections for subunit size and allelic frequency (Nei 1977, 1978). In either case, $F_{ST}$ varies inversely with interdemic gene flow, taking values from 0 to 1. At both $Acp$ and $Pgi$ loci, the $F_{ST}$ calculations show significant genetic differentiation between insects sampled from the three pine plots and from different trees within plots (Table 1.3). Wright (1951) showed that the number of migrant exchanges ($N_e m$) is inversely proportional to $F_{ST}$, so these allozyme data from black pineleaf scale indicate about 2–15 between-tree migrant exchanges per generation. In 5 of 13 cases, $F_{ST}$ estimates at the $Pgi$ locus also show significant differentiation between twigs sampled only a few meters apart on the same tree (Alstad and Corbin 1990).

$F_{IS}$ estimates the deviation of observed heterozygote numbers ($H_I$) from those expected ($H_S$) on the basis of Hardy–Weinberg equilibria (Wright 1965, 1978; with expectations adjusted for sample size and frequency biases according to the method of Nei 1977, 1978).

$$F_{IS} = \frac{H_S - H_I}{H_S}$$

Positive $F_{IS}$ values indicate a heterozygote deficiency relative to the Hardy–Weinberg expectation, and negative values an excess. In a hierarchical analysis of population genetic structure, $F_{IS}$ is conceptually equivalent to an $F_{ST}$ estimate made one step lower in the spatial hierarchy. Procedurally, however, $F_{IS}$ has more statistical power, because it compares the entire data set with a formal null model (binomial expectation), whereas $F_{ST}$ evaluates the variance among subsets. Genotypic distributions at the tree level gave $F_{IS}$ values for $Acp$ and $Pgi$ that were significantly positive in 9 and 11 of the 13 cases, respectively, demonstrating

*Table 1.3*  Hierarchical $F_{ST}$ values among plots, among trees within plots, and among twigs within trees within plots.

|  | Acp | Pgi | ME-1 |
|---|---|---|---|
| $F_{ST}$ **among plots** | 0.0233*** | 0.1123*** | 0.0527*** |
| $F_{ST}$ **among trees** |  |  |  |
| **Within Plot I** | 0.0289*** | 0.0874*** | 0.0830* |
| **Within Plot II** | 0.0241** | 0.0741*** | 0.0047 ns |
| **Within Plot III** | 0.0106* | 0.1216*** | 0.0010 ns |
| $F_{ST}$ **among twigs** |  |  |  |
| **Within Plot I** |  |  |  |
| **Within Tree 142** | 0.0000 ns | 0.0000 ns |  |
| **Within Tree 143** | 0.0000 ns | 0.0000 ns |  |
| **Within Tree 145** | 0.0033 ns | 0.0647 ns |  |
| **Within Tree 146** | 0.0170 ns | 0.0652 ns |  |
| **Within Tree 147** | 0.0305 ns | 0.0086 ns |  |
| **Within Tree 191** | 0.0418 ns | 0.1311* |  |
| **Within Plot II** |  |  |  |
| **Within Tree 152** | 0.0241 ns | 0.0474 ns |  |
| **Within Tree 153** | 0.0235 ns | 0.1022** |  |
| **Within Plot II** |  |  |  |
| **Within Tree 157** | 0.0023 ns | 0.1207** |  |
| **Within Tree 158** | 0.0298 ns | 0.0149 ns |  |
| **Within Tree 159** | 0.0032 ns | 0.1269*** |  |
| **Within Tree 160** | 0.0019 ns | 0.0000 ns |  |
| **Within Tree 161** | 0.0000 ns | 0.2826*** |  |

Sample-size limitations permitted only a two-level analysis for Me-1. Significance of deviations from 0 was calculated as $X^2 = 2NF_{ST}$, where $N$ is the number of individuals, and $df$ is 1 less than the number of subpopulations sampled (Neel and Ward 1972). Data from Alstad and Corbin 1990. * = 0.05, ** = 0.01, *** = 0.001.

heterozygote deficiencies, nonrandom mating, and pervasive substructure within host trees. $F_{IS}$ estimates from individual twigs showed little statistically significant deviation from genotypic equilibrium, suggesting that random-mating, demic units typically encompassed twigs or branches within individual host trees (Alstad and Corbin 1990). To appreciate the extraordinary structure that these data imply, it is interesting to compare them with the 1987 analysis by McCauley and Eanes of the sedentary and geographically differentiated milkweed beetle *Tetraopes tetraoph-thalmus*. Standardized genetic variances and statistical confidences for the two data sets are comparable, but the regions of McCauley and Eanes comprised the states of Virginia, Tennessee, Illinois, New Hampshire, and New York, whereas ours were portions of a single hectare on the northern edge of Spokane. Black pineleaf scale insects show demic differentiation over extremely short distances.

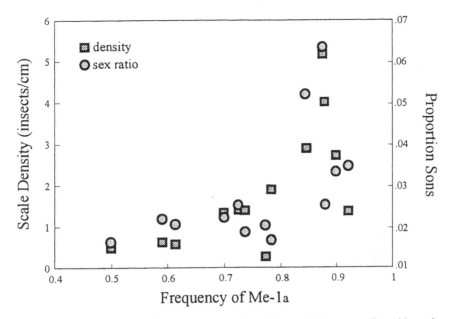

*Figure 1.5* Density and late-instar sex ratio of surviving scale insects collected in early May from 13 pines, plotted against frequencies of the most common malic enzyme allele. Density and sex-ratio estimates were made by counting insects on 30 needles per twig from five twigs sampled for the allozyme study. Redrawn after Alstad and Corbin 1990.

What about the malic enzyme locus? An analysis of population structure using Wright's *F*-statistics requires that the genetic markers be selectively neutral. To test this assumption, we compared heterozygosities and allelic frequencies for each enzyme system with two components of insect fitness, local scale density, and the surviving sex ratio. Neither allelic nor genotypic frequencies at the *Acp* and *Pgi* loci were related to these indices of scale performance, and we have no evidence that selection affected the pattern of their variation. In contrast, the density achieved by scales on different trees correlated with frequencies of the most common *Me* allele (Fig. 1.5, $n = 0\ 13$, $r = 0.652$, $p = 0.016$). There was also a significant association between the sex ratio at mating and the frequency of this same allele ($n = 13$, $r = 0.583$, $p = 0.037$; Alstad and Corbin 1990).

These correlations suggest that we have identified either a locus under selection, or, more likely, a hitchhiker linked in disequilibrium with some other selected locus. They compromise the malic enzyme system as an indicator of population structure, because selection might affect genotypic frequencies within our samples; for example, allelic frequency differences between the sexes resulting from the asymmetrical survival of males and females might inflate heterozygosities, producing Hardy–Weinberg equilibria in the $F_{IS}$ analyses despite pervasive substructuring, and causing $F_{ST}$ values for this locus to be overly conservative estimates of local differentiation (Crozier 1985; Hartl and Clark 1989). At the same time, the apparent selection on $Me$ suggests a new hypothetical mechanism of potential local adaptation. Scale densities were low on trees with the highest $Me$ variance, and increased with $Me^a$ frequencies ranging from 0.6 to 0.9, suggesting a density-dependent selection hypothesis. Figure 1.5 suggests that $Me^a$ has a competitive advantage over $Me^b$ on chlorotic needles that have been heavily damaged by scale feeding. In effect, scale insects themselves may be an important extrinsic factor, altering the foliage and the selection pressures consequently imposed on the insect herbivores.

## 1.7 Discussion

The first conclusion to be drawn from these collected data and field observations is that the neutral drift hypothesis is alive and well in black pineleaf scale insects. Two polymorphic marker loci demonstrate barriers to random mating between trees, and even between the twigs of individual trees. Black pineleaf scale populations are genetically differentiated over very short distances, and the interdemic variances suggest that only a dozen or so migrant exchanges occur between trees in each generation. Counterhypotheses based on some selection process would have to explain concordant allelic frequency variation and heterozygote deficiencies in two different marker systems while the individual twig samples stay in genotypic equilibrium; they offer a less parsimonious explanation for our data. This result is fully consistent with the viscous mating system and natural history of scale insects, which afford considerable scope for neutral evolution. Since many of the taxa in which ecologists have inferred local adaptation have sedentary habits (Mopper 1996), this may be a fairly general result.

The intrinsic local adaptation hypothesis, that insects evolve in response to genotypes of individual host trees, has become progressively less tenable as we have learned more about the pine–scale system. It is directly falsified by the reciprocal transfer results of Tables 1.1 and 1.2, which show receptor main effects but neither donor main effects nor interactions. One hypothetical explanation for this result emerges from our long-term analyses of scale population biology. Changes in density and the scale insect sex ratio within the course of a single generation demonstrate powerful selective forces; there is very heavy mortality associated nonrandomly with ploidy, and hence insect genotype. The problem with respect to intrinsic local adaptation is that these strong selective forces vary independently of plant genotype. When we first began to keep records on marked trees, we saw three consecutive years in which insect density and the surviving sex ratio increased annually on every tree, and interpreted these measures as indices of increasing local adaptation (Alstad and Edmunds 1983a); then, in the fourth year, both densities and sex ratios crashed precipitously (Alstad and Edmunds 1989). The fact that late-instar sex ratios on individual trees showed little significant correlation from year to year over a long time series suggests that unpredictable environmental changes contribute variance that may limit the scales' selection response to the constant, intrinsic characteristics of their host.

The transfer experiment reported by Edmunds and Alstad (1978) showed evidence of local adaptation, whereas the reciprocal experiment summarized in Tables 1.1 and 1.2 did not. There were extrinsic environmental variations (as well as genotypic differences) between donor and receptor trees in the 1978 experiment, but only intrinsic genotypic treatments in the later, reciprocal design. This implies that extrinsic differences between the trees at Spokane and Deer Park contributed to variation in insect survival. Allozyme data support the extrinsic local adaptation hypothesis more directly, suggesting that scales themselves can alter the selection regime. Scale feeding causes chlorotic lesions, and at moderate scale density, these lesions merge into a continuous band, affecting up to 80% of each pine needle. Under these conditions, elongation of successive annual needle crops is stunted and tree condition deteriorates (Edmunds 1973). It is not surprising that this damage might produce density-dependent changes in selection pressure. In this case, the $Me^a$ allele is at high frequency on chlorotic pines and at lower frequency on green ones, with a tree-to-tree frequency range of 40%. We do not know whether the pattern persists from generation to generation (a requirement of local adaptation; Jaenike 1981), but it seems unlikely that natural selection would cause such extreme changes in the course of a single season. These data thus suggest extrinsic local adaptation mediated by the effect of insect feeding on the tree phenotype.

In conclusion, after 20 years' field research with black pineleaf scale and ponderosa pine, the neutral drift hypothesis and the extrinsic local adaptation hypothesis remain viable. The intrinsic local adaptation hypothesis is dead. Tree genotypes do not function in isolation, and it might seem that I have killed a straw man by defining intrinsic host traits so narrowly as to exclude attributes affected by genotype–environment interaction. There are both conceptual and practical reasons for the narrowly drawn hypothesis. It is the genotype *per se* that is interesting to biologists with respect to parasites, recombination, and sex; and it is genotype *per se* that is a plant adaptation, defining the limit to broad-sense heritability of traits that influence insect performance. Finally, it is host genotype that gives us a theoretical basis for experimental designs. Transfer experiments are only instructive if we understand the pattern of selection pressures that drive local adaptation; tree genotype sets an *a priori* expectation for these spatial limits. If the genotypes of individual trees do not define the spatial distribution of the pertinent selection processes, then tree-to-tree transfer experiments yield inconsistent results. Tree-to-tree variation in insect density originally led us to an interest in local adaptation; but insect density also varies within trees (between shaded interior and sunlit peripheral branches). In retrospect, parasites, recombination, and sex lured us initially to the wrong spatial dimension, and it took awhile to straighten things out.

Can we scale up a pine to view the world? Plants exert very powerful selection pressures on insect herbivores, these pressures can be locally variable, and the panmictic neighborhood sizes of many sedentary insects are small. I continue to believe that the natural history of plant–insect interaction holds many examples of local adaptation. Experience with black pineleaf scale on ponderosa pine has shown that the spatial pattern of environmental variation sufficient to affect insect herbivores is also small, and that much local adaptation will thus be mediated by extrinsic causes. This is unfortunate for the discipline, because (1) the extrinsic hypothesis has fewer coevolutionary consequences, (2) the mechanistic diversity of extrinsic effects will confuse the theoretical bases of productive research and render the results a catalog of special cases, and (3) extrinsic variation (along with neutral evolution) will increase the noise-to-signal ratio of intrinsic local adaptation mediated by plant genotype.

## Acknowledgments

I thank G. Edmunds for introducing me to scale insects and for his long-term collaboration; S. Altizer, D. McCauley, S. Mopper, K. Oberhauser, M. Rossiter, S. Strauss, and S. Wagenius for suggesting improvements in the manuscript; J. Wooley for identification of our parasitoid; J. Davidson for the setal pattern of male and female crawlers; B. Ly and T. Pham for counting and sexing a zillion scale insects; K. Corbin for teaching me isoelectric focusing; P. Saunders for gels *ad nauseam*; K. and C. Christiansen and T. and M. Dahle for experimental use of their land at Dryden and The Dalles; E. Darley for weather records; and the National Science Foundation (DEB-7724617, DEB-8011139, BSR-8307363, BSR-9006643) for research support.

## 1.8 References

Alstad, D. N. and K. W. Corbin. 1990. Scale insect allozyme differentiation within and between host trees. *Evol. Ecol.* 4:43–56.

Alstad, D. N. and G. F. Edmunds Jr. 1983a. Selection, outbreeding depression, and the sex ratio of scale insects. *Science 220*:93–95.

Alstad, D. N. and G. F. Edmunds Jr. 1983b. Adaptation, host specificity and gene flow in the black pineleaf scale. Pp. 413–426 *in* R. F. Denno and M. S. McClure (Eds.), *Variable Plants and Herbivores in Natural and Managed Systems*. Academic Press, New York.

Alstad, D. N. and G. F. Edmunds Jr. 1987. Black pineleaf scale population density in relation to interdemic mating (Hemiptera: Diaspididae). *Ann. Entomol. Soc. Am. 80*:652–654.

Alstad, D. N. and G. F. Edmunds Jr. 1989. Haploid and diploid survival differences demonstrate selection in scale insect demes. *Evol. Ecol.* 3:253–263.

Alstad, D. N., G. F. Edmunds Jr., and S. C. Johnson. 1980. Host adaptation, sex ratio, and flight activity in male black pineleaf scale. *Ann. Entomol. Soc. Am. 73*:665–667.

Baranyovits, F. 1953. Some aspects of the biology of armoured scale insects. *Endeavour* *12*:202–209.

Beardsley, J. W. Jr. and R. H. Gonzalez. 1975. The biology and ecology of armored scales. *Annu. Rev. Entomol. 20*:47–73.

Bennett, F. D. and S. W. Brown. 1958. Life history and sex determination in the *Diaspine* scale, *Pseudaulacaspis pentagona* (Targ.) (Coccoidea). *Can. Entomol. 90*:317–324.

Brown, S. W. 1958. Haplodiploidy in the Diaspididae—confirmation of an evolutionary hypothesis. *Evolution 12*:115–116.

Brown, S. W. 1965. Chromosomal survey of the armored and palm scale insects (Coccoidea: Diaspididae and Phoenicococcidae). *Hilgardia 36*:189–294.

Brown, S. W. and H. L. McKenzie. 1962. Evolutionary patterns in the armored scale insects and their allies. *Hilgardia 33*:140–170.

Bull, J. J. 1983. *Evolution of Sex Determining Mechanisms.* Benjamin/Cummings, Menlo Park, CA.

Bulmer, M. G. and P. D. Taylor. 1980. Dispersal and the sex ratio. *Nature 284*:448–449.

Charnov, E. L. 1982. *The Theory of Sex Allocation.* Princeton University Press Monographs in Population Biology #18.

Crow, J. F. and M. Kimura. 1970. *An Introduction to Population Genetics Theory.* Harper and Row, New York.

Crozier, R. H. 1985. Adaptive consequences of male-haploidy. Pp. 201–222 *in* W. Helle and M. W. Sabelis (Eds.), *Spider Mites: Their Biology, Natural Enemies and Control* Vol. 1A. Elsevier Science Publishers, Amsterdam, The Netherlands.

Edmunds, G. F. Jr. 1973. Ecology of black pineleaf scale (Homoptera: Diaspididae). *Environ. Entomol. 2*:765–777.

Edmunds, G. F. Jr. and D. N. Alstad. 1978. Coevolution in insect herbivores and conifers. *Science 199*:941–945.

Edmunds, G. F. Jr. and D. N. Alstad. 1981. Responses of black pineleaf ccales to host plant variability. Pp. 29–38 *in* R. F. Denno and H. Dingle (Eds.), *Insect Life History Patterns.* Springer-Verlag, New York.

Edmunds, G. F. Jr. and D. N. Alstad. 1985. Malathion induced sex ratio changes in black pineleaf scale. *Ann. Entomol. Soc. Am. 70*:403–405.

Endler, J. A. 1986. *Natural Selection in the Wild.* Princeton University Press, Princeton, NJ.

Frank, S. A. 1993. Evolution of host-parasite diversity. *Evolution 47*:1721–1732.

Ferris, G. F. 1937–1955. *Atlas of the Scale Insects of North America.* Stanford University Press, Stanford, CA.

Furniss, R. L. and V. M. Carolin. 1977. *Western Forest Insects.* USDA Forest Service Michigan Publications No. 1339.

Hairston, N. G. Sr. 1989. *Ecological Experiments, Purpose, Design, and Execution.* Cambridge University Press, Cambridge, UK.

Hamilton, W. D. 1967. Extraordinary sex ratios. *Science 156*:477–488.

Hamilton, W. D. 1982. Pathogens as causes of genetic diversity in their host populations. Pp. 269–296 in R. M. Anderson and R. M. May (Eds.), *Population Biology of Infectious Diseases.* Dahlem Konferenzen, Springer-Verlag, Berlin.

Hartl, D. and A. G. Clark. 1989. *Principles of Population Genetics.* Sinauer Associates, Sunderland, MA.

Hayat, M. 1983. The genera of Aphelinidae (Hymenoptera) of the world. *Syst. Entomol.* 8:63–102.

Herre, E. A. 1985. Sex ratio adjustment in fig wasps. *Science 228*:896–898.

Jaenike, J. 1981. Criteria for ascertaining the existence of host races. *Am. Nat. 117*:830–834.

Linhart, Y. B., J. B. Mitton, K. B. Sturgeon, and M. L. Davis. 1979. An analysis of genetic architecture in populations of ponderosa pine. *Proc. Symp. on Isozymes of North American Forest Trees and Forest Insects.* USDA Forest Service, July 27, 1979. Berkeley, CA.

Linhart, Y. B., J. B. Mitton, K. B. Sturgeon, and M. L. Davis. 1981. Genetic variation in space and time in a population of ponderosa pine. *Heredity 46*:407–426.

Lively, C. M. 1987. Evidence from a New Zealand snail for the maintenance of sex by parasitism. *Nature 328*:519–521.

Luck, R. F. 1973. Natural decline of an insecticide induced outbreak of the pine needle scale, *Chionaspis pinifoliae* (Fitch) at South Lake Tahoe, California. Ph.D. dissertation, University of California, Berkeley.

Luck, R. F. and D. L. Dahlsten. 1974. Bionomics of the pine needle scale, *Chionaspis pinifoliae* and its natural enemies at South Lake Tahoe, California. *Ann. Entomol. Soc. Am.* 67:309–316.

Luck, R. F. and D. L. Dahlsten. 1975. Natural decline of a pine needle scale (*Chionaspis pinifoliae*) outbreak at South Lake Tahoe, California following cessation of adult mosquito control with malathion. *Ecology 56*:893–904.

Maynard Smith, J. 1978. *The Evolution of Sex.* Cambridge University Press, Cambridge, UK.

McCauley, D. E. and W. F. Eanes. 1987. Hierarchical population structure analysis of the milkweed beetle, *Tetraopes tetraophthalmus* (Forster). *Heredity 58*:193–201.

Michod, R. E. and B. R. Levin (Eds.). 1988. *The Evolution of Sex.* Sinauer Associates, Sunderland, MA.

Miller, D. R. and M. Kosztarab. 1979. Recent advances in the study of scale insects. *Annu. Rev. Entomol. 24*:1–27.

Mopper, S. 1996. Adaptive genetic structure in phytophagous insect populations. *Trends Ecol. Evol. 11*:235–238.

Murdoch, W. W., J. D. Reeve, C. B Huffaker, and C. E. Kennett. 1984. Biological control of olive scale and its relevance to ecological theory. *Am. Nat. 123*:371–392.

Neel, J. B. and R. H. Ward. 1972. The genetic structure of a tribal population, the *Yanomama* Indians: VI. Analysis by *F*-statistics including a comparison with the *Makiritare* and *Xavante. Genetics 72*:639–666.

Nei, M. 1977. *F*-statistic and analysis of gene diversity in subdivided populations. *Ann. Hum. Genet., London 41*:225–233.

Nei, M. 1978. Estimation of average heterozygosity and genetic distance from a small number of individuals. *Genetics 89*:583–590.

Nunney, L. 1985a. Female-biased sex ratios: Individual or group selection? *Evolution 39*:349–361.

Nunney, L. 1985b. Group selection, altruism, and structured-deme models. *Am. Nat. 126*:212–230.

Nur, U. 1967. Reversal of heterochromatization and the activity of the paternal chromosome set in the male mealy bug. *Genetics 56*:375–389.

Nur, U. 1971. Parthenogenesis in coccids. *Am. Zool. 11*:301–308.

Rice, W. R. 1983. Parent–offspring pathogen transmission: A selective agent promoting sexual reproduction. *Am. Nat. 121*:187–203.

Seger, J. and W. D. Hamilton. 1988. Parasites and sex. Pp. 176–193 *in* R. E. Michod and B. R. Levin (Eds.), *The Evolution of Sex: An Examination of Current Ideas.* Sinauer Associates, Sunderland, MA.

Stoetzel, M. B. and J. A. Davidson. 1974. Sexual dimorphism in all stages of the *Aspidiotini* (Homoptera: Diaspididae). *Ann. Entomol. Soc. Am. 67*:138–140.

Taylor, P. D. and M. G. Bulmer 1980. Local mate competition and the sex ratio. *J. Theor. Biol. 86*:409–419.

Unruh, T. R. and R. F. Luck. 1987. Deme formation in scale insects: A test with the pinyon needle scale and review of other evidence. *Ecol. Entomol. 12*:439–449.

Weir, B. S. 1990. *Genetic Data Analysis.* Sinauer Associates, Sunderland, MA.

Weir, B. and C. C. Cockerham. 1984. Estimating $F$-statistics for the analysis of population structure. *Evolution 38*:1358–1370.

Werren, J. H. 1980. Sex ratio adaptations to local mate competition in a parasitic wasp. *Science 208*:1157–1159.

Williams, G. C. 1975. Sex and evolution. *Monographs in Population Biology 8.* Princeton University Press, Princeton, NJ.

Wilson, D. S., and R. K. Colwell 1981. Evolution of sex ratio in structured demes. *Evolution 35*:882–897.

Wright, S. 1951. The genetical structure of populations. *Ann. Eugen. 15*:323–354.

Wright, S. 1965. The interpretation of population structure by $F$-statistics with special regard to systems of mating. *Evolution 19*:395–420.

Wright. S. 1978. *Evolution and the Genetics of Populations: Vol. 4. Variability within and among Natural Populations.* University of Chicago Press, Chicago, IL.

# 2

# Deme Formation in a Dispersive Gall-Forming Midge

*Peter Stiling and Anthony M. Rossi*
Department of Biology, University of South Florida, Tampa, FL

## 2.1 Introduction

Deme formation in herbivores was originally thought to result from breeding within isolated populations that are highly adapted to their host plant (Edmunds and Alstad 1978). If some herbivore genotypes have advantages over others on particular hosts, and this variation is heritable over time, then these isolated populations, or demes, should diverge from the parent population and become reproductively isolated.

It is possible that deme formation could occur over a wide range of scales. For example, within a species of polyphagous–phytophagous insect, there may exist demes for particular host-plant species (Akimoto 1990; Feder et al., Chapter 16, this volume), demes for host patches in a particular geographic area (Costa, Chapter 10, this volume), or demes for conspecific plants (such as individual trees) in the same geographic location (Komatsu and Akimoto 1995; Mopper et al. 1995). This chapter concerns deme formation at the level of conspecific plants. The evidence for deme formation at this level in nature is not extensive. Until now, only eight studies have specifically attempted to test the deme-formation hypothesis at this scale. These investigations used reciprocal transplants of insects, usually scale insects, between natal and novel plants, with subsequent surveys of insect success. Only two studies provided convincing evidence of variable insect adaptation to neighboring natal and novel host plants (Karban 1989; Mopper et al. 1995), while two others (Wainhouse and Howell 1983; Hanks and Denno 1994) provided partial evidence of deme formation at this level.

There are several theories as to which biological features of plants and herbivores promote the development of demes (see Table 2.1). First, Edmunds and Alstad (1978, 1981) suggested that deme formation may be more frequent for short-lived insects that feed on long-lived plants, where they can produce hundreds of generations on the same host individual and thus gain an advantage in the evolutionary arms race. Second, haplodiploidy, a mating system of many scale insects, may also promote the formation and maintenance of demes, because dispersing

Table 2.1  Summary of Previous Studies Concerning Deme Formation That Involved Reciprocal Transplants of Herbivores between Conspecific Plants

| Insect | Host(s) | Demes found | Distance between hosts | Distance dispersed by insect | Feeding Mode | Age of Host | Generations of insect per year | Reference |
|---|---|---|---|---|---|---|---|---|
| Nuculaspos californica | Pinus lambertiana | No | ? | 1 m? | Sucker | ? | 1 | Rice 1983 |
| Cryptococcus fagisuga | Fagus sylvatica | Yes | "Far" | 1 m? | Sucker | 35 yrs | 1 | Wainhouse and Howell 1983 (as reported in Hanks and Denno 1994) |
| Matsucoccus acalyptus | Pinus monophylla | No | much > 1 m | 1 m | Sucker | 40–100 yrs | 1 | Unruh and Luck 1987 |
| Apterothrips secticornis | Erigeron glaucus | Yes | 500 m | 1–10 m | Rasper | 10–100 yrs | 8 | Karban 1989; Karban and Strauss 1994 |
| Matsucoccus acalyptus | Pinus edulis | No<br>No | 20 m<br>5 m | 1 m?<br>1 m | Sucker | ? | 1 | Cobb and Whitman 1993, Chap. 3, this volume: DelVecchio et al. 1993 |
| Pseudaulcaspis pentagona | Morus alba | Yes<br><br>No | 300 m<br><br>5 m | 1 m<br><br>1 m | Sucker | 15 yrs | 3 | Hanks and Denno 1994 |

Table 2.1 (continued)

| Insect | Host(s) | Demes found | Distance between hosts | Distance dispersed by insect | Feeding Mode | Age of Host | Generations of insect per year | Reference |
|---|---|---|---|---|---|---|---|---|
| *Stilbosis quadri-custatella* | *Quercus geminata* | Yes | 1–10 m | > 5 m | Leaf miner | ? | 1 | Mopper et al. 1995 |
| *Blepharida rhois* | *Rhus glabra* | No | 80 m | > 10 m | Chewer | 10–100 yrs | 1 | Strauss (1997) |
| *Asphondylia borrichiae* | *Borrichia frutescens* | Yes | 1 km | > 1 km | Gall maker | ? | 6 | Stiling and Rossi (cf. this study) |

males would only have the maternal genotype (Alstad et al. 1980). However, the most common suggestion has been that lack of dispersal and reduced gene flow between hosts should promote deme formation. Most studies examining deme-formation have focused on sessile insects (Karban and Strauss 1994), especially scales (Hanks and Denno 1994), although a study by Mopper et al. (1995) on a dispersive leaf-mining moth also supported the deme formation hypothesis. That study demonstrated that differentiation of populations into demes may occur if natural selection is strong enough to overcome limited amounts of gene flow.

The present study shows the existence of demes in a moderately dispersive insect, the gall-making midge, *Asphondylia borrichiae*. *Asphondylia* larvae mature in galls that develop on the stem tips of the coastal plant *Borrichia frutescens*. Colonization of experimental patches of *Borrichia* have shown that this midge can fly at least 1–2 km. Potted, nongalled *Borrichia* that were placed on four islands that had no *Borrichia* were all colonized by the midge within a year, and three were colonized within six months (Stiling and Rossi unpublished data). The existence of demes in this system suggests that another feature of the plant–insect relationship may be important in the formation of demes—mode of feeding. Gall insects are perhaps more highly adapted to their host plant than many other insects, because the larvae are embedded within, and often modify, the tissues of their host plant. Similarly, leaf miners, such as those studied by Mopper et al. (1995), are embedded within the host plant's leaves for their entire larval period. On the other hand, sap-feeding scale insects are external feeders that may avoid the majority of noxious plant secondary compounds, which have been implicated as selective forces in adaptive deme formation (Edmunds and Alstad 1981).

Our final point is that most studies to date have focused on survival of adult or immature insects on natal versus novel plants. We present data here to show that fecundity of adult females is substantially impacted by host identity (natal vs. novel), and that this has as important an effect as either death induced by the host plant or by natural enemies. Therefore, future studies should examine host-induced changes in fecundity, not just host-induced mortality patterns.

## 2.2 Methods

We performed reciprocal transplant experiments, moving *Borrichia frutescens* among four islands off the west-central coast of Florida (see Stiling and Rossi 1995 for a map of the study area). These islands were formed from material dredged up from the ocean floor by the Army Corps of Engineers in 1960. The islands are all about the same size and are separated by regular intervals of approximately 0.5 km. The neighboring mainland, overrun by hotels, condominiums, and houses, is devoid of *Borrichia*. Although the islands were produced in 1960, no attempt by the Corps was made to vegetate them. All vegetation arrived naturally, presumably by seed. Although it is possible that midges have adapted to host plants within the last 35 years, it is also possible that midges adapted to the parent plants also colonized these islands subsequently from patches of similar or identical genetic background.

Each of our study islands contained a single, different, *Borrichia* clone (as determined by gel electrophoresis; Stiling 1994) and, compared to other islands, these all had relatively high gall densities in 1993. The frequency of galled stem terminals on native plants on the four islands, referred to as TS2, CW6, CW7, and UTB, were 23.0%, 11.5%, 8.1%, and 5.7%, respectively. Using these islands ensured that there was always an abundant source of midges to colonize the experimental plants. We assumed that the majority of galls at each site came from midges that developed from larvae feeding on native plants, not from midges that had fed on novel plants and had immigrated to the island.

Ramets were dug up in April 1993, and placed in plastic flower pots containing a common soilless rooting medium. This technique was used, rather than direct planting into the ground at each site, to minimize genotype × maternal soil interactions. During the course of the experiment, there was no root growth through the drainage holes in the pots, and we believe that the effects of the soilless mix were not swamped by the surrounding soil. Previous comparisons have shown that gall densities on natal plants in pots are not different from gall densities on wild, nonpotted plants (Stiling and Rossi 1996; $\chi^2 = 0.821$; $df = 1$; $P = 0.364$). Twenty stems were placed in each pot. Sixty pots of *Borrichia* (1,200 stems) were collected from each island. To remove extant galls from the potted plants and to further minimize maternal effects of the plants on gall densities, the stems were cut back to within 10 cm of the soil surface. Potted plants were maintained in a botanical garden on the campus of the University of South Florida until the stems had completely reflushed with leaves. In August 1993, the plants were returned to the field. Fifteen pots of *Borrichia* (300 stems) from each island were placed onto the four islands in a completely reciprocal design. Pots were randomly assigned to positions in one of six 1 × 10 grids on each island. Pots were buried in the ground, so that the potting medium was flush with the surrounding substrate. The native stems were trimmed so that about 1,200 remained on each island—roughly the same as were in the experimental grids. Pots and native patches were separated by a few meters.

In April 1994, a full year after the plants had been collected from the field, we began censusing plants for galls. We felt that maternal effects such as plant quality, induced by different soil types or salinities, would be minimized by waiting a year between plant removal and gall censusing. We recorded gall density weekly until September, and each new gall was tagged and numbered. Gall diameters were measured to the nearest 0.05 mm using dial calipers. Gall counts were taken on all experimental plants and on three patches of 200 "wild" stems (i.e., those growing naturally on an island). The fate of all marked galls was recorded. Possible fates included decay of the gall without any insect emergence (host-induced abortion), predation of the gall and its inhabitants by birds or other predators, parasitism of the midges within the galls, and successful completion of midge development to adult. Galls typically mature within 30–35 days and after rapid growth rates early in the life cycle of the midge, the growth rate reaches a plateau that corresponds to pupation of the midges (Stiling et al. 1992; Rossi and Stiling 1995). Therefore, mature gall size was taken to be the diameter at which a gall stopped growing. Once mature, each gall was picked, returned to the laboratory, and placed into a clear plastic vial to capture the midges or parasitoids as they emerged.

Twenty female midges were dissected, and wing size was measured as an estimate of adult midge size, because wing length has been found to be positively correlated with size in other species of *Asphondylia* (Freeman and Geohagen 1987). The number of eggs in each ovary were recorded, so that the relationship between fecundity and midge size could be investigated. Galls were again counted in 1995, between April and June, but no data on gall size, abortion, and levels of parasitism were recorded during this period.

Tests of partial association, using log-linear analyses, were calculated for total numbers of galls per clone per island, percent of galls aborted and percent of galls parasitized on 1994 data (Wilkinson 1989; Stiling and Rossi 1996). Log-linear analyses on totals per clone per island had to be used for abortion rates and parasitism because of the relatively low number of galls per pot. Results from log-linear analyses are very similar to those using analysis of variance (ANOVA) on the same data, at least for total gall counts, which is the only comparison that can be made with both analyses (Stiling and Rossi 1996). A two-way ANOVA, with receptor island and plant clone as main effects, was calculated for gall diameter. In all analyses, plant clone was classified as either natal (originally occurred on an island, $df = 1$) or novel (brought in from another island, $df = 3$).

## 2.3 Results

Log-linear analysis indicated that, for 1994, there was a significant interaction between plant clone (either natal or novel) and gall density, as well as between receptor island and gall density (Table 2.2). The significance appears to be largely driven by island TS2, which had more galls than other islands, and where the difference between natal and novel was the greatest (Fig. 2.1). Natal plants at CW7 also supported more galls than novel plants from other islands, but at CW6 and UTB, the reverse pattern was observed. Thus, there was a significant three-way interaction between natal–novel host and receptor island and gall density, indicating that, although natal plants accrued more galls than novel plants, the natal–novel effect was different on some islands than others. The log-linear analysis suggests that the effect of receptor island (environment) is stronger than the clonal effect (natal vs. novel genotype) on gall density.

Interestingly, data from 1995 indicated a stronger deme effect compared to 1994. For any given island, gall density was always highest on a native clone (Table 2.3), except for island CW6, where the CW7 clone did marginally better. There are two possible reasons for the differences in gall densities between natal and novel plants. Either ovipositing females lay more eggs on some plants, or there is differential mortality from egg to first instar among natal and novel clones. We have no information on egg death, and so we focused on the strength

*Table 2.2* Three-Way Log-Linear Contingency Table Analysis of Gall Numbers, Galls Aborted and Numbers of Flies Parasitized from Reciprocal Transplant Experiment in 1994

|  | Interaction | $\chi^2$ | df | P |
|---|---|---|---|---|
| Gall abundance | Natal–novel × gall density | 5.69 | 1 | < 0.02 |
|  | Receptor island × gall density | 148.15 | 3 | < 0.001 |
|  | Receptor island × natal–novel × gall density | 8.87 | 3 | 0.031 |
| Galls aborted | Natal–novel × galls aborted | 0.66 | 1 | < 0.5 |
|  | Receptor island × galls aborted | 3.04 | 3 | < 0.5 |
|  | Receptor island × natal–novel × galls aborted | 2.89 | 3 | 0.409 |
| Midges parasitized | Natal–novel × midges parasitized | 1.61 | 1 | < 0.3 |
|  | Receptor island × midges parasitized | 9.44 | 3 | < 0.05 |
|  | Receptor island × natal–novel × midges parasitized | 11.20 | 3 | 0.011 |

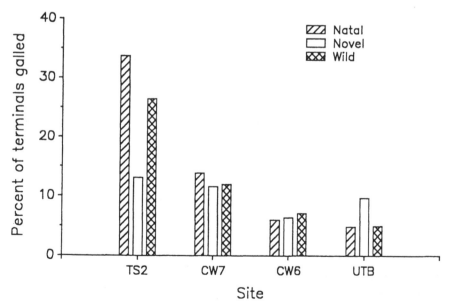

*Figure. 2.1* Gall abundance on experimental and native *Borrichia* plants on four islands: TS2, CW7, CW6, and UTB in 1994. *Novel* refers to the sum of all nonnative clones at this site. *Natal* refers to the native clone returned to its original site. *Wild* refers to native, non-transplanted plants at that site.

*Table 2.3*  Percent of Stems Galled in Reciprocal Transplant
Experiments in 1995

| Recipient island | Donor Island | | | |
|---|---|---|---|---|
|  | TS2 | CW7 | CW6 | UTB |
| TS2 | 52.4 | 20.7 | 3.3 | 25.7 |
| CW7 | 15.2 | 38.0 | 21.1 | 1.2 |
| CW6 | 13.1 | 18.4 | 16.2 | 6.0 |
| UTB | 2.6 | 1.6 | 1.5 | 3.1 |

of mortality factors acting on later stages of the midge's life cycle. While predation removed 1.9% of galls and their inhabitants on potted plants in 1994, gall
abortion and parasitism by four species of wasps accounted for 18.7% of total
galls and 39.4% of total midges killed, respectively. We cannot be certain exactly
how many midges were killed by abortion because we do not know how many
chambers were in a gall prior to abortion. Data from wild plants indicate that gall
abortion rates are lowest at the high-density sites such as TS2 and highest at the
low-density sites such as UTB (Fig. 2.2). However, data from the reciprocal
transplant experiment do not confirm this trend. Although, at every site, death due
to abortion was lower on natal plants than on novel *Borrichia* clones, the trend
was not significant (Table 2.2), perhaps because of relatively low sample sizes.

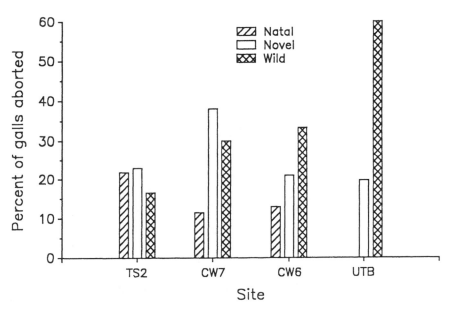

*Figure 2.2*  Gall abortion rates on experimental and native *Borrichia* plants on four islands. There were no galls aborted on natal plants at UTB.

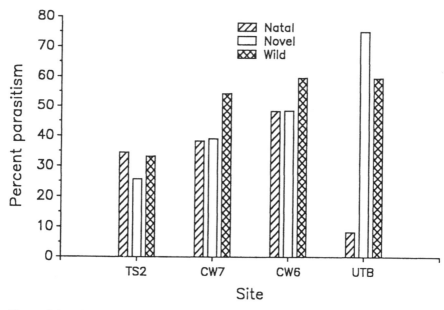

*Figure. 2.3* Midge parasitism rates on experimental and native *Borrichia* plants on four islands.

The parasitism rate on wild plants was also lower at high-density sites than at the low-density sites (Fig. 2.3). Differences in parasitism on experimental plants among islands were consistent with this trend, but there was not a significant interaction between parasitism and natal versus novel plant clone. However, there was a significant three-way interaction among clone, island, and levels of parasitism, indicating that galls on novel clones are parasitized more than galls on natal clones at some sites.

Natal versus novel host had a significant effect on gall diameter, with natal hosts producing bigger galls at every site (Table 2.4, Fig. 2.4). There was also a significant effect of receptor island (environment) on gall diameter.

Larger galls tended to produce larger midges, but the relationship is complicated by the crowding effect of multiple chambers within galls and is discussed more fully in Rossi et al. (1996). As is true of many insects (Honìk 1993), larger midges produced more eggs than smaller ones ($r = 0.927$; $n = 20$; $P < 0.001$; Fig. 2.5).

## 2.4 Discussion

In this study, host-plant genotype and environment both play an important role in influencing the density of galls on *Borrichia frutescens*, although the effects of environment are stronger than those of genotype. We have already explained the environmental effect in terms of differential nitrogen levels, plant growth charac-

*Table 2.4*  Results from ANOVA on Mature Gall Diameter from Reciprocal Transplant Experiment in 1994

| Source | Sum of square | *df* | Mean square | *F* | *P* |
|---|---|---|---|---|---|
| Natal–novel | 0.022 | 1 | 0.022 | 6.010 | 0.015 |
| Receptor island | 0.182 | 3 | 0.061 | 16.649 | < 0.001 |
| Receptor island × natal–novel | 0.058 | 3 | 0.019 | 5.282 | 0.001 |
| Error | 1.137 | 312 | 0.004 | | |

teristics, resultant gall sizes, and the effects of parasitism rates (Stiling and Rossi 1996). Galls on plants in favorable sites grew larger, and this resulted in decreased parasitism. Here, we focus on local adaptation of midges to particular host genotypes. On some islands in 1994, and all islands in 1995, galls were more numerous on natal genotypes than novel ones. Moreover, on every island, galls were always bigger on natal plants compared to novel ones. There is good evidence, therefore, that *Asphondylia* populations are locally adapted to specific host clones. Interestingly, in both years and for most factors, the trend for local adaptation seems strongest on islands that typically have the highest gall densities, such as TS2.

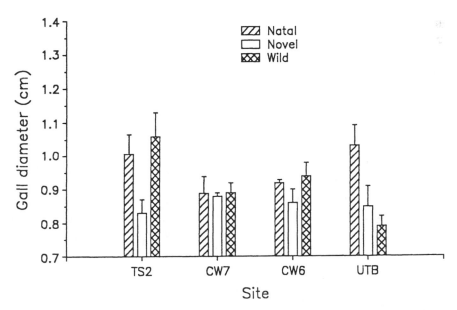

*Figure 2.4*  Gall diameter on experimental and native *Borrichia* plants on four islands.

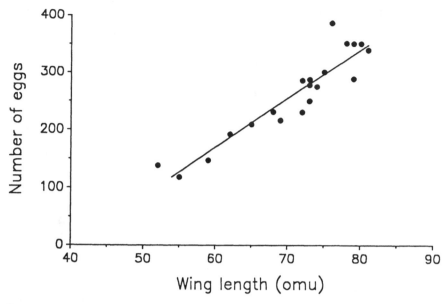

*Figure 2.5* Relationship between potential fecundity (number of eggs at eclosion) and wing length (ocular micrometer units) for *Asphondylia* reared from *Borrichia* ($r = 0.927$; $n = 20$; $P < 0.001$).

How general is deme formation in this system? We tested for the existence of demes on islands that had only single *Borrichia* clones. However, this is the exception rather than the rule. Stiling (1994) noted that on mainland patches in north Florida, the percentage of populations that are multiclonal is 83%, which is approximately the same percentage of populations we have found to be multiclonal on spoil islands in central Florida. It is still possible that adapted demes could form on multiclonal populations of *Borrichia*; however, we did not attempt to assess whether this has occurred in the current study. Interestingly, reciprocal transplants of *Borrichia* from multiclonal islands also demonstrated a significant effect of plant genotype on herbivore densities (Stiling 1994).

Most studies to date have examined the deme-formation hypothesis in terms of survival of insects either through plant- or enemy-induced mortality. In our system, the effects of natal versus novel plant genotype on parasitism and abortion rates were not significant. We believe the existence of demes in this system could be caused as much by differential midge fecundity on natal versus novel hosts as by parasitism or gall abortion. In nature, gall size ranges from < 0.8 cm at low-density sites to > 1.3 cm at high-density ones. Rossi et al. (1996) showed that potential fecundity of female midges is inversely correlated with larval crowding. Increased larval crowding results in smaller female midges, and their fecundity is strongly affected by adult size. For *Asphondylia borrichiae*, number of eggs at eclosion ranges from approximately 100–400 eggs. On average, galls on novel plants were only 88% of the size of those on novel plants. A 12% reduction in midge size (as measured by wing length) equates to > 25% decrease in egg number at eclosion (Fig. 2.5). Thus, the ability to create even slightly larger galls on native plants probably substantially increases potential midge fecundity.

We can compare the contribution of various components to the relative fitness of different insect lines using the procedure outlined by Strauss and Karban (Chapter 8, this volume); that is, we assumed the fitness of a natal insect line to be 1.0 and compared the fitness of novel populations relative to the natal population. We repeated this procedure for mortality caused by parasitism, gall abortion, and reduction in fecundity caused by a reduction in gall size on novel plants. For parasitism, the relative fitness of novel to natal lines is 0.77 + 0.41 ($SD$), for abortion 0.89 + 0.64, and for fecundity 0.75 + 0.07. Thus, the effects of gall size on the potential fecundity of *Asphondylia* were as substantial as those from either parasitism and or abortion.

Why is there good evidence of the existence of demes in this system but not others? One popular explanation, as outlined by Edmunds and Alstad (1978) and underscored by Cobb and Whitham (1993), is that multivoltine phytophagous insects that pass hundreds of generations on a long-lived host may be the best place to look for the formation of demes. Clonal plants can be very long-lived. We have observed *Borrichia* plants living for at least six years, as long as our censuses have been conducted in central Florida, and we believe that *Borrichia* probably lives much longer than this. However, we suggest that intimacy of the herbivore with its host plant could be another crucial factor that favors the formation of demes in herbivorous insects. Endophagous insects may be under stronger selection, mediated through the host plant, than exophagous feeders. Unfortunately, there are too few studies on deme formation involving gallmakers and leafminers to test this idea. Nevertheless, the existence of interactions between host genotype and insect genotype in related experiments presents the possibility of fine-scale adaptation by local populations of insects to particular host-plant genotypes. For example, Marino et al. (1993) found a significant interaction of clone and fertilizer (environment) on the larval survival of *Phytomyza ilicicola*, a leaf-mining moth on American holly, *Ilex opaca*. In addition, Quiring and Butterworth (1994) showed significant family × site effects for egg densities and herbivory by the spruce bud moth, *Zeiraphera canadensis*, which, in its early instars, mines the needles under the bud caps of white spruce, *Picea glauca*. On the other hand, Ayres et al. (1987) failed to show tree × brood interactions for external-feeding caterpillars on mountain birch, and Strauss (1990) found only weak support for the existence of genotype by environment interactions for attack of *Rhus glabra* by an externally feeding chrysomelid beetle, *Blepharida rhois*. Although Fritz (1990) demonstrated no interaction of clone and site on densities of leaf petiole stem gallers of arroyo willow, *Salix lasiolepis*, only two sites were used in his study. Taken together, these studies lend some weight to the idea that mode of feeding affects whether insect herbivores do better on certain plant genotypes in certain environments and, ultimately, whether they form demes.

## Acknowledgments

This study could not have been completed without help in the field from Kerry Hennenfent and especially Todd Bowdish and Maria Cattell. We are especially grateful to the Martin Marietta Company, Largo, Florida, for performing the nitrogen analyses. This manuscript was improved by the comments of Susan Mopper, Sharon Strauss, and two anonymous reviewers. Financial support was provided by National Science Foundation Grant DEB #9309298.

## 2.5 Summary

We performed reciprocal transplants of different *Borrichia frutescens* clones, each with relatively high numbers of galls, between four offshore islands in Pinellas County, Florida, to test whether demes of the gall midge *Asphondylia borrichiae* were adapted to specific host clones. We measured gall abundance, gall size, gall abortion rate, and levels of parasitism on experimental plants. We found a significant effect of plant clone (either natal or novel) on gall abundance and gall size, which suggests that *Asphondylia* does form demes. Gall abortion rate and levels of parasitism were not affected as much by plant clone. Deme formation in this system may be driven as much by differences in midge fecundity that are directly related to host clone as by midge death rates. Galls on natal plants were bigger than those on novel plants. Bigger galls gave rise to larger midges that had more eggs than smaller ones. On some islands, midges in bigger galls also suffered less parasitism. Finally, we believe the endophagous habit of gall insects may make them more likely to form demes than some external feeders.

## 2.6 References

Akimoto, S. 1990. Local adaptation and host race formation of a gall-forming aphid in relation to environmental heterogeneity. *Oecologia 83*:162–170.

Alstad, D. N., G. F. Edmunds Jr. and S. C. Johnson. 1980. Host adaptation, sex ratio, and flight activity in male black pineleaf scale. *Ann. Entomol. Soc. Am. 73*:665–667.

Ayres, M. P., V. J. Soumela, and S. F. Maclean. 1987. Growth performance of *Ephirrita autumnata* (Lepidoptera: Geometridae) on mountain birch: Trees, broods, and tree and brood interactions. *Oecologia 74*:450–457.

Cobb, N. S. and T. G. Whitham. 1993. Herbivore deme formation on individual trees: A test case. *Oecologia 94*:496–502.

Del Vecchio, T. A., C. A. Gehring, N. S. Cobb, and T. G. Whitham, 1993. Negative effects of insect herbivory on the ectomychorrhizae of juvenile pinyon pine. *Ecology 74*:2297–2302.

Edmunds, G. F. Jr. and D. N. Alstad. 1978. Coevolution in insect herbivores and conifers. *Science 199*:941–945.

Edmunds, G. F. Jr. and D. N. Alstad. 1981. Responses of black pineleaf scales to host plant variability. Pp. 29–38 *in* R. F. Denno and H. Dingle (Eds.), *Insect Life History Patterns.* Springer-Verlag, New York.

Freeman, B. E. and A. Geoghagen. 1987. Size and fecundity in the Jamaican gall-midge *Asphondylia boerhaaviae. Ecol. Entomol. 12*:239–249.

Fritz, R. S. 1990. Effects of genetic and environmental variation on resistance of willow to sawflies. *Oecologia 82*:325–332.

Hanks, L. M. and R. F. Denno. 1994. Evidence for local adaptation in the armored scale insect *Pseudaulacaspis pentagona* (Targioni: Tozzetti) (Homoptera: Diaspididae). *Ecology 75*:2301–2310.

Honìk, A. 1993. Intraspecific variation in body size and fecundity in insects: A general relationship. *Oikos 66*:483–492.

Karban, R. 1989. Fine scale adaptation of herbivorous thrips to individual host plants. *Nature 340*:60–61.

Karban, R. and S. Y. Strauss. 1994. Colonization of new host plant individuals by locally adapted thrips. *Ecography 17*:82–87.

Komatsu, T. and S. Akimoto. 1995. Genetic differentiation as a result of adaptation to the phenologies of individual host trees in the galling aphid *Kattenbachiella japonica. Ecol. Entmol. 20*:33–42.

Marino, P. C., H. V. Cornell, and D. H. Kahn. 1993. Environmental and clonal influences on host choice and larval survival in a leafmining insect. *J. Anim. Ecol. 62*:503–510.

Mopper, S., M. Beck, D. Simberloff, and P. Stiling. 1995. Local adaptation and agents of selection in a mobile insect. *Evolution 49*:810–815.

Quiring, D. T. and E. W. Butterworth. 1994. Genotype and environment interact to influence acceptability and susceptibility of white spruce for a specialist herbivore, *Zeiraptera canadensis. Ecol. Entomol. 19*:230–238.

Rice, W. R. 1983. Sexual reproduction: An adaptation reducing parent-offspring contagion. *Evolution 37*:1317–1320.

Rossi, A. M. and P. Stiling. 1995. Intraspecific variation in growth rate, size, and parasitism of galls induced by *Asphondylia borrichiae* (Diptera: Cecidomyiidae) on three host species. *Ann. Entomol. Soc. Am. 88*:39–44.

Rossi, A. M., P. Stiling, M. V. Cattell, and T. I. Bowdish. 1997. Evidence for the existence of host-associated races in a gall-forming midge: Trade-offs in potential fecundity. *Ecol. Entomol., in review.*

Stiling, P. 1994. Coastal insect herbivore populations are strongly influenced by environmental variation. *Ecol. Entomol. 19*:39–44.

Stiling, P. and A. M. Rossi. 1995. Coastal insect herbivores are affected more by local environmental conditions than by plant genotype. *Ecol. Entomol. 20*:184–190.

Stiling, P. and A. M. Rossi. 1996. Complex effects of genotype and environment on insect herbivores and their natural enemies on coastal plants. *Ecology, 77*:2212–2218.

Stiling, P., A. M. Rossi, D. R. Strong, and D. M. Johnson. 1992. Life history and parasites of *Asphondylia borrichiae* (Diptera: Cecidomyiidae), a gall maker on *Borrichia frutescens*. *Fla. Entomol. 75*:130–137.

Strauss, S. Y. 1990. The role of plant genotype, environment and gender in resistance to a specialist chrysomelid herbivore. *Oecologia 84*:111–116.

Strauss, S. Y. 1997. Lack of evidence for local adaptation to individual plant clones or site by a mobile specialist herbivore. *Oecologia 110*: 77–85.

Unruh, T. R., and R. F. Luck. 1987. Deme formation in scale insects: A test with the pinyon needle scale and a review of other evidence. *Ecol. Entomol. 12*:439–449.

Wainhouse, D. and R. S. Howell. 1983. Intraspecific variation in beech scale populations and susceptibility of their host *Fagus sylvatica*. *Ecol. Entomol. 8*:351–359.

Wilkinson, L. 1989. *Systat: The System for Statistics*. Systat, Inc., Evanston, IL.

# 3

# Prevention of Deme Formation by the Pinyon Needle Scale: Problems of Specializing in a Dynamic System

*Neil S. Cobb and Thomas G. Whitham*
Department of Biological Sciences, Northern Arizona University,
Flagstaff, AZ

## 3.1 Introduction

Genetic differentiation within populations has evolutionary consequences for both populations and species, and is important to issues such as local adaptation, speciation, and biodiversity (Wright 1968,1969, 1978; Lewontin 1974; Endler 1977; Brown 1979; Mitter and Futuyma 1979; Bradshaw 1984; Futuyma and Peterson 1985; Waser and Price 1985). Becoming locally adapted may allow widespread generalist species to exploit a variety of resources by forming host races (Thompson 1994); additionally, specialist herbivores may track certain host genotypes by forming biotypes (Gallun et al. 1975; Gould 1983; Service 1984; Parker 1985; Feder et al. 1988; Moran and Whitham 1988). In either case, a species may be able to increase niche breadth as a result of genetic diversity that results from local adaptation (van Valen 1965). It is therefore important to understand what mechanisms promote or prevent deme formation from occurring at different spatial levels of organization.

In 1978, Edmunds and Alstad proposed an extreme form of local adaptation, the deme formation (DF) hypothesis, whereby insect herbivores became genetically adapted to individual host trees (Edmunds and Alstad 1981; Alstad and Edmunds 1983a,b, 1987; Alstad and Corbin 1990). The ability to adapt to an individual tree could occur because (1) the host was long-lived relative to the herbivore, (2) each plant represented a discrete resource for herbivores, and (3) herbivores were relatively sessile, thus reducing gene flow in herbivore populations among trees. Support for the DF hypothesis has been mixed: Some studies found no support (Rice 1983; Unruh and Luck 1987; Cobb and Whitham 1993, Memmott et al. 1995; Kimberling and Price 1996; Strauss 1996); other studies found partial confirmation (Wainhouse and Howell 1983; Hanks and Denno 1992, 1994), while two studies found clear evidence for deme formation on individual plants (Karban 1989; Mopper et al. 1995).

A major assumption of the DF hypothesis is that resource quality within an individual tree is relatively constant in space and time; that is, if pests adapt to agricultural monocultures in space, then long-lived trees as monocultures in time should also be exposed to rapidly evolving pests. Whitham (1981) addressed this conundrum by proposing that long-lived trees are not monocultures through time, and that through diverse processes, they are highly heterogeneous resources that act to counter the evolution of virulent herbivore genotypes at the individual-tree level. There is evidence that resources utilized by herbivores exhibit considerable within-tree variation within a season (Whitham and Slobodchikoff 1981; Whitham 1983; Whitham et al. 1984; Larsson 1985; Fay and Whitham 1990) or over longer periods of time as a result of developmental resistance (Craig et al. 1988; Kearsley and Whitham 1989). These studies argue that the dynamic nature of resources within perennial plants poses a serious problem for herbivores by challenging their ability to evolutionarily track an individual host plant.

The major theme of this chapter is to examine how variable resources affect the lack of fine-scale adaptations by the pinyon needle scale, *Matsucoccus acalyptus*. Although we initially thought our studies of this scale insect would support the DF hypothesis, our subsequent experiments rejected it (Cobb and Whitham 1993). Here, we summarize those studies and explore several mechanisms that may prevent genetic differentiation from occurring among herbivore populations on individual trees. Unless otherwise noted, reference to a scale insect population refers to the collection of scales on an individual tree.

The chapter is organized into three sections. In the first section we review our evidence that refutes deme formation on individual pinyons by *M. acalyptus*. We show that scale survival is largely determined by host resistance traits, and no evidence supports the hypothesis that scales make fine-tuned adaptations to an individual tree. In the following sections, we propose mechanisms that could produce these results. First, we hypothesize that there is a feedback loop between scale-mediated changes in needle quality and changes in herbivore population size and structure that ultimately promotes gene flow among scale populations. Specifically, increasing population size promotes the dispersal of adult males and newly emerged larvae, leading to gene flow. Because the sizes of most scale populations fluctuate from year to year, a scale population may be an exporter or importer of genes, depending on the number of adult males and larvae that exist in the population in any given year. Second, we present evidence that catastrophic events can lead to population extinctions or bottlenecks on many trees. These trees are then subsequently colonized by scales from nearby trees, such that local demes are unlikely to develop. Third, we propose that selection regimes may change from one year to the next (Cobb 1990) or over longer periods time (i.e., decades; Kearsley and Whitham 1989). Both of these processes, operating at different time scales, create variable host phenotypes that make it difficult for herbivores to track their host plants over evolutionary time.

## 3.2 Pinyon–Scale System and Sampling Methods

The study site, located 33 km northeast of Flagstaff, Arizona (elevation 1,880 m), covers approximately 125 ha in a mixed *Juniperus monosperma–Pinus edulis* woodland dominated by *P. edulis*. We selected the scale–pinyon system because it contains the necessary attributes that should promote deme formation (Edmunds and Alstad 1981; Cobb and Whitham 1993). These attributes consist of a sessile scale insect with relatively poor dispersal abilities that feeds on a long-lived and chemically diverse host tree. In addition, the lack of predators, parasitoids, and competitors acting on *M. acalyptus* (Krombein et al. 1979; Cobb and Whitham 1993, and unpublished data) should result in strong plant–herbivore interactions that should promote deme formation under the scenario proposed by Edmunds and Alstad (1978). We have also commonly observed adjacent trees with extremes in scale densities ranging from zero to over 500,000. This suggests local adaptation by insect herbivores, host resistance differences, and/or some combination of both (Edmunds 1973).

The life history of *M. acalyptus* at our study site is similar to other *M. acalyptus* populations in the western United States (McCambridge and Pierce 1964; Unruh 1985). Eggs are laid March–April at the base of tree trunks and in bark crevices. Within five weeks, first-instar crawlers climb up the tree and either disperse by wind to other trees or they begin feeding. Scales insert their mouthparts through a needle stoma and suck out the contents of mesophyll tissue, and remain sessile through two feeding instars. Because scales colonize needles in the spring, but before bud break, the current year's needles generally escape attack. Scales preferentially colonize one-year-old needles over older needle cohorts. Extensive chlorosis of needles occurs as a result of scale feeding, and at high scale densities, chlorosis may lead to needle death before scales reach maturity. Such heavily attacked trees have a characteristic poodle-tail appearance in which only the current year's needles remain. Males feed until November, eclose into a mobile crawler morph, climb down the tree, and undergo a prepupal and pupal stage in plant litter at the base of the tree. Females remain on the needles as second instars throughout the winter. In March, they undergo a rapid increase in size and eclose as legged, wingless adults. At the same time in March, males emerge as winged individuals and congregate at the base of trees and mate with descending females. Males may also fly up to the foliage to mate with emerging females.

### 3.2.1 Sampling of Scales

Except where noted, sampling of scale populations occurred on small trees (mean age = 32 yrs $\pm$ 0.99 *SE*; mean height = 1.15 m $\pm$ 0.04 *SE*). Our work has concentrated on small trees, because they account for over 80% of the scale-infested trees in the population (Cobb et al. 1994). Scales were censused by collecting needles immediately after adults emerged in March. We typically collected 30–60 fascicles per tree, depending on the size of the tree. This allowed us to collect from all parts of canopies that averaged 2.41 m³. Scale remains were classified into one of four groups based on remains of carcasses or exuvia from successfully emerged adults: (1) individuals that died as first instars, (2) individuals that died as second instars, (3) individuals that eclosed into adult males, and (4) individuals that eclosed into adult females. Percent mortality was based on the number of individuals that did not pass through the second instar. We also visually estimated percentage needle chlorosis due to scale feeding based on the total amount of needle area that was not green.

To examine the DF hypothesis, we transferred scales from different sources onto their natal trees (i.e., their home trees) and novel trees (i. e., foreign trees). To prevent existing scales from confounding the performance of these experimental scale populations, the existing scales were removed. Scales were removed by hand collecting all observable eggs from the bases of trees and placing a barrier of Tanglefoot© on the tree trunk to prevent any remaining larvae from colonizing the canopy (Tanglefoot© was later removed). Transfer of scales onto trees involved either the addition of scale eggs to the base of trees or the placement of scale eggs in small containers attached to tree branches. The latter method of transfer allowed us to keep scales from several donor sources separate on the same receptor tree. Methods specific to experimental and observational data sets are described in the following sections. A more detailed description of scale removal and transfer methods is described in Cobb and Whitham (1993).

## 3.3 Experimental Tests of the Deme Formation Hypothesis

### 3.3.1 Scale Performance On "Natal" And "Novel" Trees

The DF hypothesis predicts that herbivores become increasingly adapted to the phenotype of their host tree through time. Such adaptation should be reflected by an increasing herbivore population size resulting from greater survival of superior insect genotype(s). We tested the DF hypothesis using transfer experiments in which eggs were collected from scale populations representing three different population densities (eggs from different trees were kept separate). These eggs were then reciprocally transferred onto their natal trees and onto novel foreign trees. Because the DF hypothesis predicted that scales from low-, medium-, and high-density populations represent different levels of adaptation to their natal tree, by transferring scales back onto their natal tree and onto other novel trees, we could critically test this major prediction. We only obtained one year of population density for these trees and, because we had to remove existing scales from the trees, we do not know long-term population densities for these trees. Additional scales from the three different population densities were placed on trees that had no existing populations (i.e., presumably resistant trees, or trees that had not been discovered by scales).

Confirmation of the DF hypothesis required several major outcomes. First, if pinyon needle scales form demes on individual trees, we predicted that scale mortality would be lower on natal trees compared to "novel" trees. Second, because high-density populations should reflect the highest degree of adaptation to their natal tree, they should exhibit the lowest mortality on their natal tree compared to low- and medium-density scale populations. Alternatively, we predicted that high-density populations would exhibit the lowest degree of performance when transferred onto previously uninfested trees and previously infested foreign trees.

Our experiments did not support the prediction that scale populations perform better on their natal tree (black bars) compared to novel (open bars) trees (Fig. 3.1). For example, the average mortality of the natal transfers (3 black bars) was 48.6% (± 6.8 *SE*), whereas the average mortality of scales transferred to novel trees (six open bars) was 49.3% (± 4.7 *SE*). Likewise, our results did not support the prediction that the highest density populations were more adapted to their natal tree than low- and medium-density populations, although we assumed that population sizes were relatively constant from year to year. Our later monitoring of scale population sizes showed that scale population densities can fluctuate considerably from year to year, and our designation of density classes may have only reflected the status of the populations for 1985. Regardless, when we examined only those trees with scale populations (resistant trees excluded), scales did just as well on their natal trees as they did on novel trees.

Our results showed that trees without established natural populations were resistant to scales (shaded bars; Fig. 3.1). Here, mortality was high for all donor populations and varied between 88% and 95%. We do not know whether resistance is genetically based or the result of microhabitat differences. We believe resistance is more likely to be genetically based, because resistant and susceptible trees are commonly growing adjacent to each other. The resistance of uninfested trees has been confirmed in similar transfers of scales onto uninfested trees (Del Vecchio et al. 1993; Gehring et al. unpublished manuscript) and demonstrated in other plant–herbivore systems (Memmot et al. 1995). A seven-year ongoing scale-transfer experiment involving resistant trees currently provides no evidence for local adaptation (Cobb and Whitham unpublished data). We conclude that there do not appear to be a multitude of host phenotypes that scales are adapted to; however, there are clearly intrinsic differences between trees with established populations and uninfested trees.

### 3.3.2 Performance of "Incipient" and "Established" Scale Populations

The DF hypothesis predicts that incipient populations (i.e., populations on newly colonized trees) are not adapted to their host tree and require a number of generations to adapt to a particular tree phenotype. Therefore, we would expect decreasing mortality of these incipient populations as they become more adapted to their natal tree. We also would predict that performance of incipient populations should be significantly lower than the performance of established populations that have had many generations to adapt to their natal tree.

Incipient populations were established by first removing existing scale populations in 1985 and allowing trees to be naturally recolonized. It required two years before these trees had developed large enough scale populations to estimate mortality (i.e., > 30 scales). To test the first prediction, we monitored 20 incipient populations over a six-year period to determine if mortality decreases over several

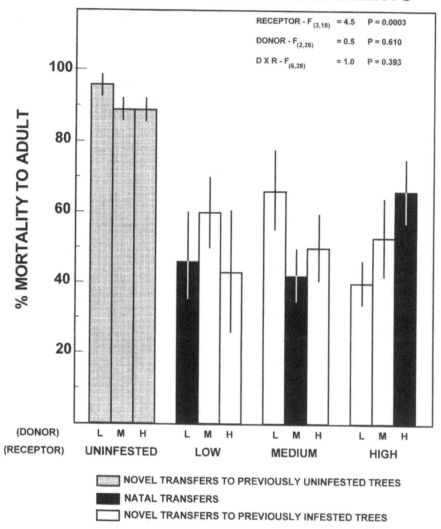

*Figure 3.1* Scale transfer experiments showing no difference in mortality between natal and novel trees that supported natural scale populations. The only significant result was increased mortality of all scale populations on trees that did not harbor scale populations before the experiment. The three scale population levels represented by donor and receptor trees are low-density (L), medium-density (M) and high-density (H) populations. Bars indicate means ± 1 *SE*.

insect generations. To test the second prediction, we compared scale mortality in incipient populations and established populations over a two-year period.

We found the opposite pattern predicted by the DF hypothesis (Fig. 3.2). As incipient populations had more years to adapt to their hosts, performance declined rather than increased over the six-year period ($F_{(4,68)} = 16.76$, $p = 0.0001$; based on profile transformation in a repeated measures analysis of variance (Pilson 1992). Mortality in 1987 was 18% but increased to 43% in 1992. Due to the increase in

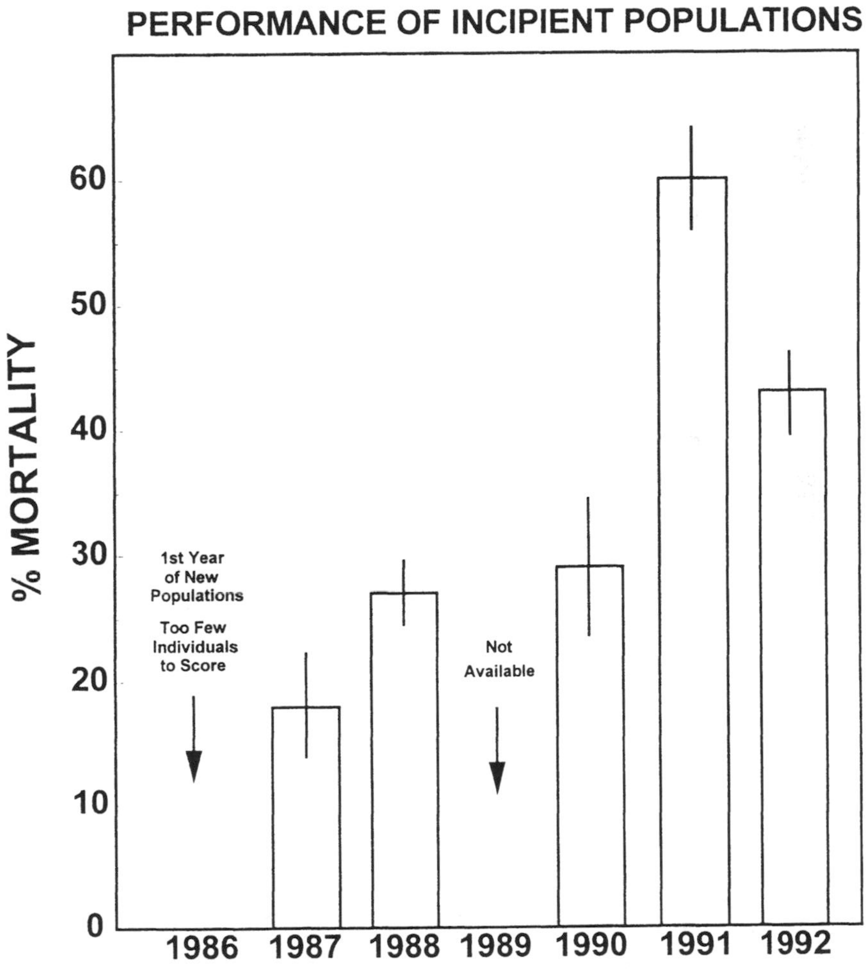

## PERFORMANCE OF INCIPIENT POPULATIONS

*Figure 3.2*  Increasing mortality in incipient (newly colonized) scale populations over a six-year period showing no adaptation to host trees through time. Bars indicate means ± 1 *SE*.

population density over the same time span, there may have been a corresponding increase in mortality resulting from density-dependent mortality. Because predators, parasites, and other herbivores have little impact on the pinyon needle scale, the observed changes in mortality cannot be due to these factors, and it is clear that scales are not becoming better adapted to their host trees. This pattern also does not reflect what was occurring in the rest of the scale populations: Established scale population densities decreased steadily from a high in 1987 (mean = 5.78 ± 0.99 scales/cm needle) to a low in 1990 (mean = 1.81 ± 0.41 scales/cm needle). This indicates that the pattern of increasing population density and mortality was unique to incipient populations.

Although the DF hypothesis predicts that established herbivore populations should perform better than incipient populations, in two separate years of analysis we found that the mortality in established scale populations did not differ from incipient populations (Fig. 3.3). Figure 3.3 only shows mortality after scale density was accounted for as a covariate; mortality was actually significantly greater in established populations when we did not account for scale density (Cobb and Whitham 1993). So, even when we account for differences in density, we observe a nonsignificant trend in which the mortality of incipient population averaged 8% lower over the two-year period compared to established populations. This demonstrates that scales do not require a transition period where they must adapt to individual tree phenotypes (Cobb and Whitham 1993).

Hanks and Denno (1994) suggested that we did not detect deme formation because of the close proximity of our experimental trees, which would have resulted in limited gene flow among scale populations. For example, although we found no difference in scale mortality between our natal and novel trees that were spaced approximately 20 m apart, Hanks and Denno (1994) found increased scale survivorship on natal trees compared to novel trees when novel trees were ≥ 300 meters from natal trees. Although we have not specifically conducted reciprocal transfer experiments to test for a distance effect, we have conducted two additional transfer experiments (Del Vecchio et al. 1993; Gehring et al. unpublished data) where our source populations and receptor trees were ≥ 300 m apart. In both of these experiments, we found that scale survival was comparable to survival in established populations. From these studies, we conclude that deme formation or local adaptation does not occur at the level of individual trees or at the local level of a few hectares.

## 3.4 Potential Mechanisms That Prevent Deme Formation

We emphasize that individual trees represent very heterogeneous resources, and that scale populations on trees are very dynamic. We propose that the dynamic nature of this system promotes gene flow among scale populations on different trees, as well as changing selection regimes that ultimately prevents demes from forming. Despite the fact that *M. acalyptus* is a relatively poor disperser com-

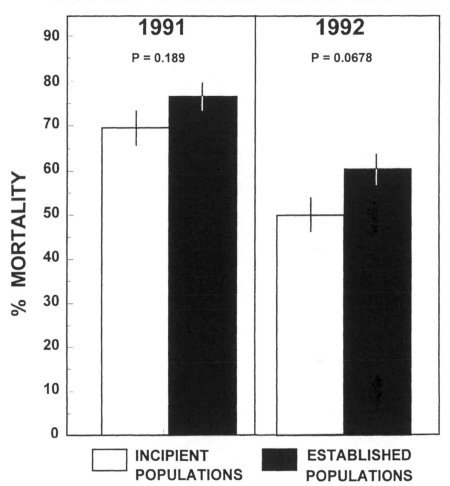

*Figure 3.3* Comparisons of mortality in incipient and established scale populations demonstrating that the performance of incipient populations was comparable to established populations. Means were adjusted for scale density, since established populations had higher densities. Incipient populations were six years old in 1991. Bars indicate means ± 1 *SE*.

pared to other insect species, Unruh (1985) has shown in a California population of *M. acalyptus* that adult male dispersal is great enough to prevent genetic differentiation and is supported by the lack of among-tree allozyme differentiation in scales. Using the same species of scale, here we focus on how gene flow may be promoted as a result of three major factors: (1) changing herbivore sex ratios, (2) density-dependent larval dispersal, and (3) climatic catastrophes that lead to extinction of populations on some trees and subsequent colonization from adjacent trees. Last, we examine how temporal changes in resource quality may prevent deme formation. We show that resources within individual trees vary annually and over longer periods of time, both of which result in changing host phenotypes that could prevent herbivores from genetically adapting to an individual tree (Whitham 1981; Whitham et al. 1984; Cobb 1990).

## 3.5 Changing Sex Ratios in Scale Populations: Effect on Gene Flow among Scale Populations

Gene flow among herbivore populations on different trees can be promoted if sex ratios are highly variable among scale populations. We propose the "male export" hypothesis, which predicts that skewed sex ratios among scale populations lead to excess males on some trees that subsequently disperse and mate with females on other trees that have relatively few males. To support this hypothesis, three criteria must be met. First, there must be significant year-to-year variation in scale sex ratios within a tree, so that some years a scale population acts as a "male exporter" and in other years the population is a "male importer." Second, there must be significant spatial variation among scale populations on trees within a single year. Third, there must be a mechanism that catalyzes this import–export process. The putative mechanism would be the production of female sex pheromones that attract surplus males from their natal tree to a female–biased tree. Unruh (1985) has demonstrated male attraction to female-bait traps in *M. acalyptus,* and several researchers have isolated sex pheromones in other species of *Matsucoccus* (Doane 1966; Young et al. 1984; Park et al. 1986).

Sex ratios vary greatly among trees in the same year, and also within the same tree from one year to the next (Fig. 3.4A). For example, on tree #18, males comprised 80% of the surviving population in 1986, 15% of the population in 1987, and 86% in 1988, whereas on tree #8, males comprised only 21% of surviving adults in 1986, 80% in 1987, and 3% in 1987. So, for any given year, some trees have relatively large numbers of surviving males, whereas other trees have very few, and the relative number of surviving males on a single tree typically fluctuates greatly from year to year.

To understand why scale sex ratios exhibit these within- and among-tree patterns of variation, it is important to understand patterns of fluctuating scale density and needle chlorosis caused by scale feeding. Both scale density (Fig. 3.4B) and percentage needle chlorosis (Fig. 3.4C) exhibit large differences among trees within a single year, and individual trees fluctuate greatly among years. Looking back at trees #18 and #8, scale densities (number of scales/cm needle) over the 1986–1988 period were 5.4, 0.7, and 5.3 for tree #18, and 1.7, 5.6, and 0.9 for tree #8. Likewise, percentage needle chlorosis during 1986–1988 for tree #8 was 93%, 32%, and 93%, respectively, whereas during the same time period for tree #8, percentage needle chlorosis was 23%, 75%, and 11%, respectively.

# TEMPORAL VARIATION

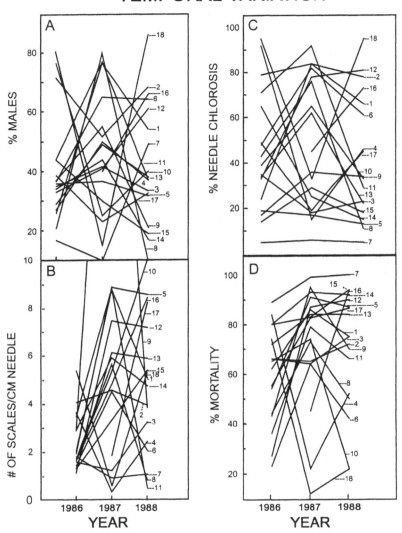

*Figure 3.4*  Population attributes of the pinyon needle scale and its affect on the host plant (i.e., needle chlorosis) vary greatly within a year among trees as well as between years on a single tree. Graphs show the year to year variation in scale sex ratio (A), scale density (B), needle chlorosis (C), and adult survival (D), over a three-year period on 18 haphazardly chosen trees. Numbers refer to individual trees.

We suggest that scale density and needle resource quality (i.e., chlorosis) acting in concert cause much of the variation in scale sex ratios on individual trees. Sixty-three percent of the variation in adult scale sex ratios on individual trees is explained by scale density and needle chlorosis (Table 3.1). Thus, female survival is reduced at high scale densities due primarily to extensive needle chlorosis caused by males and earlier instars of both sexes. Females take the brunt of the negative impact of needle chlorosis, because males stop feeding six months before females, when chlorosis is much less extensive. This results in male-biased sex ratios at high scale densities, coupled with high levels of needle chlorosis, whereas low scale densities result in female-biased sex ratios. Although percentage of needle chlorosis is more important as a predictor of sex ratios ($p < 0.001$), scale density accounts for a significant amount of variation in sex ratios ($p = 0.0201$) not accounted for by chlorosis (Table 3.1). This indicates that some unknown density-dependent factor, in addition to needle chlorosis, promotes the differential survival of males over females. Even though we do not know what this additional density-dependent factor is, there is a clear relationship between scale density, needle quality, and sex ratios.

We contend that sex-ratio variation promotes gene flow among trees, because some trees produce excess males (i.e., "male exporters"), whereas other trees produce excess females (i.e., "male importers"). To quantify the degree of variation in sex ratios of scale populations within and among trees, we performed the following analyses. We measured annual variation in sex ratios within an individual tree by calculating the absolute difference in sex ratios between two scale generations over two time periods (1986 vs. 1987, and 1987 vs. 1988). This was done for all 18 trees illustrated in Figure 3.4, and a mean absolute difference in mortality was obtained. The measure of among-tree variation was calculated in a similar manner, except the two generations compared were from different trees. We then determined whether within and among-tree variation was significantly greater than zero and comparable to each other. Analysis of this variation demonstrates two major points. First, variation within- and among trees was significantly greater than zero ($p < 0.0001$ for all four groups) for both time periods (Fig. 3.5), thus demonstrating the dynamic nature of scale populations. Second, variation within trees was comparable to among-tree variation (Fig. 3.5). Consequently, there is enough asynchrony in scale-population cycles among trees that for any given year, some scale populations are likely to be "male exporters," whereas others are "male importers." Such variation should make it very difficult for scales to adapt to individual trees. We are currently testing the "male export" hypothesis by comparing sex ratios of emerging adults and sex ratios of mating swarms, where we would predict a nonsignificant correlation due to immigrating males from other trees. We are also conducting an experiment to determine if trees with greater number of females attract more males. This we do by removing all males from populations (using sticky traps to catch males as they enter the litter in October to pupate) and measuring the number of immigrating males the following spring.

Table 3.1 Multiple Regression Results Showing the Relationship between Two Predictor Variables, % Needle Chlorosis and Scale Density (#/cm needle) and Three Response Variables, % Males of Surviving Individuals, % of Individuals That Died as First Instars, and % of Individuals That Died as Second Instars

| | Multiple Regression Results | | | INDEPENDENT VARIABLE EFFECTS | | | | | |
| | | | | % Chlorosis | | | Scale Density | | |
| DEPENDENT VARIABLE | $R^2$ | $F_{(2,139)}$ | p-value | t-value | p-value | Regression Coefficient | t-value | p-value | Regression Coefficient |
|---|---|---|---|---|---|---|---|---|---|
| SEX RATIO | | | | | | | | | |
| % Males | 0.660 | 136.3 | < 0.0001 | 14.22 | < 0.0001 | 0.595 | 2.35 | 0.0201 | 0.897 |
| MORTALITY | | | | | | | | | |
| % 1st Instars | 0.612 | 110.7 | < 0.0001 | −13.74 | < 0.0001 | −0.715 | 10.49 | < 0.0001 | 4.976 |
| % 2nd Instars | 0.631 | 120.0 | < 0.0001 | 15.27 | < 0.0001 | 0.625 | −3.42 | 0.008 | −1.276 |

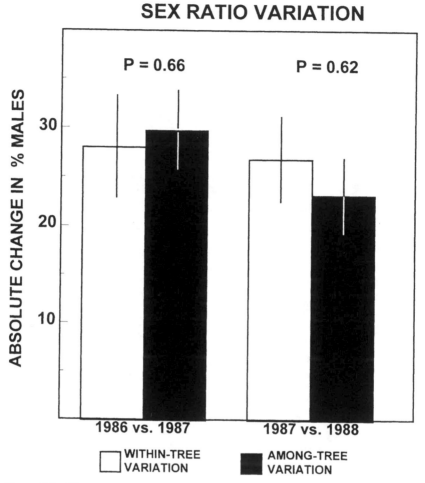

*Figure 3.5* Year-to-year variation in sex ratios on individual trees between scale generations (open bars) and among randomly paired trees (black bars). For both time periods the amount of within-tree variation is comparable to among-tree variation. Bars indicate means ± 1 *SE*.

### 3.5.1 Density-Dependent Larval Emigration Promotes Gene Flow

Another source of gene flow that would prevent deme formation is the wind dispersal of larvae. By using elevated petri plates covered with Tanglefoot© to trap emigrating larvae, we found that larval dispersal is positively density dependent (Fig. 3.6). Furthermore, the response is disproportionate; high-density populations contribute 2.5–6 times as many colonists as low-density populations.

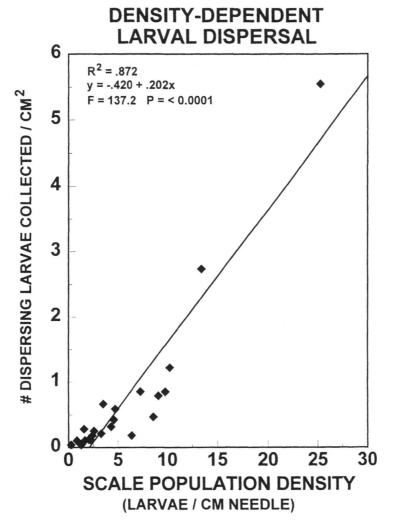

## DENSITY-DEPENDENT LARVAL DISPERSAL

$R^2 = .872$
$y = -.420 + .202x$
$F = 137.2$   $P = < 0.0001$

**# DISPERSING LARVAE COLLECTED / CM$^2$**

**SCALE POPULATION DENSITY**
**(LARVAE / CM NEEDLE)**

*Figure 3.6*   Density-dependent larval dispersal showing the disproportionate number of emigrating larvae produced by high-density scale populations.
The *x* axis denotes the number of scales/cm needle that colonized a tree; the
*y* axis denotes the number of windblown larvae captured by petri plates. Petri
plates were covered with Tanglefoot© and secured on poles that were placed
equidistant from each other and 0.25 m from the canopy edge.

The relevance of this pattern to promoting gene flow is that most scale populations on individual trees fluctuate between high- and low-density populations (Fig. 3.4B). Trees that we have monitored for three to six years have on average varied over eight-fold from low-density years (1.2 scale/cm needle) to high population density years (10.2 scale/cm needle). So the range in densities among trees shown in Figure 3.6 is representative of the range in densities exhibited by an individual tree over time. This indicates that most scale populations have the potential for producing a significant number of dispersing larvae, and that trees probably cycle between producing few or no emigrating larvae to producing relatively large numbers of emigrating larvae. We hypothesize that a scale population will vary from being a relatively important contributor to gene flow via wind-blown larvae in high-population years to producing very few larvae in low-density years.

### 3.5.2 Catastrophes That Hinder Deme Formation

El Niño–Southern Oscillation (ENSO) events can create climatic catastrophes that have severe and long-lasting effects on populations, communities, and ecosystems. These effects have been well documented for marine systems (Dayton and Tegner 1984; Glynn 1988; Glynn and Colgan 1992). In 1992, an ENSO-mediated catastrophe led to extinction of scale populations on many trees, followed by recolonization of emigrating larvae from nearby trees whose scale populations had survived (some trees have yet to be recolonized). An obvious consequence of extinction is the prevention of scales from adapting to host phenotypes. We also show that population bottlenecks are created, thereby increasing the effect of gene flow from immigrating larvae, and the effects of genetic drift, both of which would likely further hinder deme formation.

In Northern Arizona, May is typically one of the driest months of the year; however, in 1992, extensive rains occurred that were 8.5 times higher than normal. This event was associated with a massive reduction of the scale population within our study site, and scales suffered complete extinction at other sites in Northern Arizona (Christensen et al. 1995). We suspect that population reductions and extinctions were most likely due to the direct effect of rain on emerging larvae. Rain has been shown to significantly affect mortality in another scale insect (Moran and Hoffman 1987).

The effect of these rains on scale populations was variable within our study site. Extinctions occurred on 17% ($n = 23$) of infested trees; severe population crashes occurred on an additional 13% ($n = 17$), and less severe reductions on 70% ($n = 92$) of the remaining populations. Table 3.2 shows the three population responses to the 1992 ENSO event. Interestingly, the populations that went extinct after the catastrophe were only one-half the size of the other two population types one year prior to the ENSO event. This indicates that small populations of even 27,000 scales/tree were more vulnerable to extinction than populations that averaged 58,000 scales/tree. There was no difference in population size between "bottleneck" and other populations that survived prior to the rains, but there was

*Table 3.2* Impact of ENSO-Mediated Rains in May 1992. Estimated scale population sizes for populations that went extinct (EXTINCT), populations that experienced severe population reductions (BOTTLENECK), and populations that did not experience as severe population reductions as the previous two groups (REMAINING POPULATIONS). Values represent population sizes for the generation prior to the catastrophe and the generation following the catastrophe. All EXTINCT and BOTTLENECK populations were used in the calculations, 20 populations were randomly selected to represent the remaining populations.

| POPULATION TYPES | Precatastrophe Population Size | Postcatastrophe Population Size |
|---|---|---|
| EXTINCT ($n=23$) | 27,000 | Extinct* |
| BOTTLENECK ($n=16$) | 59,000 | 62 |
| REMAINING POPULATIONS ($n=20$) | 58,000 | 17,000 |

* Three remaining populations did not go extinct until the following year in 1993.

a significant difference afterwards. We believe there was a stochastic effect whereby the timing of larval emergence determined the degree to which a population was reduced. In other words, populations where peak larvae emergence coincided with heavy rain were the most decimated.

Clearly, extinction followed by prolonged establishment of a new population would deter deme formation. There may be a strong case for the "bottleneck" populations being affected as well. Bottleneck populations were operationally defined as those populations that exhibited population reductions by several orders of magnitude and had absolute population sizes below 200 individuals after the ENSO event. An estimate of the average population size for the "bottleneck" populations prior to the 1992 rains was 59,000 scales/tree compared to 62 scales/tree after the rains (Table 3.2). It is likely that the effective population was small enough to be significantly affected by immigrating larvae and males from other trees, as well as genetic drift, both of which would interfere with deme formation based on adaptation to the individual host tree. Although the effects of this ENSO event in the pinyon–scale system are not as dramatic as that observed in marine systems, where entire areas are completely decimated (Glynn 1988), such environmental variability is one important factor that, in conjunction with other factors we develop in the chapter, may prevent scales from acquiring "fine-tune" adaptation to individual host trees.

### 3.5.3 Short-Term Differential Mortality

Scale survival within an individual tree can change significantly from year to year due to density-dependent mortality. As the density of scales increases, food resource quality decreases dramatically due to chlorosis, leading to increased host-related mortality of scales. We argue that extreme variation in survivorship from one year to the next could lead to (1) differential selection of scale genotypes that survive better under conditions of high scale densities versus low densities, and/or

(2) relaxed selection against genotypes under conditions of low scale density. These scenarios could lead to the decreased ability of a scale population to genetically track its host plant. Although we do not have genetic evidence for differential selection of genotypes, differential selection at different densities has been established or inferred for several species (Begon 1984; Wall and Begon 1986; Dunham et al. 1990; Weber 1990; Bagley et al. 1994; Santos et al. 1994).

Using scale mortality as a bioassay of changing selection pressures, we show that selection pressure can vary greatly from year to year. This annual variation in scale mortality is illustrated for each tree in Figure 3.4D. For example, on tree #9 over a three-year period, the mortality of scales changed from 44% to 95% to 70%, and over the same time period, tree #18 mortality changed from 84% to 5% to 24%.

The year-to-year variation in the mortality of scales for individual trees rivals the variation we observed among trees in the population (Fig. 3.7). To compare within- and among-tree variation in scale mortality, we calculated within-tree variation as the mean absolute difference in scale mortality from one year to the next, for two time periods (1986–1987 and 1987–1988). Using the same 18 trees, we also compiled among-tree variation from one year to the next by creating random tree pairs. For example, the mortality of scales on tree #18 in 1986 was subtracted from the mortality of scales from another tree in the population in 1987 (a random pair). Figure 3.7 shows the mean absolute change in percentage mortality for the two time periods (1986 vs. 1987, and 1987 vs. 1988) within single trees and between 36 randomly paired trees (18 pairs). Both within- and among-tree variation was significantly greater than zero for both time periods ($p < 0.0001$ for all four means). We also observed the same degree of variation within trees between years as we see between randomly paired trees (Fig. 3.7), where the year-to-year variation in scale mortality for individual trees equals the variation in scale mortality among trees in the population. This high degree of year-to-year variation within a single tree would make it very difficult for scales to evolutionarily track individual trees from other trees in the population. Additionally, it is difficult to envision how deme formation could occur when variation in scale performance is as great within trees as it is among trees.

The changes in mortality are clearly related to changes in scale density and needle chlorosis (i.e., food-resource quality). We can explain over 60% of the variation in first and second instar mortality by knowing the percentage of needles that were chlorotic due to scale feeding and scale density (Table 3.1). Mortality listed in Table 3.1 is separated into the percentage of individuals that died as first instars and those that died as second instars. Sixty-one percent of the variation in the number of individuals that died as first instars is explained by needle chlorosis and scale density, although density is positively correlated, and chlorosis is negatively correlated with percentage of the population that died as first instars (see regression coefficients). The opposite is true for the percentage of individuals that died as second instars, where chlorosis is positively correlated, and

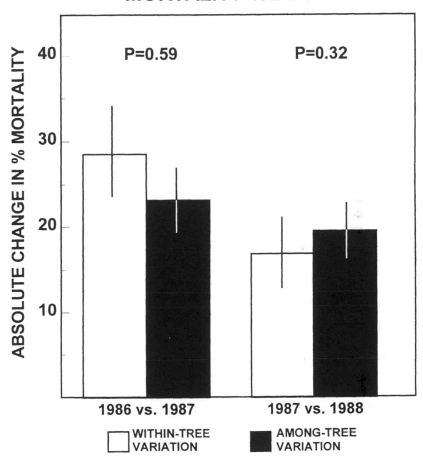

*Figure 3.7* Year-to-year variation in scale mortality on individual trees between years (open bars) and among randomly paired trees (black bars). For both time periods, the amount of within-tree variation in scale mortality is comparable to among-tree variation. Bars indicate means ± 1 *SE*.

density is actually negatively correlated. The effect of needle chlorosis on second instar death is easily understood: The second instar is the stage when most growth occurs, and it is 10–20 times larger in volume than first instars. Hence, when most of the individuals pass successfully through the first instar, they can greatly affect needle damage, because high densities of second instars can completely kill a needle before adults are ready to emerge. We do not yet understand why scale density *per se* leads to increasing numbers of individuals that die as first instars, considering that they produce very little chlorosis. Evidently, there is some un-known induced response by the plant that mediates high first instar death at high scale densities. These data indicate that at intermediate scale densities, many scales survive the first instar and subsequently die in their second instar as a result of resource depletion, whereas at high scale densities, many individuals die as first instars before extensive chlorosis occurs.

### 3.5.4 Long-term Differential Selection—Developmental Immunity

A major prediction of the DF hypothesis is that deme formation is most likely to form on long-lived hosts. Several studies, however, show that phase shifts in plant development, either genetic turning on and off of genes (ontogenetic) or re-source limitation to shoots (senescence or physiological aging) result in major changes of plant resistance traits through time (Zagory and Libby 1985; Kearsley and Whitham 1989). We propose that developmental changes in long-lived host plants occur over their life span, making them temporal habitat mosaics that are difficult for scale insects to evolutionarily track.

In support of the developmental changes in pinyons, we find that scales are al-most entirely restricted to juvenile trees (Fig. 3.8A), demonstrating that scales do not have the entire life span of the tree to form demes. A survey of trees in 1996 showed that scales infest 79% of the juvenile trees (32 yrs old), 27% of the inter-mediate aged trees (60 yrs old), and only 8% of mature trees (150 yrs old). This pattern of infestation is comparable to that found in a survey conducted in 1985 (Cobb et al. 1994). Additionally, scale density also decreases dramatically on these three age classes (Fig. 3.8B), where scale densities on juvenile trees were 3.5 and 17 times higher than intermediate-aged and mature trees, respectively. This demonstrates that even the most susceptible trees in the two older age classes do not support populations as dense as those found on juvenile trees.

These observational data indicate that scales do not progressively become more adapted to natal trees, but lose the ability to successfully attack trees as they age and become mature. Although we need to conduct transfer experiments to confirm that individual trees become more resistant with age, experiments in an-other system showed increased resistance with tree age. When gall aphids were transferred to different-aged ramets within the same cottonwood clone, a pre-dictable 10-fold change in resistance occurred over a two-year period as trees shift from juvenile to mature phases (Kearsley and Whitham 1989). Such devel-opmental changes are widespread and can affect diverse taxa associated with the host tree (Waltz and Whitham unpublished data).

## 3.6 Summary

Although the pinyon needle scale has many of the attributes of an organism that is most likely to make "fine-scale" adaptations to individual trees (Edmunds and Al-stad 1978, 1981), we found no evidence to support the DF hypothesis. We pro-

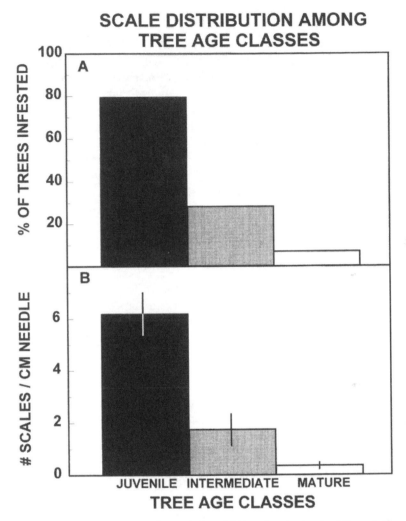

## SCALE DISTRIBUTION AMONG TREE AGE CLASSES

*Figure 3.8* The percentage of trees infested (A) and scale density (B) are both inversely related to developmental stage of the tree. This 1996 survey strongly indicates that scales do not become progressively more adapted to their host trees.

vide two sources of evidence demonstrating that scales have not acquired special adaptations to individual trees but are equally capable of attacking most juvenile trees, which make up approximately 80% of the pinyon population. First, in a reciprocal transfer experiment, scales survived just as well on novel trees that had previously supported scale populations as they did on their natal trees. Second, in comparing scale mortality of incipient scale populations with established scale populations, we found no differences in mortality, even when the effect of scale density was eliminated as a confounding variable.

Although we found no evidence for deme formation using scale mortality as a bioassay of tree resistance, we show that individual trees were either highly resistant or susceptible to scale attack. The group of juvenile trees that supported no natural scale populations was found to be highly resistant to scale attack; scales experimentally transferred to these trees suffered 91% mortality. These resistant trees make up approximately 11% of all juvenile trees. In contrast, scales experimentally transferred to trees that supported natural scale populations only experienced 50% mortality. These susceptible trees make up approximately 89% of all juvenile trees in the population. Although the presence of resistant trees should represent a selection pressure favoring scale biotype and/or deme formation, a seven-year ongoing experiment currently provides no evidence of such local adaptation.

We examined several mechanisms that could prevent deme formation by *M. acalyptus* on individual trees by increasing gene flow. First, the fluctuation of scale populations on individual trees from year to year can promote gene flow among trees. Fluctuating population size promotes the survival of males at high densities, whereas low scale densities promote the production of more females than males. Because there is asynchrony among scale populations on different trees, some trees will be "male exporters," whereas other trees will be "male importers" (male export hypothesis). Second, high population densities also leads to increased larval dispersion, which will further promote gene flow. Third, gene flow can also be promoted by catastrophic events that can lead to (1) the extinction of scales on certain trees, which are subsequently colonized by scales from adjacent populations, or (2) population bottlenecks, which would be relatively more affected by immigrants and subject to genetic drift.

Changing selection pressures can confound deme formation and select for more generalized scale genotypes. Several levels of changing selection pressures contribute to this evolutionary dilemma for scales. First, short-term variation (year to year) in scale mortality on individual trees largely brought about by changing scale densities and density-dependent selection results in selection of different scale genotypes at different scale densities, or relaxed selection at low scale densities. Second, over longer periods of time, in our case, decades, selection regimes due to developmental changes in the host plant may increase resistance to scales and overwhelm any local adaptation occurring in the herbivore population. This is supported by the fact that the percentage of trees infested by scales decreases sharply from young to mature trees. These last two conclusions still need experimentation to confirm that differential selection within a scale population is a consequence of changing resource quality, and that pinyon resistance to scale attack increases with developmental aging. Despite this, the observational patterns are certainly strong enough to justify our conclusions, and determination of differential selection is a critical avenue for further research.

Third, although less studied in this system, other studies have shown that individual trees are highly heterogeneous resources and are mosaics of resistance. In a series of papers, Whitham and his colleagues (Whitham 1981, 1983; Whitham and Slobodchikoff 1981; Whitham et al. 1984) have argued for the importance of within-tree variability in preventing herbivores from tracking their host plants. Most of these arguments were based on variation in resource quality that we have not mentioned and include differences in resource quality (1) within an individual leaf, (2) between terminal and lateral leaves, (3) among individual branches of the same developmental stage, and (4) between the lower juvenile and the upper mature portion of the canopy. Our work on the scale-pinyon system further underscores the importance of within-tree variability that can exist even in small trees (~2.5 m³ canopy volume).

In conclusion, although we find extreme variation in plant resistance traits that should favor deme formation, the variation is so fine-scale at both spatial and temporal levels that we doubt that adaptations to individual trees could occur. However, at the landscape level we believe local adaptation is possible, and we are currently pursuing these studies. For example, the pinyon needle scale is only found in a few areas within the range of pinyon pine in Arizona. Scales are restricted to either *P. edulis* populations that experience high levels of edaphic stress, or hybrid populations of *P. edulis* and *P. californiarum*. Are these populations of *M. acalyptus* the same, or have they genetically diverged, one adapted to "stressed" *P. edulis* populations and another adapted to hybrid pinyons? Furthermore, is there local adaptation among different "stressed" host populations and among different hybrid populations? It is important to know whether a species such as *M. acalyptus*, which does not form demes on individual trees, would become local adapted at larger spatial levels. Or are herbivores that exhibit fine-scale adaptation to host plants more likely to form biotypes or host races? For example, Mopper et al. (1995) found adaptation by *Stilbosis quadricustatella* at three different spatial levels: individual trees, populations of different trees in the host species, and different host species. Understanding the spatial level at which species exhibit genetic adaptations, and the factors responsible for these patterns, remains a fundamental question in evolutionary ecology.

## Acknowledgments

We thank U.S. National Forest Service and Sunset Crater National Monument personnel for their cooperation during our study. We are grateful to Greg Martinsen, Susan Mopper, and especially Sharon Strauss for insightful discussions and/or reviewing the manuscript. The research was supported by the United States Department of Agriculture Grant Nos. 91-37302-6224 and 92-37302-7854, and the National Science Foundation Grant Nos. BSR-9107042 and DEB-9408009, and Dept. of Energy Grant No. 94ER61849.

## 3.7 References

Alstad, D. N. and K. W. Corbin. 1990. Scale insect allozyme differentiation within and between host trees. *Evol. Ecol.* 4:43–56.

Alstad, D. N. and G. F. Edmunds Jr. 1983a. Selection, outbreeding depression, and the sex ratio of scale insects. *Science 220*:93–95.

Alstad, D. N. and G. F. Edmunds Jr. 1983b. Adaptation, host specificity, and gene flow in the black pineleaf scale. Pp. 413–426 *in* R. F. Denno and M. S. McClure (Eds.), *Variable Plants and Herbivores in Natural and Managed Systems.* Academic Press, New York.

Alstad, D. N. and G. F. Edmunds Jr. 1987. Black pineleaf scale population density in relation to interdemic mating (Hemiptera: Diaspididae). *Annu. Rev. Soc. Amer.* 7:532–536.

Bagley, M. J., B. Bentley, and G. A. E. Gall. 1994. A genetic evaluation of the influence of stocking density on the early growth of rainbow trout (*Onchorynchus mykiss*). *Aquaculture 121*:313–326.

Begon, M. 1984. Density and individual fitness: Asymmetric competition. Pp. 175–194 *in* B. Shorrocks (Ed.), *Evolutionary Ecology.* Blackwell Scientific Publishing, Oxford, UK.

Bradshaw, A. D. 1984. Ecological significance of genetic variation between populations. Pp. 213–228 *in* R. Dirzo and J. Sarukhan (Eds.), *Perspectives in Plant Population Biology.* Sinauer Associates, Sunderland, MA.

Brown, A. H. D. 1979. Enzyme polymorphism in plant populations. *Theor. Pop. Biol. 15*:1–42.

Christensen, K. M., T. G. Whitham, and P. Keim. 1995. Herbivory and tree mortality across a pinyon pine hybrid zone. *Oecologia 101*:29–36.

Cobb, N. S. 1990. *The role of host heterogeneity in preventing herbivore deme formation on individual trees.* M.S. thesis, Northern Arizona University, Flagstaff, AZ.

Cobb, N. S., J. B. Mitton, and T. G. Whitham. 1994. Genetic variation associated with chronic water and nutrient stress in pinyon pine. *Am. J. Bot. 81*:936–940.

Cobb, N. S. and T. G. Whitham. 1993. Herbivore deme formation on individual trees: A test case. *Oecologia 94*:496–502.

Craig, T. P., P. W. Price, K. M. Clancy, G. W. Waring, and C. F. Sacchi. 1988. Forces preventing coevolution in a three trophic level system: Willow, a gall forming herbivore, and parasitoid. Pp. 57–80 *in* K. C. Spencer (Ed.), *Chemical Mediation of Coevolution.* Academic Press, New York.

Dayton, P. K. and M. J. Tegner. 1984. Catastrophic storms, El Niño, and patch stability in a southern California kelp community. *Science 224*:283–285.

Del Vecchio, T. A., C. A. Gehring, N. S. Cobb, and T. G. Whitham. 1993. Negative effects of scale insect herbivory on the ectomychorrhizae of juvenile pinyon pine. *Ecology 74*:2297–2302.

Doane, C. C. 1966. Evidence for a sex attractant in female of the red pine scale. *J. Econ. Entomol. 59*:1539–1540.

Dunham, R. A., R. E. Brummett, M. O. Ella, and R. O. Smitherman. 1990. Genotype-environment interactions for growth of blue, channel and hybrid catfish in ponds and cages at varying densities. *Aquaculture 85*:143–151.

Edmunds, G. F. Jr. 1973. The ecology of black pineleaf scale. *Environ. Entomol. 2*:765–777.

Edmunds, G. F. Jr. and D. N. Alstad. 1978. Coevolution in insect herbivores and conifers. *Science 199*:941–945.

Edmunds, G. F. Jr. and D. N. Alstad. 1981. Responses of black pineleaf scales to plant variability. Pages 29–38 *in* R. F. Denno and H. Dingle (Eds.), *Insect Life History Patterns.* Springer-Verlag, New York.

Endler, J. A. 1977. *Geographic Variation, Speciation and Clines.* Princeton University Press, Princeton, NJ.

Fay, P. A. and T. G. Whitham. 1990. Within-plant distribution of a galling adelgid (Homoptera: Adelgidae): The consequences of conflicting survivorship, growth, and reproduction. *Ecol. Entomol. 15*:245–254.

Feder, J. L., C. A. Chilcote, and G. L. Bush. 1988. Genetic differentiation between sympatric host races of the apple maggot fly *Rhagoletis pomonella. Nature 336*:61–64.

Futuyma, D. J. and S. C. Peterson. 1985. Genetic variation in the use of resources by insects. *Annu. Rev. Entomol. 30*:217–238.

Gallun, R. L., K. J. Starks, and W. D. Guthrie. 1975. Plant resistance to insects attacking cereals. *Annu. Rev. Entomol. 20*:337–357.

Glynn, P. W. 1988. El Niño-Southern Oscillation 1982–1983: Nearshore population, community, and ecosystem responses. *Annu. Rev. Ecol. Syst. 19*:309–345.

Glynn, P. W. and M. W. Colgan. 1992. Sporadic disturbances in fluctuating coral reef environments: El Niño and coral reef development in the eastern Pacific. *Am. Zool. 32*:707–718.

Gould, F. 1983. Genetics of Plant-herbivore systems: Interactions between applied and basic studies. Pp. 599–653 *in* Denno, R. F. and M. S. McClure (Eds.), *Variable Plants and Herbivores in Natural and Managed Systems.* Academic Press, New York.

Hanks, L. M. and R. F. Denno. 1992. The role of demic adaptation in colonization and spread of scale insect populations. Pp. 393–411 *in* K. C. Kim (Ed.), *Evolution of Insect Pests: The Patterns of Variations.* John Wiley, New York.

Hanks, L. M. and R. F. Denno. 1994. Local adaptation in the armored scale insect *Pseudaulacaspis pentagona* (Homoptera: Diaspididae). *Ecology 75*:2301–2310.

Karban, R. 1987. Herbivory dependent on plant age: A hypothesis based on acquired resistance. *Oikos 48*:336–337.

Karban, R. 1989. Fine-scale adaptation of herbivorous thrips to individual host plants. *Nature 340*:60–61.

Kearsley, M. C. and T. G. Whitham. 1989. Developmental changes in resistance to herbivory: Implications for individuals and populations. *Ecology 70*:422–434.

Kimberling, D. N. and P. W. Price. 1996. Variability in grape *Phylloxera* preference and performance on canyon grape (*Vitis arizonica*). *Oecologia, 107*:553–559.

Krombein, K. D., P. D. Hurd Jr., D. R. Smith, and B. D. Burks. 1979. *Catalog of Hymenoptera in America North of Mexico,* 3 vols. Smithsonian Press, Washington, DC.

Larsson, S. 1985. Seasonal changes in the within-crown distribution of the aphid *Cinara pini* on Scots Pine. *Oikos* 45:217–222.

Lewontin, R. C. 1974. *The Genetic Basis of Evolutionary Change.* Columbia University Press, New York.

McCambridge, W. F. and D. A. Pierce. 1964. Observations on the life history of the pinyon needle scale *Matsucoccus acalyptus* (Homoptera: Coccoidea: Margarodidae). *Entomol. Soc. Amer. Ann.* 57:197–200.

Memmott, J., R. K. Day, and H. C. J. Godfray. 1995. Intraspecific variation in host plant quality: The aphid *Cinara cupressi* on the Mexican cypress, *Cupressus lusitanica. Ecol. Entomol.* 20:153–158.

Mitter, C. and D. J. Futuyma. 1979. Population genetic consequences of feeding habits in some forest Lepidoptera. *Genetics* 92:1005–1021.

Mopper, S., M. Beck, D. Simberloff, and P. Stiling. 1995. Local adaptation and agents of selection in a mobile insect. *Evolution* 49:810–815.

Moran, V. C. and J. H. Hoffmann. 1987. The effects of simulated and natural rainfall on cochineal insects (Homoptera: Dactylopiidae): Colony composition and survival on cactus cladodes. *Ecol. Entomol.* 12:61–68.

Moran, N. A. and T. G. Whitham. 1988. Predicting population fluctuations of organisms with complex life cycles: An aphid example. *Ecology* 69:1214–1218.

Park, S. C., J. R. West, L. P. Abrahmson, G. N. Lanier, and R. M. Silverstein. 1986. Cross-attraction between two species of *Matsucoccus*: Extraction, bioassay, and isolations of the sex pheromone. *J. Chem. Ecol.* 12:609–617.

Parker, M. A. 1985. Local population differentiation for compatibility in an annual legume and its host-specific fungal pathogen. *Evolution* 39:713–723.

Pilson, D. 1992. Relative resistance of goldenrod to aphid attack: Changes through the growing season. *Evolution* 46:1230–1236.

Rice, W. R. 1983. Sexual reproduction: An adaptation reducing parent–offspring contagion. *Evolution* 37:1317–1320.

Santos, M., K. Fowler, and L. Partridge. 1994. Gene–environment interaction for body size and larval density in *Drosophila melanogaster*: An investigation of effects on development time, thorax length and adult sex ratio. *Heredity* 72:515–521.

Service, P. 1984. Genotypic interactions in an aphid–host plant relationship: *Uroleucon rudbeckiae* and *Rudbeckia laciniata. Oecologia* 61:271–276.

Strauss, S. Y. 1997. Lack of evidence for local adaptation to individual plant clones or site by a mobile specialist herbivore. *Oecologia* 110:77–85.

Thompson, J. N. 1994. *The Coevolutionary Process.* University of Chicago Press, Chicago, IL.

Unruh, T. R. 1985. Insect–plant interactions of the pinyon needle scale and single leaf pinyon in Southern California. Ph.D. dissertation. University of California, Riverside.

Unruh, T. R. and R. F. Luck. 1987. Deme formation in scale insects: A test with the pinyon needle scale and a review of other evidence. *Ecol. Entomol.* 12:439–449.

Van Valen, L. 1965. Morphological variation and the width of the ecological niche, *Am Nat.* 99:377–390.

Wainhouse, D. and R. S. Howell. 1983. Intraspecific variation in beech scale populations and in susceptibility of their host *Fagus sylvatica. Ecol. Entomol.* 8:351–359.

Wall, R. and M. Begon. 1986. Population density, phenotype and mortality in the grasshopper *Chorthippus brunneus. Ecol. Entomol.* 11:445–456.

Waser, N. M. and M. V. Price. 1985. Reciprocal transplant experiments with *Delphinium nelsonii* (Ranunculaceae): Evidence for local adaptation. *Am. J. Bot.* 72:1726–1732.

Weber, K. E. 1990. Increased selection response in larger populations. I. Selection for wing-tip height in *Drosophila melanogaster* at three population sizes. *Genetics* 125:579–584.

Whitham, T. G. 1981. Individual trees as heterogeneous environments: Adaptation to herbivory or epigenetic noise? Pp. 9–28 *in* R. F. Denno and H. Dingle (Eds.), *Insect Life History Patterns.* Springer-Verlag, New York.

Whitham, T. G. 1983. Host manipulation of parasites: Within-plant variation as a defense against rapidly evolving pests. Pp. 15–41 *in* R. F. Denno and M. S. McClure, (Eds.), *Variable Plants and Herbivores in Natural and Managed Systems.* Academic Press, New York.

Whitham, T. G. and C. N. Slobodchikoff. 1981. Evolution by individuals, plant-herbivore interactions, and mosaics of genetic variability: The adaptive significance of somatic mutations in plants. *Oecologia* 49:287–292.

Whitham, T. G., A. G. Williams, and A. M. Robinson. 1984. The variation principle: Individual plants as temporal and spatial mosaics of resistance to rapidly evolving pests. Pp. 15–51 *in* P. W. Price, C. N. Slobodchikoff, and W. S. Gaud (Eds.), *A New Ecology: Novel Approaches to Interactive Systems.* John Wiley, New York.

Wright, S. 1968. *Evolution and the Genetics of Populations, Vol. 1. Genetic and Biometric Foundations.* University of Chicago Press, Chicago, IL.

Wright, S. 1969. *Evolution and the Genetics of Populations: Vol. 2. The Theory of Gene Frequencies.* University Chicago Press, Chicago, IL.

Wright, S. 1978. *Evolution and the Genetics of Populations: Vol. 4. Variability within and among Natural Populations.* University Chicago Press, Chicago, IL.

Young, B., D. R. Miller, and M. S. McClure. 1984. Attractivity of the female sex pheromone of Chinese *Matsucoccus matsumurae* (Kuwana) to males of *M. resinosae* Bean and Godwin in the United States (Margarodidae, Coccoidea, Homoptera). *Contr. Shanghai Inst. Entomol.* 4:1–20.

Zagory, D. and W. J. Libby. 1985. Maturation-related resistance of *Pinus radiata* to western gall rust. *Phytopath.* 75:1443–1447.

# 4

# Local Adaptation in Specialist Herbivores: Theory and Evidence

*William J. Boecklen*
Department of Biology, New Mexico State University, Las Cruces, NM

*Susan Mopper*
Department of Biology, University of Southwestern Louisiana, Lafayette, LA

## 4.1 Introduction

Evolutionary ecologists are interested in microgeographic genetic structure in herbivore populations and its importance to herbivore population dynamics (Slatkin 1987, Mopper 1996a, Gandon et al., Chapter 13, this volume, McCauley and Goff, Chapter 9, this volume, Peterson and Denno, Chapter 12, this volume). Two major patterns have emerged from the pioneering work by Bush (1969) and Edmunds and Alstad (1978): Polyphagous herbivores exhibit distinct "races" adapted to sympatrically distributed host species (Bush 1969; Pashley 1988; Akimoto 1990; Feder et al. 1990, Chapter 16, this volume; Via 1991; Carroll and Boyd 1992), and specialist herbivores display genetic differentiation—demes—at the spatial scale of individual conspecific host plants (McCauley et al. 1988; McPheron et al. 1988; Alstad and Corbin 1990; Komatsu and Akimoto 1995). The fine-scale partitioning of herbivore populations into demes has been implicated as an important factor promoting discontinuous distributions of herbivores within natural populations of host plants (Edmunds and Alstad 1978; Wainhouse and Howell 1983; Hanks and Denno 1994; Mopper 1996a), and has been cited as an important variable in understanding herbivore outbreaks (Wainhouse and Howell 1983).

A central question underlying studies of microgeographic genetic variation within herbivore populations is whether the variation is adaptive. In other words, are herbivores locally adapted to their hosts? Demes may form in response to natural selection or genetic drift, and therefore may be adaptive or nonadaptive. The adaptive value of a deme is its ability to survive and reproduce relative to other demes. Gene flow among local demes can counteract genetic drift (unpredictable changes in gene frequencies associated with small population size), and it can eliminate genetic structure produced by natural selection. However, if the agents of selection are sufficiently strong, structure will prevail despite substantial gene flow (Slatkin 1987; Strauss and Karban, Chapter 8, this volume). While there is convincing experimental evidence of host-race formation—local adaptation to sympatrically distributed host species (Akimoto 1990; Sandstrom 1996; Feder et al., Chapter 16, this volume), the experimental evidence for herbivore adaptation to individual host plants is less compelling and remains controversial (see Unruh and Luck 1987; Cobb and Whitham 1993; Alstad, Chapter 1, this volume).

Edmunds and Alstad (1978) provided the first experimental evidence of "host tracking" in the black pineleaf scale, *Nuculaspis californica*, on ponderosa pine, *Pinus ponderosa*. Cohorts of scales were transferred among natal and novel trees. Scales survived best on natal trees, indicating local adaptation of scale demes at the level of individual trees. However, potential flaws in the experimental design have called into question the existence of adaptive deme formation in this system (Unruh and Luck 1987; Karban 1989; Cobb and Whitham 1993; Alstad, Chapter 1, this volume).

Transfer experiments have subsequently been conducted in a number of plant-herbivore systems with mixed results. For example, Mopper et al. (1995) observed higher survival rates among leafminers transferred to natal trees than to novel trees, whereas Unruh and Luck (1987) found no significant differences in scale insect survival between natal and novel hosts. In a recent literature review, Cobb and Whitham (Chapter 3, this volume) cite two studies that provide strong evidence of adaptive deme formation by phytophagous insects (Karban 1989; Mopper et al. 1995), two studies that provide partial evidence (Wainhouse and Howell 1983; Hanks and Denno 1994), and five studies that provide no evidence (Rice 1983; Unruh and Luck 1987; Cobb and Whitham 1993; Memmott et al. 1995; Kimberling and Price 1996). In contrast, Mopper (1996a) reviewed much of the same literature and reported that 6 of 10 studies support the hypothesis of adaptive deme formation. Similarly, Stiling and Rossi (Chapter 2, this volume) list 5 of 8 studies that provide experimental evidence of local adaptation.

What exactly constitutes adaptive deme formation in herbivore populations? That different researchers can review practically the same literature yet draw such different conclusions indicates that diverse standards of evidence are used, and underscores the need for a critical review of the theory of adaptive deme formation and the evidence adduced to support and refute it. Here we attempt such a review.

## 4.2 The Theory of Adaptive Deme Formation

Ongoing investigation of adaptive deme formation has brought about several modifications and extensions of the original hypothesis proposed by Edmunds and Alstad (1978, 1981). Therefore, we begin our review with an explicit formulation of the adaptive deme formation hypothesis. Our formulation of the hypothesis is similar to Edmunds and Alstad's original construction, but is generalized to include dispersive as well as sessile insects (e.g., Ayers et al. 1987; Mopper et al. 1995; Stiling and Rossi, Chapter 2, this volume), and non-defense-related plant characteristics (see Komatsu and Akimoto 1995). See Gandon et al. (Chapter 13, this volume), Peterson and Denno (Chapter 12, this volume), and Costa (Chapter 10, this volume), for detailed discussions of the theory and evidence regarding the influence of host and herbivore behavior and dispersal patterns on gene flow and local adaptation.

We propose a formulation of the adaptive deme formation hypothesis that strongly emphasizes process (natural selection) over pattern (herbivore demes). Our goal is to redirect hypothesis testing toward mechanisms, not mere demonstrations of the presence or absence of predicted patterns. We consider adaptive deme formation to be one of several competing mechanisms that can structure herbivore populations into demes. By emphasizing *mechanism,* we avoid casting the adaptive deme formation hypothesis as an existential hypothesis, which is verifiable but not falsifiable, and therefore, outside the domain of empirical science (Popper 1959). A mechanistic construction comes with a cost—it limits the number of experimental systems that can rigorously test the hypothesis, since the pattern must be present before the underlying mechanisms can be tested (see below). Existential hypotheses have no such constraint. We recognize that these philosophical considerations might carry little weight with experimentalists, but we argue that systems lacking the appropriate conditions for deme formation cannot provide rigorous tests of the deme formation hypothesis.

### 4.2.1 The Adaptive Deme Formation Hypothesis

The spatial heterogeneity in nutritive, defensive, or phenological traits within populations of long-lived plants structures populations of short-lived, specialized herbivores into demes adapted to the phenotypes of individual hosts.

In his retrospective of the black pineleaf scale–ponderosa pine system, Alstad (Chapter 1, this volume) suggests that the adaptive deme formation hypothesis (also known as the local-adaptation hypothesis) may be divided into intrinsic and extrinsic hypotheses. The distinction is based on the nature of the plant traits driving the adaptive response of herbivore demes and represents an explicit recognition of the prominent role that environmental factors can play in plant–herbivore interactions. The intrinsic hypothesis emphasizes adaptation of herbivore demes to plant traits that are under strong genotypic control, whereas the extrinsic hypothesis is directed to traits controlled primarily by the environment or by interactions between plant genotypes and the environment. Alstad concludes that much of the original evidence inferred by Edmunds and Alstad (1978) as supporting the intrinsic (plant genotype) local-adaptation hypothesis, is more indicative of extrinsic (environmental) local adaptation.

We will not adopt a distinction between extrinsic and intrinsic factors in our construction of the adaptive deme formation hypothesis, or in our review of the experimental evidence. Interactions between herbivores and host plants are typically complex and often involve several, simultaneously covarying aspects of the host phenotype. It would be a rare case indeed in which herbivore demes are adapted to host phenotypes determined solely by genotypic effects. The subdivision of host traits into unambiguous categories that are either under strict genotypic or environment control is impractical; therefore, partitioning the adaptive deme formation hypothesis into intrinsic and extrinsic hypotheses is not operational (*sensu* Brady 1979) in natural systems.

## 4.3 Corollaries to the Adaptive Deme Formation Hypothesis

We have identified from the literature four major predictions that have assumed such a prominent role in testing the adaptive deme formation hypothesis that they may be considered corollaries, or natural consequences, of the main hypothesis.

- **Corollary 1:** Herbivore adaptation to an individual host plant is maladaptive for colonizing other conspecific host plants. Corollary 1 derives directly from Edmunds and Alstad's (1978) original test of the adaptive deme formation hypothesis and is motivated by their observation that individual host plants varied in defensive chemistry and therefore exerted different selection pressures on their herbivore colonizers. All subsequent experimental studies of the adaptive deme formation hypothesis have tested this prediction by measuring herbivore performance following transfers to natal and novel hosts plants.

- **Corollary 2:** Herbivore performance and population growth increase through time as herbivore demes become increasingly adapted to their host plants. Corollary 2 was first presented by Edmunds and Alstad (1978) as a positive correlation between scale density and the age of the host tree, which supported the adaptive deme formation hypothesis in black pineleaf scale. It has since been considered a major prediction of the theory (see Wainhouse and Howell 1983; Unruh and Luck 1987; Cobb and Whitham, Chapter 3, this volume).

- **Corollary 3:** Adaptive deme formation is facilitated when host plants are spatially isolated. Corollary 3 is based primarily on the results of Hanks and Denno (1994, Chapter 11, this volume) and Wainhouse and Howell (1983), who observed local adaptation among herbivores on spatially isolated hosts.

- **Corollary 4:** For haplodiploid herbivores, haploids (usually males) suffer higher mortality on novel hosts than diploids. Therefore, as demes adapt to the host plant, haploid survival and density rise. Corollary 4 (here referred to as "haploid handicap") is based primarily on the ponderosa pine–black pineleaf scale interaction (Alstad et al. 1980; Alstad and Edmunds 1983a, 1983b; see Unruh and Luck 1987; Alstad, Chapter 1, this volume).

## 4.4 Experimental Evidence of Adaptive Deme Formation

The adaptive deme formation hypothesis envisions a close genetic match between herbivore demes and host-plant phenotype ("herbivore tracking"), and predicts that herbivores will suffer lower relative fitness on novel hosts compared with natal plants (see Corollary 1). Artificial transfers of herbivores onto natal and novel host plants provide the primary experimental evidence adduced in support or refutation of the adaptive deme formation hypothesis (but see Komatsu and Akimoto 1995). Colonization success, survival, fecundity, and other aspects of herbivore performance are used as assays of adaptation. For example, reduced herbivore performance on novel plants compared to natal plants is indicative of local adaptation and deemed support of the adaptive deme formation hypothesis (e.g., Mopper et al. 1995), whereas, similar performance on natal and novel plants, or superior performance on novel plants, is inconsistent with the hypothesis (e.g., Unruh and Luck 1987).

Herbivore transfers are also the primary experimental evidence adduced in support or refutation of the haploid handicap hypothesis. Corollary 4 predicts high initial rates of haploid male mortality (compared to that of diploid females) following colonization of novel hosts, with increasing male survival and densities thereafter. To date, transfers of the armored scale, *Pseudaulacaspis pentagona* (Hanks and Denno 1994, Chapter 11, this volume), and the black-pineleaf scale, *Nuculaspis californica* (Alstad and Edmunds 1983a, 1983b, 1987; Alstad, Chapter 1, this volume) represent the only experimental evidence explicitly testing the haploid handicap hypothesis. Hanks and Denno (1994) inferred from generally similar male and female survival following reciprocal transfers between spatially isolated trees that ploidy did not influence male colonization success relative to females. In contrast, greater intragenerational male black pineleaf scale mortality relative to females following exposure to the insecticide Malathion resulted in an increasingly female-biased cohort (Edmunds and Alstad 1985; Alstad and Edmunds 1989). In addition, transfer experiments suggested that male pineleaf scale insects were more vulnerable to gene flow and its potential to disrupt genetic adaptations to the natal host plant (Alstad and Edmunds 1983a, 1987, Chapter 1, this volume).

Herbivore transfer experiments, like any assay used as a diagnostic test, are subject to false positives and to false negatives. A false positive occurs when herbivore populations lack local differentiation (demic structure), yet nevertheless exhibit superior performance on natal compared to novel hosts. A number of mechanisms that do not involve genetic adaptation of herbivores to individual hosts may account for this. For example, a mismatch between herbivore and host phenologies can reduce performance. In addition, maternal effects (see Unruh and Luck 1987; Rossiter, Chapter 6, this volume), physiological acclimation (Karban 1989), and environmental differences (Hanks and Denno 1994; Alstad,

Chapter 1, this volume) have been proposed as alternatives to adaptive deme formation in explaining reduced herbivore performance on novel hosts following experimental transfers.

A plurality of mechanisms can confound interpretations of transfer experiments and undermine evidence supporting the adaptive deme formation hypothesis. For example, Karban (1989) argued that the transfer experiments of Edmunds and Alstad (1978) could not distinguish between adaptation to individual hosts (adaptive deme formation) and adaptation to sites (environmental effects), because all intertree transfers involved donor trees and recipient trees from distant sites. In the same vein, Karban criticized the transfer experiments of Wainhouse and Howell (1983), because recipient hosts were not arranged in a common garden. Hanks and Denno (1994) observed in their own transfer experiments and those of Wainhouse and Howell (1983) that evidence of adaptive deme formation appeared only when recipient trees were relatively isolated from other heavily infested trees. In fact, Hanks and Denno (1994) suggested that the failure to detect adaptive deme formation in other experiments (Rice 1983; Unruh and Luck 1987; Cobb and Whitham 1993) was attributable to the close proximity of heavily infested trees to experimental trees, and argued that the "negative" results provided circumstantial evidence for the importance of isolation in adaptive deme formation (see Corollary 3). An alternative explanation is the spatial autocorrelation of environmental factors, in that isolated trees represent distinct environments, potentially confounding adaptive deme formation with environmental differences. These examples clearly illustrate that transfer experiments can provide unambiguous evidence of adaptive deme formation only if alternative mechanisms can be dismissed.

Not all sources of false positives in herbivore transfer experiments are mechanistic. Unruh and Luck (1987) described a scenario in which a statistical artifact, the misinterpretation of a significant interaction effect in an analysis of variance table (ANOVA), can result in false acceptance of the adaptive deme formation hypothesis. In transfer experiments involving reciprocal transfers between hosts, a significant interaction between the natal and novel treatments (donor and recipient hosts) is considered consistent with adaptive deme formation (see Edmunds and Alstad 1979; Wainhouse and Howell 1983; Karban 1989; Mopper et al. 1995; Stiling and Rossi, Chapter 2, this volume). In one of their experiments, Unruh and Luck (1987) observed a significant novel-by-natal-tree interaction on scale survival. However, the interaction effect was not due to higher survival of native scales compared to immigrant scales, as predicted by Corollary 1 of the adaptive deme formation hypothesis, but rather by unusually low survival of native scales on one particular tree. Unruh and Luck suggested that significant interaction effects produced by reciprocal transfer experiments must be decomposed into separate components to ascertain the source of the interaction, and to determine if the interaction is truly consistent with the adaptive deme formation hypothesis.

### 4.4.1 Problems with Power

Experimental tests of the adaptive deme formation hypothesis may also produce false negatives. A false negative obtains when demes adapted to individual hosts exist within herbivore populations, yet no reduction in herbivore performance is observed following transfers to novel hosts. In most instances, false negatives will occur because of a lack of statistical power in the experimental design. Power is the probability of rejecting the null hypothesis when it is false. The power of transfer experiments is determined jointly by sample size, variation in herbivore responses, and the magnitude of fitness differences imposed on herbivores by variation in host phenotype. The size of detectable differences is an important consideration in an experimental design, and in interpreting the results of transfer experiments (or any experiment designed to detect fitness differences), especially for those experiments that yield negative results (see Endler 1986).

If selection differences between plant hosts are small, even though they are biologically important, the typical herbivore transfer experiment may not detect them. The genetic evidence from field experiments suggests that the magnitude of interdemic variation between herbivore populations may be too small to reveal differences. For example, McPheron et al. (1988) and Alstad and Corbin (1990) reported statistically significant microgeographic genetic variation in populations of *Rhagoletis pomonella* and *Nuculaspis californica*, respectively. However, the reported $F_{st}$ values were very small, averaging only 0.010 for *R. pomonella* and 0.048 for *N. californica*. These values are probably typical of the magnitude of differentiation at the spatial scale of individual hosts. A compendium of 29 such $F_{st}$ values collected from the literature averages only 0.045, and only 0.031 if the large outlier ($F_{st} = 0.427$) is excluded (Table 4.1). It is important to note that even slight differences in selection coefficients between hosts may be sufficient, depending on the level of gene flow, to produce interdemic variation of this magnitude (see Falconer 1981). Consequently, a robust test of the adaptive deme formation hypothesis in a system with limited gene flow may require a transfer experiment that can detect as little as a 1–5% difference in herbivore performance between novel and natal hosts.

Low power is a special concern when herbivore responses are highly variable (as they so often are in natural systems) and when reciprocal transfers are conducted between hosts with little phenotypic variation. It is common to use only high density hosts in transfer experiments (e.g., Mopper et al. 1995), since sufficient numbers of herbivores must be available for reciprocal transfers. The use of high-density hosts is motivated also by a tacit assumption underlying the adaptive deme formation hypothesis (Corollary 2), namely, that high herbivore densities represent highly adapted demes (see Alstad and Edmunds 1983a; Cobb and Whitham 1993). It is arguable, however, that transfers between such hosts may be a poor strategy for detecting local adaptation. First, high-density hosts may support the least differentiated herbivore populations. They may be high-density

*Table 4.1*  Estimates of Genetic Variation in Phytophagous Insects at the Level of Individual Host Plants

| HERBIVORE | HOST | POPULATION | $F_{st}$ | SOURCE |
|---|---|---|---|---|
| *Chrysomela* | *Salix* | BPCa | 0.000 | Rank 1992 |
| *aeneicollis* | *orestera* | BPCb | 0.113 | |
| | | BPCc | 0.043 | |
| | | BPCd | 0.036 | |
| | | BPCe | 0.015 | |
| | | BPCf | 0.039 | |
| | | BPCg | 0.000 | |
| | | RCa | 0.023 | |
| | | RCb | 0.079 | |
| | | RCc | 0.165 | |
| | | SLa | 0.034 | |
| | | SLb | 0.016 | |
| | | SLc | 0.009 | |
| | | SLd | 0.000 | |
| | | SLe | 0.427 | |
| *Nuculaspis* | *Pinus* | Plot 1 | 0.066* | Alstad and Corbin |
| *californica* | *ponderosa* | Plot 2 | 0.034* | (1990) |
| | | Plot 3 | 0.044* | |
| *Plagiodera* | *Salix nigra* | 1985-2 | 0.008 | McCauley et al. |
| *versicolora* | | 1986-1 | 0.024 | (1988) |
| | | 1986-2 | 0.041 | |
| | *Salix interior* | 1986-1 | 0.015 | |
| | | 1986-2 | 0.037 | |
| *Rhagoletis* | *Crataegus* | 1982 | 0.011* | McPheron et al. |
| *pomonella* | *mollis* | 1985 | 0.009* | (1988) |
| *Rhagoletis* | *Malus* | 1985 | 0.006 | Feder et al. (1990) |
| *pomonella* | *pumila* | 1987 | 0.006 | |
| | *Crataegus* | 1985 | 0.002 | |
| | *mollis* | 1987 | 0.001 | |

All *Fst* values are composite estimates except those indicated by an asterisk, which are averages of *Fst* over all loci.

hosts simply because they are successfully colonized by herbivores with a diversity of genetic backgrounds. For example, Memmott et al. (1995) mixed together aphids collected from a variety of sources and transferred them to cypress hosts in four different infestation categories. There was a significant positive relationship between colonization success and infestation class of the host, suggesting that cypress plants varied in their resistance to herbivory. Similarly, Strauss and Karban (1994) observed that the densities of thrips were influenced by plant resistance characters rather than by local adaptation to a given host. If high-density hosts do support a composite of herbivore demes, then transfers involving such populations are unlikely to yield consistent results. Second, heavily attacked trees may represent the same adaptive peak (poor defense, high nutritional quality, coincident phenologies, etc.) within the evolutionary landscape of host plants. If a given level of herbivore infestation represents an adaptive peak, then exclusive use of high-density hosts (or matching or blocking hosts according to densities, as in Cobb and Whitham 1993; Memmott et al. 1995) may produce experiments with little phenotypic variation among hosts, low or undetectable selection gradients among hosts (see above), and experimental designs with relatively little power to detect local adaptation.

Endler (1986) described a number of reasons why selection might not be detected despite its existence, including the case when the measurement error of the trait examined is larger than the magnitude of selection. Herbivore responses measured in transfer experiments are often quite variable. For example, Karban (1989) used the number of thrips per plant as a measure of herbivore performance. This value (based on 10 replicates) after three generations had an average coefficient of variation of 90.8%. Nevertheless, Karban could detect significant effects in this experiment because of wide differences in thrips performance between treatments and a large sample size (90 host plants). Perhaps owing to strong performance differentials and/or low variance among herbivores transferred to natal and novel plants, both Mopper et al. (1995) and Hanks and Denno (1994) detected significant differences between treatments despite relatively low samples sizes. These examples indicate that herbivore density and herbivore performance can be variable in transfer experiments, even in those that exhibit significant effects. We could not provide a more general estimate of the variability of herbivore responses observed in herbivore transfer experiments because performance and density data are rarely published in tabular form.

The number and allocation of experimental units are probably the most important components affecting statistical power under experimenter control. In general, replication generates power. However, transfer experiments typically involve small numbers of herbivore cohorts (putative demes) and hosts, owing to the logistical difficulties of finding and transferring large numbers of herbivores. For example, in completely reciprocal transfer designs, Karban (1989) used only three *Apterothrips* herbivore cohorts and three *Erigeron* host clones, and Stiling and Rossi (Chapter 2, this volume) and Mopper et al. (1995) used four herbivore-by-host combinations.

Compounding the effects of small sample size is the common practice in herbivore transfer experiments of using nested designs with multiple replicates of a given herbivore cohort-by-host combination. For example, Mopper et al. (1995) used 10 replicates (cages) of each donor population on each host (although replicates were ultimately pooled within trees for the analysis), Unruh and Luck (1987) used three replicate per cohort–host combinations, and Hanks and Denno

(1994) used between six and nine branches (set of natal and novel cages) per host. Such multiple replication of herbivore cohorts within individual hosts represents pseudoreplication (*sensu* Hurlbert 1984). A preferred method would be fewer herbivore replicates within hosts, but more combinations of different herbivore–host transfers. The fact that several of these experiments, despite their low level of effective replication, detected significant differences in herbivore performance between natal and novel hosts suggests that the selection gradients created by host-plant phenotypes were relatively large.

The power of transfer experiments is determined also by the type of experimental design used. Some designs are inherently more powerful than others in detecting certain effects. For example, Cobb and Whitham (1993) conducted a reciprocal transfer experiment where hosts were blocked according to infestation levels (see also Memmott et al. 1995). The test of donor-by-receptor interaction (evidence of deme formation) was based on the number of donor and receptor blocks (N. Cobb, personal communication). A more powerful approach would have been to test the interaction using individual trees, perhaps using natural scale infestation levels as a covariate. In addition, Horton et al. (1991) described the use of repeated-measures designs in detecting local adaptation in host races, and demonstrated that these designs can offer substantial power advantages over completely randomized designs in detecting host-by-population interactions. Repeated-measures designs would appear also to be a good choice for experimental tests of the adaptive deme formation hypothesis.

For these reasons, the typical herbivore transfer experiment likely will have low power in detecting slight differences in selection coefficients between hosts. The degree to which such experiments refute the adaptive deme formation hypothesis is questionable. Experiments with low statistical power confound a lack of effect (adaptive deme formation) with an inability to detect the effect, and cannot therefore provide strong refutations of hypotheses. Unfortunately, none of the studies that have failed to support the adaptive deme formation hypothesis (Rice 1983; Unruh and Luck 1987; Cobb and Whitham 1993; Hanks and Denno 1994 [near tree transfers]; Memmott et al. 1995; Kimberling and Price 1996) provide power analyses, so it is impossible to gauge, in a statistical sense, the rigor of these experimental refutations.

Herbivore transfer experiments may generate an additional type of false negative with respect to the adaptive deme formation hypothesis. This occurs when transfers are conducted in experimental systems that lack the necessary conditions for adaptive deme formation. A number of studies have described attributes of experimental systems that make adaptive deme formation unlikely, including high levels of gene flow, weak or density-dependent selection, ontogenetic or temporal changes in host resistance, within-host heterogeneity in resources or resistance, and environmental catastrophes (see Hanks and Denno 1994, Chapter 11, this volume; Cobb and Whitham 1993, Chapter 3, this volume). In such systems, a rigorous test is not possible (see above), and failure to provide evidence of adaptive deme formation must be viewed as a trivial result (false negative), particularly if demes cannot exist in the first place, or if transfers are between hosts with little phenotypic variation. Demonstrating a lack of adaptive deme formation in these types of experimental systems may say something about the prevalence of the conditions necessary for adaptive deme formation, but it says little about the mechanism itself.

### 4.4.2 Correlative Evidence

Correlative evidence has been adduced in support or in refutation of all four corollaries of the adaptive deme formation hypothesis. With respect to Corollary 1, Alstad and Edmunds (1983b) reported a significant inverse relationship between the density of scales on a host and their colonization success following transfer to novel hosts (see also Edmunds and Alstad 1981). Alstad and Edmunds (1983b) interpreted this as support for Corollary 1 and its prediction that adaptation to a given host individual is maladaptive for colonizing new hosts.

Most of the correlative evidence gathered with respect to the adaptive deme formation hypothesis has been directed at Corollary 2 and its prediction that herbivore performance and population growth will increase as herbivore demes become better adapted to their hosts. Temporal patterns of herbivore density and performance have been examined in a variety of systems and have yielded mixed results. For example, a positive relationship between host age and herbivore density has been adduced in support of Corollary 2 (Edmunds and Alstad 1978, 1981; Wainhouse and Howell 1983), while failure to find such a relationship has been offered as refutation (Cobb and Whitham, Chapter 3, this volume). In addition, Cobb and Whitham (1993) failed to observe a decrease in rates of herbivore mortality in newly established populations (putative demes) through time and deemed this result to be inconsistent with the adaptive deme formation hypothesis.

The contention that adaptive deme formation is facilitated by host spatial isolation and has (Corollary 3) attracted relatively little attention. Alstad and Edmunds (1987) examined scale densities on twigs adjacent to and distant from neighboring trees, and observed significantly higher densities on the distant twigs, which they interpreted as evidence of outbreeding depression (breakdown of adaptive demes) at the margins of individual hosts (see also Alstad and Edmunds 1983a). At a larger spatial scale, Kimberling and Price (1996) failed to detect a significant relationship between colonization success of *Phylloxera* cohorts on grape cuttings in experimental arenas and the distance between the original grape clones in the field. To test for environmental effects on adaptive variation, Strauss (1997) examined the relative performance of *Blepharida rhois* on natal and novel plants of *Rhus glabra* as a function of the distance between hosts. There were significant environmental gradients in survivorship and development time, but no evidence that relative performance increased with distance *per se*. For larval weight, relative performance actually decreased with distance between hosts. Although Strauss (1997) and Kimberling and Price (1996) provide some correlative data, they did not measure the spatial distance between insect subpopulations, and the relationship between isolation and deme formation remains unresolved.

The ponderosa pine–black pineleaf scale system has produced all of the correlative evidence adduced in support of Corollary 4 and its prediction of increasing male bias (performance) as herbivore demes become more adapted to their hosts (Alstad et al. 1980; Alstad and Edmunds 1983a, 1983b; Alstad, Chapter 1, this volume). For example, Alstad and Edmunds (1983a) reported significant yearly increases in the relative frequency of male scales on 18 ponderosa pines sampled over a three-year period (see also Alstad et al. 1980; Alstad and Edmunds 1983b). In addition, Alstad and Edmunds (1983a) reported significantly lower male-to-female ratios for scales on twigs adjacent to neighboring pines than for scales on distant twigs (see also Alstad and Edmunds 1983b; Alstad, Chapter 1, this volume). However, Unruh and Luck (1987) and Alstad (Chapter 1, this volume) described a number of potential confounding factors and alternative explanations for these results, including sex-biased differential mortality, density-related sex-ratio shifts, and sexual dimorphism in tolerance to environmental stress.

## 4.5 Testing the Adaptive Deme Formation Hypothesis

### 4.5.1 Defining the Question

The mechanistic formulation of the adaptive deme formation hypothesis asserts that host-plant mediated natural selection produces locally adapted demes within herbivore populations. In other words, natural selection is the mechanism that produces the pattern–herbivore demes adapted to individual host plants. The hypothesis is conditioned jointly on the strength of the selection gradient created by phenotypic variation among hosts, and the degree of gene flow within the herbivore population (Strauss and Karban, Chapter 8, this volume). Plant phenotypic variation and limited herbivore gene flow create the necessary, but not sufficient, conditions for natural selection to structure herbivore populations into demes (adaptive deme formation). On the other hand, similarity among host phenotypes and/or high levels of gene flow could counteract natural selection and prevent demes from evolving. Under these conditions, adaptive deme formation is not likely to occur, and an experimental system exhibiting these characteristics is inappropriate for testing the hypothesis. Simply put, an experimental system that does not provide the necessary conditions for adaptive deme formation cannot provide a rigorous test of the adaptive deme formation hypothesis.

Therefore, the strongest possible endorsement or refutation of the hypothesis requires two conditions: First, the host-plant population must comprise phenotypically heterogeneous individuals (see Berenbaum and Zangerl, Chapter 5; Strauss and Karban, Chapter 8; and Gandon et al., Chapter 13, this volume). These traits may be concentrations of nutritive or defensive compounds, mechanical properties such as leaf size or trichome density, phenological variation in foliage production, and so on. However, they must have the potential to exert selection pressure on colonizing herbivores. Second, genetic variation in the herbivore population must be structured into demes at the spatial scale of individual host plants. Demic structure is indicative of restricted gene flow, which may arise from dispersal behavior, genetic drift, stochastic events, or natural selection. Under these conditions, experimental evidence that the observed demic structure is adaptive provides strong support of the mechanistic adaptive deme formation hypothesis. Refutation of the hypothesis is achieved when deme formation is independent of variation in host phenotype, that is, nonadaptive.

### 4.5.2 The Essential Conditions

#### 4.5.2.1 Plant Phenotypic Variation

Surprisingly few studies of adaptive deme formation have explicitly addressed plant phenotypic variation. That populations of long-lived, sexually reproducing plants are heterogeneous enough to generate differential selection pressures on herbivores is a virtual paradigm. Perhaps because this assumption is so pervasive, no experimental study of adaptive deme formation has explicitly examined variation in the phenotypic traits of host plants to which insects were transferred. Nevertheless, the assumption of plant phenotypic variation is well-founded and based on an extensive literature documenting individual variation in chemical and mechanical resistance to herbivory (see reviews in Denno and McClure 1983; Fritz and Simms 1992; Horner and Abrahamson 1992). And in a review of the deme- and host-race-formation literature, Strauss and Karban (Chapter 8, this volume) observed that the strength of selection imposed on novel insect lines by individual plants of the same species was comparable to that imposed on novel insect lines by different host plant species. One of the few studies to investigate agents of plant-mediated selection (Komatsu and Akimoto 1995) provides insight into how variation in a plant trait-budburst phenology-is associated with local adaptation in herbivores. Mopper (Chapter 7, this volume) also detected significant variation in leaf production phenology among individual oak trees, and between sites and oak species.

#### 4.5.2.2 Herbivore Population Structure

Numerous studies document fine-scale genetic structure in herbivore populations (see reviews in this volume by Peterson and Denno, Chapter 12; Costa, Chapter 10; McCauley and Goff, Chapter 9; Itami et al., Chapter 15; Thomas and Singer, Chapter 14; Feder et al., Chapter 16) but rarely in the context of adaptive deme formation (see Table 4.1). Nonetheless, the most rigorous test of the mechanistic hypothesis requires that demic structure is not only possible, but evident. Studies by McPheron et al. (1988) and McCauley and Eanes (1987) detected demic structure, although they disagree as to the underlying mechanisms. McPheron et al. (1988) propose that natural selection structures *Rhagoletis pomonella* flies on mature hawthorn trees, and McCauley and Eanes (1987) hypothesize that genetic drift/founder effects associated with ephemeral host-plant patch dynamics produce microgeographic structure in *Tetraopes tetraopthalmus* milkweed beetle populations. Without experimental manipulation, it is impossible to understand causative mechanisms.

Two unpublished studies have determined allozyme variation among herbivores collected from the same populations used in reciprocal transfer experiments testing adaptive deme formation (Strauss, unpublished data; Landau and Mopper, unpublished data). Strauss's (1977) transfer experiments with sumac flea beetles refuted the adaptive deme formation hypothesis, in contrast to Mopper et al's. (1995) transfers of oak leafminers, which supported the hypothesis. The unpublished allozyme data collected from both systems indicated local population structure, providing evidence on the one hand for nonadaptive demic structure of flea beetles (Strauss) and on the other hand, for adaptive demic structure in the oak leaf miner (Landau and Mopper).

The only published study directly addressing genetic structure in the context of adaptive deme formation detected structure in black pineleaf scale insects at the spatial scale of sites, trees within sites, and branches within trees (Alstad and Corbin 1990; Alstad, Chapter 1, this volume). Of the three polymorphic enzymes examined, two appeared neutral and were nonrandomly distributed among scales insects inhabiting different branches within the same tree. This supports the "drift hypothesis"—nonrandom, nonadaptive structure caused primarily by the high "viscosity" of the pineleaf scale mating system, which markedly restricts gene flow. However, one locus (malic enzyme—examined among, not within, trees) is correlated with scale density and sex ratio, and therefore appears to be adaptive or linked with a gene under selection. Alstad (Chapter 1, this volume) proposes that selection on malic enzyme results from scale-induced deterioration in plant quality by feeding, not from selection by intrinsic host-plant genetic traits. All of these studies indicate that herbivore genetic structure is common, even at very fine spatial scales; but its adaptive significance remains to be determined.

### 4.5.3 Testing the Hypothesis

If the essential conditions prevail, the final objective is to test the prediction (Corollary 1) that herbivores are locally adapted to individual host plants. Adaptation, or lack thereof, must be determined, and to do so requires experiments that compare herbivore fitness on natal and novel host plants. The experimenter transfers herbivores, then measures designated fitness components (depending in large part on species' life-history attributes) to test the central prediction that herbivores will realize greater fitness on their natal host plant than on a novel host plant. Many indices of performance have been employed to estimate herbivore fitness, including density, egg and larval survival, developmental rates, and fecundity (Table 4.2).

*Table 4.2*   Experimental tests of the adaptive deme formation hypothesis

| HERBIVORE | HOST | EVIDENCE | DONOR POPULATIONS | RECIPIENT HOSTS |
|---|---|---|---|---|
| *Apterothrips secticornis* | *Erigeron glaucus* | Experimental | 3 | 3 |
| | | Experimental | 3 | 3 |
| *Asphondylia borrichiae* | *Borrichia frutescens* | Experimental | 4 | 4 |
| *Blepharida rhois* | *Rhus glabra* | Experimental | 8 | 8 |
| *Cinara cupressi* | *Cupressus lusitanica* | Experimental | 8 | 4 pairs |
| *Cryptococcus fagisuga* | *Fagus sylvatica* | Experimental (1979) | 3 | 2 |
| | | Experimental (1980) | 4 | 2 |
| | | Experimental (forest trees) | 5 | 5 |
| *Daktulosphaira vitifoliae* | *Vitus arizonica* | Experimental | 4(2) | 6 |
| *Kaltenbachiella japonica* | *Ulmus davidiana* | Quantitative genetic (1991) | 6 | |
| | | Quantitative genetic (1992) | 6 | |
| *Matsucoccus acalyptus* | *Pinus edulis* | Experimental (transfers) Experimental (defaunation) | 3 density classes | 4 density classes |
| *Matsucoccus acalyptus* | *Pinus monophylla* | Experimental (SB) | 10 | 10 |
| | | Experimental (TM) | 8 | 6 |
| | | Experimental (SB to TM) | 12* | 6 |

| REPLICATION | ADAPTIVE RESPONSE VARIABLE | DEME FORMATION | COROLLARY 1 | 2 | 3 | 4 | SOURCE |
|---|---|---|---|---|---|---|---|
| 10 | Population size | + | | | | | Karban (1989) |
| | | | + | | | | |
| 2 | Population size | + | | | | | Strauss and Kar-ban (1994) |
| | | | + | | | | |
| 15 | Abundance gall size, gall abortion, parasitism | + | + | | | | Stiling and Rossi (1997) |
| 6 | Survivorship Developmental time, pupal weight, pre-dation | − | | − | | | Strauss (1997) |
| 2 | Survivorship | − | | − | | | Memmott et al. (1995) |
| 5 | Survivorship, fecundity | + | + | | | | Wainhouse and Howell (1983) |
| 5–6 | Survivorship, fecundity | | + | | | | |
| | Survivorship, fecundity | | + | | | | |
| 3 | Colonization, survivorship, fecundity | − | | − | − | | Kimberling and Price (1996) |
| 3–8 | Hatching time | + | | | | | Komatsu and Akimoto (1995) |
| 2–6 | Hatching time | + | | | | | |
| 2 | Mortality | − | | − | − | | Cobb and Whit-ham (1993) |
| 20 | Mortality | | | − | | | Cobb and Whit-ham (1997) |
| 3 | Survivorship | − | | − | | | Unruh and Luck (1987) |
| 4 | Survivorship | | − | | | | |
| 4 | Survivorship | | | | | | |
| | | | + | | | | |

Table 4.2   (continued)

| HERBIVORE | HOST | EVIDENCE | DONOR POPULATIONS | RECIPIENT HOSTS |
|---|---|---|---|---|
| Nuculaspis californica | Pinus ponderosa | Experimental | 10 | 10 |
| | | Experimental Correlative | 5 | 5 |
| | | Correlative | | |
| | | Correlative | | |
| | | Correlative | | |
| | | Correlative | | |
| Nuculaspis californica | Pinus lamber-tiana | Experimental | 10 | 65(?) |
| | | Experimental | 18(15) | 31 |
| Pseudaulacaspis pentagona | Morus alba | Experimental (near) | 10 | 5 pairs |
| | | Experimental (far) | 10 | 5 pairs |
| Stilbosis quadri-custatella | Quercus geminata | Experimental | 4 | 4 |

| REPLICATION | ADAPTIVE RESPONSE VARIABLE | DEME FORMATION | COROLLARY 1 | 2 | 3 | 4 | SOURCE |
|---|---|---|---|---|---|---|---|
| 3 | Survivorship | + | + | | | | Edmunds and Alstad (1978) |
| | | | | | | | Edmunds and Alstad (1981) |
| 4 | | (+) | − | | | | Alstad (1997) |
| 667 | Population size & age | | | + | | | Edmunds and Alstad (1978) |
| | | | | | | | Edmunds and Alstad (1981) |
| 11 pairs | Sex ratio & position | + | | | | + | Alstad and Edmunds (1983a) |
| | | | | | | | Alstad and Edmunds (1983b) |
| | | | | | | | Alstad (1997) |
| 11–18 | Sex ratio & time | + | | | | + | Alstad et al. (1980) |
| | | | | | | | Alstad and Edmunds (1983b) |
| | | | | | | | Alstad and Edmunds (1991) |
| 11 | | (+) | | | | (+) | Alstad (1997) |
| 22 pairs | Population size & position | + | | + | | | Alstad and Edmunds (1987) |
| 18 | Survivorship & population size | + | + | | | | Alstad and Edmunds (1983b) |
| | Sex ratio & population size | + | | | | + | Alstad (1997) |
| 1 | Survivorship | − | − | | | | Rice (1983) |
| 1 | Survivorship | | − | | | | |
| 6–9 | Survivorship | + | − | | + | − | Hanks and |
| 6–9 | Survivorship | | 1 | | − | − | Denno (1994) |
| 10 | Mortality Predation/parasitism | + | + | | | | Mopper et al. (1995) |

### 4.5.3.1 Experimental Design

The most powerful experimental test of the adaptive deme formation hypothesis is a transfer of herbivores between host plants. Most experiments are conducted as reciprocal transfers in which insects are collected from, and returned to, their natal host and to novel host plants. Typically, data are analyzed by a factorial analysis of variance (Hanks and Denno 1994; Strauss 1997) when the response variables are continuous, or by log-linear models (Mopper et al. 1995; Stiling and Rossi, Chapter 2, this volume) when response variables are discrete. Evidence for adaptation to individual host plants appears as an interaction effect between the natal and novel treatments, *not* between individual host plants, as Unruh and Luck (1987) caution. Horton et al. (1991) describe an alternative and potentially more powerful method of testing local adaptation by using a repeated-measures design. This approach requires certain conditions (ability to subdivide and transfer related herbivores) that may be unrealistic in most natural herbivore populations but could be an excellent alternative when herbivores occur in distinct groups of easily transferred siblings (e.g., diprionid sawflies; Mopper et al., 1990), or when they can be collected and reared in the laboratory prior to distribution to experimental treatments (example in Horton et al. 1991).

A test of Corollary 3 (adaptive deme formation facilitated by genetic isolation) can be incorporated into the design by manipulating the distance between donor and recipient plants. However, if host plants are widely dispersed, the results must be carefully interpreted because of the potential confounding effects of environmental variation (Edmunds and Alstad 1978). An additional caveat is whether host plants are isolated from all potential sources of herbivore gene flow or only isolated from other plants in the experiment. Pairwise comparisons do not eliminate the possibility of gene flow between herbivores inhabiting experimental and nonexperimental host plants.

Hanks and Denno (1994) incorporated isolation into their design and observed a significant effect: Only scale insects that were isolated from conspecific neighbors displayed differential survival consistent with adaptive deme formation. Mopper et al. (1995) transferred oak leafminers at three spatial scales: between individual host plants, between different host plant species growing sympatrically, and between host plant populations growing 60 km apart. In each comparison, leafminers transferred to natal hosts performed significantly better than those transferred to the novel hosts. Furthermore, the largest difference in herbivore performance occurred in the between-site transfer (Mopper, Chapter 7, this volume). This study supports the adaptive deme formation hypothesis, and it also indicates that isolation and environmental effects have discernible impacts on population structure. Virtually all experimental tests of the adaptive deme formation hypothesis conduct reciprocal transfers of herbivores among stationary natal and novel host plants. One exception is the study conducted by Stiling and Rossi (Chapter 2, this volume) who designed a creative alternative to the standard approaches. They reciprocally transferred host plants (*Borrichia frutescens*) among four offshore islands and monitored the performance of stem gall midges (*Asphondylia borrichiae*) native to each islands. The experiment revealed adaptive genetic structure in the stem gall populations.

### 4.5.3.2 Estimating Fitness

#### 4.5.3.2.1 Density

The use of herbivore density as a measure of local adaptation has been a fundamental assumption underlying much of the correlative evidence used to invoke adaptive deme formation (Corollary 2). However, density is an unreliable fitness component, because herbivore populations are notoriously variable over time and space, often exhibiting unpredictable explosions and collapses (Andrewartha and Birch 1954). Even when measured on the same host plants over time, herbivore populations are typically quite variable (Alstad, Chapter 1, this volume; Mopper, Chapter 7, this volume). For example, Boecklen and Price (1991) observed that herbivore densities on arroyo willow clones varied by more than an order of magnitude over three years and exhibited highly significant year-by-clone interactions (see also Boecklen et al. 1994). Although performance data suggest local adaptation to individual trees (Mopper et al. 1995), oak leafminers exhibit annual variation in density that is strongly correlated with precipitation (Mopper, Chapter 7, this volume). Even in an experimental context, herbivore density may not measure adaptation to hosts at all, but instead result from abiotic forces, predators and parasites, or demographic stochasticity. Additional factors that confound herbivore density patterns include plant resistance traits, density-dependent mortality and dispersal, maternal effects, and environmental catastrophes (Unruh and Luck 1987; Cobb and Whitham 1993, Chapter 3, this volume; Strauss and Karban 1994; Mopper, Chapter 7, this volume; Alstad, Chapter 1, this volume). We must concur with Unruh and Luck's (1987) conclusion that, except under highly controlled conditions (e.g., Karban 1989), density can be a poor measure of herbivore adaptation.

#### 4.5.3.2.2 Survival and Fecundity

The adaptive deme formation hypothesis posits that plant-mediated selection pressures produce demic structure in herbivore populations, yet this prediction has rarely been directly tested. Survival is a commonly employed fitness estimate, but the sources of mortality are seldom determined. The strongest test of Corollary 1 would be a comparison of plant-mediated performance among herbivores transferred to natal and novel trees (Mopper, Chapter 7, this volume). The studies by Komatsu and Akimoto (1995), Mopper et al. (1995), Stiling and Rossi (Chapter 2, this volume), and Strauss (1997) identified the specific agents of herbivore mortality; three of the studies support the hypothesis of plant-mediated adaptive deme formation (Table 4.2). Strauss, however, found no association between source of mortality and natal or novel treatment, although there was a significant clone effect on rate of beetle predation. Stiling and Rossi's experiment also included female fecundity in the test of the adaptive deme formation as did that of Wainhouse and Howell (1983), and Kimberling and Price (1996).

## 4.6 Conclusion

Despite the different drawbacks of individual studies, we are convinced by the body of research that adaptive deme formation is an important evolutionary phenomenon. Of the 13 separate experimental tests, 7 support and 6 refute the central hypothesis that herbivores are locally adapted at the spatial scale of individual host plants (Table 4.2). These studies share a major weakness: no *a priori* confirmation that the conditions necessary for deme formation (such as herbivore genetic structure and host heterogeneity) exist in the system. This places a burden of proof on experiments refuting adaptive deme formation to ensure the validity of the system for testing the hypothesis. But studies confirming the hypothesis also have an obligation—to eliminate the potential for other factors to confound the experimental results. Although seven years have elapsed since its publication, the Karban (1989) study remains one of the strongest tests of local adaptation, because it included a common garden experimental design and the removal of potentially confounding maternal–environmental conditioning effects (see also Komatsu and Akimoto 1995). Nonetheless, evidence for local adaptation is accumulating from insect species with very different life-history attributes. This suggests that it may be a relatively general phenomenon in phytophagous insects.

As this volume attests, adaptive deme formation has broad evolutionary implications. One of the most general issues is the coevolutionary interactions between parasites and hosts, such as the influence of sexual reproduction on host resistance against parasite and pathogen attack. If herbivores are adapted to individual plants (particularly long-lived ones), then sexually reproducing hosts should have a fitness advantage over hosts that reproduce clonally, because offspring possessing recombinant genotypes may be better equipped to escape herbivores' preadapted traits. This theory has been debated extensively (Williams 1975; Maynard Smith 1978; Hamilton 1980; Gandon et al., Chapter 13, this volume) and was tested (and supported) in the context of adaptive deme formation by Rice (1983). If herbivores undergo adaptive evolution at fine spatial scales, then the maintenance of genetically heterogeneous, outbreeding plant populations may be necessary to minimize insect outbreaks and large-scale destruction. Differentiation of insects at the spatial scale of individual plants is the first step toward reproductive isolation and speciation. It stands to reason that to fully comprehend these evolutionary processes, one must be confident in the methods by which the patterns and mechanisms are revealed.

## Acknowledgments

We thank Sharon Strauss, Karl Hasenstein, and Keli Landau for their helpful comments on the manuscript. Keli Landau provided invaluable editorial assistance with this chapter and the volume in which it appears. This research was supported by a 1996 University of Southwestern Louisiana (USL) Summer Research Fellowship, the Lafayette Parish Medical Society Endowed Professorship (1993–1999), and the following research grants: National Science Foundation BSR90-07144 (1990–1993), EPSCoR NSF/LEQSF-ADP-02 (1992–1996), LEQSF-RD-A-37 (1994–1996), and National Science Foundation (1996–1999) DEB-9632302

## 4.7 References

Akimoto, S. 1990. Local adaptation and host race formation of a gall-forming aphid in relation to environmental heterogeneity. *Oecologia 83*:162–170.

Alstad, D. N., and K. W. Corgin, 1990. Scale insect alloyme differentiation within and between host trees. *Evol. Ecol. 4*:43–56.

Alstad, D. N. and G. F. Edmunds Jr. 1983a. Selection, outbreeding depression, and the sex ratio of scale insects. *Science 220*:93–95.

Alstad, D. N. and G. F. Edmunds Jr. 1983b. Adaptation, host specificity and gene flow in the black pineleaf scale. Pp. 413–426 *in* R. F. Denno and M. S. McClure (Eds.), *Variable Plants and Herbivores in Natural and Managed Systems*. Academic Press, New York.

Alstad, D. N. and G. F. Edmunds Jr. 1987. Black pineleaf scale (Homoptera: Diaspididae) population density in relation to interdemic mating. *Ann. Entomol. Soc. Am. 80*:652–654.

Alstad, D. N. and G. F. Edmunds Jr. 1989. Haploid and diploid survival differences demonstrate selection in scale insect demes. *Evol. Ecol. 3*:253–263.

Alstad, D. N., G. F. Edmunds Jr., and S. C. Johnson. 1980. Host adaptation, sex ratio, and flight activity in male black pineleaf scale. *Ann. Entomol. Soc. Am. 73*:665–667.

Andrewartha, H. G. and L. C. Birch. 1954. *The Distribution and Abundance of Animals*. University of Chicago Press, Chicago, IL.

Ayres, M. P., J. Suomela, and S. F. MacLean Jr. 1987. Growth performance of *Epirrita autumnata* (Lepidoptera: Geometridae) on mountain birch: Trees, broods, and tree × brood interactions. *Oecologia 74*:450–457.

Boecklen, W. J., S. Mopper, and P. W. Price. 1994. Sex-biased herbivory in arroyo willow: Are there general patterns among herbivores? *Oikos 71*:267–272.

Boecklen, W. J. and P. W. Price. 1991. Nonequilibrial community structure of sawflies on arroyo willow. *Oecologia 85*:483–491.

Brady, R. H. 1979. Natural selection and the criteria by which a theory is judged. *Syst. Zool. 28*:600–621.

Bush, G. L. 1969. Sympatric host race formation and speciation in frugivorous flies of the genus *Rhagoletis* (Diptera: Tephritidae). *Evolution 23*:237–251.

Bush, G. L. 1969. Mating behavior, host specificity, and the ecological significance of sibling species in frugivorous flies of the genus *Rhagoletis* (Deptera: Tephritidae). *Am. Nat. 103*:669–672.

Carroll, S. P. and C. Boyd. 1992. Host race radiation in the soapberry bug: Natural history with the history. *Evolution 46*:1052–1069.

Cobb, N. S. and T. G. Whitham. 1993. Herbivore deme formation on individual trees: A test case. *Oecologia 94*:496–502.

Denno, R. F. and M. S. McClure. 1983. *Variable Plants and Herbivores in Natural and Managed Systems.* Academic Press, New York.

Edmunds, G. F. Jr. and D. N. Alstad. 1978. Coevolution in insect herbivores and conifers. *Science 199*:941–945.

Edmunds, G. F. Jr. and D. N. Alstad. 1981. Responses of black pineleaf scales to host plant variability. Pp. 29–39 *in* R. F. Denno and H. Dingle (Eds.), *Insect Life-History Patterns,* Springer-Verlag, New York.

Edmunds, G. F. Jr. and D. N. Alstad. 1985. Malathion induced sex ratio changes in black pineleaf scale (Homoptera: Diaspididae). *Pan. Pac. Entomol. 60*:267–268.

Endler, J. A. 1986. *Natural Selection in the Wild.* Princeton University Press, Princeton, NJ.

Falconer, D. S. 1981. *Introduction to Quantitative Genetics,* 2nd ed. Longman, Essex, UK.

Feder, J. L., C. A. Chilcote, and G. L. Bush. 1990. Regional, local and microgeographic allele frequency variation between apple and hawthorn populations of *Rhagoletis pomonella* in western Michigan. *Evolution 44*:595–608.

Fritz, R. S. and E. L. Simms. 1992. *Plant Resistance to Herbivores and Pathogens: Ecology, Evolution, and Genetics.* University of Chicago Press, Chicago, IL.

Hamilton, W. D. 1980. Sex versus non-sex versus parasite. *Oikos 35*:282–290.

Hanks, L. M. and R. F. Denno. 1994. Local adaptation in the armored scale insect *Pseudaulacaspis pentagona* (Homoptera: Diaspididae). *Ecology 75*:2301–2310.

Horner, J. D., and W. G. Abrahamson. 1992. Influence of plant genotype and environment on oviposition preference and offspring survival in a gall-making herbivore. *Oecologia 90*:323–332.

Horton, D. R., P. L. Chapman, and J. L. Capinera. 1991. Detecting local adaptation in phytophagous insects using repeated measures design. *Env. Entomol. 20*:410–418.

Hurlbert, S. H. 1984. Pseudoreplication and the design of ecological field experiments. *Ecol. Monogr. 54*:187–211.

Karban R. 1989. Fine-scale adaptation of herbivorous thrips to individual host plants. *Nature 340*:60–61.

Kimberling, D. N. and P. W. Price. 1996. Variability in grape phylloxera preference and performance on canyon grape (*Vitus arizonica*). *Oecologia 107*:553–559.

Komatsu, T. and S. Akimoto. 1995. Genetic differentiation as a result of adaptation to the phenologies of individual host trees in the galling aphid *Kaltenbachiella japonica. Ecol. Entomol. 20*:33–42.

Maynard Smith, L. 1978. *The Evolution of Sex.* Cambridge University Press, New York.

McCauley, D. E. and W. F. Eanes. 1987. Hierarchical population structure analysis of the milkweed beetle, *Tetraopes tetraophthalmus* (Forster). *Heredity 58*:193–201.

McCauley, D. E., M. J. Wade, F. J. Breden, and M. Wohltman. 1988. Spatial and temporal variation in group relatedness: Evidence from the imported willow leaf beetle. *Evolution 42*:184–192.

McPheron, B. A., D. C. Smith, and S. H. Berlocher. 1988. Microgeographic genetic variation in the apple maggot, *Rhagoletis pomonella*. *Genetics 119*:445–451.

Memmott, J., R. K. Day, and H. C. J. Godfray. 1995. Intraspecific variation in host plant quality: The aphid *Cinara cupressi* on the Mexican cypress, *Cupressus lusitanica*. *Ecol. Entomol. 20*:153–158.

Mopper, S. 1996a. Adaptive genetic structure in phytophagous insect populations. *Trends Ecol. Evol. 11*:235–238.

Mopper, S. 1996b. Temporal variability and local adaptation: A reply to Y. Michalakis. Trends *Ecol. Evol. 11*(10):431–432.

Mopper, S., M. Beck, D. Simberloff, and P. Stiling. 1995. Local adaptation and agents of selection in a mobile insect. *Evolution 49*:810–815.

Mopper, S. and D. Simberloff. 1995. Differential herbivory in an oak population: The role of plant phenology and insect performance. *Ecology 76*:1233–1241.

Mopper, S., T. G. Whitham, and P. W. Price. 1990. Plant phenotype and interspecific competition determine sawfly performance and density. *Ecology 71*:2135–2144.

Moran, N. 1981. Intraspecific variability in herbivore performance and host quality: A field study of *Uroleucon caligatrum* (Homoptera: Aphididae) and its solidago hosts (Asteraceae). *Ecol. Entomol. 6*:301–306.

Pashley, D. P. 1988. Quantitative genetics, development, and physiological adaptation in host strains of fall armyworm. *Evolution 42*:93–102.

Popper, K. R. 1959. *The Logic of Scientific Discovery.* Basic Books, New York.

Rank, N. E. 1992. A hierarchical analysis of genetic differentiation in a montane leaf beetle *Chrysomela aeneicollis* (Coleoptera: Chrysomelidae). *Evolution 46*:1097–1111.

Rice, W. R. 1983. Sexual reproduction: An adaptation reducing parent–offspring contagion. *Evolution 37*:1317–1320.

Sandstrom, J. 1996. Temporal changes in host adaptation in the pea aphid, *Acyrthosiphon pisum. Ecol. Entomol. 21*:56–62.

Service, P. 1984. Genotypic interactions in an aphid–host plant relationship: *Uroleucon rudbeckiae* and *Rudbeckia laciniata*. *Oecologia 61*:271–276.

Slatkin, M. 1987. Gene flow and the geographic structure of natural populations. *Science 236*:787–792.

Strauss, S. Y. 1997. Lack of evidence for local adaptation to individual plant clones or site by a mobile specialist herbivore. *Oecologia. 110*:77–85.

Strauss, S.Y. and R. Karban. 1994. The significance of outcrossing in an intimate plant/herbivore relationship: II. Does outcrossing pose a problem for insects adapted to individual host plants? *Evolution 48*:465–476.

Unruh, T. R. and R. F. Luck. 1987. Deme formation in scale insects: A test with the pinyon needle scale and a review of other evidence. *Ecol. Entomol. 12*:439–449.

Via, S. 1991. Specialized host plant performance of pea aphid clones is not altered by experience. *Ecology 72*:1420–1427.

Wainhouse, D. and R. S. Howell. 1983. Intraspecific variation in beech scale populations and in susceptibility of their host *Fagus sylvatica. Ecol. Entomol. 8*:351–359.

Williams, G. C. 1975. *Sex and Evolution.* Princeton University Press, Princeton, NJ.

## PART II

# Foundations of Local Adaptation: The Genetic Basis of Host-Plant Use and the Nature of Selection

# 5

# Population-Level Adaptation to Host-Plant Chemicals: The Role of Cytochrome P450 Monooxygenases

*May R. Berenbaum and Arthur R. Zangerl*
Department of Entomology, University of Illinois, Urbana, IL

## 5.1 Introduction

The ability of herbivorous insects to specialize with respect to the range of host species utilized is perhaps unrivaled by any other group of plant-feeding animals. Over 90% of the known species of herbivorous insects feed on three or fewer plant families (Bernays and Graham 1988). Indeed, this predilection for specialization may well be the principal factor involved in the tremendous diversification of this group of organisms (Ehrlich and Raven 1964). Flowering plants and the herbivores that consume them collectively comprise approximately half of the earth's biota; as such, interactions between plants and their associated herbivores have profound consequences on the structure and function of the vast majority of terrestrial ecosystems.

Until recently, the degree to which herbivorous insects specialize has been underestimated; over the past two decades, considerable evidence has accumulated that host specialization extends beyond the level of species. Populations of insect herbivores often exhibit fine-scale genetic differentiation in association with host species, even under conditions of sympatry (Mopper 1996). Such sympatric host-race formation has been documented to occur in at least four orders (Lepidoptera: *Spodoptera frugiperda*, Pashley 1988; Diptera: *Rhagoletis pomonella*, Feder et al. 1990; Homoptera: *Acrythosiphon pisum*, Via 1991; Hemiptera: *Jadera haematoloma*, Carroll and Boyd 1992). The existence of such host races illustrates the selective impact of the host plant on the physiology and behavior of the herbivore; suites of physiological and behavioral traits that are conducive to survival on one host plant may be maladaptive on a different host plant.

Among the most important ways in which plants differ in their suitability as host plants is in their allelochemical content and composition. By altering rates of food intake, interfering with food-utilization efficiency, reducing fecundity, or causing outright mortality, plant allelochemicals can profoundly affect insect fitness. Interspecific differences in chemistry can be considerable; indeed, idiosyncratic distributions across taxa are among the defining features of allelochemicals, which can be restricted to single families, genera, or even species (Hegnauer 1966–1973; Rosenthal and Berenbaum 1991). Even within species, substantial genetic variation exists for both the content and composition of allelochemicals (Berenbaum and Zangerl 1992a). Subtle differences in the chemical composition of the diet, even with respect to the relative abundance of isomeric forms of compounds sharing molecular weights and elemental composition, can have profound effects on growth, survivorship, and fitness of an insect herbivore (Berenbaum et al. 1989).

Differential adaptation to host-plant allelochemicals is thought to be largely responsible for patterns of host specialization at the species level (Ehrlich and Raven 1964). Whether such differential adaptation plays a role in host-race formation as well is a question that has received virtually no attention to date. A promising approach to the question is to examine patterns of biochemical adaptation to host-plant allelochemicals. Arthropods, in general, have a wide range of genetically based mechanisms for dealing with natural toxins. Resistance can be behaviorally based, in that certain behaviors reduce the probability of encountering a toxin (Tallamy 1986), physiologically based, in that the physical attributes of morphological structures can prevent a toxin from reaching a target site (Berenbaum 1986), or biochemically based, in that metabolic systems can alter toxin structure so as to render it biologically inactive (Brattsten 1992). In this review, we examine the nature of genetic variation in the enzymatic metabolism of toxins by insects, specifically that effected by the cytochrome P450 monooxygenases. We concentrate only on biochemically based resistance mechanisms in part, because the genetic bases for this type of resistance are often more clearly definable than for behavioral or physiological resistance, which tend to involve entire suites of genes, and in part, because an extensive (but still surprisingly incomplete) literature exists that can be used for drawing reasonably sound inferences.

## 5.2 Mechanisms of Resistance to Plant Allelochemicals

Local adaptation to host-plant chemistry, particularly in the short term, may not necessarily involve novel mutations; many traits involved in host-plant adaptation are likely to be polygenic and may be shaped by selective screening of combinations of preexisting allelic variants. Host shifts and subsequent host-race formation, however, in all probability involve novel genetic events. Genetic mechanisms of resistance to plant allelochemicals can be structural or regulatory in nature (Feyereisen 1995). Structural genes encode proteins; mutational changes in such genes can yield gene products with altered properties. In the case of metabolic enzymes associated with resistance, such altered properties include differential substrate specificity or rates of substrate turnover. Structural changes associated with resistance generally result from point mutations in coding regions. An example of such a point mutation that may have implications for adaptation to plant allelochemicals involves the *Ace* gene of *Drosophila melanogaster*. The *Ace* gene codes for the enzyme acetylcholinesterase, which is responsible for post-synaptically breaking down the neurotransmitter acetylcholine (Chapman 1982). In *D. melanogaster,* with reduced sensitivity of acetylcholinesterase to carbamate and organophosphate insecticides, the coding sequence of the *Ace* gene in resistant flies differs by five point mutations in four positions from that in susceptible flies; these changes generate an enzyme with elevated insensitivity to inhibition (Oppenorth 1984). Although, in this case, resistance developed as a result of insecticide selection, there exist plant-derived carbamate compounds with the same mode of action (e.g., physostigmine, from *Physostigma venenosum*) (Matsumura 1975), raising the possibility that target-site insensitivity derived from structural mutation is involved in adaptation to such plants.

Another form of structural change associated with resistance involves gene amplification; gene duplication events lead to an increase in the number of copies of coding sequence, allowing an organism to produce greater amounts of gene product faster than would be possible with a single-copy gene. Again, an example from the insecticide literature has potential applicability toward understanding adaptation to fine-scale differences in host-plant chemistry. Organophosphate-resistant populations of the aphid *Myzus persicae* display up to 64-fold amplification of a gene encoding an esterase that breaks down the insecticide, thus conferring resistance (Devonshire and Field 1991).

In contrast with structural changes, which affect the gene product itself, regulatory changes affect gene expression and influence the amount of gene product produced. Regulatory changes are of two sorts: Cis mutations involve alteration of regulatory elements found upstream of the coding regions of a gene, whereas trans mutations affect genes, generally on different chromosomes, that encode proteins that bind to regulatory elements. Insecticide-resistant houseflies (*Musca domestica*) display higher levels of constitutive expression of the cytochrome P450 1A1 than do susceptible houseflies. The structural gene encodes an enzyme that participates in the metabolism not only of aldrin and heptachlor but also several plant terpenoids. Although the structural gene for this enzyme maps to chromosome V, insecticide resistance (elevated constitutive expression) maps to chromosome II; resistance in this instance most likely involves production of a trans-acting factor that alters patterns of transcription in resistant flies (Cohen et al. 1994; Feyereisen et al. 1995). As well, in permethrin-resistant houseflies, although the structural gene encoding *CYP6D1*, the enzyme putatively responsible for metabolism of synthetic pyrethroids, maps to chromosome I (Tomita and Scott 1995; Liu et al. 1995), resistance maps to chromosome II, suggesting the involvement of a trans-acting factor (Scott 1996).

*A priori*, it is difficult to determine which forms of genetic change are most likely to accompany local adaptation to chemically distinct hosts. Some insight can be gained, however, from the model of Hedrick and McDonald (1980). These authors constructed a hierarchical model designed to determine conditions favorable for adaptation via regulatory genes. A component of this hierarchical approach is that, all else being equal, changes in regulatory genes have a greater impact on phenotype per locus than do changes in structural genes (or, according to their terminology, "producer gene loci"). Due to this difference in phenotypic impact, regulatory gene changes become selectively advantageous under conditions of rapid and extreme environmental change. The polygenic structural gene system, with multiple loci, each contributing a small phenotypic effect, becomes selectively advantageous under circumstances requiring "fine-tuning" adaptation or gradual environmental change, since variation around the optimum phenotype is less dramatic than it is as a result of a regulatory gene change. Thus, according to this model, in the context of adaptation to toxins, although regulatory genes may be principally involved in adaptation to radical environmental changes (e.g., the introduction of a novel chemical, such as a synthetic organic insecticide, into the environment), structural gene changes may be principally involved in fine-scale adjustments to alterations in the chemical environment (such as differential abundance of allelochemicals that occur as a series of structural analogs within a plant species).

## 5.3 Cytochrome P450 Monooxygenases and Detoxification of Plant Chemicals

Of the various and sundry biochemical mechanisms utilized by organisms to cope with plant allelochemicals, cytochrome P450 monooxygenases play a central role in virtually all plant-consuming taxa. These membrane-bound heme proteins belong to a gene superfamily with over 400 known members (Nelson et al. 1993). P450s effect a wide range of metabolic transformations; these are for the most part oxidative reactions that increase hydrophilicity and thus reduce toxicity (Gonzalez and Nebert 1990). The P450s metabolize both endogenous substrates, such as pheromones and hormones, and exogenous substrates, most notably environmental toxins such as pesticides and allelochemicals. The P450 proteins are classified into subfamilies based on levels of sequence similarity, with 40% or greater sequence identity constituting the criterion for membership in a family (designated by a number) and 60% or greater sequence identity constituting the criterion for membership in a subfamily (designated by a letter). Allelic variants at a single locus are defined as sharing greater than 98% sequence identity and are designated as numbered variants.

Genetic variation in P450s is well documented. Interestingly, such variation is more likely to occur at loci associated with xenobiotic metabolism than at loci associated with metabolism of endogenous substrates (Krynetskii 1996). In terms of how genetic variation in P450 structure and function relates to adaptation to plant allelochemicals, information derived from studies of insects is pathetically scanty. Insight can be gained, however, from reviewing the vast literature accumulated on P450-mediated metabolism of plant compounds by humans. These studies were conducted not so much from the perspective of understanding human dietary habits as from the perspective of understanding drug metabolism; a substantial number of prescription drugs in use today are plant allelochemicals or are chemically modified derivatives of plant allelochemicals.

Over 30 P450 proteins in 13 families have been characterized to date in humans; of these, members of four families play a principal role in xenobiotic metabolism. The *CYP2* family is particularly noteworthy in this regard (Goldstein and de Morais 1994; Bertilsson 1995). *CYP2D6* was among the first P450s identified to exhibit interracial and interpopulational polymorphisms in structure and function (Skoda et al. 1988; Kimura et al. 1989; Gaedigk et al. 1991). This enzyme is responsible for hydroxylating both drugs and natural products; among the drugs included are debrisoquine, propanolol, captopril, dextromethorphan, nortryptiline, and 4-methoxyamphetamine, and among the natural products are several alkaloids, including codeine, nicotine, and sparteine (Guengerich 1994). Comparisons of drug disposition among a broad cross-section of patients revealed a wide range of metabolic abilities. These metabolic differences frequently fall along racial or population lines. Whereas 5–10% of Caucasians are poor metabolizers of debrisoquine, for example, only 1% of Orientals are slow metabolizers. Pedigree analysis and genetic studies revealed that slow metabolism of debrisoquine (as well as several other important drugs) is inherited as an autosomal recessive trait. At least 11 allelic variants have been described at this locus in which mutations affect enzyme activity (Table 5.1). Among the most common of these allelic variants in "slow metabolizers" is *CYP2D6-B*, which carries several mutations, including a frame-shift point mutation. *CYP2D6-A* is considerably less

*Table 5.1* Allelic Variation at the *CYP2D6* Locus, and Phenotypic Manifestation (adapted from Krynetskii 1996)

| Allele | Activity Phenotype | Mutation |
|---|---|---|
| *CYP2D6-wt* | Normal | |
| *CYP2D6-L* | Normal | G1726 to C, C2938-T (Arg-Cys), C4268-C (Ser-Thr) |
| *CYP2D6L2* | Elevated | |
| *CYP2D6L12* | Elevated | |
| *CYP2D6A* | None | A2637 deletion |
| *CYP2D6B* | None | C188-T (Pro-Ser), G1749-C, G1934-A, T3979-C (Leu-Pro) |
| *CYP2D6D* | None | Complete deletion |
| *CYP2D6T* | None | T1795 deletion |
| *CYP2D6C* | Normal | AGA deletion |
| *CYP2D6J* | Decreased | C188-T (Pro-Ser), G1749-C, G4268-C (Ser-Thr) |
| *CYP2D6ChI* | Decreased | C188-T (Pro-ser) |

common and involves a single base-pair deletion in an exon region. In the
*CYP2D6-D* allele, the entire gene is deleted. Whereas these mutations interfere
with metabolism, there are other mutations that actually enhance metabolic rate,
giving rise to "superfast" metabolizing phenotypes that convert substrate to
metabolite at levels up to five times the normal rate. In one family of superfast
metabolizers, for example, 12-fold amplification of one allele is recorded (Johansson et al. 1993).

*CYP219* is involved in metabolism of drugs (such as mephenytoin and the anticoagulant warfarin) as well as plant allelochemicals (Krynetskii 1996); it is, for
example, the enzyme responsible for 7-hydroxylation of tetrahydrocannabinol,
the psychotropic terpenoid from *Cannabis sativa*. Interracial differences in rates
of mephenytoin metabolism are striking; whereas 3–5% of Caucasians are poor metabolizers (unable to attach a hydroxyl group to the 4-position of the S-enantiomer
of mephenytoin), between 18% and 23% of Orientals tested are poor metabolizers. Two mutations for differential activity have been identified; one point mutation creates an aberrant splice site in an exon, and a second, a guanine to adenine
switch at position 636, creates a premature stop codon, leading to the production
of a truncated polypeptide.

*CYP2E1* is important in the 7-demethylation of caffeine, an alkaloid ingested
by consumers of *Coffea arabica* (London et al. 1996). A rare variant of this enzyme, associated with very high levels of transcription, is found with differing
frequency among human populations. Whereas only 2% of African Americans in
Los Angeles displayed high rates of metabolism, 8% of Caucasians displayed
these high rates (London et al. 1996). This change is thought to be regulatory in
nature, in which a point mutation in the 5′ flanking region alters binding with a
transcription factor.

*CYP2A6* is responsible for hydroxylation of coumarin, a plant product reported
from members of the Leguminosae, as well as the conversion of nicotine (a
solanaceous alkaloid) to the metabolite cotinine. Two allelic variants, *CYP2A6v1*
and *CYP2A6v2*, have been identified at this locus (Fernandez-Salguero et al.
1995). The allelic variant *CYP2A6v2* results from a single base substitution
(thymine to adenine) and produces a defective enzyme. Individuals homozygous
for this allele are incapable of metabolizing coumarin; heterozygous individuals
display metabolic rates that are half those of normal *CYP2A6* genotypes. A second allelic variant, *CYP2A6v1*, results from the presence of point mutations in
three different exons. These variants appear with differing frequency among
human races (Table 5.2), and interracial differences in frequency distributions
pertain even in different geographic localities.

An examination of the patterns of populational differentiation in allelochemical metabolism in *CYP2* genes in humans can provide some indication as to the
ability of members of this gene family to change over evolutionary time. Analysis
of blood groups and protein characters place the time of separation between Cau-

*Table 5.2*   Allele Frequencies for the *CYP2A6* Locus According to Human Ethnic Group (adapted from Fernandez-Salguero et al. 1995).

| | Allele frequencies (%) | | |
|---|---|---|---|
| | *CYP2A6*[a] | *CYP2A6v1*[b] | *CYP2A6v2*[c] |
| Caucasian (Finnish-English) | 76–86 | 15–17 | 0–7 |
| Oriental (Japanese-Taiwanese) | 52–83 | 11–20 | 6–28 |
| African-American | 97.5 | 0 | 2.5 |

[a] African-American frequency is significantly different from all other groups.

[b] African-American frequency is significantly different from all other groups; within Oriental groups, Hapanese differ from Taiwanese and African-Americans.

[c] Finnish (0%) differ significantly from English, Taiwanese, and Japanese; Japanese differ from all other groups.

casians and Orientals at 40,000–60,000 years ago; the separation between Blacks and Caucasians/Orientals is placed at 150,000 years ago (Krynetskii 1996). Any evolutionary scenario based on human race differentiation must be interpreted with caution; notwithstanding, even these rough estimates of divergence times indicate considerable evolutionary lability. That frequency differences exist for alleles across human races is remarkable, given that it is unlikely these alleles are under strong selection; the omnivory of humans and the idiosyncratic distribution of plant toxins argue against recurrent exposure to toxic or even fitness-reducing levels of any particular compound.

For most herbivorous insects, levels of exposure to particular plant toxins are almost assuredly high enough to compromise fitness. Particularly for sedentary specialists, which may consume only a single organ of a single plant species over the course of development, levels of exposure to plant toxins can be phenomenally high. The parsnip webworm, *Depressaria pastinacella*, for example, is restricted in its diet to the reproductive structures of species in the genus *Pastinaca* and the closely related *Heracleum* (Berenbaum and Zangerl 1992b). Both of these plants produce large quantities of furanocoumarins in flowers and fruits. By virtue of its diet, the parsnip webworm can consume up to 7% of its body weight in furanocoumarins over the course of a single day (personal observation). Because furanocoumarins can be extremely toxic, particularly at high concentrations (Berenbaum et al. 1989; Berenbaum 1991a), exposure to such large quantities of these compounds is likely to place strong selection pressures on maintaining a functional detoxification system. The importance of furanocoumarin detoxification ability to the webworm in dealing with its host plants is evidenced by the fact that growth of webworms consuming host-plant tissues containing high levels of furanocoumarins is significantly correlated with the individual caterpillars' cytochrome P450-mediated detoxification capacity (Fig. 5.1).

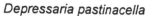

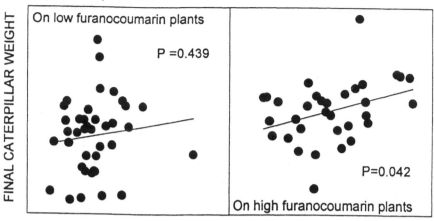

**METABOLISM OF FURANOCOUMARINS**

*Figure 5.1* Partial regression of parsnip webworm growth as a function of furanocoumarin detoxification capacity (measured *in vitro*) for larvae fed plant tissue with low
and with high furanocoumarin content. The additional independent variables (not shown)
are initial caterpillar weight and amount of tissue consumed (from data in Zangerl and
Berenbaum 1993).

## 5.4 Genetic Variation in Insect P450-Mediated Detoxification and Possible Importance in Local Adaptation to Allelochemicals

Operational challenges have made the measurement of within-species variation in
P450-mediated detoxification difficult in insects. Among other things, it is difficult to assay P450-mediated metabolism of individual insects if they are very
small (as many insects are); as well, because age, diet, experimental conditions,
and phase of the moon all appear to affect metabolism assays, isolating the portion of genetic variation from the sea of experimental and environmental variation can be challenging. Whereas enormous amounts of data are available on interspecific differences in metabolic rates (aldrin–epoxidase activities have been
measured in over 100 species of Lepidoptera alone; Neal 1987), relatively few
data are available on the range of variation within and between populations in any
single species. From this limited number of studies (Table 5.3), it does in fact appear that toxin-metabolizing P450s within a species display a range of variation
associated with genetic differentiation.

The information base on genetic variation in P450-mediated metabolism of allelochemicals is considerably thinner than is the information base on genetic variation in P450-mediated metabolism of insecticides; in fact, it appears to relate exclusively to furanocoumarin metabolism by Lepidoptera (Table 5.3A). These

*Table 5.3*    Intraspecific Variation to Cytochrone P450-Mediated Metabolism of Zenobiotics

A. Allelochemical metabolism
Black swallowtail (*Papilio polyxenes*)
*Berenbaum and Zengerl (1993)*

|  | Metabolism (nmoles/min/g caterpillar) ± *SD* | |
|---|---|---|
|  | Xanthotoxin | Angelicin |
| Maternal families | 1.  8.6 ± 3.0 | 3.5 ± 1.8 |
|  | 2.  5.3 ± 3.7 | 3.0 ± 1.4 |
|  | 3.  3.3 ± 1.1 | 2.4 ± 2.1 |
|  | 4.  4.7 ± 2.4 | 2.3 ± 1.7 |
|  | 5.  6.3 ± 2.1 | 1.9 ± 1.2 |
|  | 6.  6.2 ± 3.2 | 2.2 ± 1.7 |
| Range (fold variation) | 2.6-fold | 1.8-fold |

Parsnip webworm (*Depressaria pastinacella*)
*Berenbaum and Zangerl (1992b)*

|  | Metabolism (nmoles/min/g caterpillar) ± *SD* | | |
|---|---|---|---|
|  | Xanthotoxin | Bergapten | Sphondin |
| Full-sib families | 1.  28.9 ± 14.9 | 6.0 ± 3.1 | 9.6 ± 0.8 |
|  | 2.  39.6 ± 14.8 | 8.4 ± 3.1 | 10.1 ± 1.1 |
|  | 3.  31.8 ± 14.2 | 12.0 ± 3.8 | 12.7 ± 1.0 |
|  | 4.  19.6 ± 16.8 | 7.0 ± 5.0 | 10.2 ± 1.3 |
|  | 5.  20.9 ± 19.9 | 7.9 ± 3.7 | 8.5 ± 0.8 |
|  | 6.  30.3 ± 16.2 | 11.5 ± 3.7 | 10.3 ± 0.8 |
|  | 7.  30.7 ± 16.7 | 9.1 ± 5.0 | 11.3 ± 1.3 |
|  | 8.  15.7 ± 13.5 | 8.7 ± 5.0 | 12.2 ± 1.2 |
|  | 9.  21.5 ± 13.7 | 11.8 ± 1.1 |  |
|  | 10.  34.6 ± 12.0 |  |  |
|  | 11.  23.1 ±  6.7 |  |  |
|  | 12.  33.9 ± 17.4 |  |  |
|  | 13.  27.3 ± 17.4 |  |  |
| Range (fold variation) | 2.6-fold | 2.0-fold | 1.5-fold |

Tiger swallowtail (*Papilio glaucus*)

| | Metabolism (pmoles/mg protein/minute) | | |
|---|---|---|---|
| Population | Xanthotoxin | Bergapten | Imperatorin |
| Georgia | 20.6 ± 6.2 | 119.4 ± 31.3 | 463.8 ± 106.0 |
| Ohio | 15.0 ± 5.5 | 84.6 ± 18.9 | 251.5 ±  35.8 |
| Range (fold variation) | 1.4-fold | 1.4-fold | 1.8-fold |

Table 5.3   (continued)

---

B. Insecticide metabolism
Housefly (*Musca domestica*)
*Terriere (1968)*

| Strain | Hydroxylation of naphthalene |
|---|---|
| Milan susceptible | 88 |
| Corvallis resistant | 203 |
| Orlando resistant | 372 |
| Naphthalene resistant | 428 |
| Range (fold variation) | 4.9-fold |

*Schonbrod et al. (1968)*

| | Mmoles product/fly | |
|---|---|---|
| Strain | Hydroxylation (30 min) | Epoxidation (60 min) |
| Isolan-R | 1.08 | 1.22 |
| Grothe | 0.08 | 0.52 |
| Fc | 0.80 | 0.46 |
| Carbaryl-R | 0.71 | 0.31 |
| Naph.-R | 0.71 | 0.31 |
| DDT-R; dov | 0.46 | 0.11 |
| kdr-O; w5 | 0.37 | 0.45 |
| Trop.p; clw | 0.30 | 0.06 |
| Calif. Parathion-R | 0.31 | 0.07 |
| organotin-R; stw | 0.28 | 0.08 |
| Orlando DDT | 0.30 | 0.20 |
| Orlando-regular | 0.30 | 0.11 |
| Dield-R; w5; stw | 0.09 | 0.03 |
| Dield-R; cyw | 0.10 | 0.03 |
| Range (fold variation) | 12-fold | 41-fold |

*Hammock et al. (1977)*

| Strain | O-demethylase | Epoxidase | Hydroxylase |
|---|---|---|---|
| S-NAIDM | 0.43 | 0.04 | 0.015 |
| R-dimethoate | 2.35 | 0.13 | 0.037 |
| R-methoprene | 1.96 | 0.15 | 0.058 |
| Range (fold variation) | 5.5-fold | 3.8-fold | 3.9-fold |

*Plapp and Casida (1969)*

| | % Metabolized | | | |
|---|---|---|---|---|
| Strains | Aldrin | Allethrin | Diazinon | Baygon |
| S-stw; bwb; ocra | 32 | 29 | 15 | 8 |
| R-Baygon | 68 | 85 | 54 | 39 |
| R-Baygon; bwb; ocra | 76 | 59 | 35 | 35 |
| R-Fc | 62 | 30 | 46 | 13 |
| R-Fc; bwb; stw | 62 | 27 | 29 | 11 |
| Range (fold variation) | .4-fold | 3.1-fold | 3.6-fold | 4.9-fold |

---

*Table 5.3    (continued)*

**Drosophila melanogaster**
*Hallstrom and Grafstrom (1981)*

| Strain | Benzopyrene monoxygenase activity pmol/mg/min | PB-induced pmol/mg/min |
|---|---|---|
| Hikone R | 63 | 354 |
| Karsnas 60 | 47 | 433 |
| Berlin K | 41 | 82 |
| Range (fold variation) | 1.5-fold | 5.3-fold |

*Hallstrom (1987)*

| Strain | Nmoles formaldehyde formed/mg microsomal protein/min | | |
|---|---|---|---|
| | Aminopyrine | Ethylmorphine | Benzphetamine |
| Karsnas 60w | 1.5 + 0.3 | 1.0 + 0.2 | 1.2 + 0.2 |
| Hikone R | 2.1 + 0.2 | 0.6 + 0.1 | 0.6 + 0.1 |
| Florida 9 | 2.4 + 0.6 | 2.3 + 0.6 | 2.7 + 0.3 |
| Lausanne S | 1.4 + 0.3 | 1.0 + 0.2 | 1.4 + 0.3 |
| Canton S | 1.1 + 0.3 | 0.8 + 0.2 | 1.3 + 0.3 |
| Eth-29 | 1.5 + 0.3 | 1.4 + 0.4 | 2.0 + 0.4 |
| Berlin K | 0.9 + 0.1 | 0.7 + 0.2 | Not determined |
| Oregon R | 3.8 + 0.6 | 1.4 + 0.3 | 1.8 + 0.4 |
| Marked inversion 1 | .0 + 0.1 | 0.6 + 0.1 | Not determined |
| Range (fold variation) | 4.2-fold | 3.8-fold | 4.5-fold |

Cabbage looper fat body (*Trichoplusia ni*)
*Kuhr (1971)*

| Strain | μg carbaryl/30 min/mg protein |
|---|---|
| Blue (DDT susceptible) | 0.76 |
| Lab (DDT resistant) | 1.27 |
| Field-1968 | 6.14 |
| Range (fold variation) | 8.0-fold |

Colorado potato beetle (*Leptinotarsa decemlineata*)
*Argetine et al. (1992)*

| Strain | Abamectin metabolism nmol/min/mg protein |
|---|---|
| SS (Abameftin-susceptible) | 138 |
| AB-Fd (Abamectin-resistant) | 158 |
| AB-L (Abamectin-resistant) | 119 |
| Range (fold variation) | 1.3-fold |

studies differ from those dealing with insecticide metabolism, in that for the most part they examine variation among maternal families, rather than among genetic strains. Be that as it may, despite the exceedingly limited scope of these studies, genetic variation is readily apparent, even across as few as six maternal families. That this genetic variation is available for selection is evidenced by relatively high heritabilities for rates of furanocoumarin metabolism (alone and in combination) in at least two of these species: *Depressaria pastinacella*, in which heritabilities for bergapten and xanthotoxin metabolism were measured as 0.232 and 0.221, respectively (Berenbaum and Zangerl 1992b) and *Papilio polyxenes*, in which heritabilities for xanthotoxin metabolism was measured as 0.546 (Berenbaum and Zangerl 1993).

Not surprisingly, substantially greater genetic variation is documented for P450-mediated insecticide resistance; of considerable economic importance, such studies have been ongoing for close to 30 years. A survey of these studies (Table 5.3B) reveals strain-related variation not only in general activity levels (which can vary over 40-fold across strains) but also in patterns of substrate specificity (e.g., Hallstrom 1987).

## 5.5 Scenario for P450-Mediated Local Adaptation to Allelochemicals

The genetic basis for P450-mediated metabolism of allelochemicals by insects is perhaps most thoroughly characterized for metabolism of furanocoumarins. Furanocoumarins owe their biological activity to their ability to absorb photons of ultraviolet light energy and form an excited triplet state; the highly reactive triplet-state molecule can interact with DNA to form cross-links, with amino acids to denature enzymes and other proteins, with unsaturated fatty acids to form cycloadducts, and with ground-state oxygen to generate toxic oxyradicals that can damage many kinds of biomolecules. Thus, they are toxic to a wide range of organisms, including insects (for a review, see Berenbaum 1991a, 1995a). Furanocoumarins are classified according to their structure as either linear, with the furan ring attached at the 6,7 positions of the benz-2-pyrone nucleus, or angular, with the furan ring attached at the 7,8 positions of the benz-2-pyrone nucleus. These two groups share a common precursor, umbelliferone, but are biosynthetically distinct by virtue of the action of site-specific prenylating enzymes that initiate the attachment of the furan ring (Berenbaum 1991a). Linear furanocoumarins, known to occur in approximately 10 families, are structurally diverse and widely distributed only in two families, the Rutaceae and Apiaceae. Angular furanocoumarins are even more restricted in distribution, known only from a few genera in three families (Leguminosae, Rutaceae, and Apiaceae) and occurring with regularity only in two tribes of the Apiaceae (Murray et al. 1982).

Lepidopterans that consume foliage rich in furanocoumarins rely primarily on cytochrome P450-mediated metabolism to detoxify the compounds (Berenbaum 1995b). Lepidopterans display substantial differences not only in constitutive levels of furanocoumarin metabolism but also in the inducibility of metabolism in response to furanocoumarins (Berenbaum 1991b, 1995b). As a general rule, the level of metabolic activity against furanocoumarins corresponds to the frequency with which furanocoumarins are encountered within the host range. In the family Papilionidae, levels of constitutive P450-mediated metabolism of xanthotoxin, a linear furanocoumarin present in both umbelliferous and rutaceous hosts, are high in *P. cresphontes*, a Section IV species associated almost exclusively with furanocoumarin-containing Rutaceae, and in *P. polyxenes* and *P. brevicauda*, two Section II species associated almost exclusively with furanocoumarin-containing Umbelliferae. P450-mediated metabolism of xanthotoxin does not exist, however, in *P. troilus*, a Section III species associated exclusively with the Lauraceae, a family in which furanocoumarins are absent. Species in the family outside the genus *Papilio* that feed on plants entirely lacking furanocoumarins (*Battus philenor, Eurytides marcellus*) have no detectable ability to metabolize xanthotoxin (Cohen et al. 1992).

The genetic mechanisms underlying the metabolism of furanocoumarins are most thoroughly characterized in *P. polyxenes*, the black swallowtail, a species that feeds almost exclusively on plants containing furanocoumarins (Berenbaum 1981). From *P. polyxenes*, two cDNAs sharing over 98% sequence identity, *CYP6B1v1* and *CYP6B1v2*, were cloned, sequenced, and shown to be inducible by xanthotoxin (Cohen et al. 1992). Baculovirus-mediated expression of these cDNAs in two different lepidopteran cell lines (Sf9, Tn5) confirmed that these cDNAs encode furanocoumarin-metabolizing P450 isozymes (Ma et al. 1994). In these *in vitro* assays, the two allelic variants differed only slightly in their catalytic activity.

Northern analysis at high stringency indicated that mRNAs cross-reactive with *CYP6B1* are detectable in *P. brevicauda*; at lower stringency, more divergent mRNA transcripts could be detected in *P. cresphontes* (Section IV furanocoumarin feeder) and *P. glaucus* (Section III generalist). These three species all demonstrate xanthotoxin-inducible metabolism of xanthotoxin; papilionid species lacking this attribute also lack detectable *CYP6B1*–cross-reactive transcripts in Northern analysis (Cohen et al. 1992). As well, the three species capable of substrate-inducible metabolism of xanthotoxin—*P. glaucus, P. polyxenes,* and *P. cresphontes*—are also able to grow and develop on foliage supplemented with xanthotoxin in bioassay (Heininger 1989; Berenbaum 1991b).

A species that does not precisely fit expected patterns is *P. glaucus*, the tiger swallowtail. In contrast with most of its congeners, this Section III species only rarely utilizes furanocoumarin-containing plants as hosts; notwithstanding its infrequent exposure, *P. glaucus* displays constitutive activity against xanthotoxin that is up to 13-fold inducible (Cohen et al. 1992, this volume), a level of responsiveness comparable to that displayed by the furanocoumarin specialist *P. polyxenes* (Cohen et al. 1989). Although constitutive activities against furanocoumarins are low relative to activities displayed by furanocoumarin-consuming Section II swallowtails, when induced, their activities against some substrates are within an order of magnitude of those exhibited by furanocoumarin specialists. In fact, in laboratory bioassays, *P. glaucus* larvae are far more tolerant of furanocoumarins added to their diet than are confamilials (e.g., *Eurytides marcellus, Battus philenor*) that never encounter furanocoumarins in their diet and possess no detectable furanocoumarin metabolism (E. Heininger 1989, and personal observation).

 *P. glaucus* (comprising *P. g. glaucus + P. g. australis*) is without doubt the most polyphagous of all swallowtails in terms of numbers of plant families utilized as hosts. *P. glaucus* has been recorded on 12 genera in 7 families (Bossart and Scriber 1995a) and in the laboratory has been reared successfully on 120 species in 34 families (Scriber 1988, 1995). Hoptree (*Ptelea trifoliata*) is distinct among the host plants of *P. glaucus,* in that it is the sole rutaceous plant (and the sole furanocoumarin-containing plant) normally utilized by this species (Dreyer 1969). Tiger swallowtail populations differ throughout their range with respect to the use of hoptree as a host. Although this shrub is broadly distributed across North America, its distribution is characterized by patches of locally high population density along streambanks, forest edges, and sandy soils, interspersed throughout areas of low abundance (Ambrose et al. 1984; Bailey 1962; Bailey et al. 1970). Due to these peculiarities in the distributional range (Fig. 5.2; Scriber and Gage 1995), it is possible to identify populations of *P. glaucus* that are exposed to furanocoumarins, as well as populations whose members are unlikely ever to encounter furanocoumarins. Populational differences in the ability to utilize this host are suggested (but not conclusively demonstrated) by preliminary studies (Bossart, unpublished data) showing that populations from Georgia tend to complete larval development faster on hoptree foliage (23.7 days) than do populations in Ohio (25.2 days), where hoptrees are infrequently encountered ($p = 0.19$). Differentiation in larval development traits between these two populations is more dramatically demonstrated on another host, white ash (*Fraxinus americana*); Georgia populations complete development significantly faster on white ash (32.1 days) than do individuals from the Ohio population (39.2 days; $p = 0.001$), giving rise to a significant population $\times$ host interaction term ($p = 0.0014$, two-way ANOVA). Individual preferences among female butterflies differ as well with respect to selection of host plants for oviposition; these differences tend to reflect local availability of hosts (Bossart and Scriber 1995a). In two-choice tests with tuliptree, a host plant utilized throughout the range of the tiger swallowtail, 43% of females from Georgia, where hoptrees are abundant, preferred to oviposit on hoptree; in contrast, only 11% of the females tested preferred sweetbay, a host not naturally found in Georgia, to tuliptree.

 The differential degree to which populations of tiger swallowtails are exposed to hoptrees is associated to some degree with their furanocoumarin-metabolizing capabilities (Table 5.3). In Georgia, where hoptrees are abundant and where larvae can predictably be collected from hoptree foliage (M. Scriber personal communication), *P. glaucus* larvae are capable of rapid and efficient metabolism of all three furanocoumarins tested. In contrast, in southern Ohio, where hoptrees are infre-

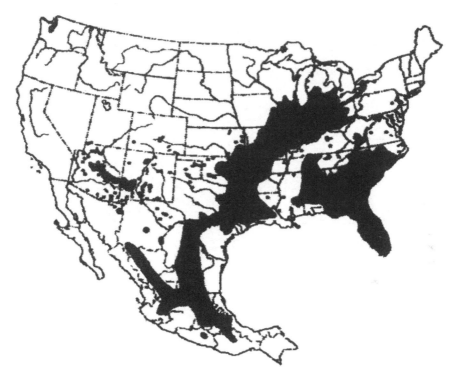

*Figure 5.2*   Geographic distribution of *Ptelea trifoliata* (hoptree) in North America (from Scriber and Gage 1995)

quent, and where preliminary evidence suggests that hoptree is a suitable, although not preferred host plant (Bossart 1993; unpublished data), rates of P450-mediated metabolism of furanocoumarins tend to be lower than they are in larvae from Georgia (e.g., imperatorin, $p < 0.11$, Fisher's–Least Significant Difference [LSD]).

Although these findings are very preliminary, they are suggestive of a P450-mediated contribution to local adaptation. Identifying and characterizing the P450s responsible for this metabolism will allow for a more powerful examination of genetic differentiation among these populations. *CYP6B4* cDNA has been cloned and sequenced from *P. glaucus*; several lines of evidence suggest that this P450 is the principal one responsible for the furanocoumarin metabolic capabilities of *P. glaucus* (Hung 1996). Not only is it over 60% identical with *CYP6B1* in the coding region but Northern analysis demonstrates that transcripts are highly induced in response to xanthotoxin (Hung 1996). Baculovirus-mediated expression of this cDNA in Transcriber (Tn5) cells demonstrates that this P450 metabolizes isopimpinellin, imperatorin, and bergapten at high levels, xanthotoxin and psoralen at intermediate levels, and angelicin, sphondin, and trioxsalen at low levels (Table 5.4).

*Table 5.4*  Metabolism of Furanocoumrins by *CYP6B4* Expressed in Tn5 Baculovirus Expression System (means with *SD* in parentheses, *n*=3)

| Substrate | Metabolism (pmol/min/mg protein) |
|---|---|
| Isopimpinellin | 217.8 (16.6)[c] |
| Imperatorin | 142.2 (43)[b] |
| Bergapten | 128.2 (7.8)[b] |
| Xanthotoxin | 50.8 (0.9)[a] |
| Psoralen | 20.9 (2.1)[a] |
| Angelicin | 12.6 (2.6)[a] |
| Sphondin | 13.5 (3.9)[a] |
| Trioxsalen | 6.7 (1.6)[a] |

From Hung 1996; values sharing the same letter are not significantly different at p < 0.05 (Fishers' *post hoc* test)

Two genes, *CYP6B4v2* and *CYP6B5v1*, which are over 99% identical in sequence, have recently been characterized from *P. glaucus*; both of these genes share over 99% identity with *CYP6B4* cDNA (Hung 1996). These two genes contain an element in their promoter region that aligns with the sequence between $-146$ and $-97$ in the *CYP6B1v3* promoter regions; in the *CYP6B1v3 P. polyxenes* gene, this element appears to function as a xanthotoxin-responsive element in *P. polyxenes* and may serve a similar function in *P. glaucus*. Although their extreme sequence similarity would indicate that *CYP6B4v2* and *CYP6B5v1* are allelic variants at the same locus, the fact that they were isolated from the same recombinant phage thus likely represents recently duplicated distinct loci within 10kb of each other (Hung 1996). Whether these two loci have diverged in function since duplication (Walsh 1995) has not yet been determined.

Recent studies indicate the presence of genetic variation in *CYP6B* genes in the tiger swallowtail. In a preliminary screen of Restriction Fragment Length Polymorphisms (RFLP) variation across populations (Fig. 5.2), Hung (unpublished) obtained polymorphisms characterizing *P. glaucus* from four populations: Georgia, Illinois, Ohio, and Florida (these caterpillars were reared from eggs laid by females collected in the field and kindly sent to us by J. M. Scriber). For RFLP analysis, 40 µg genomic DNA of individual larvae from different populations was digested by restriction enzymes (SalI, BamHI + HindIII or EcoRI + HindIII) The digested DNA fragments were separated by agarose gel electrophoresis, blotted on nylon membrane, and probed with *CYP6B4* at high stringency (30% formamide; 50°C). Genomic DNA isolated from larvae and digested with SalI yielded four distinct fragments present in Georgia, Florida, and Illinois. Because *CYP6B4* does not have a SalI sight, these four fragments likely represent distinct alleles. Cuts with EcoRI and HindIII clearly show differences among individuals from different populations in the distribution of allelic fragments (Fig. 5.3).

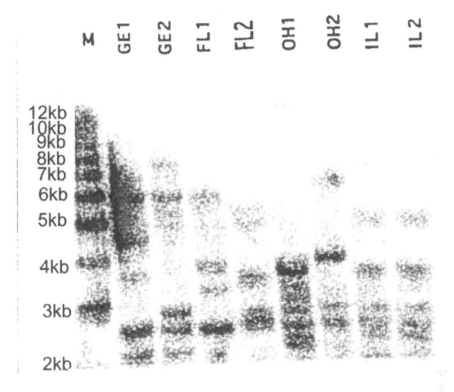

*Figure 5.3* Southern blot analysis of *Papilio glaucus* DNA after digestion with three sets of restriction endonucleases and hybridization with 32P-labeled *CYP6B4* cDNA probe (GE = Georgia, FL = Florida, OH = Ohio, IL = Illinois).

Whether these allelic differences represent functional differences in fura-nocoumarin metabolism, resulting from local adaptation, has yet to be estab-lished. It must be stated, most emphatically, that this study is in an inchoate stage. Many important questions have not yet been addressed; it is far from clear, for ex-ample, how much gene flow occurs between populations and even putative sub-species of *P. glaucus* (e.g., Scriber 1986). The extent to which furanocoumarin metabolism is exclusively the province of *CYP6B4* and *CYP6B5* in this species is not yet established; if, in fact, there are other gene products involved in their me-tabolism, genetic variation in *CYP6B4* and *CYP6B5* may not have a pronounced effect on furanocoumarin metabolism rates. If work can proceed to the point that functional variation in these loci can be shown to map onto populations that differ in the frequency with which furanocoumarins are encountered and the degree to which larvae can tolerate them, the tiger swallowtail may then provide an example as to how genetic differentiation in allelochemical metabolism contributes to the evolution of host-associated populations.

## 5.6 Conclusions

The existence of locally adapted populations of herbivorous insects is often assumed to be *prima facie* evidence that genetic differentiation and adaptation are rapid forms of evolution that can take place at fine spatial scales. As well, evidence exists (in the form of locally adapted demes) that insects have the potential to adapt to an individual plant's unique chemical profile. Without elucidating the selective mechanisms that generate the outcome, however, any understanding of the process of local adaptation remains necessarily incomplete. If this review serves no other useful purpose, it can at least illustrate the fact that investigations of selective mechanisms are conspicuous by their absence (there are, as far as we can tell, *no* studies that can justifiably be called complete) and are sorely needed in order to advance this field. Tools and techniques are now available to allow investigators to pursue more mechanistic studies of local adaptation, particularly with respect to host-plant chemistry: to reconstruct the evolutionary process and to define its spatial limits more precisely. Although such studies may be difficult, expensive, and time-consuming, they have the potential to provide powerful insights into a process that otherwise might prove difficult to explain.

## Acknowledgments

We thank Chien-Fu Hung for valuable discussion and editors Susan Mopper and Sharon Strauss for patience deserving of sainthood. The preparation of this manuscript, as well as much of the work described in it, was supported by National Science Foundation Grant No. DEB 95-09826.

## 5.7 References

Ambrose, J. D., P. G. Kevan, and R. M. Gadawski. 1984. Hop tree (*Ptelea trifoliata*) in Canada: Population and reproductive biology of a rare species. *Can. J. Bot. 63*:1928–1935.

Argentine, J. A., J. M. Clark, and H. Lin. 1992. Genetics and biochemical mechanisms of abamectin resistance in two isogenic strains of Colorado potato beetle. *Pest. Biochem. Physiol. 44*:191–207.

Bailey, V. L. 1962. Revision of the genus *Ptelea* (Rutaceae). *Brittonia 14*:1–45.

Bailey, V. L., S. B. Herlin, and H. E. Brown. 1970. *Ptelea trifoliata* ssp. *trifoliata* (Rutaceae) in deciduous forest regions of eastern North America. *Brittonia 22*:346–358.

Berenbaum, M. R. 1981. Effects of linear furanocoumarins on an adapted specialist insect (*Papilio polyxenes*). *Ecol. Entomol. 6*:345–351.

Berenbaum, M. R. 1986. Target-site insensitivity in plant insect interactions. Pp. 257–272 *in* L. Brattsten and S. Ahmad (Eds.), *Molecular Mechanisms in Insect–Plant Associations.* Plenum Press, New York.

Berenbaum, M. R. 1991a. Coumarins. Pp. 221–249 *in* G. Rosenthal and M. Berenbaum (Eds.), *Herbivores: Their Interactions with Secondary Plant Metabolites,* Vol. 1. Academic Press, New York.

Berenbaum, M. R. 1991b. Comparative processing of allelochemicals in the *Papilionidae* (Lepidoptera). *Arch. Insect Biochem. Physiol. 17*:213–222.

Berenbaum, M. R. 1995a. Phototoxicity of plant secondary metabolites: Insect and mammalian perspectives. *Arch. Insect Biochem. Physiol. 29*:119–134.

Berenbaum, M. R. 1995b. Metabolic detoxification of plant prooxidants. Pp. 181–209 *in* S. Ahmad (Ed.), *Oxidative Stress and Antioxidant Defense in Biology.* Routledge, Chapman & Hall, New York.

Berenbaum, M. R. and A. R. Zangerl. 1992a. Genetics of secondary metabolism and herbivore resistance in plants. Pp. 415–438 *in* G. Rosenthal and M. Berenbaum, (Eds.), *Herbivores: Their Interactions with Secondary Plant Metabolites,* Vol. 2, 2nd ed. Academic Press, San Diego, CA.

Berenbaum, M. R. and A. R. Zangerl. 1992b. Genetics of physiological and behavioral resistance to host furanocoumarins in the parsnip webworm. *Evolution 46*:1373–1384.

Berenbaum, M. R. and A. R. Zangerl. 1993. Furanocoumarin metabolism in *Papilio polyxenes:* Genetic variability, biochemistry, and ecological significance. *Oecologia 95*:370–375.

Berenbaum, M. R., A. R. Zangerl, and K. Lee. 1989. Chemical barriers to adaptation by a specialist herbivore. *Oecologia 80*:501–506.

Bernays, E. and M. Graham. 1988. On the evolution of host specificity in phytophagous arthropods. *Ecology 69*:886–892.

Bertilsson, L. 1995. Geographical/interracial differences in polymorphic drug oxidation. *Clin. Pharmcokinet. 29*:192–209.

Bossart, J. L. 1993. Differential selection and adaptation in different host environments: Genotypic and phenotypic variation in host use traits in the tiger swallowtail butterfly, *Papilio glaucus* (Laws.). Ph.D. dissertation, Michigan State University, East Lansing, MI.

Bossart, J. L. and J. M. Scriber. 1995a. Genetic variation in oviposition preference in tiger swallowtail butterflies: Interspecific, interpopulation and interindividual comparisons. Pp. 183–193 *in* J. M. Scriber, Y. Tsubaki, and R.C. Lederhouse (Eds.), *Swallowtail Butterflies: Their Ecology and Evolutionary Biology.* Scientific Publishing, Gainesville, FL.

Brattsten, L. B. 1992. Metabolic defenses against plant allelochemicals. Pp. 175–242 *in* G. Rosenthal and M. Berenbaum (Eds.), *Herbivores: Their Interactions with Secondary Plant Metabolites,* Vol. 2. Academic Press, San Diego, CA.

Carroll, S. P. and C. Boyd. 1992. Host race radiation in the soapberry bug: Natural history with the history. *Evolution 46*:1052–1069.

Chapman, R. 1982. *The Insects: Structure and Function.* Elsevier, New York.

Cohen, M. R., M. R. Berenbaum, and M. A. Schuler. 1989. Induction of cytochrome P450–mediated detoxification in the black swallowtail. *J. Chem. Ecol. 15*:2347–2355.

Cohen, M. B., J. F. Koener, and R. Feyereisen. 1994. Structure and chromosomal localization of *CYP6A1*, a cytochrome P450-encoding gene from the housefly. *Gene 146*:267–272.

Cohen, M. B., M. A. Schuler, and M. R. Berenbaum. 1992. A host-inducible cytochrome P450 from a host-specific caterpillar: Molecular cloning and evolution. *Proc. Natl. Acad. Sci. USA 89*:10920–10924.

Devonshire, A. L. and L. M. Field. 1991. Gene amplification and insecticide resistance. *Annu. Rev. Entomol. 36*:1–23.

Dreyer, D. L. 1969. Coumarins and alkaloids of the genus *Ptelea*. *Phytochem. 8*:1013–1020.

Ehrlich, P. R. and P. R. Raven. 1964. Butterflies and plants: A study in coevolution. *Evolution 18*:586–608.

Feder, J. L., T. A. Hunt, and G. L. Bush. 1990. The effect of climate, host plant phenology, and host fidelity on the genetics of apple and hawthorn infesting races of *Rhagoletis pomonella*. *Entomol. Exp. Appl. 69*:117–135.

Fernandez-Salguero, P., S. M. G. Hoffman, S. Cholerton, H. Mohrenweiser, H. Raunio, A. Rautio, O. Pelkonen, J.-D. Huang, W. E. Evans, J. R. Idle, and F. J. Gonzalez. 1995. A genetic polymorphism in coumarin 7–hydroxylation: Sequence of the human *CYP2A* genes and identification of variant *CYP2A6* alleles. *Am. J. Hum. Genet. 57*:651–660.

Feyereisen, R. 1995. Molecular biology of insecticide resistance. *Tox. Lett. 82/83*:83–90.

Feyereisen, R., J. F. Andersen, F. A. Carino, M. B. Cohen, and J. F. Koener. 1995. Cytochrome P450 in the housefly: Structure, catalytic activity, and regulation of expression in an insecticide-resistant strain. *J. Pestic. Sci. 43*:233–239.

Gaedigk, A., M. Blum, R. Gaedigk, M. Eichelbaum, and U. A. Meyer. 1991. Deletion of the entire cytochrome P450 *CYP2D6* gene as a cause of impaired drug metabolism in poor metabolizers of the debrisoquine/sparteine polymorphism. *Am. J. Hum. Genet. 48*:943–950.

Goldstein, J. A. and S. M. F. de Morais. 1994. Biochemistry and molecular biology of the human *CYP2C* subfamily. *Pharmacogenetics 4*:285–299.

Gonzalez, F. J. and D. W. Nebert. 1990. Evolution of the P450 gene superfamily. *Trends in Genetics 6*:182–186.

Guengerich, F. P. 1994. Catalytic selectivity of human cytochrome P450 enzymes: Relevance to drug metabolism and toxicity. *Tox. Lett. 70*:133–138.

Hallstrom, I. 1987. Genetic variation in cytochrome P450-dependent demethylation in *Drosophila melanogaster*. *Biochem. Pharmacol. 36*:2279–2282.

Hallstrom, I. and R. Grafstrom. 1981. The metabolism of drugs and carcinogens in isolated subcellular fractions of *Drosophila melanogaster*: II. Enzyme induction and metabolism of benzo[a]pyrene. *Chem.-Biol. Interactions 34*:145–159.

Hammock, B. D., S. M. Mumby, and P. W. Lee. 1977. Mechanisms of resistance to the juvenoid methoprene in the housefly *Musca domestica* (Laws.). *Pestic. Biochem. Physiol. 7*:261–272.

Hedrick, P. W. and J. F. McDonald. 1980. Regulatory gene adaptation: An evolutionary model. *Heredity 45*:83–97.

Hegnauer, R. 1966–1973. *Chemotaxonomie der Pflanzen*. Birkhauser Verlag, Basel, France.

Heininger, E. 1989. Effects of furocoumarin and furoquinoline allelochemicals on host-plant utilization by Papilionidae. Ph.D. dissertation, University of Illinois at Urbana–Champaign, IL.

Hung, C. F. 1996. Isolation and characterization of cytochrome P450s from *Papilio polyxenes* and *Papilio glaucus*. Ph.D. dissertation, University of Illinois at Urbana–Champaign, IL.

Hung, C. F., H. Prapaipong, M. R. Berenbaum, and M. A. Schuler. 1995. Differential induction of cytochrome P450 transcripts in *Papilio polyxenes* by linear and angular furanocoumarins. *Insect Biochem. Mol. Biol. 25*:89–99.

Johannson, I., E. Lundqvist, L. Bertilsson, M.-L. Dahl, F. Sjoqvist, and M. Ingelman-Sundberg. 1993. Inherited amplification of an active gene in the cytochrome P450 *CYP2D* locus as a cause of ultrarapid metabolism of debrisoquine. *Proc. Natl. Acad. Sci. USA 90*:11825–11829.

Kimura, S., M. Umeno, R. C. Skoda, U. A. Meyer, and F. J. Gonzalez. 1989. The human debrisoquine 4–hydroxylase (*CYP2D*) locus: Sequence and identification of the polymorphic *CYP2D6* gene, a related gene, and a pseudogene. *Am. J. Hum. Gen. 45*:889–904.

Krynetskii, E. Y. 1996. Polymorphism of drug-metabolizing enzymes: Gene structure and enzyme activity (a review). *Molec. Biol. 30*:17–23.

Kuhr, R. J. 1971. Comparative metabolism of carbaryl by resistant and susceptible strains of the cabbage looper. *J. Econ. Entomol. 64*:1373–1378.

Liu, N, T. Tomita, and J. G. Scott. 1995. Allele-specific PCR reveals that *CYP6D1* is on chromosome 1 in the housefly, *Musca domestica. Experientia 51*:164–167.

London, S. J., A. K. Daly, J. Cooper, C. L. Carpenter, W. C. Navidi, L. Ding, and J. R. Idle. 1996. Lung cancer risk in relation to the *CYP2E1 RsaI* genetic polymorphism among African-Americans and Caucasians in Los Angeles County. *Pharmacogenetics 6*:151–158.

Ma, R., M. B. Cohen, M. R. Berenbaum, and M.A. Schuler. 1994. Black swallowtail (*Papilio polyxenes*) alleles encode cytochrome P450s that selectively metabolize linear furanocoumarins. *Arch. Biochem. Biophys. 310*:332–340.

Matsumura, F. 1975. *Toxicology of Insecticides*. Plenum Press, New York.

Mopper, S. 1996. Adaptive genetic structure in phytophagous insect populations. *Trends Ecol. Evol. 11*:235–238.

Murray, R. D. H., J. Mendez, and S. A. Brown. 1982. *The Natural Coumarins: Occurence, Chemistry, and Biochemistry*. John Wiley, New York.

Neal , J. J. 1987. Ecological aspects of insect detoxication enzymes and their interaction with plant allelochemicals. Ph.D. dissertation, University of Illinois at Urbana–Champaign, IL.

Nelson, D, T. Kamataki, D. J. Waxman, F.P. Guengerich, R. W. Estabrook, R. Feyereisen, F. J. Gonzalez, M. J. Coon, I. C. Gunsalus, O. Gotoh, K. Okuda, and D. W. Nebert. 1993. The P450 superfamily: Update on new sequences, gene mapping, accession numbers, early trivial names of enzymes, and nomenclature. *DNA Cell Biol. 12*:1–51.

Oppenorth, F. J. 1984. Biochemistry of insecticide resistance. *Pestic. Biochem. Physiol. 22*:187–193.

Pashley, D. P. 1988. Quantitative genetics, development, and physiological adaptation in host strains of fall armyworm. *Evolution 42*:93–102.

Plapp, F. W. Jr. and J. E. Casida. 1969. Genetic control of housefly NADPH-dependent oxidases: Relation to insecticide chemical metabolism and resistance. *J. Econ. Entomol. 62*:1174–1179.

Rosenthal, G. and M. R. Berenbaum. 1991. *Herbivores: Their Interactions with Secondary Plant Metabolites,* Vol. 1. Academic Press, San Diego, CA.

Schonbrod, R. D., M. A. Q. Khan, L. C. Terriere, and F. W. Plapp Jr. 1968. Microsomal oxidases in the housefly: A survey of fourteen strains. *Life Sciences 7*:681–688.

Scott, J. G. 1996. Cytochrome P450 monooxygenase-mediated resistance to insecticides. *J. Pestic. Sci. 21*:241–245.

Scriber, J. M. 1986. Origins of the regional feeding abilities in the tiger swallowtail butterfly: Ecological monophagy and the *Papilio glaucus australis* subspecies in Florida. *Oecologia 71*:94–103.

Scriber, J. M. 1988. Tale of the tiger: Beringial biogeography, binomial classification, and breakfast choices in the *Papilio glaucus* complex of butterflies. Pp. 241–301 *in* K. C. Spencer (Ed.), *Chemical Mediation of Coevolution.* Academic Press, New York.

Scriber, J. M. 1995. Overview of swallowtail butterflies: Taxonomic and distributional latitude. Pp. 3–8 *in* J. M. Scriber, Y. Tsubaki and R. C. Lederhouse (Eds.), *Swallowtail Butterflies: Their Ecology and Evolutionary Biology.* Scientific Publishing, Gainesville, FL.

Scriber, J. M. and S. H. Gage. 1995. Pollution and global climate change: Plant ecotones, butterfly hybrid zones and changes in biodiversity. Pp. 319–344 *in* J. M. Scriber, Y. Tsubaki and R. C. Lederhouse (Eds.), *Swallowtail Butterflies: Their Ecology and Evolutionary Biology.* Scientific Publishing, Gainesville, FL.

Skoda, R. C., F. J. Gonzalez, A. Demierre, and U. A. Meyer. 1988. Two mutant alleles of the human cytochrome P-450db1 gene (*P450C2D1*) associated with genetically deficient metabolism of debrisoquine and other drugs. *Proc. Natl. Acad. Sci. USA 85*:5240–5243.

Tallamy, D., 1986. Behavioral adaptations in insects to plant allelochemicals. Pp. 273–300 *in* L. B. Brattsten and S. Ahmad (Eds.), *Molecular Aspects of Insect–Plant Associations.* Plenum Press, New York.

Terriere, L. C. 1968. The oxidation of pesticide: The comparative approach. Pp. 175–196 *in* E. Hodgson (Ed.), *The Enzymatic Oxidation of Toxicants.* North Carolina State University Press, Raleigh, NC.

Tomita, T. and J. G. Scott. 1995. cDNA and deduced protein sequence of *CYP6D1*—the putative gene for a cytochrome P450 responsible for pyrethroid resistance in a housefly. *Insect Biochem. Mol. Biol. 25*:275–283.

Via, S. 1991. Specialized host plant performance of pea aphid clones is not altered by experience. *Ecology 72*:1420–1427.

Walsh, J. B. 1995. How often do duplicated genes evolve new functions? *Genetics 139*:421–428.

Zangerl, A. R. and M. R. Berenbaum. 1993. Plant chemistry and insect adaptations to plant chemistry as determinants of hostplant utilization patterns. *Ecology 74*:47–53.

# 6

# Assessment of Genetic Variation in the Presence of Maternal or Paternal Effects in Herbivorous Insects

*MaryCarol Rossiter*
Institute of Ecology
University of Georgia, Athens, GA

## 6.1 Introduction

A long-standing issue in evolutionary ecology has been the evolution of diet breadth in herbivorous insects. Because the history of an insect species' ecological circumstance is built into current patterns of host-plant utilization, we have attempted to speculate how and why insects eat particular host-plant species using phylogenetic, geographic, and life history relationships among extant insect taxa. These approaches have produced some robust generalizations about host-use patterns among herbivores (e.g., coevolutionary arms race) but provide less information on the microevolutionary processes involved (but see Mitter and Futuyma 1983). Investigations of the relationship between preference and performance (Futuyma 1983; Singer et al. 1989), the role of nonnutritional factors in host use (Rossiter 1987; Bernays and Graham 1988; Hunter 1992), and the physiology of host utilization (e.g., Martin et al. 1987) have permitted inferential but speculative conclusions about the microevolution of plant–insect relationships. About 15 years ago, a new approach to the empirical study of herbivore evolution was taken up by Rausher (1984) and Via (1984) in response to the work of quantitative biologists who extended the theoretical genetic basis of quantitative genetics, developed by Fisher (1918), Wright (1921), and Haldane (1932) for the improvement of crops through artificial selection, to the study of evolutionary change in wild populations under natural selection (e.g., Lande 1979; Lande and Arnold 1983). Their quantitative genetics (QG) approach was particularly appealing for its ability to make estimates of genetic variation in insect populations under variable host environments. This foundation provided a general experimental approach for the study of microevolutionary aspects of the evolution of host utilization.

It is important to remember that QG has its origin in statistics rather than genetics. Although there is a strong theoretical foundation for equating statistical and biological inference, statistical reasoning still dominates the discipline (Barton and Turelli 1989). Biological inference is based on the approximate equation of multilocus Mendelian inheritance and properties of the multivariate normal distribution, an equation that requires many biological assumptions (Bulmer 1980). When the assumptions of a QG model can be met, the approach provides an excellent method for partitioning the phenotypic variation associated with a given morphological, physiological, or behavioral life-history trait into its genetic and environmental components, based on the familial relationships among measured individuals. However, in the study of quantitative inheritance in natural populations (versus domestic stock), some of the assumptions that we make because of experimental constraints may not be warranted. And, at this point in time, we do not know the extent to which the adoption of these assumptions about natural populations reduces or destroys the integrity of predictions about evolutionary change.

This chapter examines one of the assumptions commonly made in the study of quantitative traits in natural populations, namely, that maternal and paternal effects are absent or, if present, do not bias estimates of genetic variation, genetic correlation, response to selection, or evolutionary trajectory of herbivore traits. The examination of this assumption is critical, as there is growing evidence that inherited environmental effects are widespread (Roach and Wulff 1987; Labeyrie 1988; Mousseau and Dingle 1991a; Boggs 1995; Rossiter 1996; Bernardo 1996), that their presence can modify population dynamics, even causing destabilization (Rossiter 1994; Ginzburg and Taneyhill 1994), and that their presence can modify the possibility and trajectory of evolutionary change (Riska et al. 1985; Riska 1991; Kirkpatrick and Lande 1989; Boggs 1990; Lande and Kirkpatrick 1990). In this chapter, I examine an assumption that is commonly made in quantitative genetic studies of natural populations: that maternal and paternal effects will not bias the conclusions we draw about the magnitude of genetic variation or correlation. To do this, the sources of inherited environmental effects, also called parental, maternal or paternal effects, depending on context (Rossiter 1996 and see below), will first be described. Then, by way of example, I will use a popular QG model, the paternal half-sib design, to examine the different ways in which maternal or paternal effects can interfere with assessments of genetic variation. This discussion will be expanded to include measurement of genetic variation under a multiple-environment condition. Finally, I will discuss the benefits and shortcomings of several experimental designs that aim either to maximize the ability to partition parental effects from genetic effects or minimize the confounding influence of parental effects on estimation of genetic variation.

## 6.2 Impact of Inherited Environmental Effects on Phenotypic Variation

Inherited environmental effects are those components of the phenotype that are derived from either parent, apart from nuclear genes. This phenomenon is so named because the cross-generational transfer of anything beyond the parent's genome is received as an environmental rather than a genetic component of the offspring phenotype. Although the environmental effect is inherited (like money), it is not heritable, the latter indicating the transmission of genetic material. However, the capacity of a parent to deliver a parental effect may itself have a genetic basis. In such a case, one would say that the pathways involved represent a heritable trait upon which natural selection may act, if genetic variation for the ability to generate a parental effect is present.

Inherited environmental effects arise as the product of parental genes, parental environment, or their interaction. Inherited environmental effects can include contributions that reflect abiotic, nutritional, and other ecological features of the parental environment. Their impact on offspring will be positive or negative depending on the nature of the contribution and the ecological context in which the offspring exist (Rossiter 1996). Within the confines of this definition, I will use the more abbreviated "parental effect" or specify "maternal" or "paternal" when the parental source of the inherited environmental effect is known.

The cross-generational nature of parental effects makes it difficult to separate genetic and inherited environmental components of offspring phenotype. To be assured that estimates of genetic variation are not confounded by these cross-generational effects, QG analysis must take into account their contribution to phenotypic variation in offspring (Rossiter 1994, 1996). However, the experimental designs used in most studies do not account for inherited environmental effects and assume that they are not a confounding influence.

Figure 6.1 is based on a QG model described by Eisen and Saxton (1983) that accounts for sources of phenotypic variation arising from parental effects when expressed in a variable offspring environment. I have added sources of offspring phenotypic variation owing to a variable parental environment. This model illustrates the complexity of assessing genetic variation in natural populations that express parental effects. See Table 6.1A and B for abbreviated definitions of the terms used in Figure 6.1. There are eight potential sources of offspring phenotypic variation, $P_{o(t+1)}$, which arise from genetic (G) or environmental (E) causes originating in either the parental (m) or offspring (o) generation. Of these eight, sources 4–8 represent the contribution of parental effects to offspring phenotype. Let us say that we wish to estimate genetic variation for a trait in an herbivore population. Normally, we would make the assumption that parental effects are not present and ignore the potential contribution of sources 4–8. However, the influence of these sources on offspring phenotype has been empirically demonstrated for a number of plant and animal species (see below and further references in Rossiter 1996, p. 455).

In Figure 6.1, we can see that a parental effect may be realized through several nonexclusive pathways. First, a parental effect can arise from some aspect of the parental performance phenotype, $P_{m(t)}$, for example, maternal body size (Cowley et al. 1989), maternal age (Bridges and Heppell 1996), and maternal behavior

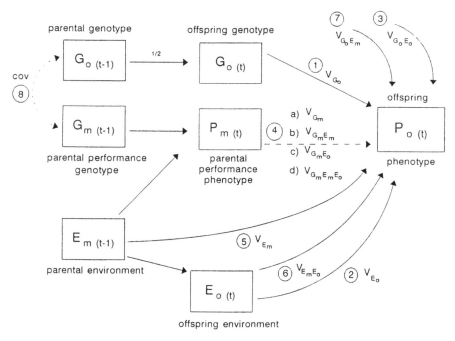

*Figure 6.1* The components of offspring phenotype ($P_o$) expressed in time $t$, deriving from the direct contribution of nuclear genes by one parent ($G_o$), a time-lagged presentation of the parental environment ($E_m$), a time-lagged expression of parental performance genes ($G_m$), and their interaction with the parental environment to produce the parental performance phenotype ($P_m$), plus the offspring's own environment ($E_o$). For simplicity of presentation, $G$ indicates additive genetic effects with dominance and epistasis (nonadditive genetic variation $VNA$) assumed to be negligible. The numbered sources indicate possible routes of contribution to the offspring phenotype; Source 4a = $VG_m$, 4b = $VG_mE_m$, etc.; see Table 6.1 for a full description of variables. This diagram can represent contributions by either mother of father (from Rossiter 1996).

(Roosenburg 1996). Components of the parental performance phenotype that influence the offspring phenotype may be due to the parental genotype alone, $G_m$ (source 4a; e.g., Carmona et al. 1994), the adjustment of maternal genotypic expression by the parental environment, $G_m \times E_m$ (source 4b; e.g., Keena et al. 1995), or by the offspring environment, $G_m \times E_o$ or $G_m \times E_m \times E_o$ (sources 4c and 4d; e.g., likely in Fox et al. 1995; Futuyma et al. 1993; Groeters and Dingle 1988; Kobayashi 1990; Watson and Hoffmann 1995).

To illustrate the different ways in which the parental performance phenotype can adjust offspring phenotype, consider a hypothetical population of herbivores that

*Table 6.1A*   Definitions of Selected Terms Used in This Chapter

---

**Sources of variation in offspring phenotype, $P_{o(t+1)}$**, as seen in Figure 6.1, arise from any to all of the following: $P_{o(t+1)} = G_{o(t)} + E_{o(t)} + G_{o(t)} \times E_{o(t)} + P_{m(t)} + E_m + E_m \times E_o + G_o \times E_m + G_m G_o$, where $G_{o(t)}$: nuclear genes provided by the parent (*source 1*)

$E_{o(t)}$: offspring environment (*source 2*)

$G_{o(t)} \times E_{o(t)}$: interaction between offspring genotype and offspring environment (*source 3*)

$P_{m(t)}$: parental performance phenotype; can include the following sources of variation in $P_{o(t+1)}$:

    $G_m$: the component of $P_{m(t)}$ due to parental genes alone (*source 4a*)

    $G_m \times E_m$: the component of $P_{m(t)}$ due to the adjustment of parental genetic expression by the parental environment (*source 4b*)

    $G_m \times E_o$: the component of $P_{m(t)}$ due to the adjustment of parental genetic expression by the offspring environment (*source 4c*)

    $G_m \times E_m \times E_o$: the component of $P_{m(t)}$ due to a three-way interaction between parental genes, parental environment and offspring environment (*source 4d*)

$E_m$: direct effects of the parental environment (*source 5*)

$E_m \times E_o$: interaction between parental and offspring environments (*source 6*)

$G_o \times E_m$: adjustment of offspring genotypic expression by direct effects of the parental environment (*source 7*)

$cov G_m G_o$ or $cov(G_m \times E_o)(G_o \times E_o)$: genetic correlation between the parental performance trait (e.g., pupal weight) and the same or a different trait expressed in the next generation (e.g., offspring pupal weight or offspring development time; *source 8*)

---

*B*   Selected Quantitative Genetics Terms Used in the Text and Tables 6.2–6.6.

---

$V_A$: additive genetic variation

$V_{NA}$: nonadditive genetic variation due to epistasis and dominance

$V_{I(A)}$: variation due to interaction between additive genetic variation and offspring environment

$V_{I(NA)}$: variation due to interaction between nonadditive genetic variation and offspring environment

$V_{ew}$: variation among progeny due to microenvironmental effects

$V_{ec1}$: variation arising from parental effects that adjust mean dam values or increase resemblance among siblings

$V_{ec2}$: variation among progeny due to parental effects that decrease resemblance among siblings

---

uses two host species, A and B, but to varying degrees in any given generation, depending on host plant quality and herbivore population density. The genotype of the parent, and the host species used, affects some parental performance trait such as adult body size, egg weight, or mobility. The parental performance trait is identified as such when it influences some attribute of offspring phenotype (e.g., early survival, development rate, or reproductive output). To the extent that the parental genotype dictates the value of the parental performance trait, regardless of which host species is used, offspring phenotypic variation includes $G_m$ (source 4a). If the parental host species modifies the impact of the parental genotype on egg size, then offspring phenotypic variation includes $G_m \times E_m$ (source 4b). Likewise, if the host species used by the offspring adjusts the impact of the nongenetic parental contribution to growth potential, then $G_m \times E_o$ (source 4c) will contribute to offspring phenotypic variation. Finally, when the sequence of parent–offspring host use (e.g., host species A-A, A-B, B-A, or B-B) determines the extent to which the parental effect is able to adjust offspring phenotype, then $G_m \times E_m \times E_o$ (source 4d) will be present.

Offspring phenotype, $P_{o(t +1)}$, can also be influenced by the parental environment, independent of the parental performance phenotype, that is, beyond any influence of the parent's genotype. For example, mothers who grew up on host species A are able to sequester and transmit an antimicrobial plant compound (e.g., Dussourd et al. 1988; Boppre and Fischer 1994) in a quantity that is independent of maternal pupal weight, but dependent on, say, plant quality. This means that $E_m$ (source 5) will be a component of offspring phenotype, $P_{o(t+1)}$. When the sequence of parent–offspring host use (e.g., host species A-A, A-B, B-A, or B-B) determines the extent to which the parental effect adjusts offspring phenotypic expression, independent of parental genotype, then $E_m \times E_o$ (source 6) is a component of offspring phenotype, $P_{o(t +1)}$. For example, efficacy of the antimicrobial compound, and thus parameters of offspring vigor, are adjusted according to which host species is eaten by offspring. This means that $E_m \times E_o$ (source 6) will be a component of offspring phenotype. Finally, if the efficacy of the antimicrobial compound is adjusted according to the offspring genotype, then $G_o \times E_m$ (source 7) will contribute to $P_{o(t +1)}$. For insects, there are empirical examples for source 5 (Islam et al. 1994) and source 6 (Groeters and Dingle 1987, 1988; Gould 1988; Kobayashi 1990; Rossiter 1991a; Futuyma et al. 1993; Keena et al. 1995; Fox et al. 1995; Watson and Hoffman 1995).

### 6.2.1 Cross-Generational Genetic Covariance

Finally, offspring phenotype, $P_{o(t +1)}$, can be influenced by a genetic correlation between the parental performance trait and the same or a different trait expressed in the next generation. The contribution of this covariance to offspring phenotype is designated as source 8: $\mathrm{cov}G_m\, G_o$ or $\mathrm{cov}(G_m \times E_o)(G_o \times E_o)$. It is important to realize that the most commonly used QG models make the assumption that this type of genetic covariance is negligible, although good empirical support for this covariance exists (e.g., Dickerson 1947; Willham 1963; Bondari et al. 1978; Eisen and Saxton 1983).

A cross-generational genetic covariance component has two features. Unlike the variance components (sources 1–7 in Fig. 6.1) whose value is always positive, covariance components can be either positive or negative. When the cross-generational genetic covariance is positive, parental effects can move phenotypic expression toward the maximum value and may be able to sustain that phenotypic value for many generations. The end product of this positive covariance is to give the appearance of a genetic effect.

When the cross-generational genetic covariance is negative, it has been suggested that there will be an oscillation in the phenotypic value from one generation to the next. Janssen et al. (1988) hypothesized that the generational oscillation for age of reproductive maturity in springtail populations (*Orchesella cincta*) was due to $-\mathrm{cov}G_mG_o$, whose presence was confirmed in a QG study. Cross-generational negative genetic covariance has been demonstrated for both pupal weight and family size in a flour beetle, *Tribolium castaneum* (Bondari et al. 1978). Whether $\mathrm{cov}G_mG_o$ is positive or negative, its expression can alter the trajectory of character evolution relative to the action of selection (Kirkpatrick and Lande 1989) and can bias heritability estimates (inflation or reduction), depending on its sign (Riska et al. 1985; Atchley and Newman 1989).

The second feature of cross-generational genetic covariance is that it can include paternal as well as maternal performance traits. Since I know of no published data that attempt to measure $\mathrm{cov}G_mG_{o[pat]}$, consider a hypothetical example where adult body size has a genetic basis, adult males deliver some metabolic precursor (protein, RNA, hormone) to their mates in a quantity that is related to their own body size, and the precursor influences offspring metabolism with a net cumulative effect of influencing adult body size. Species-specific developmental physiology will determine the relationship between paternal body size and the quantity of precursor provided. When the relationship between male body size and quantity of precursor transmitted is inverse, then negative genetic covariance from a paternal source, $-\mathrm{cov}G_mG_{o[pat]}$ will adjust offspring phenotype. By contrast, if there is a positive relationship between male body size and quantity of precursor transmitted, then positive genetic covariance from a paternal source, $+\mathrm{cov}G_mG_{o[pat]}$ will adjust offspring phenotype.

In light of this information, the interesting and important questions are these: Is it valid to assume that no cross-generational genetic covariance exists? And, when present, how will this covariance modify estimates of genetic variation and inferences of evolutionary consequence? The answers to these questions are discussed below but, in short, cross-generational genetic covariance will bias the estimates of variance components, heritability, genetic correlation, and evolutionary trajectory under a given selection regime. For example, $\delta^2_{sire}$ will be inflated if $\mathrm{cov}G_mG_{o[pat]}$ is positive and diminished if it is negative. Specific details about how this can influence genetic analysis will be described next with reference to a simple half-sib mating design under constant environment, then amplified for a model that includes multiple-offspring environments.

## 6.3 The Position of Inherited Environmental Effects in Quantitative Genetics Models

To facilitate an understanding of how inherited environmental effects can influence the outcome of QG analysis, a simple QG mating plan, the paternal half-sib

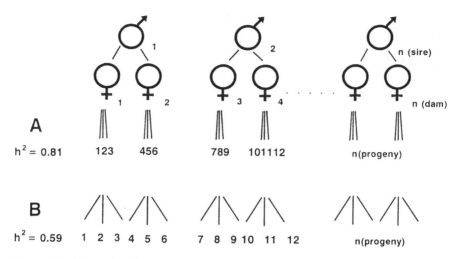

*Figure 6.2*  Paternal half-sib mating design wherein each sire is mated with two unique and unrelated dams, and three offspring from each dam are measured for some quantitative trait. As within-family resemblance changes due to maternal or paternal effects, the $h^2$ value changes: greater within-family resemblance (case A) due to parental effects yields a greater $h^2$ value, relative to a situation where there is greater within-family variation (case B; values based on data in the Appendix).

design, will be used for illustration (see Fig. 6.2A). Here, each father (SIRE) in the sample population mates with each of two unrelated mothers (DAM nested within SIRE) to produce an F1 generation (PROGENY nested within DAM). The total phenotypic variation in the F1 trait is partitioned by nested analysis of variance into variation associated with *sire*, DAM(SIRE), and PROGENY(DAM); (Table 6.2). As described in Falconer (1989), the values of observed variance components ($\delta^2_{sire}$, $\delta^2_{dam}$, $\delta^2_{progeny}$) are calculated, and these values are used to estimate the causal components of total phenotypic variance.

The theoretical genetics basis of this paternal half-sib model tells us how to decipher the causal from these observed components of variance in order to esti-

*Table 6.2*  Partitioning of Variance in a Nested Paternal Half-Sib Design

| | Components of Variance | |
|---|---|---|
| Source of Variation | Observed | Causal |
| SIRE | $\sigma^2_{sire}$ | $\frac{1}{4}V_A + \mathbf{V}_{Ec1[pat]} +/- \mathbf{cov}\mathbf{G}_m\mathbf{G}_{o[pat]}$ |
| DAM (Sire) | $\sigma^2_{dam}$ | $\frac{1}{4}V_A + \frac{1}{4}V_{NA} + \mathbf{V}_{Ec1[mat]} +/-$ $\mathbf{cov}\mathbf{G}_m\mathbf{G}_{o[mat]}$ |
| PROGENY (Dam) | $\sigma^2_{error}$ | $\frac{1}{2}V_A + \frac{3}{4}V_{NA} + \mathbf{V}_{Ec2[pat].[mat]} + V_{Ew}$ |

mate the relative magnitude of genetic and environmental variation (Falconer 1989). The observed variance component associated with SIRE, $\delta^2_{sire}$, holds ¼ of $V_A$, the total additive genetic variation. Additive genetic variation ($V_A$) is variation upon which selection may act (see Table 6.2; Table 6.1B includes abbreviated definitions of these variance components). When using this equation, we always make the assumption that there are no paternally originated, inherited environmental effects (paternal effects). Should they be present, their variance, $V_{Ec1[pat]}$, will be associated with the SIRE term (Table 6.2) as will any paternally originated, cross-generational genetic covariance, $+/- \text{cov}G_m G_{o[pat]}$. The presence of $V_{Ec1[pat]}$ or $+\text{cov}G_m G_{o[pat]}$ will inflate the value of the *sire* variance component and, therefore, bias estimates of heritability and genetic correlation.

The variance component associated with DAM, $\delta^2_{dam}$, also holds ¼ $V_A$ as well as ¼ $V_{NA}$, the nonadditive genetic variation, which is the genetic variation due to dominance relationships among alleles at the same locus and interactions between different loci (epistasis), plus $V_{Ec1[mat]}$, the maternal effects (Table 6.2). Finally, the variance component associated with PROGENY, $\delta^2_{within}$, holds ½ $V_A$, ¾ $V_{NA}$, plus $V_{Ec2}$, a parental effect that causes differences among siblings and $V_{Ew}$, which is the microenvironmental variation that causes differences among siblings.

$V_{Ec1}$, whether of paternal or maternal origin, increases the resemblance among siblings through a nongenetic effect. For example, a favorable maternal diet results in the production of propagules of the largest possible size. When environmental influence produces greater conformity in offspring phenotype, the phenomenon is called a *Type I maternal (or paternal) effect* (Rossiter et al. 1990). $V_{Ec2}$ reduces the resemblance among siblings through a nongenetic effect. For example, a mother provisions eggs differentially according to their position in the ovariole, and the quality of her diet influences the degree of variation in the provisioning of her eggs (Rossiter et al. 1993). This phenomenon generates diversification and is called a *Type II maternal (or paternal) effect.*

Once the variance components are calculated, we can determine the relative contribution of genes and environment to a population's phenotype and, with information on selection pressure, can make predictions about short-term evolutionary change (e.g., Lande and Kirkpatrick 1990). A formal measurement of evolutionary potential is narrow-sense heritability, $h^2$, which is based on the relative magnitude of additive genetic variation ($VA$) to total phenotypic variation ($VP$). Heritability, which ranges from zero to 1, can be calculated a number of ways (Falconer 1989). Using a half-sib design, $h^2 = 4\delta^2_{sire} / \delta^2_{total}$, based on the assumption that $\delta^2_{sire}$ is equivalent to ¼ $V_A$, and $\delta^2_{total}$ is equivalent to $V_P$. Consequently, estimates of $h^2$ and other population genetic descriptors (e.g., additive genetic correlation), will be biased whenever $\delta^2_{sire}$ includes $V_{Ec1[pat]}$ or $\text{cov}G_m G_{o[pat]}$. Biological examples of $V_{Ec1}$, $\text{cov}G_m G_o$, and $V_{Ec2}$ will be given in the next section, along with a description of how their presence will bias the inferences made from QG studies.

## 6.4 Impact of Parental Effects on Quantitative Genetic Analysis

### 6.4.1 $V_{Ec1}$, Type I Parental Effects Increase Family Resemblance

$V_{Ec1}$ is the variation arising from parental effects that adjust mean DAM values or increase resemblance among siblings. For example, the average quantity of yolk protein provided to progeny is set by maternal food quality, and amount of yolk dictates body size at hatch. Here, $V_{Ec1}$ is maternal in origin, $V_{Ec1[mat]}$. The origin can be paternal as well, for example, when the average amount of defensive compound provided by a father to offspring determines the likelihood that progeny will survive through the egg stage, $V_{Ec1[pat]}$. (Note: $V_{Ec1}$ can also arise from shared environmental experiences that are not provided by the parent, such as photoperiod during embryonation in a univoltine species. In empirical work, interference from this latter source of $V_{Ec1[external]}$ can usually be eliminated with appropriate design, and so it will not be discussed further.)

$V_{Ec1}$ may arise from any or all of the following sources shown in Figure 6.1: source 4a ($VG_m$), source 4b ($VG_mE_m$), source 5 ($VE_m$), or source 7 ($VG_oE_m$). $V_{Ec1}$, whether of a paternal or maternal source, increases the resemblance among siblings. Consequently, as the value of $V_{Ec1[pat]}$ increases, the value of $\delta^2_{sire}$ increases, thereby causing inflation of the heritability estimate. This is particularly critical, as there is growing evidence that fathers can make important environmental contributions to offspring phenotype in insect species (Boggs 1995), and that these contributions inflate estimates of $\delta^2_{sire}$ (Lacey 1996).

Likewise, as the value of $V_{Ec1[mat]}$ increases, the value of $\delta^2_{dam}$ increases. When of maternal origin, $V_{Ec1}$ acts to reduce the heritability value because it causes a relative reduction in the magnitude of $\delta^2_{sire}$, along with an increase in $\delta^2_{total}$ (Table 6.3, cases D1–D3). The presence of $V_{Ec1[mat]}$ is often inferred when the value of $\delta^2_{dam}$ is greater than $\delta^2_{sire}$, and $V_{NA}$ is assumed absent (Falconer 1989; Webb and Roff 1992).

### 6.4.2 $V_{Ec2}$, Type II Parental Effect: Decrease Family Resemblance

$V_{Ec2}$ may arise from any or all of the following sources shown in Figure 6.1: source 4a ($VG_m$), source 4b ($VG_mE_m$), source 5 ($VE_m$), or source 7 ($VG_oE_m$). $V_{Ec2}$, whether of a paternal or maternal origin, decreases the resemblance among siblings. $V_{Ec2}$ is the among-sibling component of variation due to parental effects. For example, full-siblings differ from one another due to variation in yolk provisioning by the mother ($V_{Ec2[mat]}$; e.g., Wellington 1965; Rossiter et al. 1990, 1993), or half-siblings differ from one another due to differences in the quantity or quality of a defensive agent contributed by the father ($V_{Ec2[pat]}$; an interesting prospect but not yet studied). By increasing the magnitude of within-family variation, $V_{Ec2}$ increases $\delta^2_{total}$, which results in the reduction of $h^2$ (remember, $h^2 = 4$ $\delta^2_{sire} / \delta^2_{total}$): see Table 6.3, cases P1–P4 and Figure 6.2B; results are based on the data set provided in the Appendix. In real terms, this means that Type II parental

*Table 6.3*  Demonstration of the Impact of Parental Effects on Estimates of Heritability.

| Hypothetical experiment | Variance Components | | | | Heritability $h^2$ | Relative difference among dams(sire) |
|---|---|---|---|---|---|---|
| | $\sigma^2_{sire}$ | $\sigma^2_{dam}$ | $\sigma^2_{error}$ | $\sigma^2_{total}$ | | |
| D1 | 0.79 | 2.88 | 0.01 | 3.68 | 0.86 | less |
| D2 | 0.54 | 3.38 | 0.01 | 3.93 | 0.55 | |
| D3 | 0.27 | 3.92 | 0.01 | 4.20 | 0.26 | more |
| | $\sigma^2_{sire}$ | $\sigma^2_{dam}$ | $\sigma^2_{error}$ | $\sigma^2_{total}$ | $h^2$ | Relative difference among progeny(dam) |
| P1 | 0.79 | 2.88 | 0.01 | 3.68 | 0.86 | less |
| P2 | 0.79 | 2.76 | 0.36 | 3.91 | 0.81 | |
| P3 | 0.79 | 2.48 | 1.21 | 4.48 | 0.71 | |
| P4 | 0.79 | 2.03 | 2.56 | 5.38 | 0.59 | more |

From D1–D3, the average difference among dams within sire ($\sigma^2_{dam}$) increases due to maternal effects $V_{ec1}$ or $+cov_{(GmGo)}$. From P1–P4, the difference among progeny ($\sigma^2_{error}$) increases due to maternal or paternal effects $V_{ec2}$. See Appendix for data tables.

effects decrease the opportunity for natural selection to discriminate on the basis of genotype, because they reduce the correlation between phenotype and genotype (Rossiter 1991b; Carriere 1994).

### 6.4.3  CovG$_m$G$_o$, Cross-Generational Genetic Correlation: + or −

CovG$_m$G$_o$ can be positive or negative, and can arise from either parent (source 8 in Fig. 6.1). Unlike $V_{Ec1}$, the possibility of its action is not widely acknowledged by empiricists who use QG models to estimate genetic variation and correlation in natural populations. Given the intense nature of gathering data to make estimates of genetic variation for quantitative traits, most empiricists find it necessary to utilize the standard QG models with their simplifying assumptions and whichever statistical protocols the "experts" deem in current vogue. This practical approach defies the cautions of theoreticians and discussions of the ability of covG$_m$G$_o$ to adjust and even destabilize the evolutionary trajectory of a trait (e.g., Dickerson 1947; Willham 1963, 1972; Bondari et al. 1978; Riska et al. 1985; Kirkpatrick and Lande 1989).

What, then, are the consequences to QG analysis when cross-generational genetic covariance is present but unaccounted for? When covG$_m$G$_{o[mat]}$ is positive, the associated increase in the value of $\delta^2_{dam}$ causes a reduction in the heritability estimate for the same reason that $V_{Ec1}$ reduces the estimate of $h^2$ (see above).

When covG$_m$G$_{o[mat]}$ is negative, there will be a corresponding reduction in the value of $\delta^2_{dam}$, producing two effects. First, the value of $h^2$ will increase through an apparent reduction in total phenotypic variance (the denominator in the calculation of $h^2$). Second, the likelihood of getting a significant SIRE effect will increase, because, in a nested model, $\delta^2_{dam}$ is the denominator of the $F$-test for $\delta^2_{sire}$. This significance test is important, because it is often done in lieu of calculating confidence intervals or standard errors to test the validity of heritability estimate when data sets are (and they usually are) unbalanced.

When $\text{cov}G_m G_{o[pat]}$ is positive, there will be an associated increase in the value of $\delta^2_{sire}$ and the estimate of $h^2$ will be inflated. When $\text{cov}G_m G_{o[pat]}$ is negative, there will be a corresponding reduction in the value of $\delta^2_{sire}$ and the estimate of $h^2$.

Perhaps the most frustrating aspect of $\text{cov}G_m G_o$ is the difficulty in ascertaining whether it is expressed by the population under study, that is, whether it is valid to make the assumption that it does not exist. Unfortunately, when $\text{cov}G_m G_o$ is present but not accounted for, all estimates of QG variation and predictions about trajectories of character evolution are inaccurate and, possibly, misleading. Although studies that directly measure $\text{cov}G_m G_o$ have been done for mammals (e.g., Willham 1963; Atchley and Newman 1989; Southwood and Kennedy 1990), I know of no such measurement for an herbivorous insect species (but see Bondari et al. 1978 with *Tribolium*).

Despite the lack of protocols to test for the presence of negative cross-generational genetic covariance, it can be inferred in two ways. First, consider the distribution of the causal components of variance among the observed variance components in a nested analysis of variance (ANOVA; Table 6.2). Since $\delta^2_{sire}$ and $\delta^2_{dam}$ each hold $\frac{1}{4} V_A$, $\delta^2_{sire}$ should always be less than or equal to $\delta^2_{sire}$ (which can also include nonadditive genetic variation and common environment effects), assuming adequate sample size. However, this is not always the case. For example, results from a study on genetic variation in *Collembola* showed that the value of $\delta^2_{dam}$ was considerably less than the value of $\delta^2_{sire}$ (Posthuma et al. 1993), providing a strong indication of negative cross-generational genetic covariance (or the presence of a paternal effect that would inflate the value of $\delta^2_{sire}$).

Second, the presence of cross-generational genetic covariance is supported (although not confirmed) by a negative parental effect. To understand a negative parental effect, consider the hypothetical scenarios in Figures 6.3A and 6.3B. The biological features of these species are that the parent provides an environment that influences offspring phenotype. In Figure 6.3-A1, maternal body size dictates brooding temperature in the nest. Nest temperature is inversely correlated with offspring body size at birth (Fig. 6.3-A2), because warmer nests result in smaller offspring compared to cooler nests. The impact of size at birth is maintained through development, leading to a negative correlation between maternal and offspring body size (Fig. 6.3-A3), because big females produce offspring that are relatively smaller due to the maternal influence of nest temperature. This pattern establishes the presence of a negative parental effect and suggests the possible presence of $\text{cov}G_m G_o$.

The second hypothetical scenario (Fig. 6.3B) is better suited for considering an ephemeral impact of the environment on cross-generational genetic covariance.

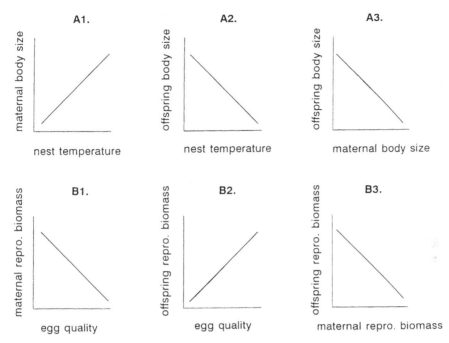

*Figure 6.3*   Two examples of a negative parental effect: the relationship between a parental environmental contribution (nest temperature or egg quality), and a trait (body size or total reproductive biomass) expressed by the parent (A1, B1) and offspring (A2, B2) is of opposite sign; this can produce a negative correlation (A3, B3) between the same trait expressed in two generations, suggesting the possibility of negative genetic covariance.

Here, environmental quality influences total reproductive biomass (average propagule size × fecundity). Total reproductive biomass is inversely related to egg quality (Fig. 6.3-B1), as is known for some organisms (e.g., Roff 1992; Cummins 1986). In turn, there is a positive correlation between egg quality and total reproductive biomass of offspring (Fig. 6.3-B2), because the influence of initial individual quality is partially maintained through adult development (e.g., Rossiter 1991b). As a consequence, the relationship between maternal and offspring total reproductive biomass is negative (Fig. 6.3-B3). This pattern establishes the presence of a negative parental effect and suggests the possible presence of $covG_mG_o$.

Examples of these scenarios have been inferred from empirical work. For example, Falconer (1965) found that female mice selected for production of larger litters had litters that were made up of offspring whose final body size (and litter size) were relatively smaller and vice versa. He inferred that the underlying mechanism was competition for resources (milk) during early development, and this produced the negative correlation between litter size in mothers and daughters. Rossiter (1991a) found that defoliation of leaves in the parental gypsy moth diet was inversely correlated with reduced pupal weight of parents but greater pupal weight of their offspring. She inferred that an adjustment in egg quality in response to perceived poor diet quality led to these results.

Once the presence of a negative parental effect is established, we need to ask: Does the negative maternal effect represent negative genetic covariance? The answer to this question will depend on whether the parental "environmental" variable (e.g., nest temperature or egg quality) has a genetic component that is expressed in both generations. For example, if the negative correlation in Figure 6.3-B3 is based on an inverse correlation between the genes that control egg quality and genes that control response to egg quality (e.g., set point for metabolic rate), negative genetic covariance would be confirmed. Unfortunately, little is known about how to quantifying cross-generational genetic covariance. What information does exist focuses primarily on mammals, animals in which the partitioning of prenatal and postnatal parental effects provides an avenue for quantification (e.g. Riska et al. 1985; Atchley and Newman 1989; Cowley 1991).

## 6.5 Quantitative Genetics Studies Involving Variable Environments

The multiple environment model of Table 6.4 is based on a paternal half-sib design, where the response of offspring from each full-sib mating is measured in more than one environment. Phenotypic variation of offspring is partitioned as for Table 6.2, with added representation from variation due to offspring environment (F1HOST) and any associated gene-environment interactions: F1HOST × SIRE and F1HOST × DAM. Respectively, these terms represent additive genetic variation and nonadditive genetic–common environment variation for diet breadth plasticity (e.g., Via 1984). The associated causal components of variance are labeled $VI(A)$ and $VI(NA)$ to indicate additive and nonadditive genetic interaction effects.

*Table 6.4* Partitioning of Variance in a Nested Paternal Half-Sib Design Where Offspring Are Represented in Multiple Environments (see text for definitions)

| Source of Variation | Components of Variance | |
|---|---|---|
| | Observed | Causal |
| SIRE | $\sigma^2_{sire}$ | $\frac{1}{4}V_A + V_{Ec1[pat]} +/- \mathbf{cov}G_mG_{o[pat]}$ |
| DAM (Sire) | $\sigma^2_{dam}$ | $\frac{1}{4}V_A + \frac{1}{4}V_{NA} + V_{Ec1[mat]} +/-$ $\mathbf{cov}G_mG_{o[mat]}$ |
| F1HOST | $\sigma^2_{F1host}$ | $V_{F1host}$ |
| F1HOST × SIRE | $\sigma^2_{F1host*sire}$ | $V_{I(A)} + V_{I(Ec1[pat])} +/- \mathbf{cov}(G_mE_o)(G_oE_o)_{[pat]}$ |
| F1HOST × DAM(Sire) | $\sigma^2_{F1host*dam}$ | $V_{I(NA)} + V_{I(Ec1[mat])} +/-$ $\mathbf{cov}(G_mE_o)(G_oE_o)_{[mat]}$ |
| PROGENY (Dam) | $\sigma^2_{error}$ | $\frac{1}{2}V_A + \frac{3}{4}V_{NA} + V_{Ec2[pat].[mat]} + V_{Ew}$ |

When we expand the paternal half-sib design to include multiple-offspring environments, additional inherited environmental effects may surface (shown in boldface, Table 6.4), owing to interactions between the offspring environment (i.e., F1HOST) and parental effects. These interactions are summarized in the causal component of variance, $VI_{(Ec1)[pat]}$ or $_{[mat]}$, which includes any or all of the following sources shown in Figure 6.1: source 4c $(VG_mE_o)$, source 4d $(VG_mE_mE_o)$, and source 6 $(VE_mE_o)$. If the F1 environment-parental effect interaction comes from the father, $VI_{(Ec1[pat])}$, it will be associated with $\delta^2_{Fhost*sire}$; if from the mother, $VI_{(Ec1[mat])}$ will be associated with $\delta^2_{Fhost*dam}$ (Table 6.3). In the same way, the cross-generational genetic covariance component, $cov(G_mE_o)(G_oE_o)_{[pat]}$ or $cov(G_mE_o)(G_oE_o)_{[mat]}$, source 8 of Figure 6.1, will be associated with the source parent. This G × E covariance is the least considered type of inherited environmental effect (Rossiter 1996) in both a theoretical and empirical realms (but see Eisen and Saxton 1983).

## 6.6 What to Do about Inherited Environmental Effects

We can see from the previous section that as experimental designs become more complex, the ability to separate and measure the impact of inherited environmental effects is reduced even further. It is this inherent complexity that encourages us to accept the working assumption that inherited environmental effects do not exist or are, at the least, not strong enough to seriously bias estimates of heritability, genetic correlation, or evolutionary trajectory. If this assumption is not accepted, we are left with two challenges: (1) to minimize the opportunity for maternal or paternal effects to bias results in both lab and field experiments, realizing that any "minimization" scheme may itself introduce a bias; and (2) to continue with theoretical and empirical efforts to develop new methods to quantify the contribution of maternal or paternal effects to phenotypic variation and response to selection.

### 6.6.1 Minimizing the Influence of Inherited Environmental Effects

There are several approaches to minimizing the confounding influence of maternal or paternal effects on estimates of genetic variation. First, when sample populations are collected as pupae or adults for immediate use (i.e., genetic variation among their offspring is measured), then sample wild populations should be collected to maximize inclusion of individuals whose experience, collectively, represents the greatest diversity of the population's microenvironmental circumstance. For example, collect pupae or adults from different host species, at different times (from first to last laid within a season), from different microhabitats (e.g., height or humidity), and so on. Collection criteria will vary according to the biology and ecology of the species under study. Second, when sample populations are collected as eggs or at early larval stages (using the same recommendations as discussed), one can rear that generation in the lab under a common environment in an attempt to equilibrate any environmental component of a maternal or paternal effect. The downside of this approach is that there is an opportunity for selection (on ability to survive the controlled conditions) to adjust the genotypic frequencies by the time that genetic variation is measured (on grandchildren of the wild type). It is important to remember that this method will not eliminate environmentally based parental effects, but rather move toward an equilibration of their expression across genotypes in a single environment. These methods to minimize the influence of maternal or paternal effects on estimates of genetic variation may be for naught if their expression is due to $-\mathrm{cov}G_m G_o$ $_{[mat]}$ or $V_{I(Ecl)}$ $_{[pat]}$, e.g., $\delta^2_{dam}$ is much lower than $\delta^2_{sire}$. Under such circumstances, it is necessary to characterize the maternal or paternal effect involved and include it as a component in the genetic analysis.

## 6.6.2 Characterizing the Influence of Inherited Environmental Effects

A number of researchers studying processes of change in natural populations have recently taken up the study of parental effects (reviewed in Rossiter 1996) in response to the growing realization of their potential importance to population dynamics and character evolution (e.g., Kirkpatrick and Lande 1989; Mousseau and Dingle 1991b; Cowley and Atchley 1992; Carriere 1994; Rossiter 1994, 1995). As is often the case in science, much of the empirical work has been initiated in response to unusual experimental results on some other topic. The greatest progress has been made by plant biologists due, at least in part, to the greater amenability of plant reproductive biology to the crossing designs used in agricultural breeding. Insects, unlike plants and cattle, have fewer reproductive options (e.g., no cloning, inability of a genotype to act as both male and female, generally only one breeding season), making them unsuitable for many of the QG designs that are used to study parental effects. Consequently, the challenge is to develop new approaches in this arena of study. Here are a few suggestions for what might be done in the meantime.

Determine whether the parental environmental experience influences the estimate of heritability or genetic correlation. Here are two ways to begin this experiment. One is to randomly distribute members of the wild-collected insect population across the parental environments you have chosen to mimic, for example, Host Species 1 and 2, hosts naturally used by at least some of the population (RANDOM ASSIGNMENT METHOD). Alternatively, select a number of families randomly, for example, 50–100 from a wild-collected insect population and distribute representatives from each family to each parental environment (FAMILY ASSIGNMENT METHOD). After rearing this parental generation, adults are mated according to the experimental design chosen (e.g., paternal half-sib mating as in Fig. 6.2) within each parental host environment (e.g., males and females from Host A are mated with one another). Selected life-history traits are measured on their offspring.

The rearing conditions of the offspring will depend on the experimental design: Offspring of all matings are reared in one environment (REAR IN SINGLE ENVIRONMENT METHOD), or offspring of all matings are reared in each of several environments (REAR IN MULTIPLE ENVIRONMENTS METHOD). In the former case, the experiment is more manageable but ignores the possibility that the magnitude of maternal or paternal effect expression in the field is dependent on the quality of the offspring's environment (i.e., it ignores sources 4c, 4d, or 6 in Fig. 6.1, which can only be measured against the backdrop of environmental variation).

When multiple offspring environments are used (REAR IN MULTIPLE ENVIRONMENTS METHOD), there is the inevitable problem of feasibility relative to requisite sample size. In both cases, there is the issue of choosing the most appropriate or representative environment(s) to test in. The best guideline is to chose the environment(s) that most likely influences the trait under study (e.g., host quality for body size, temperature, or photoperiod for propensity for diapause, density for behavioral traits), or the environment(s) that most population members experience in the wild.

Once the experimental design is decided upon, there is the issue of data analysis and what assumptions are made. Given that it is not possible to clone wild insects (yet), the appropriate statistical analysis is that shown in Table 6.5A, where SIRE is nested within PARENTAL ENVIRONMENT (called PARHOST, as host species was chosen as the environmental variable for this example). This analysis is the most conservative, and it suffers from the fact that any interaction between SIRE and PARHOST (source 7) will be subsumed into $\delta^2_{sire}$, thereby inflating the estimate of heritability. In general, linear models, the variance due to such an interaction term will be associated with the variance of the nesting term (Neter et al. 1985). As an alternative, one can use the statistical analysis shown in Table 6.5B under two conditions: (1) the family assignment method is employed, and (2) we make the biological assumption that the products of half-sib matings within a parental host represent genetic replicates from each parental environment. This assumption is based on the fact that members of each parental environmental treatment group collectively represent the same set of genotypes (i.e., siblings occur in each parental environment). This assumption was made by Goodnight (1988), who measured the impact of parental density on offspring life history.

Since the offspring environment influences the magnitude of parental-effect expression (Rossiter 1996), the experimental design can be extended to include both variable-parental and variable-offspring environments. An example of the analysis for this design is given in Table 6.6A, which presents the conservative nested model, and Table 6.6B, which presents the crossed design requiring, as for the model in Table 6.5B, use of the family-assignment method and the biological assumption that products of half-sib matings within a parental host represent genetic replicates from each parental environment.

*Table 6.5A*  Partitioning of Variance in a Nested Paternal Half-Sib Design for Multiple Parental Environments

| Sources of Variation | Components of Variance | Sources of Inherited Environmental Effects in Causal Components |
|---|---|---|
| PARHOST | $\sigma^2_{parhost}$ | (?) $V_{Ec1[pat]}$ + (?) $V_{Ec1[mat]}$ from sources 5 & 7 |
| SIRE(Par-host) | $\sigma^2_{sire}$ | $\frac{1}{4}V_A$ + (?) $V_{Ec1[pat]}$ from sources 4a & 4b +/− $covG_mG_{o[pat]}$ |
| DAM (Sire) | $\sigma^2_{dam}$ | $\frac{1}{4}V_A$ + $\frac{1}{4}V_{NA}$ + (?) $V_{Ec1[mat]}$ from sources 4a & 4b +/− $covG_mG_{o[mat]}$ |
| PROGENY (Dam) | $\sigma^2_{error}$ | $\frac{1}{2}V_A$ + $\frac{3}{4}V_{NA}$ + $V_{Ec2[pat].[mat]}$ + $V_{Ew}$ |

*B*  Partitioning of Variance in a Crossed Design under the Assumption that Matings within Parental Host Represent Replicates of Genotype in Each Parental Host Environment (i.e., nesting is not necessary)

| Sources of Variation | Components of Variance | Sources of Inherited Environmental Effects in Causal Components |
|---|---|---|
| PARHOST | $\sigma^2_{parhost}$ | (?) $V_{Ec1[pat]}$ + (?) $V_{Ec1[mat]}$ from source 5 |
| SIRE | $\sigma^2_{sire}$ | $\frac{1}{4}V_A$ + (?) $V_{Ec1[pat]}$ from sources 4a & 4b +/− $covG_mG_{o[pat]}$ |
| DAM (Sire) | $\sigma^2_{dam}$ | $\frac{1}{4}V_A$ + $\frac{1}{4}V_{NA}$ + (?) $V_{Ec1[mat]}$ from sources 4a & 4b +/− $covG_mG_{o[mat]}$ |
| PARHOST × SIRE | $\sigma^2_{parhost*sire}$ | (?) $V_{Ec1[pat]}$ + (?) $V_{Ec1[mat]}$ from source 7 |
| PARHOST × DAM | $\sigma^2_{parhost*dam}$ | (?) $V_{Ec1[pat]}$ + (?) $V_{Ec1[mat]}$ from source 7 |
| PROGENY (Dam) | $\sigma^2_{error}$ | $\frac{1}{2}V_A$ + $\frac{3}{4}V_{NA}$ + $V_{Ec2[pat].[mat]}$ + $V_{Ew}$ |

For illustration, parental host species (PARHOST) is the source of parental environmental variation. See Figure 6.1 for more information about specific sources; $V_A$ and $V_{NA}$ from source 1; $covG_mG_o$ from source 8; "?" means that the proportion of representation from causal component is unknown.

As you can see from Tables 6.5 and 6.6, the inclusion of a PARHOST term allows separation of some, but not all, of the potential sources of variation due to inherited environmental effects. Many sources are still associated with other model terms, some of which will produce a bias in estimates of heritability (as discussed earlier). Since this bias challenges the integrity of all heritability estimates and associated interpretations from theoretical models and empirical data, it is clear that some important work is yet to be done.

## 6.7  Closing Remarks

There is good empirical evidence that inherited environmental effects are widespread, and that the magnitude of their expression can be influenced by an organism's environmental experience (Rossiter 1996). Inherited environmental effects

*Table 6.6A*  Partitioning of Variance in a Nested Paternal Half-Sib Design with Multiple Parental and Offspring Environments

| Sources of Variation | Components of Variance | Sources of Inherited Environmental Effects in Causal Components |
|---|---|---|
| PARHOST | $\sigma^2_{parhost}$ | (?) $V_{Ec1[pat]}$ + (?) $V_{Ec1[mat]}$ from sources 5 & 7 |
| SIRE(Parhost) | $\sigma^2_{sire}$ | $\frac{1}{4}V_A$ + (?) $V_{Ec1[pat]}$ from sources 4a & 4b +/− cov$G_m G_{o[pat]}$ |
| DAM (Sire) | $\sigma^2_{dam}$ | $\frac{1}{4}V_A$ + $\frac{1}{4}V_{NA}$ + (?) $V_{Ec1[mat]}$ from sources 4a & 4b +/− cov$G_m G_{o[mat]}$ |
| F1HOST | $\sigma^2_{f1host}$ | $V_{f1host}$ from source 2 |
| F1HOST × PARHOST | $\sigma^2_{f1host*parhost}$ | $V_{I(E)}$ from source 6 |
| F1HOST × SIRE | $\sigma^2_{f1host*sire}$ | $V_{I(A)}$ from source 34a + 4a $V_{I(Ec1[pat])}$ from sources 4c & 4d +/− cov$(G_m E_o)(G_o E_o)_{[pat]}$ |
| F1HOST × DAM + A41 | $\sigma^2_{f1host*dam}$ | $V_{I(NA)}$ from source 3 + $V_{I(Ec1[mat])}$ +/− cov$(G_m E_o)(G_o E_o)_{[mat]}$ |
| PROGENY (Dam) | $\sigma^2_{error}$ | $\frac{1}{2}V_A$ + $\frac{3}{4}V_{NA}$ + $V_{Ec2[pat],[mat]}$ + $V_{Ew}$ |

*B*  Partitioning of Variance in Crossed Design under the Assumption That Matings within Parental Host Represent Replicates of Genotype in Each Parental Host Environment (i.e., Nesting Is Not Necessary)

| Sources of Variation | Components of Variance | Sources of Inherited Environmental Effects in Causal Components |
|---|---|---|
| PARHOST | $\sigma^2_{parhost}$ | (?) $V_{Ec1[pat]}$ + (?) $V_{Ec1[mat]}$ from source 5 |
| F1HOST | $\sigma^2_{f1host}$ | $V_{f1host}$ from source 2 |
| F1HOST × PARHOST | $\sigma^2_{f1host*parhost}$ | $V_{I(E)}$ from source 6 |
| SIRE | $\sigma^2_{sire}$ | $\frac{1}{4}V_A$ + (?) $V_{Ec1[pat]}$ from sources 4a & 4b +/− cov$G_m G_{o[pat]}$ |
| DAM (Sire) | $\sigma^2_{dam}$ | $\frac{1}{4}V_A$ + $\frac{1}{4}V_{NA}$ + (?) $V_{Ec1[mat]}$ from sources 4a & 4b +/− cov$G_m G_{o[mat]}$ |
| PARHOST × SIRE | $\sigma^2_{parhost*sire}$ | (?) $V_{Ec1[pat]}$ + (?) $V_{Ec1[mat]}$ from source 7 |
| PARHOST × DAM | $\sigma^2_{parhost*dam}$ | (?) $V_{Ec1[pat]}$ + (?) $V_{Ec1[mat]}$ from source 7 |
| F1HOST × SIRE | $\sigma^2_{f1host*sire}$ | $V_{I(A)}$ from source 3 + $V_{I(Ec1[pat])}$ from sources 4c & 4d +/− cov$(G_m E_o)(G_o E_o)_{[pat]}$ |
| F1HOST × DAM | $\sigma^2_{f1host*dam}$ | $V_{I(NA)}$ from source 3 + $V_{I(Ec1[mat])}$ +/− cov$(G_m E_o)(G_o E_o)_{[mat]}$ |
| PROGENY (Dam) | $\sigma^2_{error}$ | $\frac{1}{2}V_A$ + $\frac{3}{4}V_{NA}$ + $V_{Ec2[pat],[mat]}$ + $V_{Ew}$ |

For illustration, parental host species (PARHOST) and offspring host species (F1HOST) are the sources of environmental variation. See Figure 6.1 for more information about specific sources; $V_A$ and $V_{NA}$ from source 1; all covariance (cov) values from source 8; "?" means that the proportion of representation from causal component is unknown.

*Appendix:*   Data Used to Generate Results Shown in Table 6.3.

Columns 3–9 represent the hypothetical results of seven experiments. From cases D1 to D3, the mean difference between dams nested within sire increases, as would occur with an increasing contribution from $V_{ec1[mat]}$ or $+ covG_mG_o$; differences among progeny within dam remain constant. In cases P1–P4, the differences among dams within sire remains constant (based on the differences seen in D1) but the difference among progeny within dam increases from P1 to P4, as would occur with an increasing contribution from $V_{ec2[pat]\ or\ [mat]}$.

| Sire | Dam | D1 | D2 | D3 | P1 | P2 | P3 | P4 |
|------|-----|-----|-----|-----|-----|-----|------|------|
| 1 | 1 | 2.7 | 2.6 | 2.5 | 2.7 | 2.2 | 1.7 | 1.2 |
| 1 | 1 | 2.8 | 2.7 | 2.6 | 2.8 | 2.8 | 2.8 | 2.8 |
| 1 | 1 | 2.9 | 2.8 | 2.7 | 2.9 | 3.4 | 3.9 | 4.4 |
| 1 | 2 | 5.1 | 5.2 | 5.3 | 5.1 | 4.6 | 4.1 | 3.6 |
| 1 | 2 | 5.2 | 5.3 | 5.4 | .2 | 5.2 | 5.2 | 5.2 |
| 1 | 2 | 5.3 | 5.4 | 5.5 | 5.3 | 5.8 | 6.3 | 6.8 |
| 2 | 3 | 3.7 | 3.6 | 3.5 | 3.7 | 3.2 | 2.7 | 2.2 |
| 2 | 3 | 3.8 | 3.7 | 3.6 | 3.8 | 3.8 | 3.8 | 3.8 |
| 2 | 3 | 3.9 | 3.8 | 3.7 | 3.9 | 4.4 | 4.9 | 5.4 |
| 2 | 4 | 6.1 | 6.2 | 6.3 | 6.1 | 5.6 | 5.1 | 4.6 |
| 2 | 4 | 6.2 | 6.3 | 6.4 | 6.2 | 6.2 | 6.2 | 6.2 |
| 2 | 4 | 6.3 | 6.4 | 6.5 | 6.3 | 6.8 | 7.3 | 7.8 |
| 3 | 5 | 4.7 | 4.6 | 4.5 | 4.7 | 4.2 | 3.7 | 3.2 |
| 3 | 5 | 4.8 | 4.7 | 4.6 | 4.8 | 4.8 | 4.8 | 4.8 |
| 3 | 5 | 4.9 | 4.8 | 4.7 | 4.9 | 5.4 | 5.9 | 6.4 |
| 3 | 6 | 7.1 | 7.2 | 7.3 | 7.1 | 6.6 | 6.1 | 5.6 |
| 3 | 6 | 7.2 | 7.3 | 7.4 | 7.2 | 7.2 | 7.2 | 7.2 |
| 3 | 6 | 7.3 | 7.4 | 7.5 | 7.3 | 7.8 | 8.3 | 8.8 |
| 4 | 7 | 5.7 | 5.6 | 5.5 | 5.7 | 5.2 | 4.7 | 4.2 |
| 4 | 7 | 5.8 | 5.7 | 5.6 | 5.8 | 5.8 | 5.8 | 5.8 |
| 4 | 7 | 5.9 | 5.8 | 5.7 | 5.9 | 6.4 | 6.9 | 7.4 |
| 4 | 8 | 8.1 | 8.2 | 8.3 | 8.1 | 7.6 | 7.1 | 6.6 |
| 4 | 8 | 8.2 | 8.3 | 8.4 | 8.2 | 8.2 | 8.2 | 8.2 |
| 4 | 8 | 8.3 | 8.4 | 8.5 | 8.3 | 8.8 | 9.3 | 9.8 |
| 5 | 9 | 6.7 | 6.6 | 6.5 | 6.7 | 6.2 | 5.7 | 5.2 |
| 5 | 9 | 6.8 | 6.7 | 6.6 | 6.8 | 6.8 | 6.8 | 6.8 |
| 5 | 9 | 6.9 | 6.8 | 6.7 | 6.9 | 7.4 | 7.9 | 8.4 |
| 5 | 10 | 9.1 | 9.2 | 9.3 | 9.1 | 8.6 | 8.1 | 7.6 |
| 5 | 10 | 9.2 | 9.3 | 9.4 | 9.2 | 9.2 | 9.2 | 9.2 |
| 5 | 10 | 9.3 | 9.4 | 9.5 | 9.3 | 9.8 | 10.3 | 10.8 |
| 6 | 11 | 2.9 | 2.8 | 2.7 | 2.9 | 2.4 | 1.9 | 1.4 |
| 6 | 11 | 3.0 | 2.9 | 2.8 | 3.0 | 3.0 | 3.0 | 3.0 |
| 6 | 11 | 3.1 | 3.0 | 2.9 | 3.1 | 3.6 | 4.1 | 4.6 |
| 6 | 12 | 5.3 | 5.4 | 5.5 | 5.3 | 4.8 | 4.3 | 3.8 |
| 6 | 12 | 5.4 | 5.5 | 5.6 | 5.4 | 5.4 | 5.4 | 5.4 |
| 6 | 12 | 5.5 | 5.6 | 5.7 | 5.5 | 6.0 | 6.5 | 7.0 |
| 7 | 13 | 3.9 | 3.8 | 3.7 | 3.9 | 3.4 | 2.9 | 2.4 |

*Appendix:* (*continued*)

| Sire | Dam | D1 | D2 | D3 | P1 | P2 | P3 | P4 |
|------|-----|-----|-----|-----|-----|------|------|------|
| 7 | 13 | 4.0 | 3.9 | 3.8 | 4.0 | 4.0 | 4.0 | 4.0 |
| 7 | 13 | 4.1 | 4.0 | 3.9 | 4.1 | 4.6 | 5.1 | 5.6 |
| 7 | 14 | 6.3 | 6.4 | 6.5 | 6.3 | 5.8 | 5.3 | 4.8 |
| 7 | 14 | 6.4 | 6.5 | 6.6 | 6.4 | 6.4 | 6.4 | 6.4 |
| 7 | 14 | 6.5 | 6.6 | 6.7 | 6.5 | 7.0 | 7.5 | 8.0 |
| 8 | 15 | 4.9 | 4.8 | 4.7 | 4.9 | 4.4 | 3.9 | 3.4 |
| 8 | 15 | 5.0 | 4.9 | 4.8 | 5.0 | 5.0 | 5.0 | 5.0 |
| 8 | 15 | 5.1 | 5.0 | 4.9 | 5.1 | 5.6 | 6.1 | 6.6 |
| 8 | 16 | 7.3 | 7.4 | 7.5 | 7.3 | 6.8 | 6.3 | 5.8 |
| 8 | 16 | 7.4 | 7.5 | 7.6 | 7.4 | 7.4 | 7.4 | 7.4 |
| 8 | 16 | 7.5 | 7.6 | 7.7 | 7.5 | 8.0 | 8.6 | 9.0 |
| 9 | 17 | 5.9 | 5.8 | 5.7 | 5.9 | 5.4 | 4.9 | 4.4 |
| 9 | 17 | 6.0 | 5.9 | 5.8 | 6.0 | 6.0 | 6.0 | 6.0 |
| 9 | 17 | 6.1 | 6.0 | 5.9 | 6.1 | 6.6 | 7.1 | 7.6 |
| 9 | 18 | 8.3 | 8.4 | 8.5 | 8.3 | 7.8 | 7.3 | 6.8 |
| 9 | 18 | 8.4 | 8.5 | 8.6 | 8.4 | 8.4 | 8.4 | 8.4 |
| 9 | 18 | 8.5 | 8.6 | 8.7 | 8.5 | 9.0 | 9.5 | 10.0 |
| 10 | 19 | 6.9 | 6.8 | 6.7 | 6.9 | 6.4 | 5.9 | 5.4 |
| 10 | 19 | 7.0 | 6.9 | 6.8 | 7.0 | 7.0 | 7.0 | 7.0 |
| 10 | 19 | 7.1 | 7.0 | 6.9 | 7.1 | 7.6 | 8.1 | 8.6 |
| 10 | 20 | 9.3 | 9.4 | 9.5 | 9.3 | 8.8 | 8.3 | 7.8 |
| 10 | 20 | 9.4 | 9.5 | 9.6 | 9.4 | 9.4 | 9.4 | 9.4 |
| 10 | 20 | 9.5 | 9.6 | 9.7 | 9.5 | 10.0 | 10.5 | 11.0 |

represent an environmental contribution to offspring phenotype, arising from the parental environment, expression of the parental genes, and interaction of the two. Because this environmental contribution is transmitted simultaneously with the genetic contribution (nuclear genes), it is often difficult, and sometimes impossible, to separate from genetic contributions to offspring phenotype. With an understanding of the ways in which inherited environmental effects can contribute to phenotypic variation (e.g., Fig. 6.1), it is possible to evaluate where these sources of variation will show up in a statistical genetic analysis of quantitative traits (e.g., Tables 6.2, 6.4, 6.5, 6.6), then to use this information to develop better experimental approaches to estimate genetic variation.

Because inherited environmental effects are generally assumed absent in studies of natural populations, we have only recently begun to appreciate the multiple ways in which they can bias estimates of heritability or genetic correlation. In half-sibling analysis, for example, we look for statistical separation among sires as an indication that additive genetic variation exists. However, lack of statistical separation among sires ($h^2$ is not distinguishable from zero) need not mean an absolute absence of additive genetic variation. Instead, it may indicate that other effects (e.g., parental effects or offspring environment) are masking our ability to distinguish a difference among sires. Comparable to the statistical effect, these circumstantial, environmental effects mask the genotype from exposure to selection through adjustment of the phenotype, which is, in the end, the object of selection. At present, we recognize the ability of the current environment (e.g., offspring environment) to adjust phenotype in a way that reduces its correlation with genotype, but we have been slow to recognize that the previous environment (delivered as a maternal or paternal effect) can make a comparable adjustment of phenotype. Finally, the external environment can influence the magnitude and sign of inherited environmental effects. This suggests that we need to develop a dynamic perspective on the expression of additive genetic variation, which includes both spatial and temporal aspects of environmental heterogeneity. Thinking about the associated empirical work is . . . troublesome, but logistic difficulties do not remove the need to understand how inherited environmental effects function in both an ecological and evolutionary context. We currently face the challenge of developing new experimental approaches to test for and quantify the impact of inherited environmental effects in natural populations of insect species on the assessment of genetic variation. The following work indicates that such an effort is underway: Goodnight (1988), Groeters and Dingle (1988), Rossiter et al. (1990, 1993), Rossiter (1991a, 1991b), Futuyma et al. (1993), Posthuma et al. (1993), Carriere (1994), Fox et al. (1995), Watson and Hoffman (1996).

As work in this area progresses, we can begin to evaluate whether inherited environmental effects can influence the rate of local adaptation or deme formation, an influence that is certainly plausible, given the ability of maternal and paternal effects to adjust the exposure of the phenotype to natural selection. The theoretical groundwork laid by Kirkpatrick and Lande (1989) demonstrates this potential.

## Acknowlegments

I thank Susan Mopper and Sharon Strauss for inviting my contribution to the ongoing exploration of insect population genetics. I also thank them for insightful comments and suggestions to improve the chapter. I appreciated the comments of several anonymous reviews as well. This work was supported by a grant from the National Science Foundation, DEB-9629735.

## 6.8 References

Atchley, W. R. and S. Newman. 1989. A quantitative-genetics perspective on mammalian development. *Am. Nat. 134*:486–512.

Barton, N. H. and M. Turelli. 1989. Evolutionary quantitative genetics: How little do we know? *Annu. Rev. Genet. 23*:337–370.

Bernardo J. 1996. Maternal effects in animal ecology. *Am. Zool. 36*:83–105.

Bernays, E. and M. Graham. 1988. On the evolution of host specificity in phytophagous arthropods. *Ecology 69*:886–892.

Boggs, C. L. 1990. A general model of the role of male-donated nutrients in female insects' reproduction. *Am. Nat. 136*:598–617.

Boggs, C. L. 1995. Male nuptial gifts: Phenotypic consequences and evolutionary implications. Pp. 215–242 *in* S. R. Leather and J. Hardie (Eds.), *Insect Reproduction.* CRC Press, New York.

Bondari, K. R., L. Willham, and A. E. Freeman. 1978. Estimates of direct and maternal genetics correlations for pupa weight and family size of *Tribolium. J. Anim. Sci. 47*:358–365.

Boppre, M. and O. W. Fischer. 1994. Zonocerus and Chromolaena in West Africa. Pp. 108–126 *in* S. Krall and H. Wilps (Eds.), *New Trends in Locust Control.* GTZ, D-Eschborn.

Bridges, T. S. and S. Heppell. 1996. Fitness consequences of maternal effects in *Streblospio benedicti* (Annelida: Polychaeta). *Am. Zool. 36*:132–146.

Bulmer, M. G. 1980. *The Mathematical Theory of Quantitative Genetics.* Oxford, Clarendon, UK.

Carmona, M. J., M. Serra, and M. R. Miracle. 1994. Effect of population density and genotype on life-history traits in the rotifer *Brachionus plicatilis* O.F. Mueller. *J. Exper. Mar. Biol. Ecol. 182*:223–235.

Carriere, Y. 1994. Evolution of phenoptypic variance: Non-Mendelian parental influences on phenotypic and genotypic components of life-history traits in a generalist herbivore. *Heredity 72*:420–430.

Cowley, D. E. 1991. Prenatal effects on mammalian growth: Embryo transfer results. Pp. 762–779 *in* E. C. Dudley (Ed.), *The Unity of Evolutionary Biology.* Dioscorides Press, Portland, OR.

Cowley, D. E. and W. R. Atchley. 1992. Quantitative genetic models for development epigenetic selection and phenotypic evolution. *Evolution 46*:495–518.

Cowley, D. E., D. Pomp, W. R. Atchley, E. J. Eisen, and D. Hawkins-Brown. 1989. The impact of maternal uterine genotype on postnatal growth and adult body size in mice. *Genetics 122*:193–204.

Cummins, C. P. 1986. Temporal and spatial variation in egg size and fecundity in *Rana temporaria. J. Anim. Ecol. 55*:303–316

Dickerson, G. E. 1947. Composition of hog carcasses as influenced by heritable differences in rate and economy of gain. *Iowa Agric. Exp. Station Res. Bull. 354*:492–524.

Dussourd, D. E, K. Ubik, C. Harvis, J. Resch, J. Meinwald, and T. Eisner. 1988. Biparental defensive endowment of eggs with acquired plant alkaloid in the moth *Utetheisa ornatrix. Proc. Natl. Acad. Sci. USA 85*:5992–5996.

Eisen, E. J. and A. M. Saxton. 1983. Genotype by environment interactions and genetic correlations involving two environmental factors. *Theor. Appl. Genet. 67*:75–86.

Falconer, D. S. 1965. Maternal effects and selection response. Pp. 763–774 *in* S. J. Geerts (Ed.), *Genetics Today: Proceedings of the XI International Congress on Genetics,* Vol. 3. Pergamon, Oxford, UK.

Falconer, D. S. 1989. *Introduction to Quantitative Genetics,* 3rd ed. Longman, London.

Fisher, R A. 1918. The correlation between relatives on the supposition of Mendelian inheritance. *Trans. Royal Soc. Edinburgh 52*:399–433.

Fox, C. W., K. J. Waddell, and T. A. Mousseau. 1995. Parental host plant affects offspring life histories in a seed beetle. *Ecology 76*:402–411.

Futuyma, D. J. 1983. Selective factors in the evolution of host choice by insects. Pp. 227–244 *in* S. Ahmed (Ed.), *Herbivorous Insects: Host Seeking Behavior and Mechanisms.* Academic Press, New York.

Futuyma, D. J., C. Herrmann, S. Milstein, and M. C. Keese. 1993. Apparent transgenerational effects of host plant in the leaf beetle *Ophraella notulata* (Coleoptera: Chrysomelidae). *Oecologia 96*:365–372.

Ginzburg, L. R. and D. E. Taneyhill. 1994. Population cycles of forest Lepidoptera: a maternal effect hypothesis. *J. Anim. Ecol. 63*:79–92.

Goodnight, C. J. 1988. Population differentiation and the transmission of density effects between generations. *Evolution 42*:399–403.

Gould, F. 1988. Stress specificity of maternal effects in *Heliothis virescens* (Lepidoptera: Noctuidae) larvae. Pp. 191–197 *in* T. S. Sahota (Ed.), *Paths from a Viewpoint: The Wellington Festschrift on Insect Ecology.* CS Holling, *Mem. Entomol. Soc. Canada No. 146:* Entomol. Soc. Can., Ottowa, Canada.

Groeters, F. R. and H. Dingle. 1987. Genetic and maternal influences on life history plasticity in response to photoperiod by milkweed bugs (*Oncopeltus fasciatus*). *Am. Nat. 129*:332–346.

Groeters, F. R. and H. Dingle. 1988. Genetic and maternal influences on life history plasticity in milkweed bugs (*Oncopeltus*): Response to temperature. *J. Evol. Biol. 1*:317–333.

Haldane, J. B. S. 1932. *The Causes of Evolution.* Longmans and Green, London, UK.

Hunter, M. D. 1992. A variable insect-plant interaction: The relationship between tree budburst phenology and population levels of heribvores among trees. *Ecol. Entomol. 16*:91–95.

Islam, M. S., P. Roessingh, S. J. Simpson, and A. R. McCaffery. 1994. Effects of population density experienced by parents during mating and oviposition on the phase of hatchling desert locusts, *Schistocerca gregaria. Proc. Royal Soc. Lond. B 257*:93–98.

Janssen, G. M., G. DeJong, E. N. G. Joose, and W. Scharloo. 1988. A negative maternal effect in springtails. *Evolution 42*:828–834.

Keena, M. A., T. M. O'Dell, and J. A. Tanner. 1995. Phenotypic response of two successive gypsy moth (Lepidoptera: Lymantriidae) generations to environment and diet in the laboratory. *Ann. Entomol. Soc. Am. 88*:680–689.

Kirkpatrick, M. and R. Lande. 1989. The evolution of maternal effects. *Evolution 43*:485–503.

Kobayashi, J. 1990. Effects of photoperiod on the induction of egg diapause of tropical races of the domestic silkworm, *Bombyx mori,* and the wild silkworm, *B. mandarina. Jap. Agric. Res. Quart. 23*:202–205.

Labeyrie, V. 1988. Maternal effects and biology of insect populations. *Mem. Entomol. Soc. Canada 146*:153–170 (in French).

Lacey, E. P. 1996. Parental effects in *Plantago lanceolata* L. I.: A growth chamber experiment to examine pre-and post-sygotic temperature effects. *Evolution 50*:865–878.

Lande, R. 1979. Quantative genetic analysis of multivariate evolution applied to brain:body allometry. *Evolution 34*:402–416.

Lande, R. and S. Arnold. 1983. The measurement of selection on correlated traits. *Evolution 37*:1210–1226.

Lande, R. and M. Kirkpatrick. 1990. Selection response in traits with maternal inheritance. *Gen. Res. 55*:189–198.

Martin, J. S., M. M. Martin, and E. A. Bernays. 1987. Failure of tannic acid to inhibit digestion or reduce digestibility of plant protein in gut fluids of insect herbivores: Implications for theories of plant defense. *J. Chem. Ecol. 13*:605–621.

Mitter, C. and D. J. Futuyma. 1983. An evolutionary-genetic view of host-plant utilization by insects. Pp. 427–459 *in* R. F. Denno and M. S. McClure (Eds.), *Variable Plants and Herbivores in Natural and Managed Systems*. Academic Press, New York.

Mousseau, T. A. and H. Dingle. 1991a. Maternal effects in insect life histories. *Annu. Rev. Entomol. 36*:511–534.

Mousseau, T. A. and H. Dingle. 1991b. Maternal effects in insects: Examples, constraints, and geographic variation. Pp. 745–761 *in* E. C. Dudley (Ed.), *The Unity of Evolutionary Biology*. Dioscorides Press, Portland, OR.

Neter, J., W. Wasserman, and M. H. Kutner. 1985. *Applied Linear Statistical Models*. Richard Irwin, Inc., Homewood, IL.

Posthuma, L., R. F. Hogervorst, E. N. G. Joosse, and N. M. Van Straalen. 1993. Genetic variation and covariation for characteristics associated with cadmium tolerance in natural populations of the springtail *Orchesella cincta* (L.). *Evolution 47*:619–631.

Rausher, M. D. 1984. Trade-offs in performance on different hosts: Evidence from within- and between-site variation in beetle *Deloyala guttata*. *Evolution 38*:582–595.

Riska, B. 1991. Maternal effects in evolutionary biology: Introduction to the symposium. Pp. 719–724 *in* E. C. Dudley (Ed.), *The Unity of Evolutionary Biology*. Dioscorides Press, Portland, OR.

Riska, B., J. J. Rutledge, and W. R. Atchley. 1985. Covariance between direct and maternal genetic effects in mice, with a model of persistent environmental influences. *Genet. Res. Camb. 45*:287–297.

Roach, D. A. and R. D. Wulff. 1987. Maternal effects in plants. *Annu. Rev. Ecol. Syst. 18*:209–235.

Roff, D. A. 1992. *The Evolution of Life Histories: Theory and Analysis*. Chapman & Hall, New York.

Roosenburg, W. M. 1996. Maternal condition and nest site choice: An alternative for maintenance of environmental sex determination. *Am. Zool. 36*:157–168.

Rossiter, M. C. 1987. Use of a secondary host, pitch pine, by non-outbreak populations of the gypsy moth. *Ecology 68*:857–868.

Rossiter, M. C. 1991a. Environmentally based maternal effects: A hidden force in insect population dynamics. *Oecologia 87*:288–294.

Rossiter, M. C. 1991b. Maternal effects generate variation in life history: Consequences of egg weight plasticity in the gypsy moth. *Func. Ecol. 5*:386–393.

Rossiter, M. C. 1994. Maternal effects hypothesis of herbivore outbreak. *BioScience* 44:752–763.

Rossiter, M. C. 1995. Impact of life history evolution on population dynamics: Predicting the presence of maternal effects. Pp. 251–275 *in* N. Cappuccino and P. W. Price (Eds.), *Population Dynamics: New Approaches and Synthesis*. Academic Press, San Diego, CA.

Rossiter, M. C. 1996. Incidence and consequences of inherited environmental effects. *Annu. Rev. Ecol. Syst.* 27:451–476.

Rossiter, M. C., D. L. Cox-Foster, and M. A. Briggs. 1993. Initiation of maternal effects in *Lymantria dispar*: Genetic and ecological components of egg provisioning. *J. Evol. Biol.* 6:577–589.

Rossiter, M. C., W. G. Yendol, and N. R. Dubois. 1990. Resistance to *Bacillus thuringiensis* in the gypsy moth (Lepidoptera: Lymantriidae): Genetic and environmental causes. *J. Econ. Entomol.* 83:2211–2218.

Singer, M. C., C. D. Thomas, H. L. Billington, and C. Parmesan. 1989. Variation among conspecific insect populations in the mechanistic basis of diet breadth. *Anim. Behav.* 37:751–759.

Southwood, O. I. and B. W. Kennedy. 1990. Estimation of direct and maternal genetic variance for litter size in Canadian Yorkshire and Landrace swine using an animal model. *J. Anim. Scien.* 68:1841–1847.

Via, S. 1984. The quantitative genetics of polyphagy in an insect herbivore: I. Genotype-environment interaction in larval performance on different host plant species. *Evolution* 38:881–895.

Watson, M. J. O. and A. A. Hoffmann. 1995. Cross-generation effects for cold resistance in tropical populations of *Drosophila melanogaster* and *D. simulans*. *Austr. J. Zool.* 43:51–58.

Watson, M. J. O. and A. A. Hoffman. 1996. Acclimation, cross-generation effects, and the response to selection for increased cold resistance in *Drosophila*. *Evolution* 50:1182–1192.

Webb, K. L. and D. A. Roff. 1992. The quantitative genetics of sound production in *Gryllus firmus*. *Anim. Behav.* 44:823–832.

Wellington, W. G. 1965. Some maternal influences on progeny quality in the western caterpillar, *Malacosoma pluviale* (Dyar). *Can. Entomol.* 97:1–14.

Willham, R. L. 1963. The covariance between relatives for characters composed of components contributed by related individuals. *Biometrics* 19:18–27.

Willham, R. L. 1972. The role of maternal effects in animal breeding: III. Biometrical aspects of maternal effects in animals. *J. Anim. Sci.* 35:1288–1293.

Wright, S. 1921. Systems of mating. *Genetics* 6:111–178.

# 7

# Local Adaptation and Stochastic Events in an Oak Leaf-Miner Population

*Susan Mopper*
Department of Biology, University of Southwestern Louisiana, Lafayette, LA

## 7.1 Introduction

There is strong evidence for local genetic variation and demic structure in phytophagous insect populations (McCauley and Eanes 1987; McPheron et al. 1988; Gittman et al. 1989; Rank 1992; and see Chapter 9 in this volume by McCauley and Goff; Chapter 12 by Peterson and Denno; and Chapter 14 by Thomas and Singer). Research in agricultural systems indicates that a major cause of this structure is genetic variation in resistance among individual host plants in a population (Denno and McClure 1983). Genetic variation within natural plant populations is also well established (Hiebert and Hamrick 1983; Plessas and Strauss 1986; Mopper et al. 1991; Berg and Hamrick 1995; Strauss and Karban 1994a, 1994b; Berenbaum and Zangerl, Chapter 5 this volume), but perhaps because of inherently greater spatial and temporal complexity, its relation to insect population structure is not as well understood as in managed systems (Michalakis et al. 1993).

In natural systems, populations of phytophagous insects usually occur at low to intermediate densities. Because outbreaks are relatively rare despite seemingly plentiful food, ecologists have pondered the forces that influence insect population biology. Hairston et al. (1960) argued that because plants are abundant and largely intact, phytophagous insects are limited not by food supply, but by natural enemies. In contrast, the "coevolutionary arms race" hypothesis, proposed that natural selection by herbivores produced adaptive chemical defenses in plants, which in turn led to the evolution and diversification of insect herbivores (Ehrlich and Raven, 1964).

Although the "coevolutionary arms race" theory elucidated important macroevolutionary forces influencing plants and insects, it was also relevant at the population level, because it could explain patterns of local variation. The "arms race" hypothesis stimulated the hypothesis of "plant chemical defense" in which plants were no longer viewed as abundant and unlimited food supplies but rather as vessels of diverse noxious and toxic chemical compounds that prevented consumption by herbivores (Feeny 1976; Rhoades and Cates 1976). Lawton and McNeil (1979) supported this perspective in their review of mechanisms that influence populations of plant-feeding insects, but also reemphasized the importance of natural enemies (and coined the popular term "the world is green" in homage to the Hairston et al. 1960 paper). This paper is one of the earliest formulations of a multitrophic theory describing insect population regulation. This synthetic approach toward insect population biology continued into the 1980s and 1990s with the advancement of concepts such as "tritrophic-level interactions," "top-down versus bottom-up," and "trophic cascade" regulatory forces, which dominate the field of insect evolutionary ecology today (Hunter 1992a; Hunter and Price 1992).

The growing awareness of genetic variation among and within insect and plant populations (Ehrlich and Raven 1969; Hamrick 1976; Hedrick et al. 1976), in conjunction with the acceptance of the host plant as the primary force in the ecological and evolutionary dynamics of phytophagous insects, set the stage for the phytocentric "local adaptation hypothesis" (Edmunds and Alstad 1978), which elucidated evolutionary ecology at the population level. This hypothesis predicts that because of spatial variation in plant chemical defenses, insect herbivores must adapt to the traits of individual trees. Adaptation to one host reduced ability to colonize a different host. Hence, deme formation of short-lived insects inhabiting long-lived plants produced adaptive genetic structure at very fine spatial scales.

At least 13 field experiments have independently tested the local adaptation hypothesis in natural insect populations, and the results are about equally divided in support for and refutation of the theory (see reviews in Mopper 1996a; and Boecklen and Mopper, Chapter 4, this volume). The results of these studies are paradoxical, in a sense, because the presence or absence of local genetic structure is unrelated to insect dispersal ability and the potential for gene flow among host plants (see review in Stiling and Rossi, Chapter 2, this volume). Some dispersive insects exhibit local adaptation (Mopper et al. 1995; Komatsu and Akimoto 1995; Stiling and Rossi, Chapter 2, this volume), whereas much more sessile insects do not (Unruh and Luck 1987; Cobb and Whitham 1993).

The paradox may arise from viewing local adaptation and deme structure from a phytocentric perspective. In some herbivorous insect populations, host plants may be the principle agents of genetic structure, but selection by natural enemies could also be important. Interaction between the plant and the enemy trophic levels has been overlooked in the adaptive deme formation literature, despite its potentially strong effect on herbivore fitness (Price et al. 1980). A fundamental principle of the local adaptation hypothesis is strong selection by host plants on colonizing insects, but it has rarely been explicitly tested. Although insect survival is the most commonly employed index of adaptation (Table 4.2 in Boecklen and Mopper, Chapter 4, this volume), most studies do not identify the specific sources of mortality (e.g., plants or enemies). Using total survival as an index of adaptation can lead to erroneous assumptions about population genetic structure and conceal potentially important evolutionary forces. For example, Mopper et al. (1995) tested the local adaptation hypothesis by measuring total survival of leaf miners transferred to natal and novel host plants. Leaf miners survived at roughly equivalent rates regardless of the host type, but decomposition of mortality into separate categories indicated that host plants and natural enemies played opposing roles in the genetic structure of the leaf-miner population.

This chapter presents the oak leaf-miner system in which I have worked for over a decade as a model to examine the ecological forces that influence population biology and produce genetic structure. *Stilbosis quadricustatella* is a lepidopteran leaf miner that occurs at high densities on its primary host, *Quercus geminata* (sand-live oak). Census of population densities since 1985 reveals distinct spatial and temporal trends. Field transfer experiments and genetic data indicate genetic structure in *S. quadricustatella* leaf miners at three spatial scales: among individual trees within a population, between two sympatric host-plant species, and between two isolated populations. Based on work by Peter Stiling, Dan Simberloff, and myself, I will discuss how both natural selection and stochastic forces influence the population biology of this phytophagous insect.

## 7.2 Natural History and Experimental Methods

*Stilbosis quadricustatella* (Cham.) is a univoltine leaf miner (Lepidoptera: Cosmopterigidae) common to north Florida. Adults are winged, but juveniles are restricted to a single leaf throughout larval development. Moths emerge from pupation in the spring and mate but do not feed. Females deposit single, minute eggs (0.2 mm diameter) amid dense trichomes on the lower leaf surface. Leaves contain from one to five eggs, which hatch in about two weeks. Failure to successfully excavate a mine leads to high larval mortality in the first instar, and late-instar larvae are preyed on by ants, birds, and hymenopteran parasitoids (Mopper et al. 1984, 1995). Late-instars also suffer plant-mediated mortality because of incomplete development in attached (Connor and Beck 1993) or prematurely abscised leaves (Simberloff and Stiling 1987; Mopper and Simberloff 1995). In the fall, larvae emerge from the mine, drop to the ground, and pupate in the soil and leaf litter beneath the tree.

*S. quadricustatella* miners occur at high densities on *Quercus geminata* (sand-live oak). Populations of these shrubby trees occur in sandy soils in the interior and Gulf Coast of north Florida. Bud break occurs in late April, and most leaves are fully expanded when leaf miners oviposit. We conducted experiments in an inland (Lost Lake) and coastal (Alligator Point) population of *S. quadricustatella* separated by 60 km. The inland site is a small monotypic stand of *Q. geminata*; the coastal site is an intermixed woodland of *Q. geminata* and *Q. myrtifolia* (myrtle oak), which is also attacked by *S. quadricustatella* leaf miners. We have censused individual trees at the inland Lost Lake population since 1985, and conducted the transfer experiments described here in 1991 and 1992. Genetic analysis of protein markers was completed in 1997. Detailed life-history and experimental methods are reported in Mopper et al. (1984), Mopper et al. (1995), and Mopper and Simberloff (1995).

## 7.3 Spatial and Temporal Variation in the Leaf-Miner Population

### 7.3.1 Density

Densities of *S. quadricustatella* vary widely among individual oak trees. In some trees, 30–40% of all leaves are attacked by leaf miners, yet other trees suffer intermediate or low levels of attack (Fig. 7.1). Relative densities are consistent over time; the same trees are repeatedly attacked or avoided year after year. Adult leaf-miners rarely oviposit on some trees, despite their close proximity to heavily infested neighbors. Adults are winged and vagile; therefore, dispersal ability does not prevent colonization of neighboring trees.

### 7.3.2 Plant Phenology

Spatial variation in the leaf-miner population is influenced by a plant temporal trait–leaf budburst phenology. Rates of *S. quadricustatella* herbivory were significantly correlated with the timing of leaf production in the spring. In a multivariate model, leaf phenology explained 61% of the variation in leaf-miner densities on individual trees (Mopper and Simberloff 1995). With the addition of leaf area, the model explained a total of 74% of the variation in leaf-miner density among individual trees (arcsine $Y = 0.485 + 0.022$ (leaf area) $- 0.055$ (leaf phenology), $F_{2,7} = 9.8$, $p = 0.009$). In contrast to the more common observation of early-phenology trees being the most heavily attacked by herbivores, (see Hunter 1992b and references therein), the late-phenology *Q. geminata* trees suffered the heaviest leaf-miner attack.

We also detected significant variation in budburst phenology between host–plant species, and between the inland and coastal sites (Fig. 7.2, and Mopper and Simberloff 1995). Our results support the contention that seasonal variation in leaf production has a profound impact on insect population dynamics and evolutionary ecology (Varley and Gradwell 1963; Askew 1961; Holliday 1977; Thompson and Price 1977; Thompson 1978; Auerbach and Simberloff 1984; Faeth 1990, 1991; Hunter 1990; Hunter and Price 1992; Connor et al. 1994). Host–plant phenology is a potentially strong agent of natural selection and reproductive isolation because insect feeding and oviposition must be synchronized with budburst and leaf abscission patterns (Crawley and Akhteruzzaman 1988; Auerbach 1991; Faeth et al. 1981; Van Dongen et al. 1997). Plant phenology may be particularly critical for endophagous insects—those that feed and reside within the host plant.

One feature shared by locally adapted dispersive insects is endophytic larval development (Table 4.2, Boecklen and Mopper, Chapter 4, this volume). Insects such as leaf miners (Mopper et al. 1995), gall midges (Stiling and Rossi, Chapter 2, this volume), and gall aphids (Komatsu and Akimoto 1995) are intimately associated with their host, because they are confined to the same plant tissue throughout larval development. Such intimacy may magnify the selection pressures imposed by host plants and natural enemies to a degree not experienced by externally feeding herbivores. Komatsu and Akimoto (1995) observed significant variation among individual trees in budburst phenology, to which aphids were lo-

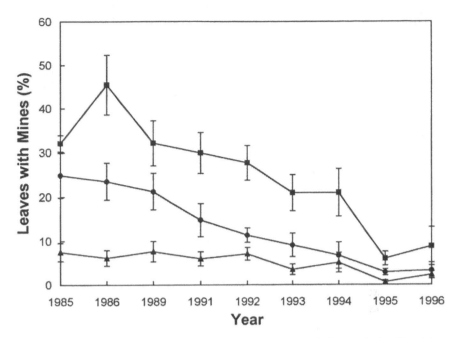

*Figure 7.1*   Annual variation in leaf miner density at inland Lost Lake population. Densities differ consistently among light (▲, $\mu = 0.0509 \pm 0.0006$), intermediate (●, $\mu = 0.1307 \pm 0.0072$, and heavily (■, $\mu = 0.2487 \pm 0.0151$) infested trees ($F_{2.24} = 11.7$, $p = 0.0003$).

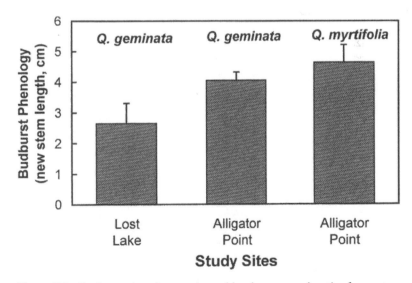

*Figure 7.2*   Budburst phenology estimated by the average length of new stem growth of *Q. geminata* at Lost Lake and Alligator Point, and *Q. myrtifolia* at Alligator Point ($F_{2.12} = 4.7$, $p = 0.03$).

cally adapted via synchronized egg hatching. The strength of the interaction between an herbivore and its host plant, and the potential for deme formation in a population, may be determined largely by the mode of feeding (Stiling and Rossi, Chapter 2, this volume).

### 7.3.3 Stochastic Events

Precipitation and catastrophic events produce spatial structure in the *S. quadricustatella* leaf-miner population by causing local extinctions and sharp reductions in the populations of leaf miners inhabiting individual host plants. The inland *S. quadricustatella* population is in steady decline, from an average of 20% of total *Q. geminata* leaves mined in 1985, to only 4% total leaves mined in 1996 (Fig. 7.1). The lowest year was 1995, when densities averaged 3%, and even the historically heavily attacked trees experienced only 6% infestation.

Much of the decline in the Lost Lake *S. quadricustatella* population since 1985 is explained by a strong positive correlation between density and precipitation, which explains 63% of the annual variation in leaf-miner densities (Fig. 7.3).

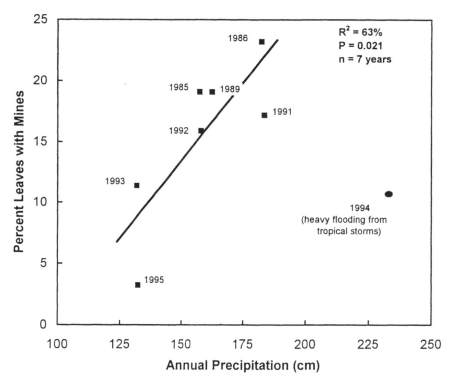

*Figure 7.3* Correlation between average annual leaf miner densities and total annual precipitation since 1985 ($R^2 = 63\%$, $p = 0.021$). The year of Hurricane Andrew, 1994, is not included in the regression.

There is a large body of literature documenting the relationship between phytophagous insect densities and precipitation, which can have beneficial or detrimental effects on fitness (see review in Waring and Cobb 1992). *S. quadricustatella* leaf miners occur at higher densities in wetter years; perhaps a change in weather conditions will reverse the decade-long downward trend.

Although moderate levels of precipitation have a positive effect on oak leaf miners, excessive amounts can reduce population densities. In 1994, numerous tropical storms produced 228 cm of rainfall, one of the largest amounts recorded in north Florida. Larvae suffered the highest plant-mediated mortality in 1994 (Fig. 7.4), and densities in the following year dropped to an all-time low of 4% total leaves mined (Fig. 7.1). Many trees remained standing in water for extended periods through the fall of 1994, which may have contributed to the decline of the leaf-miner population by preventing larval pupation in the soil and leaf litter beneath the trees.

Five Lost Lake study trees have died since 1985, causing the extinction of their leaf-miner populations, and leaf miners have gone extinct on one living tree. In September 1996, a forest fire severely burned and possibly killed about 20% of our study trees. This will undoubtedly increase the number of leaf-miner extinctions associated with individual trees in the population. Stochastic events such as hurricanes, droughts, fires, and human impacts can supersede ambient levels of natural

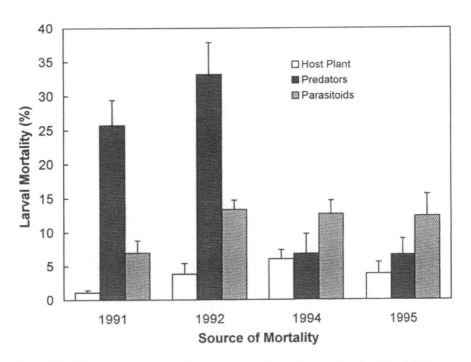

*Figure 7.4* Magnitude and source of leaf miner morality in four years on the same eight trees.

selection and produce nonadaptive demic structure through founder effects and genetic drift (Slatkin 1987: Thomas et al. 1996; Ebert and Hamilton 1996). Such unpredictable forces can dramatically change the evolutionary path of locally
adapted populations (Thomas et al. 1996), prevent local adaptation entirely (Cobb
and Whitham, Chapter 3, this volume), or drive locally adapted demes to extinction (Gandon et al., Chapter 13, this volume).

## 7.4 Local Adaptation

Our experiments indicate that *S. quadricustatella* leaf miners are locally adapted
at three spatial scales: to individual *Q. geminata* trees at Lost Lake, to *Q. geminata* and *Q. myrtifolia* host species at Alligator Point, and to the Lost Lake study
site (Fig. 7.5; Mopper et al. 1995). One might expect differentiation of between-
host-plant populations separated by 60 km, but variation among neighboring *Q.
geminata* trees (Lost Lake), and between two intermixed host species (Alligator
Point), is surprising because the winged adult moths could disperse readily among
trees in close proximity. Genetic analyses confirm the experimental data: genetic
structure at fine spatial scales (Landau and Mopper, unpublished data).

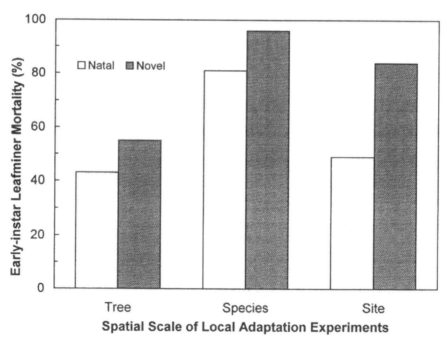

*Figure 7.5* *S. quadricustatella* local adaptation at three spatial scales. Log-linear tests of
partial correlation: $\chi^2_{tree} = 7.9$, $p = 0.005$: $\chi^2_{host\ species} = 9.5$, $p = 0.001$; $\chi^2_{site} = 70.95$, $p = 0.00001$.

### 7.4.1 Agents of Selection for Local Adaptation

Very little is known about the forces that produce local adaptation in phytophagous insects. To understand how natural selection operates on local levels, the mechanisms underlying differential selection must be identified, which requires the decomposition of total mortality into separate sources. Early-instar leaf-miner mortality is primarily plant mediated and is significantly higher on novel host plants (Fig. 7.5). Evidence from our system indicates that budburst phenology may influence performance of leaf miners inhabiting natal and novel trees. There were significant differences in average rates of new stem development: The inland *Q. geminata* were the slowest to produce new foliage, and the coastal *Q. myrtifolia* were the fastest (Fig. 7.2). This result supports recent observations that patterns of differential herbivory and/or local adaptation are associated with synchrony between budburst and larval eclosion (Crawley and Akhteruzzaman 1988; Auerbach 1991; Hunter 1992b; Komatsu and Akimoto 1995).

When we decompose late-instar mortality, a surprising pattern emerges: Natural-enemy mortality was highest among leaf miners transferred to natal trees. This compensated for the reduction in plant-mediated mortality on natal and novel host plants (Fig. 7.6). However, our local adaptation experiments were conducted in a

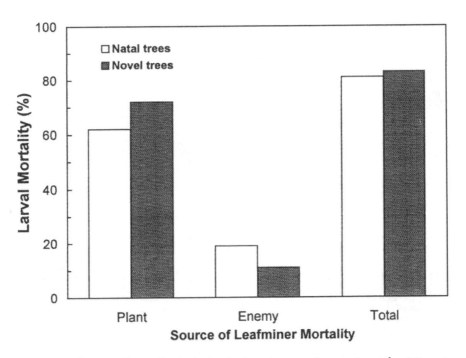

*Figure 7.6*  Sources of mortality in the local adaptation experiment: plant ($\chi^2 = 5.7, p < 0.025$), natural enemy ($\chi^2 = 6.8, P < 0.01$), and total mortality $\chi^2 = 0.328, p >> 0.05$) of miners transferred to natal and novel trees.

year when predation was extremely high and late-instar plant-mediated mortality relatively low (Fig. 7.4); conducting the experiments in a different year may have produced a stronger host-plant effect.

### 7.4.2 Local Adaptation and Spatial Variation

Does local adaptation contribute to the pattern of differential herbivory in the Lost Lake *S. quadricustatella* population? According to the local adaptation hypothesis, performance and population growth should increase through time, as herbivores become adapted to their host plants (Edmunds and Alstad 1978; Cobb and Whitham, Chapter 3, this volume). Therefore, the pattern of heavy and light infestation rates of trees at Lost Lake could indicate differential adaptation to individual oak trees.

We predicted that if high-density trees support locally adapted demes, and low-density trees support recent nonadapted colonizers, resident leaf miners should perform better on heavily attacked than on lightly attacked trees. We tested this prediction in 1991 and 1992 by comparing leaf-miner survival on heavily ($\mu = 32 \pm 6\%$, $n = 6$) and lightly ($\mu = 7 \pm 2\%$, $n = 6$) infested trees and found that mortality rates were roughly similar within years (Fig. 7.7; Mopper and Simberloff

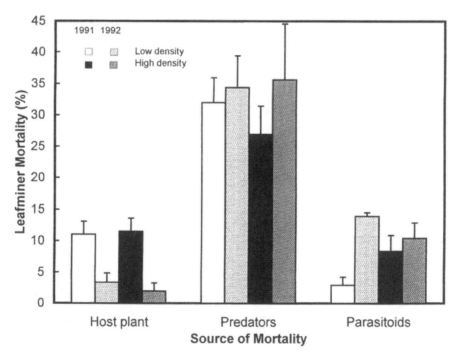

*Figure 7.7* Rates of mortality on heavily and lightly infested trees in 1991 and 1992. There were no significant differences between density categories (all $p$'s $\geqslant 0.10$). Redrawn after Mopper and Simberloff 1995.

1995). Additional studies in this population indicate that mortality is not density dependent at the level of individual trees (Simberloff and Stiling 1987; Stiling et al. 1991). Therefore, density *per se* does not indicate degree of resident leaf-miner adaptation. These findings support the argument made by Boecklen and Mopper (Chapter 4, this volume) that density is an unreliable index of local adaptation.

## 7.5 Conclusion

We detected local adaptation in a dispersive insect at three spatial scales: individual trees, host-plant species, and sites. The selection pressures associated with different sites appear stronger than those associated with neighboring conspecific trees and different but sympatrically distributed host-plant species (Fig. 7.5). The mortality differential between natal and novel treatments was 35% in the between-site comparison, 15% in the host-species comparison, and 13% in the individual-tree comparison. This is consistent with the degree of phenological variation between treatments, because the greatest difference in budburst phenology occurred between the inland and coast sites (Fig. 7.2).

Our local adaptation experiment supports the theory that insects can adapt to individual plants, but it also suggests that adapted insects are *more* apparent to natural enemies. One cost associated with adaptation to host-plant traits could be greater vulnerability to natural enemies. If adapted leaf miners are more efficient consumers than recent colonists, larval development may accelerate, thereby increasing their relative visibility to natural enemies. Greater metabolic efficiency and digestion of plant secondary compounds may also increase relative chemical apparency to natural enemies such as hymenopteran predators and parasitoids, which rely on kairomones to locate prey. Preferential attack of natal insects that are locally adapted to plant traits would provide a relative fitness advantage to recent colonizers that are less efficient metabolizers of host-plant tissues. Therefore, natural enemies could prevent local adaptation in some systems. Since few studies (Mopper et al. 1995; Stiling and Rossi, Chapter 2, this volume; Strauss 1997) determined the magnitude of natural-enemy mortality when testing the local adaptation hypothesis, their role in promoting or preventing genetic structure remains unclear.

One condition necessary for the evolution of locally adapted demes is genetic isolation. Therefore, most experimental tests of the local adaptation hypothesis have centered on relatively nondispersive organisms such as highly sessile scale or thrips insects which are spatially isolated (e.g., Edmunds and Alstad 1978; Karban 1989; Cobb and Whitham 1993; Hanks and Denno 1994). The eight tests of local adaptation in sessile insects are evenly divided in support and rejection of the hypothesis. Surprisingly, three out of five tests using a dispersive insect model detected local adaptation (Table 4.2, Boecklen and Mopper, Chapter 4, this volume).

When populations are isolated by large distances relative to insect dispersal abilities, as are the Lost Lake and Alligator Point leaf miners, or by huge expanses of water as in the *Asphondylia borrichia* gall midge (Stiling and Rossi, Chapter 2, this volume), gene flow between them may be virtually nonexistent, allowing local differentiation between populations. But when gene flow is potentially high, as among neighboring trees within a population or sympatrically distributed host plant species, adaptive differentiation is less likely. Temporal variation, as in differential leaf phenology, could be a strong isolating mechanism. Although strong selection can counterbalance the effects of gene flow (Slatkin 1987), adaptive structure within populations may be also conditional on certain genetic or behavioral attributes. Adults should be philopatric, yet dispersive enough to colonize new host plants. Assortative mating and/or selective oviposition on natal trees are probably essential for adaptive deme formation in the vagile *S. quadricustatella* leaf miners. Fidelity to the natal host may be possible only with linkage or pleiotropy of genes promoting physiological adaptations, reproductive isolation, and philopatry (Gould 1993; Slatkin 1987; Hanks and Denno 1993; Teale et al. 1994).

Local adaptation may occur only under optimal environmental conditions, which are influenced by the frequency and magnitude of stochastic and catastrophic events. Because the host is mortal, insect demes continuously arise and dissolve via local colonization and extinction. Therefore, dispersal is essential to maintain population continuation. In the initial stages of host colonization, good dispersers that are poorly adapted to the host plant predominate but eventually decline in abundance relative to philopatric residents that become efficient at exploiting host-plant tissue. Eventually, resources will deteriorate as densities approach carrying capacity, and colonization of new hosts will be necessary. Neighboring plants may be more similar by descent to the source-pool host than are distant plants (particularly clonal species); therefore, insects colonizing adjacent plants may be partially preadapted to the new hosts.

Understanding the dynamics of local adaptation in phytophagous insect populations may require a metapopulation approach (Gandon et al. 1996). A metapopulation is a spatially structured collection of local populations or demes that undergo extinction and colonization events that affect genetic variation and evolution (Levins 1970; Hastings and Harrison 1994). In the dynamics of a metapopulation of locally adapted demes, there is a tension between selection for good dispersers and for good residents (Olivieri et al. 1995), which can explain the sometimes paradoxical relationship between dispersal and genetic structure. Stable insect populations with long-lived invariant host plants and few natural enemies may best be described by simple models. Stepping-stone models of colonization and extinction (Kimura and Maruyama 1971), in which colonization occurs only between neighboring host plants, are probably most applicable to sessile organisms such as black-pineleaf scale (Alstad, Chapter 1, this volume). On the other hand, island models, in which the sources of new colonizers of vacant host plants are randomly distributed (Slatkin 1977), may be appropriate for highly dispersive insects. Metapopulation theory is relatively flexible. Complex models can include abiotic and biotic forces such as stochastic events, genetic variation, and the migration of parasites and host plants (Gilpin and Hanski 1991; Hastings and Harrison 1994; Olivieri et al. 1995; Gandon et al., Chapter 13, this volume; Thomas and Singer, Chapter 14, this volume). The structure of a population is integral to its evolutionary ecology. Populations composed of small semi-isolated groups evolve on average at faster rates than single panmictic assemblages (Gould 1993). Elucidating the abiotic and biotic forces that produce adaptive and nonadaptive genetic structure at local levels will reveal mechanisms underlying the evolutionary processes at larger spatial and temporal scales.

## Acknowledgments

My gratitude to Keli Landau, Sharon Strauss, and Anita Evans for their thoughtful review of this manuscript. Keli Landau provided essential editorial assistance with this manuscript and the volume in which it appears. I thank Karl Hasenstein for his insight, inspiration, and considerable efforts on this chapter and the entire volume. This research was supported by a 1996 USL Summer Research Fellowship, the Lafayette Parish Medical Society Endowed Professorship (1993–1999), and the following research grants: National Science Foundation BSR90-07144 (1990–1993), EPSCoR NSF/LEQSF-ADP-02 (1992–1996), LEQSF-RD-A-37 (1994–1996), and National Science Foundation (1996–1999) DEB-9632302.

## 7.6 References

Alstad, D. N. and K. W. Corbin. 1990. Scale insect allozyme differentiation within and between host trees. *Evol. Ecol. 4*:43–56.

Askew, R. R. 1961. On the biology of the inhabitants of oak galls of Cynipidae (Hymenoptera) in Britain. *Trans. Soc. Brit. Entomol. 14*:237–268.

Auerbach, M. 1991. Relative impact of interactions within and between trophic levels during an insect outbreak. *Ecology 72*:1599–1608.

Auerbach, M. and D. Simberloff. 1984. Responses to atypical leaf production patterns. *Ecol. Entomol. 9*:361–367.

Berg, E. E. and J. L. Hamrick. 1995. Fine-scale genetic structure of a turkey oak forest. *Evolution 49*:110–120.

Burdon, J. J., A. M. Jarosz, and G. C. Kirby. 1989. Patterns and patchiness in plant-pathogen interactions—causes and consequences. *Annu. Rev. Ecol. Syst. 20*:119–136.

Cobb, N. S. and T. G. Whitham. 1993. Herbivore deme formation on individual trees: A test case. *Oecologia 94*:496–502.

Connor, E. F., R. H. Adams-Manson, T. Carr, and M. W. Beck. 1994. The effects of host plant phenology on the demography and population dynamics of the leaf-mining moth. *Cameraria hamadryadella* (Lepidoptera: Gracillariidae). *Ecol. Entomol. 19*:111–120.

Connor, E. F. and M. W. Beck. 1993. Density-related mortality in *Cameraria hamadryadella* (Lepidoptera: Gracillariidae) at epidemic and endemic densities. *Oikos 66*:515–525.

Crawley, M. J. and M. Akhteruzzaman. 1988. Individual variation in the phenology of oak trees and its consequences for herbivorous insects. *Funct. Ecol.* 2:409–415.

Denno, R. F. and M. S. McClure. 1983. *Variable Plants and Herbivores in Natural and Managed Systems.* Academic Press, New York.

Dodge, K. L. and P. W. Price. 1991. Eruptive versus noneruptive species: A comparative study of host plant use by a sawfly, *Euura exiguae* (Hymenoptera: Tenthredinidae) and a leaf beetle, *Disonycha pluriligata* (Coleoptera: Chrysomelidae). *Environ. Entomol.* 20:1129–1133.

Ebert, D. and W. D. Hamilton. 1996. Sex against virulence: The coevolution of parasitic diseases. *Trends Ecol. Evol. 11*:79–82.

Edmunds, G. F. and D. N. Alstad. 1978. Coevolution in insect herbivores and conifers. *Science 199*:941–945.

Ehrlich, P. R. and P. H. Raven. 1964. Butterflies and plants: A study in coevolution. *Evolution 18*:585–608.

Ehrlich, P. R. and P. H. Raven. 1969. Differentiation of populations. *Science 165*:1228–1232.

Faeth, S. H. 1990. Aggregation of a leafminer, *Cameraria* sp. nov. (Davis): Consequences and causes. *J. Anim. Ecol. 59*:569–586.

Faeth, S. H. 1991. Effect of oak leaf size on abundance, dispersion, and survival of the leafminer *Cameraria* sp. (Lepidoptera: Gracillariidae). *Environ. Entomol. 20*:196–204.

Faeth, S. H., E. F. Connor, and D. Simberloff. 1981. Early leaf abscission: A neglected source of mortality for folivores. *Am. Nat. 117*:409–415.

Feeny, P. 1976. Plant apparency and chemical defense. *Rec. Adv. Phytochem. 10*:1–40.

Gandon, S., Y. Michalakis, and D. Ebert. 1996. Temporal variability and local adaptation. *Trends Ecol. Evol. 11*:431.

Gilpin, M. E. and I. Hanski (Eds.). 1991. *Metapopulation Dynamics: Empirical and Theoretical Investigations.* Academic Press, London.

Gittman, S. I., T. Wilson, and L. A. Weigt. 1989. Microgeographic genetic variation in the *Enchenopa binotata* complex (Homoptera: Membracidae). *Ann. Entomol. Soc. Am.* 82:156–165.

Gould, F. 1993. The spatial scale of genetic variation in insect populations. Pp. 67–88 *in* K. C. Kim and B. A. McPheron (Eds.), *Evolution of Insect Pests: Patterns of Variation.* John Wiley, New York.

Hairston, N. G., F. E. Smith, and L. B. Slobodkin. 1960. Community structure, population control and competition. *Am. Nat. 44*:421–425.

Hamrick, J. L. 1976. Variation and selection in western montane species: II. Variation within and between populations of white fir on an elevational transect. *Theor. Appl. Genet. 47*:27–34.

Hanks, L. M. and R. F. Denno. 1993. The role of demic adaptation in colonization and spread of scale insect populations. Pp. 393–411 *in* K. C. Kim and B. A. McPheron (Eds.), *Evolution of Insect Pests. Patterns of Variation.* John Wiley, New York.

Hanks, L. M. and R. F. Denno. 1994. Evidence for local adaptation in the armored scale insect *Pseudaulacaspis pentagona* (Homoptera: Diaspididae). *Ecology 75*:2301–2310.

Hastings, A. and S. Harrison. 1994. Metapopulation dynamics and genetics. *Annu. Rev. Ecol. Syst. 25*:167–188.

Hedrick, P. W., M. E. Ginevan, and E. P. Ewing. 1976. Genetic polymorphism in heterogeneous environments. *Annu. Rev. Ecol. Syst. 7*:1–32.

Hiebert, R. D. and J. L. Hamrick. 1983. Patterns and levels of genetic variation in Great Basin bristlecone pine, *Pinus longaeva*. *Evolution 37*:302–310.

Holliday, N. J. 1977. Population ecology of the winter moth (*Operophtera brumata*) on apple in relation to larval dispersal and time of budburst. *J. Appl. Ecol. 14*:803–814.

Hunter, M. D. 1990. Differential susceptibility to variable plant phenology and its role in competition between two insect herbivores on oak. *Ecol. Entomol. 15*:401–408.

Hunter, M. D. 1992a. Interactions with herbivore communities mediated by the host plant: The keystone herbivore concept. Pp. 287–325 *in* M. D. Hunter, T. Ogushi, and P. W. Price (Eds.), *Effects of Resource Distribution on Animal–Plant Interactions*. Academic Press, San Diego, CA.

Hunter, M. D. 1992b. A variable insect–plant interaction: The relationship between tree budburst phenology and population levels of insect herbivores among trees. *Ecolog. Entomol. 16*:91–95.

Hunter, M. D. and P. W. Price. 1992. Playing chutes and ladders: Heterogeneity and the relative roles of bottom-up and top-down forces in natural communities. *Ecology 73*:724–732.

Karban, R. 1989. Fine-scale adaptation of herbivorous thrips to individual host plants. *Nature 340*:60–61.

Kimura, M. and T. Maruyama. 1971. Patterns of neutral variation in a geographically structured population. *Genet. Res. 18*:125–131.

Komatsu, T. and S. Akimoto. 1995. Genetic differentiation as a result of adaptation to the phenologies of individual host trees in the galling aphid *Kaltenbachiella japonica*. *Ecolog. Entomol. 20*:33–42.

Lawton, J. H. and S. McNeil. 1979. Between the devil and the deep blue sea: On the problem of being a herbivore. *Symp. Brit. Ecol. Soc. 20*:223–244.

Levins, R. 1970. Extinction. *Lect. Math. Life Sci. 2*:75–107.

McCauley, D. E. 1991. The effect of host plant patch size variation on the population structure of a specialist herbivore insect, *Tetraopes tetraopthalmus* (Forster). *Evolution 45*:1675–1684.

McCauley, D. E. and W. F. Eanes. 1987. Hierarchical population structure analysis of the milkweed beetle, *Tetraopes tetraophthalmus* (Forster). *Heredity 58*:193–201.

McPheron, B. A., D. Courtney Smith, and S. H. Berlocher. 1988. Genetic differences between host races of *Rhagoletis pomonella*. *Nature 336*:64–66.

Michalakis, Y., A. W. Sheppard, V. Noel, and I. Olivieri. 1993. Population structure of a herbivorous insect and its host plant on a microgeographic scale. *Evolution 47*:1611–1616.

Mopper, S. 1996a. Adaptive genetic structure in phytophagous insect populations. *Trends Ecol. Evol. 11*:235–238.

Mopper, S. 1996b. Temporal variability and local adaptation—a reply to Gandon et al. *Trends Ecol. Evol. 11*:431–432.

Mopper, S., M. Beck, D. Simberloff, and P. Stiling. 1995. Local adaptation and agents of mortality in a mobile insect. *Evolution 49*:810–815.

Mopper, S., S. H. Faeth, W. J. Boecklen, and D. S. Simberloff. 1984. Host-specific variation in leafminer population dynamics: Effects on density, natural enemies and behaviour of *Stilbosis quadricustatella* (Lepidoptera: Cosmopterigidae). *Ecolog. Entomol. 9*:169–177.

Mopper, S., J. B. Mitton, T. G. Whitham, N. S. Cobb, and K. M. Christensen. 1991. Genetic differentiation and heterozygosity in pinyon pine associated with herbivory and environmental stress. *Evolution 45*:989–999.

Mopper, S. and D. Simberloff. 1995. Differential herbivory in an oak population: The role of plant phenology and insect performance. *Ecology 76*:1233–1241.

Mopper, S. and T. G. Whitham. 1992. The plant stress paradox: Effects on pinyon sawfly fecundity and sex ratios. *Ecology 73*:515–525.

Olivieri, I., Y. Michalakis, and P. Gouyon. 1995. Metapopulation genetics and the evolution of dispersal. *Am. Nat. 146*:202–228.

Price, P. W., C. E. Bouton, P. Gross, B. A. McPheron, J. N. Thompson, and A. E. Weiss. 1980. Interactions among three trophic levels: Influence of plants on interactions between herbivores and natural enemies. *Annu. Rev. Ecol. Syst. 11*:41–65.

Plessas, M. E. and S. H. Strauss. 1986. Allozyme differentiation among populations, stands, and cohorts in Monterey pine. *Can. J. For. Res. 16*:1155–1164.

Rank, N. E. 1992. A hierarchical analysis of genetic differentiation in a montane leaf beetle *Chrysomela aeneicollis* (Coleoptera: Chrysomelidae). *Evolution 46*:1097–1111.

Rhoades, D. F. and R. G. Cates. 1976. Toward a general theory of plant antiherbivore defense. *Rec. Adv. Phytochem. 10*:168–213.

Simberloff, D. S. and P. Stiling. 1987. Larval dispersion and survivorship in a leaf-mining moth. *Ecology 68*:1647–1657.

Slatkin, M. 1977. Gene flow and genetic drift in a species subject to local extinction. *Theor. Pop. Biol. 12*:253–62.

Slatkin, M. 1987. Gene flow and the geographic structure of natural populations. *Science 236*:787–792.

Stiling, P., D. S. Simberloff, and B. V. Brodbeck. 1991. Variation in rates of leaf abscission between plants may affect the distribution patterns of sessile insects. *Oecologia 88*:367–370.

Strauss, S. Y. 1997. Lack of evidence for local adaptation to individual plant clones or site by a mobile specialist herbivore. *Oecologia 110*:77–85.

Strauss, S.Y. and R. Karban. 1994a. The significance of outcrossing in an intimate plant/herbivore relationship: I. Does outcrossing provide an escape for progeny from herbivores adapted to the parental plant? *Evolution 48*:454–464.

Strauss, S. Y. and R. Karban. 1994b. The significance of outcrossing in an intimate plant–herbivore relationship: II. Does outcrossing pose a problem for thrips adapted to the host-plant clone? *Evolution 48*:465–476.

SYSTAT for Windows. 1992. Version 5 Edition. SYSTAT, Inc., Evanston, IL.

Teale, S. A., J. B. Hager, and F. X. Webster. 1994. Pheromone-based assortative mating in a bark beetle. *Anim. Behav. 48*:569–578.

Thompson, J. N. 1978. Within-patch structure and dynamics in *Pastinaca sativa* and resource availability to a specialized herbivore. *Ecology 59*:443–448.

Thompson, J. N. 1988. Evolutionary ecology of the relationship between oviposition preference and performance of offspring in phytophagous insects. *Entomol. Exp. Appl 47*:3–14.

Thompson, J. N. and O. Pellmyr. 1991. Evolution of oviposition behavior and host preference in Lepidoptera. *Annu. Rev. Entomol. 36*:65–89.

Thompson, J. N. and P. W. Price. 1977. Plant plasticity, phenology, and herbivore dispersion: Wild parsnip and the parsnip webworm. *Ecology 58*:1112–1119.

Thomas, C. D., M. C. Singer, and D. A. Boughton. 1996. Catastrophic extinction of population sources in a "source-pseudosink" butterfly metapopulation. *Am. Nat., 148*:957–975.

Unruh, T. R. and R. F. Luck. 1987. Deme formation in scale insects: A test with the pinyon needle scale and a review of other evidence. *Ecol. Entomol. 12*:439–449.

Van Dongen, S., T. Backeljau, E. Matthysen, and A. Dhondt. 1997. Synchronization of hatching date with budburst of individual host trees (*Quercus robur*) in the winter moth (*Operophtera brumata*) and its fitness consequences. *J. Anim. Ecol. 66*:113–121.

Varley, G. C. and G. R. Gradwell. 1968. Population models for the winter moth. Pp. 132–142 *in* T. R. E. Southwood (Ed.), *Insect Abundance.* Symposium Royal Society London, London, England.

Waring, G. L. and N. S. Cobb. 1992. The impact of plant quality on herbivore population dynamics: The case of plant stress. Pp 167–227 *in* E. A. Bernays (Ed.), *Insect–plant Interactions.* CRC Press, Boca Raton, FL.

# 8

# The Strength of Selection: Intraspecific Variation in Host-Plant Quality and the Fitness of Herbivores

*Sharon Y. Strauss*
Section of Evolution and Ecology

*Richard Karban*
Department of Entomology, University of California at Davis, Davis, CA

## 8.1 Introduction

To date, the body of evidence that supports fine-scale local adaptation by herbivores to individual host-plant phenotypes has been found in insects that are relatively sedentary and specialized. At the extreme, such species can consist of genetically divergent populations that are each adapted to neighboring individual plant phenotypes. While such cases have been documented, there are also other instances in which local adaptation has not been shown to occur (Cobb and Whitham 1993, Strauss 1997). Local adaptation is a result of differing selective regimes imposed by different host-plant individuals. Conditions favoring local adaptation can be offset, however, by factors that tend to homogenize subpopulations genetically (e.g., gene flow among populations of herbivores on these plants). A more general issue is: How likely is it to find fine-scale local adaptation to individual plants in herbivorous insects? In this chapter, we explore the strength of selection imposed on insect populations by intraspecific variation in the host plant. We try to relate this value to the strength of other effects, such as nongenetic parental effects that could influence insect performance. Finally, we ask how much gene flow would be required to homogenize these populations genetically, thus preventing local adaptation.

It would be remiss to address questions about local adaptation to individual host plants without first documenting that the ability of insects to deal with intraspecific variation in host-plant quality has a genetic basis (see Berenbaum and Zangerl, Chapter 5, this volume). While much evidence supports a genetic basis for insect adaptation to interspecific variation in host-plant quality (Futuyma et al. 1984; Via 1990 for review; Pashley et al. 1995; Scriber et al. 1991a), much less work has focused on genetic variation in herbivores to deal with intraspecific differences in host-plant quality. Lindroth and Weisbrod (1991) revealed strong family differences in gypsy moth performance on diets with phenolic glycosides extracted from aspen leaves. There was a significant positive correlation between phenolic glycosides and esterase activity in larvae, as well as a diet $x$ family interaction in inducibility of esterase activity. Similarly, significant family effects were found in black swallowtail larvae (*Papilio polyxenes*) for the metabolism of xanthotoxin and angelicin furanocoumarins (Berenbaum and Zangerl 1993). Broad sense heritability (i.e., estimates that include nonadditive genetic variation) for detoxification of these compounds separately and in combination fell between 0.276 and 0.546. At least two alleles exist at an inducible *CYP6B1* locus that encodes for a xanthotoxin-metabolizing cytochrome P450 monooxygenase in black swallowtails (Cohen et al. 1992), and thereby indicate a genetic basis for furanocoumarin metabolism in this species. Berenbaum and Zangerl (1993) have also documented genetic variability in metabolism of furanocoumarins within populations of parsnip webworm, a specialist insect. Thus, there is growing evidence that not only is there genetic variation among insect host races to deal with interspecific differences among host plants, but also genetic variation exists within insect populations for feeding on conspecific host plants that vary in quality.

If selection as a result of variation among host plants drives local adaptation by insects to individual plants (but see Mopper, Chapter 7, and Stiling and Rossi, Chapter 2, this volume), then a prerequisite for finding such local adaptation is that the ability to thrive on one plant phenotype precludes an ability to thrive on others. A good example of such a phenomenon is described by Komatsu and Akimoto (1995), who show a strong genetic component in aphid emergence times on elm trees. The timing of emergence of each aphid genotype is closely correlated with budburst phenology of the natal tree. Aphids transferred to (or migrating to) nonnatal plant phenotypes would not fare well, regardless of whether the new host tree leafed out later (in which case no food would be available at aphid emergence), or earlier (when leaves would be at an inappropriate stage developmentally for proper gall formation) than the natal tree. Local adaptation based on selection from host phenology is thus likely to constrain success on other host-plant individuals. Phenology appears to play a large part in local adaptation by aphids and treehoppers to their host plants (Komatsu and Akimoto 1995; Wood and Guttman 1983), and by *Rhagoletis* fruit flies to new host species (see Feder et al., Chapter 16, this volume).

Trade-offs with respect to detoxification of secondary compounds are harder to document. There are relatively few studies of secondary-compound metabolism that have linked detoxification of compounds to specific enzymes (but see Nitao 1989; Scriber et al. 1991b). It is even more difficult to determine such mechanisms for all the combinations of compounds present in a plant (Berenbaum and Zangerl 1993). Thus, if adaptation is primarily to secondary chemistry, it is not clear whether there are trade-offs in ability to detoxify compounds of different host plants within the same species. For example, an insect genotype that is effective at detoxifying secondary chemicals could perform well on all plants of that species and not become locally adapted to any individual plant. Trade-offs would result in differential efficiency of insect genotypes on plants with different ratios or profiles of defensive compounds. Alternatively, if metabolism of secondary compounds is costly, then local adaptation might be expressed through a poly-morphism in investment in detoxification metabolism. Less-invested insects might perform better on less-defended plants, whereas insect genotypes with high investment in detoxification systems would fare better on more defended plants. Whatever the basis, trade-offs must occur in abilities of different insect genotypes to use different individual host plants in order for locally adapted demes to occur.

This chapter focuses on five aspects of the biology of local adaptation by in-sects to host plants:

1. **What is the relative fitness of adapted versus nonadapted insect popu-lations on individual host plants?** About eight species of insects exhibit fine-scale local adaptation to individual host plants. One important ques-tion is, how great are the fitness differences among subpopulations when individuals are reared on their natal and on novel plant phenotypes? This information can give us an idea of the strength of selection imposed by host-plant quality, and ultimately an idea of the likelihood of finding local adaptation in these and other systems.

2. **How much gene flow would be required to homogenize insect popula-tions, given the average strength of selection imposed by variation in host-plant quality?** Ultimately, we want to know how generally likely it would be to find local adaptation in herbivorous insect populations. In this part of the chapter, we will explore how much gene flow might be required to prevent local adaptation from occurring, given the average strength of selection as calculated in the present review.

3. **How do fitness differences resulting from nongenetic parental effects compare to the fitness differences that are attributable to genetic fac-tors?** To date, only two studies testing for local adaptation by insects to host plants have examined the possible confounding factors of nongenetic parental effects. What is the evidence for parental effects? How large are these effects, and could they be responsible for patterns that appear to be local adaptation?

4. **When local adaptation occurs, does the "natal" insect line always out-perform all "novel" insect lines?** To test whether local adaptation has oc-curred, investigators typically compare the average performance of insects on their natal plants to the average performance of insects that originated from other plants. However, averaging over novel lines could conceal the fact that some "novel" insect lines outperform the "natal" line, despite the fact that, *on average*, natal insects do better than novel ones. If there are novel lines that can outperform the natal line, then some insects could be pre-adapted to use an individual host plant. The question is, how often do we find this to be the case? If we find it often, does this tell us anything about the dynamics of local adaptation within these species and the processes that maintain differentiated populations?

5. **How does the magnitude of the coefficient of selection differ in studies of intraspecific variation in host plants versus interspecific variation?** Do host races (conspecific insects that specialize on different host-plant species) have greater differences in relative fitness when they are reared on natal versus novel species than insects that are adapted to individual plants and reared on different individuals of a single plant species?

## 8.2 Estimating the Strength of Selection Imposed by Intraspecific Variation in Host-Plant Quality on Herbivorous Insects

When local adaptation occurs, the average fitness of "natal" insects (those reared on the plant on which they were laid) will exceed the average fitness of insects originating from other nonnatal plants, hereafter referred to as "novel" insects. To calculate the strength of selection, or coefficient of selection, we assumed the fitness of the natal insect line to be 1, and compared the fitness of novel subpopulations relative to the natal subpopulation. Thus, if survivorship is assumed to equal fitness and is equal to 0.50 for the natal line, and 0.10 for the novel line, then the relative fitness of that novel line is 0.20 and the coefficient of selection is 0.80 (1.0−0.20).

Table 8.1 lists studies of local adaptation by herbivores to individual host plants. We included only those studies for which we could acquire data on the performance of each novel insect lineage/subpopulation, as well as the natal lineage. In all of these studies, on average, natal insect subpopulations outperformed individuals from novel subpopulations. In the third column of Table 8.1, we present the average relative fitness of novel lines with respect to natal insect lines. Data for this column were obtained from published papers or from authors who generously contributed additional data to us. For example, in a study using four trees for rearing insects and four insect populations (one from each tree) in a full reciprocal transfer experiment, there are three values of relative fitness per tree (the fitness of insects from the novel trees relative to the fitness of the natal insects). Thus there would be 12 values from which to calculate mean relative fitness of novel insects versus natal insects. There is some degree of pseudoreplication here, since usually the same insect lines are transferred to all possible hosts. However, in order for local adaptation to be present, there have to be trade-offs in how lines perform across individual trees, so there is likely to be little overall correlation in relative performance (or fitness) by a single line across all sources.

In most cases, we do not have strict measures of fitness, but measures of some correlates of fitness. Wherever possible, we tried to avoid traits that had more dubious correlations with fitness (i.e., development time), and we have most confidence in measures of population growth rates that incorporate both viability and fecundity components of fitness. In cases for which we find local adaptation, we expect the mean relative fitness of novel insects to be significantly less than 1. From Table 8.1, we find that the average relative fitness overall of novel insect lines with respect to natal lines is 0.63 (± 0.15 *SD*; *n* = 6). The corresponding

*Table 8.1* Mean Relative Fitness and Range of Novel Insect Genotypes with Respect to Natal Genotypes in Systems Documenting Local Adaptation to Individual Host-Plant Phenotypes.

| Insect | Plant | Overall Mean Relative Fitness with Respect to Natal Insect Line (± S.D.) | Range of Relative Fitness with Respect to Natal Line (# times Natal lines outperformed Novel lines) | Trait | Value of Trait in Natal Line | # Insect Lines Used per Host Plant | # Host Plants Used | Citation |
|---|---|---|---|---|---|---|---|---|
| *Asphondylia borrichiae* | *Borrichia frutescens* | .51 ± .30 (n = 12) | .06–1.14 (1) | % gall success | 11.3% | 4 | 4 | Stiling & Rossi, Chapter 2 this volume |
| *Cryptococcus fagisuga* | *Fagus sylvatica* | .79 ± .78 (n = 20) | 0–1.50 (7) | % survival to fecund adults | 0.92% | 5 | 5 | Wainhouse & Howell (1983) |
| *Euphydryas editha* | *Pedicularis semibarbata* | .69 ± .33 (n = 11) | 0–1.22 | Larval survivorship to pupation | 55% | 12 | 2 "types" | Ng (1988) |
| *Apterothrips apteris* | *Erigeron glaucus* | .42 ± .11 (n = 6) | .33–58 | Population growth | N/A | 3 | 3 | Karban (1989) |
| *Stilbosis quadri-custatella* | *Quercus geminata* | .53 ± .16 (n = 12) | .27–.91 | % survivorship | 49% | 4 | 3 | Mopper et al. (1995) |
| *Pseudalacapsis pentagona* | *Morus alba* | .84 ± .29 (n = 10) | .28–1.38 (2) | % survivorship | 16% | 10 | 10 | Hans and Denno (1994) |

coefficient of selection is therefore 0.37, and ranges from 0.16–0.58. By convention, $s$ is considered small if it is less than 0.10 (M. Turelli personal communication), so the strength of selection estimated from these studies could be considerable, if these values reflect primarily genetic differences among lines.

One question that remains is whether selection imposed by host plants is much stronger in systems where local adaptation is present versus those in which it has been looked for, but not found (e.g., Unruh and Luck 1987, Rice 1983, Hanks and Denno 1994, Strauss 1997). Unfortunately, the data necesary for a comparison similar to the one in Table 8.1. were unavailable. Three studies, however, do provide us with an opportunity to examine this issue. In a series of reciprocal transfer experiments, Hanks and Denno (1994) found evidence for local adaptation to natal trees by scale insects when trees were at least 0.25 km apart. However, scales were not differentially adapted to neighboring trees. We compared the magnitude of $s$ in these two experiments ($n = 10$ in both cases). Where there was no local adaptation by scales (near trees), we always compared fitness with respect to the best-performing insect line, regardless of tree of origin (natal or novel). We found that the magnitude of $s$ was the same in both experiments ($s = 0.32 \pm 0.04$ with no local adaptation, and $s = 0.33 \pm 0.07$ with local adaptation). Since the authors examined the same insect lineages at both distance scales and at the same time, this study represents a good test of whether fitness differences among lines are greater in cases when we find local adaptation versus when we do not. From the data presented in Alstad, Table 1.1 (Chapter 1, this volume) in which no local adaptation by scales was found, the magnitude of $s = 0.35$ ($\pm 18$, $n = 20$). This estimate is comparable to that from the Hanks and Denno study, and is almost identical to the value of $s$ found in the six studies that do show local adaptation by insects. Finally, in a study by Strauss (1997) on chrysomelid beetles that feed on sumac clones, the coefficient of selection was smaller than that estimated in Table 8.1; $s = 0.13$ ($\pm 0.07$ $SD$, n = 28). This result was based on pupal weights of lines of insects transferred to eight sumac clones. Thus, we have few examples, and they are inconclusive concerning the relative strength of selection exerted by host plants on insects in cases where we do and do not find local adaptation.

## 8.3 Gene Flow and the Coefficient of Selection

Gene flow can potentially swamp any possibility for genetic divergence among subpopulations, even in the face of strong selection. According to the island model of gene flow, each subpopulation of insects both contributes and receives migrant individuals from other subpopulations. Individuals are assumed to be derived from each subpopulation equiprobably, and one locus is assumed responsible for the trait causing local adaptation. The fraction of zygotes with immigrant parents prior to selection is defined as $m$. Under the island model, if $m$ is less (generally) than the coefficient of selection, then this rate of migration would not be sufficient to homogenize insect subpopulations (Haldane 1930; Nagylaki 1978).

Thus, in the locally adapted subpopulations from Table 8.1, on average, more than one-third of the zygote population would have to have a migrant parent in order for gene flow to overcome effects of selection.

Few estimates of $m$ have been made in studies documenting local adaptation; in addition, it may be a mistake to assume that migrants from all subpopulations have equiprobable mating success. Karban and Strauss (1993) have shown that only two-thirds of thrips migrants colonizing new plants reproduced, and this number was a conservative estimate, because the experimental plants used were highly susceptible to infestation ($s$ expected to be very low). In addition, the more isolated a plant from a source subpopulation (another plant), the lower the rate of reproduction of by thrips migrants. Thus, migrants from more distant sources might contribute even less to the gene pool per individual than migrants from closer subpopulations (plants).

Reduction in fitness as a result of longer dispersal distances by herbivores may be due to the costs of dispersal or to greater differences between environments. For example, in many natural plant populations, neighboring plants are more closely related to one another than are more distant individuals (e.g., Schaal 1975; Linhart et al. 1981; Schnabel and Hamrick 1990). Herbivores adapted to one plant may therefore be better able to reproduce on a nearby, related plant than on a less related, more distant plant. In addition, other environmental variables to which an herbivore may be locally adapted (e.g., soil moisture, etc.), are often autocorrelated spatially. Not only will neighboring subpopulations exchange more migrants, but also plants with large populations will contribute more migrants to the migrant pool. These larger populations may arise because the host plant is generally susceptible to insect attack, or because there is a locally adapted deme of insects on that plant. If susceptible plants are generally rare in plant populations (e.g., Strauss and Karban 1994) but nonetheless are the source of many migrants, then we might expect colonists originating from these plants to have low fitness on more abundant, resistant plant genotypes. In addition, if insects are locally abundant because they are adapted to a particular plant, then they may not necessarily be effective colonists of new plants. For all these reasons, it is highly unlikely that genes from migrants are equally likely to come from all subpopulations, and that gene flow is easily related to the number of migrants entering a population.

The approximation of $m < s$ to maintain polymorphism in the population under the island model is generally robust to recessivity–dominance of the adaptive alleles (Nagylaki 1978). However, the number of loci involved, as well as the additivity of the effects of these loci, may also influence the magnitude of $m$ relative to $s$ for the maintenance of differentiated subpopulations. In a recent paper, Slatkin (1995) showed that for a particular form of epistasis, the degree of linkage between loci, as well as the number of loci involved in the trait, strongly affected the magnitude of $m$ required to maintain differentiated subpopulations. In this model, increasing numbers of loci and decreasing linkage among these loci, resulted in much lower values of $m$ required to homogenize populations than in the single-locus model. These results may, however, not be robust to other patterns of trait inheritance or epistasis. The bottom line is that the genetic basis of the traits underlying local adaptation must be known before we can estimate how much gene flow would be required to maintain genetically heterogeneous subpopulations.

What do we know about the underlying genetic basis of traits responsible for local adaptation to host plants? Huettel and Bush (1972) found results consistent with single-locus control of host-plant choice in tephritid flies. At least two alleles are involved in the detoxification of xanthotoxins by black swallowtail butterflies (Cohen et al. 1992). In contrast, several unlinked autosomal loci with some dominance were implicated in the oviposition preference behavior of *Drosophila tripunctata* (Jaenike 1989). Finally, Thompson (1988) concluded that one or more loci on the X chromosome were responsible for oviposition choice in swallowtail butterflies. We know even less about the genetics of insect performance on host plants (see Berenbaum and Zangerl, Chapter 5, this volume for review). In some agricultural systems, there seems to be good evidence for gene-for-gene resistance mechanisms. In these cases, ability to use host plants by insects may be determined by only one or few loci with many alleles (Smith 1989). Clearly, we are far from being able to estimate with confidence exactly how much gene flow would be sufficient to homogenize insect herbivore populations. Given the current estimates of the magnitude of $s$ in these systems, it is certainly possible that selection could maintain differentiated populations in the face of gene flow (Alstad et al. 1991).

## 8.4 How Does the Magnitude of Fitness Differences Caused by Nongenetic Parental Effects Compare to That Attributable to Genetic Traits?

The previous discussion assumes that the difference in relative fitness we see in experiments on local adaptation reflect underlying genetic differences among populations. However, environmental factors such as maternal and paternal diet may also influence offspring performance. Most studies of fine-scale local adaptation have not attempted to control for maternal effects on offspring performance. Two notable exceptions are Karban (1989) and Komatsu and Akimoto (1995). In the former study, insects were reared for two generations on a common host plant in the greenhouse before being transferred to cuttings in a common garden in the field. In the latter, detailed crossing experiments showed no differences in aphid offspring performance as a function of the tree of origin of the maternal or paternal aphid line. Thus, both studies provide strong evidence for local adaptation based on genetic rather than parental environmental factors. Rossiter's Chapter 6 in this volume gives a detailed discussion of parental effects and how they may bias our estimates of heritability of traits. For the purposes of this chapter, we review the small amount of data available that might document the strength of such parental effects and relate that to the coefficient of selection described earlier.

Both maternal and paternal nongenetic effects have been documented in insect performance (e.g., Boggs and Gilbert 1979; Gould 1988; Futuyma et al. 1993); thus, we refer to these effects as "parental" as opposed to the more typical "maternal" terminology. While it is expected that there may be strong maternal effects in characters such as egg provisioning, it is likely that paternal effects occur as well, often as a result of contributions of nutrients, minerals, and enzymes in ejaculate or spermatophores (e.g., Boggs and Gilbert 1979; Fox 1993). Experiments to document parental effects separate from genetic effects are extremely difficult to conduct and require extensive rearing of parents and offspring in multiple environments (see Rossiter, Chapter 6, this volume for detailed experimental design and analysis). These types of experiments have rarely been done to examine the strength of parental effects based on *inter*specific variation in host-plant quality, let alone for *intra*specific variation in quality. Why should we care about such effects? If insects are sedentary and parental diet influences larval growth, then superior performance of natal lines in reciprocal transfer experiments may be caused by parental diet rather than by any genetically based adaptation. Parental effects could be a problem for any experiment that collects samples from subpopulations as eggs or larvae and uses insects from these collections directly in experimental manipulations (the protocol for many published studies testing for local adaptation in herbivorous insects). It thus becomes imperative to know how often such effects occur, and the magnitude of these effects.

A few studies suggest that host-plant quality fed on by parents affects the performance of their offspring. Most of these studies use artificial diets to assess effects and thus are subject to artifacts that accompany such diets. For example, Gould (1988) showed that tobacco budworm larvae weighed approximately 2.5 times more when only females were raised on a diet without quercetin, a tannin found in oaks and cotton, than when both parents were raised on quercetin. In contrast, when only males were raised on standard quercetin-free diet, larvae weighed 55 times more than when both parents were raised on quercetin. These diets may have been unrealistic, as concentrations of compounds were about twice as high as was typical in host cotton leaves; however, it is difficult to ascertain what a wet weight equivalent in artificial diet would be for similar effective concentrations in fresh leaves (Gould personal communication). In addition, the diet treatments represented a qualitative difference in resources (either presence or absence of compounds), something that is probably not typical of intraspecific variation in plant quality. While this study likely does not reflect a reliable estimate of the magnitude of differences caused by intraspecific variation in plant quality, it clearly indicates that diet of both parents can influence offspring performance.

Several other studies of parental effects come a little closer to estimating how natural variation in plant quality affects herbivore performance. In the only study that addresses the role of *intra*specific variation in host-plant quality through parental effects on offspring, Rossiter (1991b) reared families of gypsy moths on conspecific trees that varied naturally in their levels of defoliation and also in levels of secondary compounds. Larvae collected as eggs from these families were then reared on a standard artificial diet. The greatest difference in relative pupal size (which is correlated with egg production in females) of siblings whose parents were raised on the highest versus lowest quality plants was 0.75 for daughters and 0.84 for sons, and the average relative difference in weight over all families was about 0.87 ( $\pm$ 0.06 *SD*) for sons and 0.83 ( $\pm$ 0.075 *SD*) for daughters (average values were calculated by these authors from data in Figures 6.1 and 6.2 in Rossiter 1991b). These differences are generally lower than relative fitness differences among lines in systems showing local adaptation (see Table 8.1).

Other studies that examine parental effects on larval performance consider consequences of feeding on different host plant *species* by parents. These studies have had mixed results on the size of such effects. Rossiter (1991a) finds that parental diets of different species of host trees have either no effect or strong effects on egg mass. Fox et al. (1994) show generally small effects of parental diet of different seeds types on larval weevil performance. In contrast, Futuyma et al. (1993) found strong effects of parental host-plant diet on larval performance. Those authors reared chrysomelid beetle parents and larvae in different combinations on two host plants. Feeding on one host-plant species (ragweed) decreased the percent of larvae surviving, regardless of whether the female or the male parent fed on that plant species. Relative fitness (compared to fitness of larvae when larvae and both parents were reared on the better plant) was, on average, 0.78, when either parent was raised on ragweed.

When parental and genetic effects have been compared in the same study, genetic effects have generally been found to be stronger than environmental ones. For example, Fox (1993) found that the largest proportion of variance in egg size due to maternal effects was 5.1%, much smaller than the 60–70% of variance explained by additive genetic effects.

In conclusion, the scant data suggest that parental effects could have a range of effects on offspring performance. The effects of parental diet on offspring performance can vary, and could cause one to conclude that there is no evidence for local adaptation, when local adaptation may indeed be occurring. In the only study that has addressed parental effects owing to intraspecific variation in host quality, the impact on herbivore fitness was somewhat smaller than the average difference in relative fitness between natal and novel insect lines. Finally, in the two studies of local adaptation that have attempted to control for parental effects, strong evidence for local adaptation was still present. Despite these results, it is clear that future studies of local adaptation by herbivores must include adequate controls for parental effects in the experimental design.

### 8.5 In Populations Where We Find Local Adaptation, is the "Natal" Insect Line Always the Best Performing Line?

While we may find that, on average, natal insects outperform novel insects on host plants, we should also be interested in whether novel lines exist that outperform natal insects. In four of the six cases in Table 8.1, at least one novel line had a greater mean relative fitness (>1) than the natal line (column 4). Errors around these mean values were unavailable, so we do not know how many times novel lines differed significantly in their performance over natal lines. (If the error for the performance of all novel lines combined is a reasonable estimate of the error for individual lines, then each of these values is significantly different from one.)

The result that four of six studies had novel lines superior to natal lines is somewhat surprising, considering that the sample size of insect lines–individual host examined is always small (3–12), usually because of the labor intensiveness of these experiments. Thus, even relatively poorly sampled insect subpopulations contain insects that outperform the natal line. If we assume that performance differences have a genetic basis, then we must consider what the high incidence of superior lines indicates about the mechanisms involved in local adaptation.

If there were fairly high migration rates among subpopulations, then we would expect that most host plants would have "sampled" the available insect genotypes, and that strong selection by the host plant would have favored the most fit lineages. Under this scenario, we expect a low incidence of novel insect lines (those with new mutations) that could outperform the natal line. Since we do not find this to be the case (based on Table 8.1), we explore what mechanisms might result in the prevalence of novel lines that are better adapted to using a host plant than the natal line. These mechanisms can be placed into two basic categories: those that invoke imperfect opportunities for matching insect and plant genotypes when host plant quality is the main selective pressure (e.g. low dispersal rates), and those that suggest that selective agents other than plant quality are also important for insect fitness; a related hypothesis is that this result is an artifact of how we measure fitness in these organisms.

When the primary selective agent is host-plant quality, then potential explanations for the superior performance of novel lines are that: (1) insects are very sedentary with extremely low dispersal rates, and/or (2) host–insect associations are fairly recent. Both of these mechanisms are based on the idea that there has not been enough time to sample all combinations of insect–host-plant genotypes; (3) female choice in oviposition sites limits the exposure of all insect genotypes to all plant types, and (4) alleles that are present, but selectively neutral in the natal environment, are expressed and confer high fitness in some novel environments.

If other agents of selection are more important to insect fitness than performance on host plants, then we can include the fact that: (5) differential predation on novel insect lines could overwhelm effects of larval performance. Finally, (6) adaptation to plant phenology or other plant attributes is more important than adaptation to leaf quality. Since experimenters tend to measure only adaptation to the latter, we may not be including important aspects of insect fitness in our assessments of relative fitness.

### 8.5.1 Low Dispersal Limits Sampling of Best Plant Phenotype–Insect Genotype Combinations.

In at least two systems showing local adaptation, low migration rates could impede sampling of the best insect genotypes for a given plant phenotype. In *Apterothrips apteris*, migration rates were documented to be 1.47 individuals/year/rosette (or about 0.20 individuals/generation/rosette) in a system in which host plants are known to exert strong effects on thrips fitness (Karban 1989; Karban and Strauss 1993; Table 8.1). Similarly, Alstad and Corbin (1990) found population differentiation in scale insects even at the level of twigs within individual trees, a consequence of the extremely low dispersal of crawler scales from their eclosion site. In such cases, it is possible that limited rates of colonization prevent exposure of plants to the best adapted insect lines. Such extremely low levels of migration are not likely to occur as a general phenomenon in insect–plant systems.

### 8.5.2 History of Individual Infestations

In most studies of local adaptation, information about the history or duration of the infestation on the plant is lacking. If infestations are relatively recent, then there may not have been sufficient time to sample all the available insect genotypes, and many novel lines could potentially outperform natal insects.

### 8.5.3 Oviposition Preferences of Females Limits Sampling of All Insect Genotype–Plant Genotype Combinations

Female preferences in oviposition sites may also prevent all insect genotypes from being exposed to all plant genotypes. For example, *Euphydryas editha* butterfly populations are polymorphic for females that do/do not discriminate among individual *Pedicularis* host plants (Ng 1988). In this case, plants preferred by discriminating females also supported better growth of their larval progeny. There are, however, many other documented cases in which female preference in oviposition site is not correlated with larval performance (for review, see Thompson and Pellmyr 1991; Via 1991). Discrepancies between oviposition preference and larval performance could result in superior performance of novel lines of insects (placed on plants by the experimenter), since past selection by host plants may have been acting on only a subset of the available insect genotypes (as a result of discrimination by ovipositing females).

### 8.5.4 Alleles at Loci That Are Effectively Neutral with Respect to Use of Some Host Individuals Could Confer High Fitness on Other Host Plants

Weis et al. (1992) proposed that the extent to which trade-offs exist in the ability to use different host-plant species (or in this case, host-plant individuals) depends on the degree to which the genetic basis of host-plant use is correlated among plant genotypes. Suppose that loci that enable efficient use of one host-plant individual are different from those allowing efficient use of other individuals. If alleles at loci affecting performance on one plant are neutral with respect to use of a second plant, then selection would operate independently on these loci. Under this scenario, insects could be preadapted to a novel plant as a result of neutral variation at a locus that was unimportant for use of the natal host plant. In order to assess the likelihood of this scenario, we need to know much more about the underlying genetic basis of host-plant use (see Section 8.3).

### 8.5.5 Selection Imposed by Natural Enemies Is as Important as Performance on Plants

Novel insect lines that perform better than natal lines may also be differentially susceptible to natural enemies. The role of enemies in local adaptation to host plants has received relatively little attention (but see Mopper et al. 1995; Stiling and Rossi, Chapter 2, and Mopper, Chapter 7, this volume), despite growing evidence that enemies respond strongly to chemical signals resulting from wounding of host plants, and may be sensitive to differences among individual plant phenotypes (Vet and Dicke 1992; Strauss 1997). While enemies attack natal lines of leaf miners more than novel lines (Mopper et al. 1995), Stiling and Rossi (this volume) have strong evidence that both parasitism and predation rates are higher on novel lines of gallers than on natal insects. In this case, local adaptation by galling midges to *Borrichia* plants may be regulated as much by enemies as by midge gall initiation or abortion on novel plants (see Stiling and Rossi this volume). This result is not surprising, especially in galler systems. First, gall size has been shown to be influenced by both plant and insect genotype (Weis and Abrahamson 1986). Second, gall size directly influences both fecundity of gallers and risk of predation—often in in opposite directions (Weis and Kapelinski 1994). In these systems, complex interactions between plant and insect genotypes, coupled with opposing selection pressures on gall size, may make galler performance on natal host plants a simplistic measure of local adaptation. If plant effects are mediated by other factors, there may be shifting successions of "best adapted" insect genotypes, the identities of which may vary with environmental conditions (e.g., high vs. low predator densities, or predators that are more or less effective with respect to attributes of gall size).

### 8.5.6 Other Plant Attributes, Such as Plant Phenology, Are Important to Local Adaptation

Finally, plant phenology or other plant traits involved in oviposition or larval success may also be important components of local adaptation by herbivores. These aspects of adaptation to host plants that affect herbivore fitness are not addressed in reciprocal transplant studies focusing strictly on larval growth and survivorship on host-plant tissue (the typical experiment). It is possible that larval performance on leaf tissue might not be completely correlated with these other adaptations that enable better plant use. In this case, the novel lines that outperform natal lines, based solely on ability to feed on host plant tissue, may not be better adapted to the host plant over all attributes.

In conclusion, closer inspection of the data supporting local adaptation by herbivorous insects to host plants suggests that identifying selection pressures leading to local adaptation is not as straightforward as might initally appear. The presence of many "preadapted" insects to novel host plants may indicate that the ability to feed and grow successfully on the natal host does not encompass all of the aspects of local adaptation by herbivores, however, dispersal or other behavioral factors could simply be limiting exposure of all insect genotypes to all plant genotypes. While some of the factors contributing to the existence of novel lines that outperform natal lines are difficult to test experimentally (e.g., the history of infestation, but see Cobb and Whitham, Chapter 3, this volume), several of these (the importance of predators, adaptation to other plant qualities, and oviposition preferences of females) can be addressed fairly easily. Elaboration on the basic reciprocal transfer experiment may help us understand better some of the underlying agents of selection in these systems.

## 8.6 How Does the Magnitude of the Coefficient of Selection Differ in Studies of Intraspecific Variation in Host Plants versus Interspecific Variation?

One interesting question is how the strength of selection differs between cases in which there is local adaptation to different plant species (i.e., the formation of host races vs. to individuals of the same host plant). One might expect that the coefficient of selection would be even greater against insects using novel host-plant *species* than against insects using novel plants that are within the same species. In Table 8.2, we summarize 12 studies that have examined the performance of lines of insects that are locally adapted to using different host-plant species. This is not a comprehensive review of such studies, but rather a sample of studies that have addressed this question, have demonstrated local adaptation with respect to insect performance, and that presented data in a way that we could easily analyze. We also did not include more than one study per insect species. It should be noted that there are many potentially confounding variables in such comparisons. For example, species exhibiting local adaptation to individual host plants have generally been monophagous insects. In contrast, by definition, a species with host races is a more polyphagous feeder. In addition, phylogenetic differences among groups could also contribute to differences in the estimates of the strength of selection, if some groups are more sensitive to variation in plant quality, as a rule.

In Table 8.2, natal lines are insects raised on the host-plant species that is naturally used by that population; novel lines are biotypes adapted to use a different host-plant species. Relative fitness is again considered with respect to the natal population, and we present the range of relative fitnesses as well as the mean. We

Table 8.2  Relative Fitness of Host Races (or Biotypes) on Other Host Species Compared to Natal Host Species

| Insect | Host Plants | Trait | Mean Relative Fitness and Range with Respect to Natal Race | # Insect Lines/Host | # Host Plant Spp. | Citation |
|---|---|---|---|---|---|---|
| Alsophila pomeraria | Oak<br>Maple | Dry wt. of prepupa | .74<br>1.47 | 1 | 2 | Futuyma et al. (1984) |
| Deloyala guttata | Ipomoea pandurata<br>Ipomoea purpurea | Fecundity | 1.52<br>(.86–3.55)<br>.49<br>(.25–.86) | 7 | 2 | Rausher (1984) |
| Acyrthosiphon pisum | Alfalfa<br>Clover | Longevity | .49<br>(.42–.55) | 4 | 2 | Via (1989) |
| Metopolophium dirhodum | Oat<br>Barley<br>Wheat | Population growth rate after 12 d. | .95<br>(.83–1.07) | 1 | 3 | Weber (1985b) |
| Sitobion avenue | Oat<br>Barley<br>Wheat | Population growth rate after 12 d. | .95<br>(.81–1.12) | 1 | 3 | Weber (1985a) |
| Stator limbatus | Acacia greggii<br>Cercidium microphyllum | % survivorship | .79<br>1.04 | 53 | 2 | Fox et al. (1994) |
| Stilbosis quadricustatella | Quercus myrtifolia<br>Q. geminata | % survivorship | .20<br>(0–1.00) | 5 | 2 | Mopper et al. (1995) |

Table 8.2 (continued)

| Insect | Host Plants | Trait | Mean Relative Fitness and Range with Respect to Natal Race | # Insect Lines/Host | # Host Plant Spp. | Citation |
|---|---|---|---|---|---|---|
| *Enchenopa binotata* | *Juglans nigra* *Ptelea trifoliata* *Robinia pseudoacacia* *Cercis canadensis* *Viburnum prunifolium* *Celastrus scandens* | % survivorship | .07 (0–.30) | 1 | 6 | Wood and Guttman (1983) |
| *Tribolium castaneum* | Wheat5, wheat, rice, corn, oat | Pupal wt. | .93 (.86–.98) | 2 | 5 | Via & Conner (1995) |
| *Spodoptera frugiperda* | Corn Rice | % survivorship | .92 (.54–1.3) | 1 | 2 | Pashley (1988) |
| *Leptinotarsa decemlineata* | *Solanum sarrachoides* *S. rostratum* *S. tuberosum* | % survivorship | .70 (.41–.98) | 1 | 3 | Horton et al. (1988) |

are interested in the average coefficient of selection in these studies and how that compares to the coefficient of selection from Table 8.1. Finally, we also calculated the coefficient of selection only for the subset of data in which the relative fitness of novel lines was less than the natal line (to get an estimate of the strength of selection when there was evidence for local adaptation in performance).

We found that, overall, the average relative fitness in cases of host-race formation was 0.77 ($\pm$ 0.38, *SD*); if we include only those subsets of host plants for which there was evidence for local adaptation (i.e., removing maple from Futuyma et al. 1993, and *I. pandurata* from Rausher 1984, etc.), we find an average relative fitness of novel lines to be 0.67 ($\pm$ 0.30, *SD; n* = 12). The corresponding coefficients of selection are 0.23 and 0.32, respectively. Counter to what we might have expected, these coefficients of selection are of the same magnitude as those documented for intraspecific differences in selective regimes among host-plant individuals.

Aside from the caveats discussed earlier, there are a number of issues that make these comparisons not wholly satisfying. For example, clearly there are many host species on which these biotypes have a fitness of zero. These are often not included in experiments, nor are they in the table. The point here was to ask whether there was any suggestion that interspecific variation in host-plant quality was much greater than intraspecific variation. To answer this question, the most appropriate comparison was to compare relative fitnesses of biotypes when there was evidence for local adaptation to the natal host-plant species. When we do this, in this small, but comparable sample to Table 8.1, we find no hint that this is the case. Mopper et al. (1995) as well as Wood and Guttman (1983) have documented average selective regimes against novel biotypes that were a lot more extreme than any documented in the within host-species comparisons among insect lines. So under some circumstances, biotypes do experience stronger selection from host-plant species than is generally present as a result of intraspecific variation. However, a comparable number of cases have shown that there is no evidence for fitness differences on different hosts, that is, no advantage to specialization by biotypes with respect to host-plant performance (e.g., Scriber et al. 1991a, 1991b; Fox and Caldwell 1994). This basic result was obtained in an earlier, much more extensive review of this topic (Futuyma and Peterson 1985). Again, these results point strongly to the possibility that in many cases, other factors aside from performance on host plants may be important in local adaptation by biotypes (see also discussion in Futuyma and Peterson 1985).

## 8.7 Conclusions

We have found that our estimate of the coefficient of selection against novel insect lines in cases where local adaptation has been documented is quite large, if it reflects primarily genetic differences among insect subpopulations. These values are consistent with the maintenance of genetically differentiated populations even in the face of gene flow from nonadapted populations under the island model. In these systems, it is clear that gene flow would need to be very extreme in order to override these selective pressures, unless (1) the single locus model is not a good descriptor of the underlying genetic basis of adaptive traits, and (2) epistasis and polygenic traits do greatly decrease the magnitude of $m$ required to homogenize populations, as proposed by Slatkin (1995). In general, we know so little about both the genetic basis of adaptive traits in these systems and the nature of potential epistatic effects that getting an estimate of $m$ relative to $s$ in which we can place much confidence is still in the somewhat distant future.

We do not know whether systems in which local adaptation occurs represent cases in which selection imposed by host plants on insects is extreme. In the best opportunity we had to test this idea, we found that the relative fitness of scale insects derived from different trees did not differ in the case where local adaptation to individual tree was found versus when it did not occur (Hanks and Denno 1994). We need to accrue more data on the relative fitness of insects from different sources in systems where no local adaptation occurs before we can really assess this point.

We find a number of cases in which novel insect lines are as good as, or better than, the natal line in using individual plants. This outcome suggests that either (1) not all insect genotypes are being exposed to all plants (as a result of low migration, of female oviposition preferences or of recent histories of infestations) and other factors such as selective predation by natural enemies are removing lineages that could fare better on these plants; or (2) adaptation is occurring to other plant traits, such as phenology, that are not being addressed in the ways in which we conduct our tests for local adaptation. This result points to our need to consider a multiplicity of mechanisms underlying local adaptation and to question the adequacy of only assessing larval performance on host-plant tissue as an indicator of local adaptation. Both within- and between-species comparisons have examples in which local adaptation may not be strictly measured based solely on larval performance (Thompson and Pellmyr 1991; Mopper, Chapter 7, and Stiling and Rossi, Chapter 2, this volume).

Contrary to our expectations, the strength of selection from host-plant species against races was of comparable magnitude to that against subpopulations that were locally adapted to individual plant phenotypes. As with within-host-plant species differences, phenological differences among species can also greatly limit the degree of gene exchange among races that must match these hosts in their own phenologies (Wood and Guttman 1983). Thus, it is still possible that, overall, the strength of selection exerted by host plants of different species on insect biotypes is greater than that exerted by intraspecific variation in host quality when all traits that are important to local adaptation are measured.

In summary, both with respect to locally adapted demes of insects associated with individual host plants and host races using different host-plant species, the strength of selection exerted by plants on herbivorous insects can be very strong.

Of interest would be a larger number of studies examining similar effects in herbivorous species that have no genetic structure associated with plant phenotypes. Such studies will enable us to determine whether life-history traits such as dispersal, or other factors such as natural enemies, are more important than selection exerted by host plants in determining the genetic structure found in natural insect populations.

## Acknowledgments

We thank Dieter Ebert, Susan Mopper, and Art Weis for their cogent comments, and M. Turelli and A. Weis for helpful discussion. This work was supported by National Science Foundation Grant No. HRD-91-03471 to S.Y. Strauss.

## 8.8 References

Alstad, D. N. and K. W. Corbin. 1990. Scale insect allozyme differentiation within and between host trees. *Evol. Ecol. 4*:43–56.

Alstad, D. N., S. C. Hotchkiss, and K. W. Corbin. 1991. Gene flow estimates implicate selection as a cause of scale insect population structure. *Evol. Ecol. 5*:88–92.

Berenbaum, M. R. and A. R. Zangerl. 1992. Genetics of behavioral and physiological resistance to host furanocoumarins in the parsnip webworm. *Evolution 46*:1373–1384.

Berenbaum, M. R. and A. R. Zangerl. 1993. Furanocoumarin metabolism in *Papilio polyxenes*: Biochemistry, genetic variability and ecological significance. *Oecologia 95*:370–375.

Bernays, E. and M. Graham.1988. On the evolution of host specificity in phytophagous arthropods. *Ecology 68*:886–892.

Boggs, C. L. and L. E. Gilbert. 1979. Male contribution to egg production in butterflies: Evidence for transfer of nutrients at mating. *Science 206*:83–84.

Cobb, N. S. and T. G. Whitham. 1993. Herbivore deme formation on individual trees: A test case. *Oecologia 94*:496–502.

Cohen, M. B., M. A. Schuler, and M. R. Berenbaum. 1992. A host-inducible ctychrome P450 from a host-specific caterpillar: Molecular cloning and evolution. *Proc. Natl. Acad. Sci. 89*:10920–10924.

Dicke, M., P. van Baarlen, R. Wessels, and H. Dijkman. 1993. Herbivory induced systemic production of plant volatiles that attract predators of the herbivore: Extraction of endogenous elicitor. *J. Chem. Ecol. 19*:581–599.

Edmunds, G. F. and D. N. Alstad. 1978. Coevolution in insect herbivores and conifers. *Science 199*:941–945.

Fox, C. W. 1993. Maternal and genetic influences on egg size and larval performance in a seed beetle (*Callosobruchus maculatus*): Multigenerational transmission of a maternal effect? *Heredity 73*:509–517.

Fox, C. W. and R. L. Caldwell. 1994. Host-associated fitness trade-offs do not limit the evolution of diet breadth in the small milkweed bug *Lygaeus kalmii* (Hemiptera: Lygaeidae). *Oecologia 97*:382–389.

Fox, C. W. and H. Dingle. 1994. Dietary mediation of maternal age effects on offspring perfromance in a seed beetle (Coleoptera: Bruchidae). *Funct. Ecol.* 8:600–606.

Fox, C. W., L. A. McLennan, and T. A. Mousseau. 1995. Male body size affects lifetime female reproductive success in a seed beetle. *Anim. Behav.* 50:281–284.

Fox, C. W., K. J. Waddell, and T. A. Mousseau. 1994. Host-associated fitness vaiation in a seed beetle (Coleoptera: Bruchidae): Evidence for local adaptation to a poor quality host. *Oecologia* 99:329–336.

Futuyma, D. J., R. P. Cort, and I.van Noordwijk. 1984. Adaptation to host plants in the fall cankerworm (*Alsophila pometaria*) and its bearing on the evolution of host affiliation in phytophagous insects. *Am. Nat.* 123:287–296.

Futuyma, D. J., C. Hermann, S. Milstein, and M. C. Keese. 1993. Apparent transgenerational effects of host plant in the leaf beetle *Ophraella notulata* (Coleoptera: Chrysomelidae). *Oecologia* 96:365–372.

Futuyma, D. J. and S. C. Peterson. 1985. Genetic variation in use of resources by insects. *Annu. Rev. Entomol.* 30:217–238.

Gould, F. 1988. Stress specificity of maternal effects in *Heliothis virescens* (Boddie) (Lepidoptera: Noctuidae) larvae. *Mem. Ent. Soc. Can.* 146:191–197.

Haldane, J. B. S. 1930. A mathematical theory of natural and artificial: Part IV. Isolation. *Proc. Cambridge Phil. Soc.* 26:220–230.

Hanks, L. M. and R. F. Denno. 1994. Local adaptation in the armored scale insect *Pseudaulacaspis pentagona* (Homoptera: Diaspididae). *Ecology* 75:2301–2310.

Horton, D. R., J. L. Capinera, and P. L. Chapman. 1988. Local differences in host use by two populations of the Colorado potato beetle. *Ecology* 69:823–831.

Huettel, M. D. and G. L. Bush. 1972. The genetics of host selection and its bearing on sympatric speciation in *Procecidochares* (Diptera: Tephritidae). *Entomol. Exp. Appl.* 15:465.

Jaenike, J. 1989. Genetics of oviposition site preference in *Drosophila tripunctata*. *Heredity* 59:363–369.

Karban, R.1989. Fine-scale adaptation of herbivorous thrips to individual host plants. *Nature* 340:60–61.

Karban, R. and S. Y. Strauss. 1993. Colonization of new host plant individuals by locally adapted thrips. *Ecography* 17:82–87.

Komatsu, T. and S. Akimoto. 1995. Genetic differentiation as a result of adaptation to the phenologies of individual host trees in the galling aphid *Kaltenbachiella japonica*. *Ecol. Entomol.* 20:33–42.

Lindroth, R. L. and A. V. Weisbrod. 1991. Genetic variation in response of the gypsy moth to aspen phenolic glycosides. *Biochem. Syst. Ecol.* 19:97–103.

Linhart, Y. B., J. B. Mitton, K. B. Sturgeon, and M. L. Davis. 1981. Genetic variation in a population of ponderosa pine. *Heredity* 46:407–426.

Mopper, S., M. Beck, D. Simberloff, and P. Stiling. 1995. Local adaptation and agents of selection in a mobile insect. *Evolution* 49:810–815.

Nagylaki, T. 1978. The geographical structure of populations. Pp. 588–624 *in* S. A. Levin (Ed.), *Studies in Mathematics*, 16th ed. (Pt. II). Mathematical Association of America Washington, DC.

Ng, D. 1988. A novel level of interactions in plant–insect systems. *Nature 334*:61–62.

Nitao, J. K. 1989. Enzymatic adaptation in a specialist herbivore for feeding on furanocoumarin-containing plants. *Ecology 70*:629–635.

Pashley, D. P. 1988. Quantitative gentics, development and physiological adaptation in host strains of the fall armyworm. *Evolution 42*:93–102.

Pashley, D. P., T. N. Hardy, and A. M. Hammond. 1995. Host effects on developmental and reproductive traits in fall armyworm strains (Lepidoptera: Noctuidae). *Ann. Entomol. Soc. Am. 88*:748–755.

Rausher, M. D. 1984. Trade-offs in performance on different hosts: Evidence from within- and between-site variation in the beetle *Deloyola guttata*. *Evolution 38*:582–595.

Rice, W. R. 1983. Sexual reproduction: an adaptation reducing parent-offspring contagion. *Evolution 37*:1317–1320.

Rossiter, M. C. 1991a. Maternal effects generate variation in life history: Consequences of egg weight plasticity in the gypsy moth. *Funct. Ecol. 5*:386–393.

Rossiter, M. C. 1991b. Environmentally based maternal effects: A hidden force in insect population dynamics. *Oecologia 87*:288–294.

Rossiter, M. C., D. L. Cox-Foster, and M. A. Briggs. 1993. Initiation of maternal effects in *Lymantria dispar*: Genetic and ecological components of egg provisioning. *J. Evol. Biol. 6*:577–589.

Schaal, B. A. 1975. Population structure and local differentiation in *Liatris cylindrica*. *Am. Nat. 109*:511–528.

Schnabel, A. and J. L. Hamrick. 1990. Comparative analysis of population genetic structure in *Quercus macrocarpa* and *Q. gambelii* (Fagaceae). *Syst. Bot. 15*:240–251.

Scriber, J. M., R. L. Lindroth, and J. K. Nitao. 1991a. Toxic phenolic glycosides from *Populus*: Physiological adaptations of the western North American tiger swallowtail butterfly, *Papilio rutulus* (Lepidoptera: Papilionidae). *Great Lakes Entom. 24*:173–180.

Scriber, J. M., J. Potter, and K. Johnson. 1991b. Lack of physiological improvement in performance of *Callosamia promethea* larvae on local host plant favorites. *Oecologia 86*:232–235.

Sierra, J. R., W. D. Weggen, and H. Schmid. 1976. Transfer of cantharidin during copulation from adult male to the female *Lytta vesicatoria* (Spanish flies). *Experientia 32*:142–144.

Slatkin, M. 1985. Gene flow in natural populations. *Annu. Rev. Ecol. Syst. 16*:393–430.

Slatkin, M. 1995. Epistatic selection opposed by immigration in multiple locus genetic systems. *J. Evol. Biol. 8*:623–633.

Smith, C. M. 1989. *Plant Resistance to Insects: A Fundamental Approach.* John Wiley, New York.

Strauss, S. Y. 1997. Lack of evidence for local adaptation to individual plant clones or site by a mobile specialist herbivore. *Oecologia, 110*:77–85.

Strauss, S. Y. and R. Karban. 1994. The significance of outcrossing in an intimate plant/herbivore relationship: I. Does outcrossing provide an escape for progeny from herbivores adapted to the parental plant? *Evolution 48*:454–464.

Thompson, J. N. 1988. Evolutionary genetics of oviposition preference in swallowtail butterflies. *Evolution 42*:1223–1234.

Thompson, J. N. and O. Pellmyr. 1991. Evolution of oviposition behavior and host preference in Lepidoptera. *Ann. Rev. Ent. 36*:65–89.

Unruh, T. R. and R. F. Luck. 1987. Deme formation in scale insects: A test with the pinyon needle scale and a review of other evidence. *Ecol. Entomol. 12*:439–449.

Vet, L. E. M. and M. Dicke, 1992. Ecology of infochemical use by natural enemies in a tritrophic ontext. *Ann. Rev. Ent. 37*:141–172.

Via, S. 1989. Field estimation of variation in host plant use between local populations of the pea aphids from two crops. *Ecol. Entomol. 14*:357–364.

Via, S. 1990. Ecological genetics and host adaptation in herbivorous species: The experimental study of evolution in natural and agricultural systems. *Ann. Rev. Entomol. 35*:421–446.

Via, S. and J. Conner. 1995. Evolution in heterogeneous environments: Genetic variablity within and across different grains in *Tribolium castaneum. Heredity 74*:80–90.

Wainhouse, D. and R. S. Howell. 1983. Intraspecific variation in beech scale populations and in susceptiblity of their host *Fagus sylvaticus. Ecol. Entomol. 8*:351–359.

Weber, G. 1985a. On the ecological genetics of *Sitobion avenae* (F.) (Hemiptera: Aphididae). *Zang. Ent. 100*:100–110.

Weber, G. 1985b. On the ecological genetics of *Metopolophium dirhodum* (Walker) (Hemiptera: Aphididae). *Zang. Ent. 100*:451–458.

Weis, A. E. and W. G. Abrahamson. 1986. Evolution of host–plant manipulation by gallmakers: Ecological and genetic factors in the *Solidago-Eurosta* system. *Am. Nat. 127*:681–695.

Weis, A. E. and W. L. Gorman. 1990. Measuring selection on reaction norms: An exploration of the *Eurosta-Solidago* system. *Evolution 44*:820–831.

Weis, A. E. and A. Kapelinksi. 1994. Variable selection on *Eurosta*'s gall size: II. A path analysis of the ecological factors behind selection. *Evolution 48*:734–745.

Weis, A. E., W. G. Abrahamson, and M. C. Andersen. 1992. Variable selection on *Eurosta*'s gall size: I. The extent and nature of variation in phenotypic selection. *Evolution 46*:1674–1697.

Wood, T. K. and S. I. Guttman. 1983. *Enchenopa binotata* complex: Sympatric speciation? *Science 220*:310–312.

# PART III
# Life History, Behavior, and Genetic Structure

# 9

# Intrademic Genetic Structure and Natural Selection in Insects

*David E. McCauley and Peter W. Goff*
Department of Biology, Vanderbilt University, Nashville, TN

## 9.1 Introduction

*Genetic structure* can be defined as the manner in which genetic variation is distributed within and among individuals grouped at hierarchical spatial scales. Classically, genetic structure has been regarded as an interdemic phenomenon (Wright 1931). In that view, species are subdivided into some number of more or less randomly mating subunits, or demes, that are connected to one another by some pattern of gene flow. An individual is characterized by its genotype, a deme, by its allele frequency. Genetic structure can then be quantified as some function of the variance among demes in allele frequency. Defined in this manner, genetic structure arises owing to the diversifying effects of genetic drift and founding events as they operate independently in the various demes, or from spatial variation in selection pressures. It is limited by the gene flow that occurs when individuals migrate among populations. By definition, interdemic structure persists across generations, though it can be modified.

Over the past 25 years, there have been numerous studies that have quantified the interdemic genetic structure found in a great diversity of taxa (Wright 1978; Slatkin 1985; Avise 1994), including many insect species (McCauley and Eanes 1987; Roderick 1996; Peterson and Denno, Chapter 12, this volume). Until the mid-1980s, electrophoretic protein variants (allozymes) were the primary genetic markers; recently, DNA markers have been used in increasing proportion (Avise 1994). Studies of genetic structure seem to be motivated by two primary objectives. The first is to use the distribution of genetic variation as an indirect measure of gene flow (Slatkin 1985) or population history (Avise 1994). The second is to evaluate the potential response to selection. Clearly, the response to selection depends on the magnitude and distribution of available genetic variation within and among demes. Since interdemic structure is controlled in large part by levels of gene flow, its magnitude might be expected to be relatively low in many insect species, especially flying insects with high dispersal power. Indeed, electrophoretic studies of insects often show low levels of interdemic structure, especially when compared to less vagile taxa, such as terrestrial snails or salamanders, though this trend is by no means universal (Wright 1978; McCauley and Eanes 1987; Roderick 1996; Peterson and Denno, Chapter 12, this volume).

Though genetic structure is usually considered an interdemic phenomenon, it can also be found within breeding groups, at least during part of the life cycle. The nonrandom spatial association of individuals within populations has been termed *intrademic* (Wade 1978) or *trait-group* (Wilson 1979) structure. This form of population structure is ephemeral in that groups form within demes for only part of the life cycle, disband before reproduction, and are formed anew the next generation. Groups can consist of random collections of individuals or collections of genetically similar individuals, as when relatives live in close proximity to one another. Intrademic genetic structure can be quantified as allele frequency variance among groups within demes. In this chapter we will address the hypothesis that in insects, intrademic genetic structure is often of greater magnitude than interdemic genetic structure and thus more likely to influence selection. We will first review the theoretical literature needed to compare the two forms of structure. We then consider the various life-history attributes that lead to intrademic structure, including their potential adaptive significance. Next, we illustrate how intrademic structure might influence various frequency-dependent selective processes. Finally, we conclude by presenting some thoughts on studying the significance of intrademic structure in natural systems.

## 9.2 Intrademic versus Interdemic Structure: Patches to Populations

### 9.2.1 Statistical Measures of Genetic Structure

Interdemic genetic structure is often quantified using Wright's *F*-statistics (Wright 1931). In a simple system, the genetic structure for some locus is represented by three parameters, $F_{is}$, $F_{st}$, and $F_{it}$ (see Table 9.1 for a summary of terms used in this chapter). They represent deviations from the distribution of selectively neutral genetic variance under complete panmixia. $F_{st}$ represents the among-population component of genetic variance (subpopulations to total; subpopulations = demes). If each subpopulation, or deme, has an allele frequency $p_i$, then one formulation of $F_{st}$ is

$$F_{st} = Vp_i/p \, (1 - p) \tag{1}$$

where $Vp_i$ is the variance in allele frequency among subpopulations, and $p$ is the mean allele frequency averaged across subpopulations. $F_{is}$ represents the effects of nonrandom mating within subpopulations on the distribution of genetic variance among individuals within subpopulations. It is usually used as a measure of inbreeding. $F_{it}$ is a measure of the total deviation from panmixia (individuals to total) such that

$$(1 - F_{it}) = (1 - F_{is})(1 - F_{st}) \tag{2}$$

*Table 9.1*   Definitions of terms and symbols used in this chapter.

| | |
|---|---|
| Deme | a collection of interbreeding individuals, a local subpopulation |
| F Statistics | statistics developed by Sewall Wright to quantify the degree to which alleles are distributed nonrandomly among and within population subunits (see below) |
| $F_{ip}$ | relates to the distribution of alleles among individuals within patches |
| $F_{is}$ | relates to the distribution of alleles among individuals within subpopulations or demes |
| $F_{it}$ | relates to the distribution of alleles among individuals within the total area studied |
| $F_{ps}$ | relates to the distribution of alleles among patches within subpopulations or demes |
| $F_{pt}$ | relates to the distribution of alleles among patches within the total area studied |
| $F_{st}$ | relates to the distribution of alleles among subpopulations or demes within the total area studied |
| Nm | the number of migrants moving between populations |
| FCPB | false Colorado potato beetle, *Leptinotarsa juncta* |
| Intrademic Genetic Structure | allele frequency variance among patches within demes as measured by $F_{ps}$ |
| Interdemic Genetic Structure | allele frequency variance among demes as measured by $F_{st}$ |
| IWLB | imported willow leaf beetle, *Plagiodera versicolora* |
| Metapopulation | collection of demes subject to frequent local extinctions but replaced by frequent colonization, a population of populations |
| Patch | a physical subdivision of a deme due to a fragmented distribution of favorable habitat or resources |
| PSB | the average frequency of phenotype B from the perspective of phenotype B, its subjective frequency |
| r | coefficient of relatedness, the probability that alleles drawn from two individuals will be identical by descent owing to a common ancestor |
| $\bar{r}$ | the average coefficient of relatedness considering all possible pairs of individuals within each of several groups |

The estimation of $F$-statistics from natural populations requires various modifications of these equations to account for the sampling variance that arises when a subset of all individuals are sampled from each of a subset of all populations, and for loci with more than two alleles (Weir 1990).

$F_{st}$ can range from zero (no interdemic structure) to one (fixed allelic differences among demes). It is often estimated from allozyme data, with information from several loci combined into a summary statistic (Weir 1990). Typically, estimates are made from several loci and the estimated $F_{st}$ is compared against the null hypothesis that $F_{st} = 0.0$ (i.e., no population structure). Tests of the null hypothesis can be conducted on a locus-by-locus basis using such methods as the $G$-test of independence (Sokal and Rohlf 1981) or by including information from all loci simultaneously by using each locus as a replicate and computing a standard error around the average using bootstrapping or jackknifing methods (Weir 1990).

Allozyme-based estimates of $F_{st}$ have been conducted for a variety of insect species. One problem with comparing estimates of $F_{st}$ obtained for different species by different authors is that studies vary widely in the spatial scale at which the sampling was conducted. Despite this caveat, it is generally the case that $F_{st}$ estimates in insects fall in the range 0.01–0.20, even when samples are taken from populations distributed over the major portion of the species' range (McCauley and Eanes 1987; Roderick 1996; Peterson and Denno, Chapter 12, this volume). Thus, while most insect species show statistically significant interdemic genetic structure, it is often of a relatively small magnitude (i.e., $F_{st} < 0.10$).

One could define population subunits within demes as well, especially if critical resources were patchy in distribution. For example, one could subdivide a population according to the distribution of an herbivorous insect species among individuals of its host plant, the distribution of a carrion-feeding insect among carcasses, or the distribution of an aquatic insect among ephemeral pools. In insects, the distribution of genetic variation within and among habitat patches often depends on the pattern of oviposition. Consider a continuum of oviposition strategies. At one extreme, each patch is visited by a different female, who deposits all of her eggs in that patch. Different patches are occupied by different full sib groups. At the other extreme, each female visits numerous patches, depositing one egg per patch. Because patches are visited by numerous females, each individual living in a given patch has a different set of parents. In the first case, among-patch variance in allele frequency will be relatively high, owing to the kin relationships of the individuals in each patch. In the second case, variance in allele frequency will represent the binomial sampling of the population at large that occurs when patches are occupied by a finite number of randomly selected individuals. Allele frequency variance will be minimal unless the number of eggs per patch is small. Intermediate cases depend on the number of females contributing eggs to a patch and the degree to which females are multiply inseminated.

In species in which immatures are much less mobile than adults, an individual may be restricted to the same habitat patch for much of its preadult life, interacting with the small subset of the local population that is restricted to that patch. Upon maturing, however, adults could disperse over a much wider area, encountering potential mates originating from numerous habitat patches. Thus, the population structure defined by interactions among immatures is much more fine scaled than the population structure defined by breeding interactions. It is the spatial distribution of the breeding pool that defines interdemic structure; any population structure that results from the temporary patchy distribution of individuals within demes would be considered intrademic (Fig. 9.1; see also Costa, Chapter 10, this volume). In many insect species, intrademic structure should be greater than interdemic structure, because winged adults are more mobile during their search for mates and oviposition sites than are nonwinged immatures searching for food.

If individuals were sampled from numerous patches within a deme, one could define a term $F_{ps}$ based on the allele frequency variation among patches within subpopulations and $F_{ip}$ based on the distribution of genetic variation among individuals within patches. Furthermore, if the sampling were done in a hierarchical fashion, with several patches sampled from within each of several demes, then the total variance among patches, $F_{pt}$, could be partitioned into intrademic and interdemic components such that

$$(1 - F_{pt}) = (1 - F_{ps})(1 - F_{st}) \tag{3}$$

Weir (1990) and Queller and Goodnight (1989) provide methods for hierarchical population structure analysis.

The genetic consequences of kin associations are often expressed as $r$, the coefficient of relatedness (Hamilton 1964a, 1964b). This parameter can be viewed as the probability that two individuals share alleles that are identical by descent, owing to a common ancestor. It is well known that in noninbreeding diploid organisms, full-sibs display $r = 0.5$, half-sibs $r = 0.25$, and unrelated individuals $r = 0.0$ Wade (1982) and others describe the average relatedness, $\bar{r}$, associated with more complex mixtures of individuals. For example, when a female uses the sperm of two males to fertilize her eggs in equal proportion, her offspring have an $\bar{r}$ equal to 0.375, the average of full- and half-sibs (i.e., the average of 0.5 and 0.25). A set of immatures attributable in equal proportion to two nonrelated mothers, each mated to a different male, would have $\bar{r} = 0.25$ (i.e., the average of the within-family relatedness, 0.5, and the between-family relatedness, 0.0). Within populations, the relationship between $F_{ps}$ and average relatedness ($\bar{r}$) is rather simple (Michod 1980).

$$\bar{r} = 2F_{ps}/(1 + F_{is}) \tag{4}$$

For example, with random mating ($F_{is} = 0$), and patches occupied by full-sibs ($\bar{r} = 0.50$), $F_{ps} = 0.25$. Thus, in insects, fairly common life-history phenomena can result in intrademic genetic structure of greater magnitude than is typically seen interdemically, even when interdemic measures are taken across the species' range.

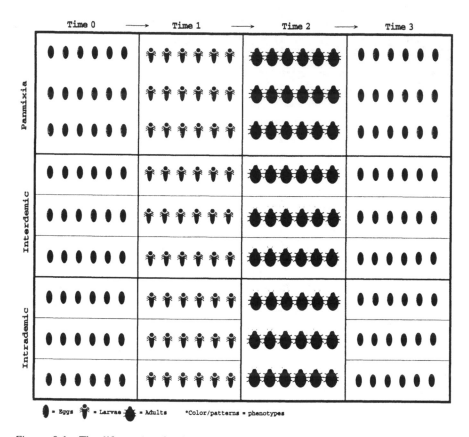

*Figure 9.1* The life cycle of a hypothetical insect under three population structures. Under panmixia there is a random association of phenotypes (colors) at all life stages. There is no egg clustering, and both immatures and adults move freely. Under interdemic structure, segregation of phenotypes by habitat patch occurs at all life stages and extends across generations owing to the lack of gene flow. Under intrademic structure, phenotypes are segregated by patch in the immatures, but adults disperse and interbreed. Patches are colonized anew, and structure is reestablished in the next generation as a consequence of egg clustering.

(There may appear to be a paradox if $F_{ps} > 0$ and $F_{is} = 0$. However, when

$$(1 - F_{is}) = (1 - F_{ps})(1 - F_{ip}) \tag{5}$$

$F_{ip}$ is often negative. Within families, alleles are not distributed into genotypes at random but according to the laws of Mendel. For example, with Mendelian crosses, some full-sib groups will consist of all heterozygotes and, averaged over all families, $F_{ip} = -0.33$.)

As stated earlier in this chapter, one motivation for studying genetic structure is to estimate levels of gene flow. Under a set of simplifying assumptions such as Wright's island model, the interdemic form of $F_{st}$ is a function of $Nm$, the number of migrants moving between populations (Slatkin 1985; Roderick 1996). Most importantly, it is assumed (1) that populations persist long enough for the forces of genetic drift and gene flow to come to equilibrium, and (2) that migrants are free to move among all subpopulations, not just near neighbors. While these assumptions are rarely met in nature, there does seem to be a good correspondence across insect taxa between estimates of $F_{st}$ and apparent dispersal ability (Peterson and Denno, Chapter 12, this volume). For example, $F_{st}$ values estimated for a variety of agricultural pest insects are typically less than 0.10 (McCauley and Eanes 1987). This might be expected, in that one life-history characteristic that preadapts insects to be agricultural pests is strong dispersal ability.

The use of $F_{ps}$ to estimate $Nm$ makes little sense, since the reality of intrademic structure is far removed from the assumptions of the island model. Intrademic genetic structure and $F_{ps}$ could be related to demography in another way, however. In its simplest form, intrademic structure is similar to metapopulation structure, with the founding and extinction of demes occurring every generation (McCauley 1993). Ovipositing females and their mates could be considered founders. The $F_{ps}$ of immatures distributed among habitat patches in an intrademic structure is equivalent to the $F_{st}$ among a set of newly founded populations in a metapopulation structure. The "migrant pool" form of colony formation in metapopulation models (Slatkin 1977; Wade and McCauley 1988) is equivalent to random mating among the parents whose eggs are distributed into habitat patches. Namely,

$$F_{st} = \tfrac{1}{2}k \text{ or } F_{ps} = \tfrac{1}{2}k \qquad (6)$$

where $k$ is the number of founders, assuming each founder contributes an equal number of offspring to a patch. When patches are occupied by full-sib groups, each patch is founded by two individuals (the male and female parents) and both equations 4 and 6 predict $F_{ps} = 0.25$. Thus, $F_{ps}$ could be used to estimate the effective number of parents represented in a group of immatures. Pamilo (1983) and Costa and Ross (1993) have each used a form of this approach to estimate the average effective number of genomes within social groups. Since different modes of founding can generate the same genetic structure (Wade and McCauley 1988; Whitlock and McCauley 1990), this would be most useful in the context of a good understanding of the natural history of the system.

## 9.2.2 The Natural History of Group Formation

In this section, we will discuss the types of natural-history traits that lead to intrademic structure at some point in the insect life cycle. In particular, we will examine life-history traits that are likely to predispose insects to aggregation at the egg, and immature and adult life-history stages. Whenever possible, real-world examples will be framed with reference to the theory of intrademic structure. However, this section also refers to examples of species whose life history does not conform exactly to intrademic structure in its simplest form, but rather to a structure intermediate to intra- and interdemic. One obvious omission of this section will be discussion of the eusocial insects, which are well-known examples of insects with both group-living and dispersal life stages. The eusocial literature is extensive and well known.

The most obvious life-history trait associated with intrademic structure is a behavior known as *egg clustering* (Stamp 1980; Courtney 1984; Parker and Courtney 1984). In insects, the success and/or behavior of a mother's offspring can be greatly influenced by her oviposition behavior (Davis and Pedigo 1989; Mopper and Simberloff 1995). In many insect species, a female will oviposit a number of eggs in close proximity to one another, usually on or near the food resource (e.g., a host plant) required by her offspring. This is well illustrated by the oviposition behavior seen in many chrysomelid beetles, for example the imported willow leaf beetle (IWLB), *Plagiodera versicolora* (Wade 1994) and the false Colorado potato beetle (FCPB), *Leptinotarsa juncta* (McCauley 1992). Females of both these species cluster up to 20 eggs on single leaves of their respective host plants (Fig. 9.2). Clustering of eggs on the host plant has been recorded in several orders of insects, including Lepidoptera (Ashiru 1988; Derrick and Showers 1990), Coleoptera (see above), Hemiptera, and Hymenoptera. As discussed earlier, the consequences of egg clustering for genetic structure depend on the number of females ovipositing in a resource patch and whether clutches display multiple paternity.

Egg clustering is thought to have evolved owing to what we will call "fecundity-enhancing" or "survivorship-enhancing" factors (Parker and Courtney 1984). Fecundity effects include environmental factors such as spatial and/or temporal heterogeneity of resources and predation rates that influence the probability that a female will locate a succession of oviposition sites. If favorable sites are rare and patchily distributed, the female may be under strong selective pressure to cluster her eggs on the first resource patch found, given that she may die before finding another oviposition site. Fecundity-enhancing egg clustering is thought to be advantageous to the female in terms of net number of offspring, even though her offspring may incur fitness costs due to the negative effects of crowding (increased apparency to predators, competition, etc.).

Clustering that enhances offspring survivorship can occur even when the female has a high probability of encountering additional resource patches. In this case, the clustering behavior maximizes offspring survivorship, because there is some advantage to her offspring living in proximity to one another, such as group predator defense. In some cases, the mother will remain with, and defend, her offspring (Tallamy 1984; Tallamy and Wood 1986).

Once the eggs hatch, intrademic structure may or may not persist, depending on the behavior of the immatures. One of the most obvious reasons for immatures to

*Figure 9.2*   Eggs of the false Colorado potato beetle, *Leptinotarsa juncta*, clustered on the underside of a leaf of the host plant, *Solanum carolinense*. Members of this egg clutch are full- or half-siblings and will remain together on this host plant for several instars after hatching (Photo by P. Goff)

remain in proximity to one another is inertia. By *inertia* we mean that clustering at the egg-stage leads to an association among immatures, simply because it will take them some time to disperse. If immatures remain relatively immobile/sedentary, inertia could lead to *passive* aggregations (i.e., collections of individuals restricted to the same resource patch, but not necessarily seeking one another out).

In some cases, intrademic structure is maintained as a consequence of active aggregation behavior by immatures (Eickwort 1981; Costa and Pierce 1996). Members of such aggregations can display behaviors such as trail following, coordinated feeding and defenses, and active recruitment of dispersed individuals to an existing aggregation. This is seen in IWLB, in which clutch mates are not only restricted to the same willow tree for their larval life, but also actively aggregate onto the same willow leaf (Wade 1994). In many cases of larval aggregation there is interinstar variation in the amount of aggregation (active or passive), and in many cases, aggregation at early instars precedes dispersal, though it is also common for insects to be associated with a single resource patch (plant, carcass, pool, dung pile, etc.) for the entire larval/immature stage, whether by active or passive means.

Insects can also display a clumped distribution at the adult stage. Patchy resources and/or low mobility can facilitate adult aggregation behavior (Steiner 1990). Adult aggregations are especially well known among some homopterans (Naranjo and Flint 1995; Tonhasca et al. 1994; Tandon and Veeresh 1989). However, even among relatively mobile taxa, adults do sometimes form aggregations. As with the previous life-history stages, the aggregations can have spatial and/or temporal components. One of the better known causes of adult aggregations is mating. The primary cues for such aggregations are pheromones (Singh 1993).

Adult aggregation is only likely to result in significant intrademic genetic structure if individuals have been associated since birth, because a nonrandom association of genotypes is most likely when relatives live in proximity to one another. Such aggregations are not true intrademic structure unless the adults disperse prior to reproducing, since genetic continuity across generations on the same resource patch would be considered interdemic structure. One well-known life history that leads to a population structure intermediate to strictly intra- or interdemic is seen in fig wasps (Herre 1985). Here, one or more females oviposit into a fig; the larvae mature within the fruit and could be considered a group. Unlike the cases discussed earlier, mating occurs before dispersal. However, because oviposition occurs after females disperse, there is no genetic continuity within resource units, so the individuals occupying a fig would not be considered a deme. Thus, the patchy nature of resources, combined with egg clustering and/or aggregation behavior, can result in kin associations and distinct nonrandom associations of genotypes at very local spatial scales.

### 9.2.3 Hierarchical Analysis of Natural Populations of Insects

Measures of $\bar{r}$, $F_{ps}$, or both, have been taken intrademically for a variety of insect taxa (Avise 1994). Often this is done in order to evaluate the impact of breeding system on relatedness in eusocial, social, or subsocial insects. When this is combined with a larger study of geographic variation in a hierarchical analysis, $F_{ps}$ defined among groups within demes is usually considerably greater than $F_{st}$ seen among demes. For example, in IWLB, females cluster eggs on the leaves of the host plant, and larvae remain aggregated in feeding groups for the first several instars (Wade 1994). McCauley et al. (1988) collected numerous groups of larvae from each of several geographical locations. A locality consisted of a grove of willow trees. Allozyme genotypes were taken at three loci, and allele frequencies were portioned among individuals within groups, among groups with localities, and among localities. $F_{ps}$, as defined among larval groups within localities, varied from 0.12 to 0.30 ( then ranges from 0.2 to 0.6), depending on the locality and the time of year at which collections were made. Relatedness values of less than 0.5 were due to frequent multiple paternity within egg clutches. In contrast, $F_{st}$ defined among localities separated by up to 1,000 km was 0.06.

McCauley et al. (1988) also analyzed population structure based on variation among samples collected from individual willow trees within some localities. In two of eight cases, significant among-tree, within-locality population structure was evident. $F_{ps}$ defined among trees within one locality was as high as 0.098. Similarly, Rank (1992) studied the adult population structure of another willow-feeding chrysomelid, *Chrysomela aeneicollis*, in California. He also found significant variation in allozyme allele frequency among samples taken from different trees within willow groves. In several cases, the magnitude of $F_{ps}$ defined among trees within localities was considerably greater than that seen among localities separated by several kilometers. In both species of willow beetles, the among-tree variance was probably due to sampling the offspring of a limited number of females that had oviposited on each tree at the beginning of the season and should be considered intrademic.

Among-tree, within-locality allele frequency variation has also been demonstrated in the apple maggot, *Rhagoletis pomonella* (Diptera: Tephritidae). McPheron et al. (1988) sampled flies from hawthorn trees spaced less than 200 m from one another in a park in Illinois. Small, but significant, allele frequency variation among trees was demonstrated for several allozyme loci ($F_{ps} = 0.01$), despite evidence that adult flies often leave their natal tree. Because of the dispersal potential of adults, the among-tree genetic structure could be considered intrademic. While small, allele frequency variance among trees within this locality was nearly equal to that seen among collections taken from hawthorn trees separated by up to 1,000 km in the eastern United States ($F_{st} = 0.03$; McPheron 1990).

The larvae of the eastern tent caterpillar, *Malacosoma americanum*, (Lepidoptera: Lasiocampidae) aggregate in "tents" constructed in the branches of their host tree, often wild cherry. Costa and Ross (1993, 1994, see also Costa, Chapter 10, this volume) conducted a hierarchical study of population structure and relatedness in *M. americanum*. Average relatedness defined at the level of individual tents ranged from 0.38 to 0.50, depending on the point in the life cycle relatedness is measured (Costa and Ross 1993). Thus, $F_{ps}$ defined by variance in allele frequency among tents within clusters of host tree must be on the order of 0.20–0.25. In contrast, $F_{st}$ defined by allele frequency variance among clusters of trees distributed from Georgia to New York was approximately 0.05 (Costa and Ross 1994).

Hierarchical population structure has been particularly well studied in the eusocial insects. For example, Pamilo (1983 and references therein) studied the population genetic structure of four species of ants in the genus *Formica* (Hymenoptera: Formicidae) in Finland. These eusocial species inhabit nests that are produced by one or more reproductive females. Nearby nests may or may not exchange workers with each other (forming colonies) and are distributed locally into demes. Demes are separated by intervening unfavorable habitat. In two of the species studied by Pamilo (1983), demes were distributed among islands, and in the other two they were distributed more continuously on the mainland. Average relatedness at the nest level ranged from a high of 0.75 in one species (expected in a haplodiploid system when a nest is founded by one, singly inseminated, reproductive female) to 0.33 in a species in which nests with multiple reproductive females are common. Among demes separated by hundreds of meters to a few kilometers, estimates of $F_{st}$ ranged from 0.04 to 0.09, depending on the species. Again, intrademic structure at the level of nests is much more pronounced than interdemic structure.

Finally, in a study of the population structure of the eusocial termite *Reticulitermes flavipes* (Isoptera: Rhinotermidae), Reilly (1986, 1987) partitioned allozyme variation among colonies separated by as little as a few meters within localities, and among localities, defined at various spatial scales. Colony structure could be considered intermediate between intra- and interdemic, in that some mating occurs within colonies, but colonies are ephemeral. Variation among localities could be considered interdemic. $F_{ps}$ among colonies within one state park in Tennessee was 0.48, whereas $F_{st}$ among localities distributed over the southeastern United States was 0.12.

In summary, while the number of studies in which intra- and interdemic structure were both measured is limited, there are several well-documented cases in which the magnitude of intrademic structure measured over a small area is of the same order or larger than measures of interdemic structure based on collections taken over much larger areas.

## 9.3 Genetic Structure and Selection

### 9.3.1 Theory: Models of Population-Structure-Dependent Selection

The impact of genetic structure on the response to selection can be viewed from several perspectives. Since the response to natural selection is generally proportional to the amount of genetic variation present (Fisher 1958), a large amount of interdemic structure can limit the response to selection within populations, since, by definition, most genetic variation is distributed among, rather than within, populations. However, other classes of selection can be facilitated by genetic structure. A variety of population-structure-dependent models of selection have been developed (Wilson 1983). Included in these are intra- and interdemic group selection (Wade 1978, 1985; Goodnight et al. 1992), trait-group selection (Wilson 1979, 1980), family-structured kin selection (Wade 1979; Michod 1980, 1982), and shifting-balance evolution (Wright 1978). One unifying feature of these models is that selection is driven, in part, by the differential productivity of groups, whether groups are defined intra- or interdemically (Wade 1985). A second feature is that the among-group variance in productivity must be correlated to among-group genetic variance underlying the trait in question. Finally, fitness must be influenced by epistatic interactions among loci (Wright 1978) or frequency-dependent interactions among individuals (Uyenoyama and Feldman 1980; Michod 1982). This chapter will focus on cases of frequency-dependent fitness.

If fitness is determined by local rather than global frequency, then population structure becomes important when there is frequency variation among the relevant population subunits. It can be said that fitness is context dependent (Goodnight et al. 1992), where the context is the local genetic environment, recognizing that genetic environments vary among demes or patches according to the genetic structure. The concept of context-dependent selection is best known in reference to the evolution of altruistic behaviors (Wilson 1980) but can be extended to any ecological or behavioral interaction that generates frequency-dependent selection.

One way to evaluate the impact of genetic structure on the response to selection is to compare the average fitness of a trait with and without genetic structure, using the concept of subjective frequency (Wilson 1980). Consider a simple case in which there are two forms of trait. Type A has an absolute fitness $F_A$ that is constant at 1. Type B has a fitness $F_B = x + p_B$, where $x$ is a constant such that $x < 1$, and $p_B$ is the frequency of type B. In the absence of population structure, trait Type B cannot invade a population because it will be less fit than Type A when rare. Suppose, however, that there is population structure and that fitness is a function of local, rather than global, frequency. In that case, Type B could be locally common in a few population subunits, even when globally rare. The fitness of Type B, averaged across population subunits, is a function of its subjective frequency, $p_{SB}$, or the frequency with which it interacts with other B types. That is,

$$p_{SB} = p_B + V_B/p_B \qquad (7)$$

where $p_B$ is the arithmetic average frequency of B, and $V_B$ is the variance among local population subunits in the frequency of B (Wilson 1980). Thus, the evolution of a positively frequency-dependent trait will be facilitated by population structure to the degree that $V_B >$ zero. The equations could be rewritten, however, to illustrate how genetic structure could inhibit the evolution of traits whose fitness is negatively frequency dependent. For example, if $x > 1$, but $F_B = x - p_B$, trait B would have its highest fitness when rare and would increase in frequency until $F_B = F_A$. Population structure would limit any advantage to $F_B$ when rare to the degree that $p_{SB}$ was greater than $p_B$.

The term *population subunits* used above could apply to either demes or patches within demes. The likelihood that either intrademic or interdemic structure would influence the nature of selection depends on several things. Most importantly, there must be ecological or behavioral interactions that generate frequency-dependent selection. This will be discussed. Second, the relevant genetic structure is that which exists at the spatial scale at which frequency determines fitness. The relevant level of population subdivision for evaluating individual fitness is that at which the fitnesses of the individuals that comprise one subunit are not a function of the genetic composition of any other subunit (Wilson 1980). A subunit could be the patch as defined variously within demes, the entire deme, or some larger collection of demes, depending on the forces driving selection.

Once the appropriate level of population structure is defined for a given frequency-dependent phenomenon, its influence on selection is proportional to the magnitude of the among-group component of genetic variance (Wade 1985), that is, to the degree that $p_{SB} > p_B$. This can be measured by $F$-statistics, assuming the genetic structure of the genetic markers is roughly equivalent to the genetic variation underlying the selected phenotype. Recall that in many insects, intrademic structure ($F_{ps}$) is greater than interdemic structure ($F_{st}$) arguing that, from a genetic perspective, intrademic structure is more likely to influence the response to selection.

Finally, if the differential productivity of groups is an important component of this process, traits favored by selection must spread (Queller 1992; Wilson et al. 1992). Under interdemic structure, this implies that some demes must export more migrants or colonists than others. Because the magnitude of interdemic genetic structure is greatest when migration rates are low, the dual requirements of meaningful genetic structure and interconnectedness among demes would seem contradictory. However, the significance of interdemic selection remains an open question, since some experimental evidence suggests that the differential productivity of demes can be important under even a relatively limited genetic structure (Wade and Goodnight 1991). The potential conflict between limited gene flow and differential productivity does not exist with intrademic selection. Since groups are defined for only part of a generation, any group that is particularly successful immediately exports its genotypes to the deme at large.

Figure 9.3a summarizes how selection might act under three population structures. This schematic illustrates positive, frequency-dependent selection on body color. The fitness of the "white" individuals is always 0.5; the fitness of the "black" individuals is equal to their local frequency (i.e., high fitness when they are common, selected against when they are rare, usual assumptions when social traits such as altruism are modeled). The results of this model of selection are shown for three types of population structure: panmixia, interdemic, and intrademic group structure. In panmixia, the rare phenotype is selected against and will be quickly eliminated. With interdemic structure, local demes will rapidly become fixed for either phenotype, depending on whichever is initially most common. In the absence of gene flow, the global allele frequency would never change. Given the same initial population distribution of phenotypes among groups, intrademic structure will give very different results. Because of dispersal, and subsequent random mating of adults, the phenotypes will be redistributed among groups every generation. This redistribution of phenotypes allows the effects of selection to influence the entire population. Fig. 9.3b illustrates how the impact of population structure on fitness depends on the life stage at which selection occurs. Intrademic structure has no influence on fitness when selection occurs after dispersal.

While theory can help organize our thinking concerning the evolutionary consequences of various forms of population structure, it leaves open the questions of what types of traits are likely to be subject to frequency-dependent selection and at what spatial scale frequency-dependent fitnesses are likely to be defined. These are essentially natural history questions.

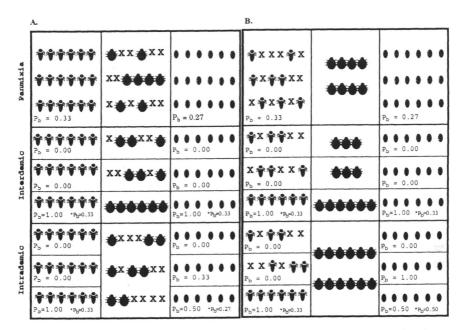

*Figure 9.3 (a,b)* An example of frequency-dependent selection on body color in a hypothetical insect living under the three population structures illustrated in Figure 9.1. Survivorship of the dark morph is equal to its local frequency. Survivorship of the light morph is constant at 0.5. In Figure 9.3a (left) the selection occurs at the adult stage. The response to selection is the same with panmictic and intrademic structure, because selection occurs in the mobile phase. In Figure 9.3b (right) selection acts on the larval stage. Under panmictic structure, the frequency of the dark morph declines because it is in the minority. Under interdemic structure there can be no response to selection within demes, because there is no within-deme variation. Variation in survivorship among demes does not influence evolution, because there is no migration. Selection favors the dark morph under intrademic structure, because it is in high frequency in one group ($Pb = 1$), even if in the minority globally (\*Pb = 0.33) and because oviposition following random mating exports the dark morph to new patches.

### 9.3.2 Reality: How Do Ecological Interactions Determine Fitness in Nature?

Frequency-dependent selection usually implies that fitness is a consequence of the direct or indirect effects that conspecifics have on one another's fitness. Direct effects might range from negative interactions, such as cannibalism, to positive interactions, such as trophallaxis. Indirect effects often mediate interactions with other species and can be negative, such as the transmission of disease, or positive, such as group predator defense. Two lines of inquiry might be pursued in order to assess what kinds of ecological scenarios are likely to generate the types of fitness effects that are influenced by population structure. The first is to ask what evolutionary advantages result from group living. The assumption is that we can learn about how something evolved by asking how it is currently adaptive. The second is to study selection in action by attempting to partition variation in individual fitness into within- and among-group components (Wade 1985; Goodnight et al. 1992).

Intrademic genetic structure would enhance the evolution of traits displaying positive frequency-dependent fitness; that is, the more common a trait is in a group, the higher the survivorship of individuals expressing that trait. Correlations between group size and survivorship are fairly common observations in studies of insects (McCauley 1994). The most commonly hypothesized advantage to group living and aggregation behavior in insects is predator defense (Vulinec 1990; Gross 1993). Defenses include warning coloration, noxious chemical displays, and aggressive responses to the approach of predators, and can be displayed by eggs, immatures, and adults. The evolution of these defenses would be positively frequency dependent if the individual contributions of group members reinforce one another.

Eggs are usually the life stage most vulnerable to predation. Attacks can come from predators, cannibals, parasites, parasitoids, and pathogens. If eggs are clustered, it might seem that an attack on one member of the aggregation would put all the others at high risk (compared to randomly placed or hyperdispersed eggs). Since egg clustering is a common occurrence in insects, however, there must be instances in which the clustering of eggs reduces predation. Stamp (1980) and Courtney (1984) have discussed some of these mechanisms. Most common appears to be reinforcement of aposematic coloration. In order for selection to act on egg-clumping behavior by females, there must be a genetic component to variation in the behavior, as shown by Del-Solar and Ruiz (1992).

Group defense is well established in the immatures of several insect species (Vulinec 1990). Immatures can be aggregated simply because they were clumped as eggs, or because they actively seek out one another after hatching. In either case, this can be an advantage in the presence of predators. Defense against predators by aggregated immatures usually takes the form of chemical, sometimes behavioral, and rarely morphological adaptations. Larval IWLB, for example, use host-derived chemicals to defend against predators (Wade 1994). Larger aggregations are more efficient at defending against predators, especially in early instars when individuals can produce very little of the chemical. Defense of grouped immatures against predators can also result from maternal behavior in some species in which the adult female remains with her eggs after they hatch (Tallamy 1984).

McCauley (1992, 1994) studied the effects of intrademic structure on fitness in FCPB. Females of this species practice egg clustering on the host plant, *Solanum carolinense*. Group members have an average relatedness of 0.38 owing to frequent multiple paternity within egg clutches (McCauley 1992). Upon hatching, larvae tend to remain on the same host plant until they disperse to pupate in the ground. Thus, intrademic structure defined at the level of the host plant is a consequence of the oviposition behavior of females plus the sedentary nature of larvae. McCauley (1992) noted that FCPB larvae suffer up to 60% mortality owing to a tachinid fly parasitoid, *Myiopharus doryphorae,* and that parasitism rates varied among groups to a greater degree than predicted by chance. In addition to remaining on their natal host plant (a form of passive aggregation), FCPB larvae sometimes actively aggregate by feeding head to head on the same leaf. Groups vary with regard to the degree that this active aggregation is expressed.

McCauley (1994) suggested that aggregation behavior could evolve by positive frequency-dependent selection in that individuals with the propensity to aggregate must encounter one another in order to express this behavior. If so, aggregation behavior would be facilitated by intrademic structure because, with structure, aggregators could be locally common in some groups, even if rare overall. McCauley was able to demonstrate selection at the phenotypic level by showing a correlation between the degree of active aggregation and avoidance of parasitism in studies of both naturally occurring and manipulated groups (Fig. 9.4).

The other major advantage attributed to group living is that it somehow enhances feeding efficiency. For example, Breden and Wade (1987) showed a positive relationship between group size and larval growth in IWLB. Similar results have also been shown in a sawfly (Ghent 1960) and a lepidopteran (Lawrence 1990). Ribeiro (1989) has demonstrated both developmental and defensive benefits to nymphal aggregation in an aposematic hemipteran. However, in this case, cannibalism by older nymphs encourages dispersal after the third instar. Interestingly, fourth- and fifth-instar nymphs reaggregate during molting.

The fitness of traits that display negative frequency-dependent fitness would be limited by intrademic genetic structure. One such trait is cannibalistic behavior. Cannibalism is a common phenomenon among insect taxa (Elgar and Crespi 1992, and references therein). In species with egg clustering, it often involves cannibalism by newly emerged larvae of more slowly developing clutchmates, which are their full- and half-sibs (Stevens 1992). Advantages to this type of cannibalism are usually nutritional (Stevens 1992). However, there is a cost to cannibalism in terms of inclusive fitness if the victim is genetically related to the cannibal. Thus, the evolution of this within-group cannibalistic behavior would have to be evaluated in the context of intrademic genetic structure. Studies of IWLB demonstrate how complex the selection can be (Breden and Wade 1989). Intraclutch egg cannibals gain a size and survivorship advantage over their surviving clutchmates. However, because victims are relatives of cannibals, and because subsequent group-size advantages are reduced by the loss of group members who are victims, the overall average fitness of cannibals is limited by the genetic structure. While this might predict that cannibalism

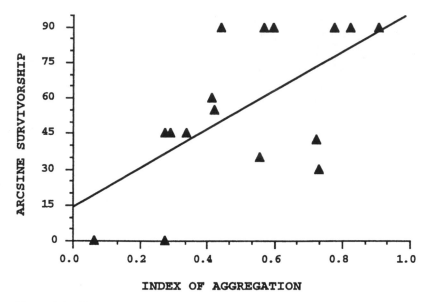

*Figure 9.4*   The relationship between survivorship (1 − percent parasitized by a tachinid fly) and aggregation behavior in 17 naturally occurring larval groups of the false Colorado potato beetle, *Leptinotarsa juncta.* Survivorship is arcsine transformed; the index of aggregation is adjusted for overall group size (McCauley 1994). The slope of the regression line is significantly different from zero ($p < 0.05$).

rates would be higher in groups with lower relatedness, no such relationship was found (Goff and Stevens 1995).

The evolutionary literature is replete with speculation as to why there should be an advantage to carrying a rare genotype or phenotype. Included are advantages in the face of predators and parasites that have adapted to overcome the defenses of the common type, and advantages when resource partitioning is possible. How often these mechanisms operate in insects remains to be seen, but it seems they would be limited by intrademic structure. One interesting example of how a form of intrademic structure influences fitness is in characters that determine the sex ratio. It is well known that, all other things being equal, a 1:1 sex ratio should be maintained by negative frequency-dependent selection (Fisher 1958). However, female-biased sex ratios are known in insects, especially species in which males are derived from unfertilized eggs, such as Hymenoptera, Thysanoptera, and a few families of Homoptera and Coleoptera (Wrensch and Ebbert 1993). In many cases, the female bias is found in species with clear intrademic structure (but with mating prior to dispersal), a condition that limits the rare advantage (Hamilton 1967; Wilson and Colwell 1981; Wrensch and Ebbert 1993, and references therein).

## 9.4 Conclusion

This chapter argues that a nonrandom spatial distribution of genotypes can often occur within demes, owing to a combination of patchy resources, sedentary or actively aggregating immatures, and the oviposition habits of females. This nonrandom distribution is termed *intrademic genetic structure*. It is argued further that in insects, intrademic structure is often considerably greater in magnitude than the interdemic structure that is the standard fare of experimental population genetics. This is because in many insect species, dispersal distances in adults can be orders of magnitude greater than the dispersal distances seen in immatures, and because there are various mechanisms whereby kin associations can be found in the more sedentary immatures.

It is becoming increasingly clear that ecological processes are influenced by genetic diversity (Real 1994). The impact of genetic diversity would depend, in part, on its spatial organization (i.e., on genetic structure), since it is the genetic similarity or dissimilarity of behaviorally or ecologically interacting individuals that is important. For example, the epidemiology of infectious agents such as parasites can be strongly dependent on the genetic diversity found in the host population (Hamilton 1980). Diseases or parasites are thought to spread more easily among genetically similar individuals. Stated from an evolutionary perspective, genetic structure can have important fitness consequences for traits in which fitness is determined by frequency-dependent interactions with others, as ARE implicit or explicit in well-known group and kin selection models. When fitness is influenced by local allele frequency, genetic structure enhances the evolution of traits with positive frequency-dependent fitness because, by definition, with genetic structure, traits can be locally common even when globally rare. This increases the average fitness of all bearers of the trait relative to the case with no structure. In the same vein, structure limits the evolution of traits with negative frequency-dependent fitness because, on average, it reduces local rarity.

These results discussed here have several other implications for the study of insect population biology, two of which will be discussed here. The first concerns some rather obvious consequences of ignoring intrademic structure when designing and interpreting a study of interdemic structure. Intrademic structure can be confounded with interdemic structure when samples are taken from natural populations for genetic analysis. For example, if all the individuals sampled from a given geographic locality belong to the same kin group, then variance among localities might reflect among-patch intrademic variance as much or more than true geographic variation. From equation 3, sampling without regard to trait-group membership measures $F_{pt}$, not $F_{st}$. $F_{pt}$ is most likely to be greater than $F_{st}$ when samples are taken prior to dispersal. Similarly, observed variation in levels of herbivory among host plants that might be ascribed to physiological or genetic differences among plants could, in fact, be due to spatially structured variation in the herbivorous insects that feed on them, or an interaction between the two effects (Alstad, Chapter 1, this volume).

The second implication is concerned with the evolution and adaptive significance of characteristics often associated with group-living herbivorous insects, such as coordinated predator defense and/or feeding. In many cases, the evolution of these characteristics should be interpreted in the context of existing intrademic structure. It is intriguing, however, to consider the interaction between selection pressures *leading to* the evolution of intrademic structure and those *resulting from* intrademic structure. How do traits associated with the formation of intrademic structure, such as egg clustering or aggregation behavior, become associated with those whose evolution is favored under intrademic structure, such as aposematic coloration? Consider the following: Intrademic structure often results from the egg-clustering behavior of females and can be viewed as a maternal effect (Cheverud and Moore 1995). In that sense, selection on the clustering behavior is both direct and indirect, because it is mediated both by how clustering influences the number of eggs laid in the mother's lifetime (a function of her genotype), and by how it influences the survivorship of the offspring (a function of how the mother's genotype determined the clustering, and how the offspring genotypes determine survivorship in the context of the group). Several authors have emphasized how maternal effects can lead to a counterintuitive response to subsequent selection acting on the mother's offspring (Kirkpatrick and Lande 1989; Cheverud and Moore 1995). This would be especially complex in the context of intrademic structure, because the offspring continue to influence one another's fitness. As a possible example, we have found that FCPB females cluster eggs on the host plant. Eggs range in color from light yellow to dark red, with the color variation being almost entirely among clutches (and hence mothers). Preliminary studies indicate that egg color influences the survivorship of eggs to hatching in the face of predation and cannibalism, with darker red eggs being more successful (Goff unpublished data). Egg color is almost certainly a maternal trait. If egg color variation has a genetic basis, it is a function of the maternal genotype but presents its fitness effects in the next generation. Thus, the evolution of the egg-clustering behavior, the egg defensive compounds, and the larval aggregation behavior mentioned earlier are interconnected in some way. It would be interesting to know the sequence in which these traits arose and how the evolution of each trait influenced the others.

In closing, it is our feeling that the ecological and evolutionary consequences of intrademic genetic structure have not received adequate attention, especially in the noneusocial insects.

## 9.5 References

Ashiru, M. O. 1988. The frequency distribution of eggs and larvae of *Anaphe venata* Butler (Lepidoptera: Notodontidae) on *Triplochiton scleroxylon* K. Schum. *Ins. Sci. Appl.* 9:587–592.

Avise, J. C. 1994. *Molecular Markers, Natural History, and Evolution.* Chapman & Hall, New York.

Breden, F. and M. J. Wade. 1987. An experimental study of the effect of group size on lar-val growth and survivorship in the imported willow leaf beetle, *Plagiodera versicolora* (Coleoptera: Chrysomelidae). *Environ. Entomol. 16* (5):1082–1086.

Breden, F. and M. J. Wade. 1989. Selection within and between kin groups of the imported willow leaf beetle. *Am. Nat. 134*:35–50.

Cheverud, J. M. and A. J. Moore 1995. Quantitative genetics and the role of the environment provided by relatives in behavioral evolution. Pp. 67–100 *in* C. R. B. Boake (Ed.), *Quantitative Genetic Studies of Behavioral Evolution.* University of Chicago Press, Chicago, IL.

Costa, J. T. III and N. E. Pierce. 1997. Social evolution in the Lepidoptera: Ecological con-text and communication in larval societies. *In* J. C. Choe and B.J. Crespi (Eds.), *Social Competition and Cooperation in Insects and Arachnids: Volume II. Evolution of Social-ity.* Cambridge University Press, Cambridge, UK. Pp. 402–442.

Costa, J. T. III and K. G. Ross. 1993. Seasonal decline in intracolony genetic relatedness in eastern tent caterpillars: Implications for social evolution. *Behav. Ecol. Sociobiol. 32*:47–54.

Costa, J. T. III and K. G. Ross. 1994. Hierarchical genetic structure and gene flow in macrogeographic populations of the eastern tent caterpillar (*Malacosoma americanum*). *Evolution 48* (4):1158–1167.

Courtney, S. P. 1984. The evolution of egg clustering by butterflies and other insects. *Am. Nat. 123*:276–281.

Davis, P. M. and L. P. Pedigo. 1989. Analysis of spatial patterns and sequential count plans for stalk borer (Lepidoptera: Noctuidae). *Environ. Entomol. 18* (3):504–509.

Del-Solar, E. and G. Ruiz. 1992. Behavioral analysis of the choice of oviposition site by single females of *Drosophila melanogaster* (Diptera: Drosophilidae). *J. Insect Behav. 5* (5):571–581.

Derrick, M. E. and W. B. Showers. 1990. Relationship of adult European corn borer (Lep-idoptera: Pyralidae) in action sites with egg masses in the corn field. *Environ. Entomol. 19*:1081–1085.

Eickwort, G. C. 1981. Presocial insects. Pp. 199–280 *in* H. R. Hermann (Ed.), *Social In-sects,* Vol. II. Academic Press, New York.

Elgar, M. A. and B. J. Crespi. 1992. *Cannibalism: Ecology and Evolution among Diverse Taxa.* Oxford University Press, New York.

Fisher, R. A. 1958. *The Genetical Theory of Natural Selection.* Dover Publications, New York.

Ghent, A. W. 1960. A study of the group-feeding behavior of larvae of the jack pine sawfly, *Neodiprion pratti banksianae* Roh. *Behaviour 16*:110–148.

Goff, P. W. and L. Stevens. 1995. A test of Hamilton's rule: Cannibalism and relatedness in beetles. *Anim. Behav. 49*:545–547.

Goodnight, C. J., J. M. Schwartz, and L. Stevens. 1992. Contextual analysis of models of groups selection, soft selection, hard selection and the evolution of altruism. *Am. Nat. 140*:743–761.

Gross, P. G. 1993. Insect behavioral and morphological defenses against parasitoids. *Annu. Rev. Entomol. 38*:251–273.

Hamilton, W. D. 1964a. The genetical evolution of social behaviour, I. *J. Theor. Biol. 7*:1–16.

Hamilton, W. D. 1964b. The genetical evolution of social behaviour, II. *J. Theor. Biol.* 7:17–52.

Hamilton, W. D. 1967. Extraordinary sex ratios. *Science 156*:477–488.

Hamilton, W. D. 1980. Sex versus non-sex versus parasite. *Oikos 35*:282–290.

Herre, E. A. 1985. Sex ratio adjustment in fig wasps. *Science 228*:896–898.

Kirkpatrick, M. and R. Lande. 1989. The evolution of maternal characters. *Evolution 43* (3):485–503.

Lawrence, W. S. 1990. The effects of group size and host species on development and survivorship of a gregarious caterpillar *Halisidota caryae* (Lepidoptera: Arctiidae). *Ecol. Entomol. 15*:53–62.

McCauley, D. E. 1992. Family structured patterns of mortality in the false Colorado potato beetle. *Ecol. Entomol. 17*:142–148.

McCauley, D. E. 1993. Evolution in metapopulations with frequent local extinction and recolonization. *Oxf. Surv. Evol. Biol. 9*:109–134.

McCauley, D. E. 1994. Intrademic group selection imposed by a parasitoid–host interaction. *Am. Nat. 144* (1):1–13.

McCauley, D. E. and W. F. Eanes. 1987. Hierarchical population structure analysis of the milkweed beetle, *Tetraopes tetraophthalmus* (Forster). *Heredity 58*:193–210.

McCauley, D. E., M. J. Wade, F. Breden, and M. Wohltman. 1988. Spatial and temporal variation in group relatedness: Evidence from the imported willow leaf beetle. *Evolution 42*:184–192.

McPheron, B. 1990. Genetic structure of apple maggot fly (Diptera: Tephritidae) populations. *Ann. Entomol. Soc. Am. 83* (3):568–577.

McPheron, B., D. C. Smith, and S. H. Berlocher. 1988. Microgeographic genetic variation in the apple maggot *Rhagoletis pomonella*. *Genetics 119*:445–451.

Michod, R. 1980. Evolution of interactions in family structured populations: Mixed mating models. *Genetics 96*:275–296.

Michod, R. 1982. The theory of kin selection. *Annu. Rev. Ecol. Syst. 13*:23–55.

Mopper, S. and D. Simberloff. 1995. Differential herbivory in an oak population: The role of plant phenology and insect performance. *Ecology 76* (4):1233–1241.

Naranjo, S. E. and H. M. Flint. 1995. Spatial distribution of adult *Bemisia tabaci* (Homoptera: Aleyrodidae) in cotton and development and validation of fixed-precision sampling plans for estimating population density. *Environ. Entomol. 24* (2):261–270.

Pamilo, P. 1983. Genetic differentiation within subdivided populations of *Formica* ants. *Evolution 37*:1010–102.

Parker, G. A. and S. P. Courtney. 1984. Models of clutch size in insect oviposition. *Theor. Pop. Biol. 26*:27–48.

Queller, D. C. 1992. Does population viscosity promote kin selection? *Trends Ecol. Evol.* 7:322–4.

Queller, D. C. and K. F. Goodnight. 1989. Estimating relatedness using genetic markers. *Evolution 42* (2):258–275.

Rank, N. E. 1992. A hierarchical analysis of genetic differentiation in a montane leaf beetle *Chrysomela aeneicollis* (Coleoptera: Chrysomelidae). *Evolution 46* (4):1097–1111.

Real, L. 1994. *Ecological Genetics.* Princeton University Press, Princeton, NJ.

Reilly, L. M. 1986. Measurements of population structure in the termite *Reticulitermes flavipes* (Kollar) (Isoptera: Rhinotermitidae) and comparisons with the eusocial Hymenoptera. Ph.D. dissertation, Vanderbilt University, TN.

Reilly, L. M. 1987. Measurements of inbreeding and average relatedness in a termite population. *Am. Nat. 130*:339–349.

Ribeiro, S. T. 1989. Group effects of aposematism in *Jadera haematoloma* (Hemiptera: Rhopalidae). *Ann. Entomol. Soc. Am. 82*:466–475.

Roderick, G. K. 1996. Geographic structure of insect populations: Gene flow, phylogeography and their uses. *Annu. Rev. Entomol. 41*:325–352.

Singh, K. 1993. Evidence of male and female of coffee bean weevil, *Araecerus fasciculatus* (Deg.) (Coleoptera: Anthribidae) emitted aggregation and sex pheromones. *Crop Res. 6* (1):97–101.

Slatkin, M. 1977. Gene flow and genetic drift in a species subject to frequent local extinctions. *Theor. Pop. Biol. 12*:253–262.

Slatkin, M. 1985. Gene flow in natural populations. *Annu. Rev. Ecol. Syst. 16*:393–430.

Sokal, R. R. and J. F. Rohlf. 1981. *Biometry.* W. H. Freeman, New York.

Stamp, N. E. 1980. Egg deposition in butterflies: Why do some species cluster their eggs rather than deposit them singly? *Am. Nat. 15*:367–380.

Steiner, M. Y. 1990. Determining population characteristics and sampling procedures for the western flower thrips (Thysanoptera: Thripidae) and the predatory mite *Amblyseius cucumeris* (Acari: Phytoseiidae) on greenhouse cucumber. *Environ. Entomol. 19* (5):1605–1613.

Stevens, L. 1992. Cannibalism in beetles. Pp. 156–174 *in* M. A. Elgar and B. J. Crespi (Eds.), *Cannibalism: Ecology and Evolution among Diverse Taxa.* Oxford University Press, New York.

Tallamy, D. W. 1984. Insect parental care. *Bioscience 34*:20–24.

Tallamy, D. W. and T. K. Wood. 1986. Convergence patterns in subsocial insects. *Annu. Rev. Entomol. 31*:369–390.

Tandon, P. L. and G. K. Veeresh. 1988. Inter-tree spatial distribution of *Coccus viridis* (Green) on mandarin. *Intern. J. Trop. Agric. 6* (3–4):270–275.

Tonhasca, A. J., J. C. Palumbo, and D. N. Byrne. 1994. Distribution patterns of *Bemisia tabaci* (Homoptera: Aleyrodidae) in cantaloupe fields in Arizona. *Environ. Entomol. 23* (4):949–954.

Uyenoyama, M. K. and M. W. Feldman. 1980. Theories of kin and group selection: A population genetics perspective. *Theor. Pop. Biol. 17*:380–414.

Vulinec, K. 1990. Collective security: Aggregation by insects as a defense. Pp. 251–288 *in* D. L. Evans and J. O. Schmidt (Eds.), *Insect Defenses: Adaptive Mechanisms and Strategies of Prey and Predators.* State University of New York Press, Albany, NY.

Wade, M. J. 1978. A critical review of the models of group selection. *Quart. Rev. Biol.* 53:101–114.

Wade, M. J. 1979. The evolution of social interactions by family selection. *Am. Nat.*113:399–417.

Wade, M. J. 1982. The effect of multiple inseminations on the evolution of social behaviors in diploid and haplodiploid organisms. *J. Theor. Biol.* 95:351–368.

Wade, M. J. 1985. Hard selection, soft selection, kin selection, and group selection. *Am. Nat.* 125:61–73.

Wade, M. J. 1994. The biology of the imported willow leaf beetle (*Plagiodera versicolora*, Laicharting). Pp. 541–547 *in* P. H. Jolivet, M. L. Cox, and E. Petitpierre (Eds.), *Novel Aspects of the Biology of the Chrysomelidae.* Kluwer Academic Publishers, The Netherlands.

Wade, M. J. and C. J. Goodnight. 1991. Wright's shifting balance theory: An experimental study. *Science 253*:1015–1018.

Wade, M. J. and D. E. McCauley. 1988. Extinction and recolonization: Their effects on the genetic differentiation of local populations. *Evolution 42*:995–1005.

Weir, B. S. 1990. *Genetic Data Analysis.* Sinauer Associates, Sunderland MA.

Whitlock, M. C. and D. E. McCauley 1990. Some population genetic consequences of colony formation and extinction: Genetic correlations within founding groups. *Evolution 44*:1717–1724.

Wilson, D. S. 1979. Structured demes and trait-group variation. *Am. Nat. 113*:606–610.

Wilson, D. S. 1980. *The Natural Selection of Populations and Communities.* Benjamin/Cummings, Menlo Park, CA.

Wilson, D. S. 1983. The group selection controversy: History and current status. *Annu. Rev. Ecol. and Syst. 14*:159–187.

Wilson, D. S. and R. K. Colwell. 1981. Evolution of sex ratio in structured demes. *Evolution 35*:882–897.

Wilson, D. S., G. Pollock, and L. A. Dugatin. 1992. Can altruism evolve in purely viscous populations? *Evol. Ecol. 6*:331–341.

Wrensch, D. L. and M. A. Ebbert. 1993. *Evolution and Diversity of Sex Ratio in Insects and Mites.* Chapman & Hall, New York.

Wright, S. 1931. Evolution in Mendelian populations. *Genetics 16*:97–159.

Wright, S. 1978. *Evolution and the Genetics of Populations: Vol 4. Variability within and among Natural Populations.* University of Chicago Press, Chicago, IL.

# 10

# Social Behavior and Its Effects on Colony- and Microgeographic Genetic Structure in Phytophagous Insect Populations

*James T. Costa*
Museum of Comparative Zoology, Harvard University, Cambridge, MA

## 10.1 Introduction

*Population genetic structure* is defined in terms of deviation from panmixis, or random mating, and explicitly refers to nonrandom spatial association of alleles. Nonrandom associations may arise in several ways as a natural consequence of the interplay among behavioral, ecological, and biogeographic factors. At large (macrogeographic) spatial scales, genetic differences between subpopulations may be maintained by natural selection or result from genetic drift associated with isolation by distance and attenuated gene flow. The causes of *microgeographic* structure—here defined as structure at the spatial scale of individual host plants and localized host plant groups—are more varied. Spatial genotypic association could arise from the joint effects of reproductive output and dispersal whereby physical association is the outcome of environmental constraints such as predation pressure and lack of suitable habitat. On the other hand, behaviors promoting association of related individuals can also produce highly patchy, localized units of genetic structure.

This chapter addresses the genetic dynamics at and within microgeographic (intrademic) spatial scales in family-structured populations, and will consider certain insect herbivore groups that engage in social interaction, exhibiting behaviorally mediated family structure. Behavioral interactions are central to such widely disparate phenomena as mate choice, mating frequency, dispersal, host selection, territoriality, and kin interactions. Social systems offer compelling examples of how behavior and natural history can determine population genetic structure and gene flow, parameters of central importance to modes of colonization and host-plant specialization. Moreover, spatial and temporal variance in population genetic parameters may be especially large in populations of social insects as a result of the unique behavioral cycle of association and dispersal characteristic of societies. Socially mediated population genetic patterns and dynamics are akin to those found in highly patchy populations; such populations may exhibit striking localized genetic structure persisting for a fraction of the organism's life cycle (see McCauley and Goff, Chapter 9, this volume). Sociality thus represents a significant, though often neglected, phenomenon shaping microgeographic genetic parameters in some insect herbivore populations.

The chapter is divided into four sections that integrate the following themes of sociality and the genetics of social herbivore populations: (1) alternative estimators of genetic structure at spatial scales in socially structured populations: (2) defining characteristics of societies and the life history and ecology of representative herbivorous social groups: (3) behavioral and ecological determinants of genetic structure at microgeographic and colony spatial scales; and (4) the interrelationship of behavioral ecology and population genetics, focusing on the importance of assessing these jointly to better understand patterns of host use and dispersal in herbivore populations. The behavioral and ecological framework of this chapter complements the explicitly theoretical approach of McCauley and Goff (Chapter 9, this volume).

## 10.2 Inbreeding and Relatedness Coefficients: Different Questions for Different Spatial Scales

Genetic structure is quantified in various ways, all of which index the relative degree of within- and between-group genetic variance. Structure at highly localized scales (such as that of the colony) is commonly expressed in terms of genetic relatedness, whereas structure at higher scales is expressed in terms of inbreeding. Although these describe essentially the same thing, namely, relative genetic similarity, different methods have historically been derived for analyzing structure at different spatial scales. At the scale of populations or subpopulations, efforts have focused on charting the degree and distribution of genetic variation in an effort to correlate patterns with evolutionary forces such as selection, drift, and gene flow (Slatkin 1987), interfacing with broader evolutionary questions pertaining to adaptation and speciation (Otte and Endler 1989; Avise 1994). At the other end of the spatial scale, investigations into "micro-microgeographic" population genetic patterns—the realm of colonies and family interactions—were motivated by students of social biology in an effort to test theories explaining the evolution of cooperation and altruism.

Many evolutionary thinkers suspected that family structure could explain the evolution of individual fitness-reducing traits (e.g., sterile castes) found in many insect societies (Darwin 1859; Fisher 1930; Haldane 1932; Williams and Williams 1957; Williams 1966). Hamilton (1964, 1972) provided the most comprehensive treatment of the problem, formalized as inclusive fitness theory. At the heart of this theory (also known as kin selection theory) is the observation that an individual's total fitness stems from two sources: *directly* from personal reproductive effort ($W_x$ for fitness of individual x) and *indirectly* from the reproductive effort of relatives ($W_y$ for fitness of individual $y$), as weighted by the degree of relatedness between individuals $x$ and $y$ ($r_{xy}$) (reviews by Hamilton 1972; West-

Eberhard 1975; Michod 1982; Page 1986). A reduction in direct fitness may thus be offset by indirect fitness, as formally expressed by the well-known relationship $W_x + (r_{xy} \cdot W_y) > 0$, according to which behaviors reducing personal fitness can evolve as long as indirect fitness, mediated by relatedness, is positive.

The most sophisticated insect societies contain behaviorally and morphologically specialized castes comprised of nonreproductive individuals that have foraging (workers) or defensive (soldiers) functions. Empirical assessment of the genetic structure and sex ratios of insect societies is central to testing the inferential power of the genetic models of Hamilton and others in explaining the evolution of these traits (Michod and Anderson 1979). Reproductive altruism is but one example of the behavioral traits favored by inclusive fitness. In fact, from a population genetic perspective, indirect fitness may arise whenever kin structure exists in a social or cooperative context (West-Eberhard 1975; Michod 1982). Thus, societies exhibiting a range of cooperative behaviors short of reproductive altruism are also worthy of genetic analysis.

Inclusive fitness theory has long provided a framework for empirical studies of insect societies, and several alternative analytical methods for assessment of colony genetic structure have been developed (e.g., Pamilo and Crozier 1982; Pamilo 1984, 1989, 1990; Wilkinson and McCracken 1985; Queller and Goodnight 1989) and used to probe the genetic structure of such complex social organisms as ants (e.g., Pamilo 1982, 1983; Crozier et al. 1984; Ross et al. 1988), wasps (e.g., Lester and Selander 1981; Ross 1986; Ross and Matthews 1989; Strassmann et al. 1989; Queller et al. 1992), bees (e.g., Laidlaw and Page 1984; Schwarz 1986), and termites (Reilly 1987). Because kin selection is essentially a population-genetic phenomenon (Uyenoyama and Feldman 1980; Uyenoyama 1984; Michod 1982), several theorists have explored the spread of "altruistic alleles" in an explicitly nonkin context using models of trait-group or interdemic selection (Wilson 1975, 1977, 1983; Maynard Smith 1976; Grafen 1984; Wade 1985; Queller 1991). In general, however, so-called altruistic and cooperative traits are nearly always associated with kin-structured populations.

Since kin groups exist within a broader population, they offer an opportunity to examine genetic relationships and their causes in a localized area centered on the colony. As spatial scale increases from the colony to the surrounding population, observed genetic similarity (an index of genetic structure) grades from identity by descent (IBD) to identity by state (IBS; Fig. 10.1). In population genetic terms, *identity* refers to *allelic* identity; two apparently identical alleles sampled from within or between two populations could be the same by common ancestry, in which case they are said to be identical by descent, or they could be the same by convergence, in which case they are said to be identical in sharing the same allelic state. The spatial gradation from prevailing IBD to prevailing IBS may be abrupt or diffuse, and is mediated by a host of ecological and behavioral characters interacting in a complex manner unique to each species and type of society.

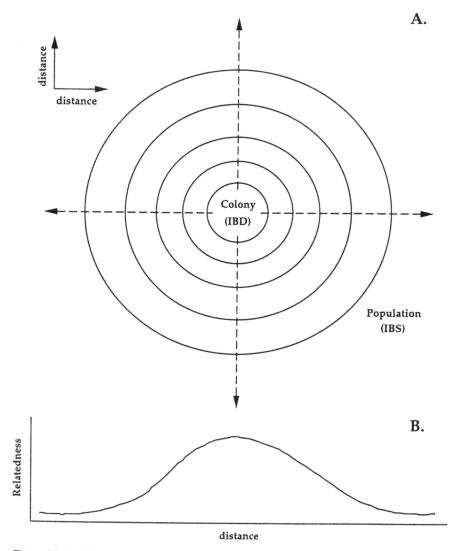

*Figure 10.1* Diagram of genetic relationships prevailing at colony (intrademic) and micro-geographic (interdemic) spatial scales. A. The center of the diagram marks the position of a family-structured colony, where structure is attributable to identity by descent (IBD). With increasing spatial scale, prevailing genetic similarity grades from IBD to identity by state (IBS). B. Spatial relatedness profile for diagram (A). Relatedness is by definition high where genetic similarity arises from common descent (IBD); the given profile is arbitrary, and the extent of local spatial dominance of family structure for a given species is a function of local population density, host use, dispersal, mating behavior, and kin discrimination.

The interdependence of microgeographic and colony-level relationships underscores the value of describing population genetic patterns in a hierarchical fashion. Both inbreeding and genetic relatedness, for example, can be expressed in global and local terms: Total or global inbreeding or relatedness can be partitioned into a within-group (local) component and a between-group component (Wright 1951; Pamilo 1984). Thus, in the case of localized, socially mediated genetic structure, global relatedness stems from relatedness of individuals *within* as well as *among* social groups within demes (Pamilo 1984, 1989; see Fig. 10.2). In Figure 10.2, among-group relatedness ($r_{ST}$) can be linked to highly localized dispersal, just as Wright's among-group variance parameter for subpopulations ($F_{ST}$) provides an estimator for gene flow ($Nm$) among populations by the relationship $Nm = (1 - F_{ST})/4F_{ST}$ under an island model (Wright 1951; see extensive discussion in McCauley and Goff, Chpater 9, this volume). For example, Costa and Ross (1993), drawing on relationships developed by Ross (1993), expressed the genetically effective number of family groups ($F_e$) making up colonies of eastern tent caterpillars using the expression $F_e = (r_s - r_{ns})/(r_t - r_{ns})$, where $r_s, r_{ns}$, and $r_t$ are empirically determined estimates of relatedness between siblings ($s$), nonsiblings in colonies on a common hostplant ($ns$), and nestmates at any point in time ($t$), respectively. Note that the $r_{ns}$ term is among-group relatedness, corresponding to $r_{ST}$ above. Shifts over time in intracolony relatedness revealed small-scale, among-colony migration.

## 10.3 Insect Sociality

Social interaction occurs in many insect and arachnid taxa, including embiids, aphids, termites, cockroaches, beetles, ants, bees and wasps, thrips, butterflies and moths, spiders, and mites (Wilson 1971; Eickwort 1981; Crespi and Choe 1997). These societies vary in demographics, size, and complexity. They include extended families and simple cohorts of siblings, species with and without morphological or behavioral castes, species living in or on their food and active foragers, and they are trophically diverse, encompassing predators, herbivores, and omnivores. In response to such diversity, an elaborate hierarchical classification (based on presence or absence of overlapping generations, cooperative brood care, and reproductive division of labor) has developed since the first formal attempts to categorize these societies (Batra 1966; Michener, 1953, 1958, 1969; Wilson 1971). In recent years, several authors have sought to resolve the many semantic and conceptual problems associated with this classification (e.g., Gadagkar 1994; Crespie and Yanega 1995; Sherman et al. 1995; Costa and Fitzgerald 1996; Wcislo 1997).

Herbivores constitute a relatively small proportion of social insects, but their colony composition and social interactions vary widely. Social herbivores occur in six major taxa: butterflies and moths (Lepidoptera), sawflies (Hymenoptera: Symphyta), aphids (Hemiptera: Homoptera), thrips (Thysanoptera), beetles (Coleoptera),

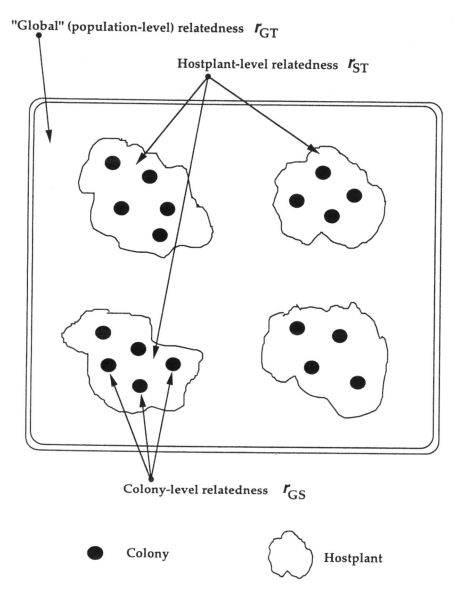

**"Global" (population-level) relatedness  $r_{GT}$**

**Hostplant-level relatedness  $r_{ST}$**

**Colony-level relatedness  $r_{GS}$**

● **Colony**    ⬡ **Hostplant**

*Figure 10.2* Hierarchical spatial scales and relatedness partitioning. Global relatedness ($r_{GT}$) stems from a within-colony relatedness component ($r_{GS}$) and a between-colony relatedness component ($r_{ST}$), treating the population as subdivided by host plant. In the absence of subdivision, ($r_{GT}$) corresponds to within-colony relatedness (based on after Pamilo 1989).

and grasshoppers (Dictyoptera: Orthoptera). Many other insect herbivores exhibit parental care in the form of provision or defense of juveniles (see Rossiter, Chapter 6, this volume), but here I will restrict the term sociality to colonial species in which members exhibit reciprocal communication and cooperation for group defense, reproduction, and/or host use. While most social herbivores would be termed pre-, sub-, or eusocial (Michener 1969; Wilson 1971), I will avoid this classification, since many of its terms lack rigor and are under revision, as indicated earlier. In this chapter I will focus only on phytophagous and sap-feeding herbivores, not the gregarious/social insects feeding on wood, pollen, or nectar (e.g., passalid beetles, termites, wood roaches, bees, web spinners, etc.).

## 10.4 Life History and Ecology of Herbivorous Social Insects

This chapter is not intended to provide a comprehensive treatment of the behavioral ecology and population genetics of social herbivores. Rather, the groups described here were chosen to serve as contrasting examples of social structure in insect herbivore populations, and only those elements of mating, dispersal, and foraging behavior relevant to population genetic patterns are discussed. For economy of space, four of the six taxa mentioned will be treated in detail, including the social aphids, thrips, caterpillars, and sawflies (Table 10.1). These groups were selected to represent a cross-section of life-history, genetic, and behavioral traits. For brevity, I will not cover the several species of chrysomelid beetles and acridid grasshoppers that have gregarious immatures with group defense and foraging (Eickwort 1981; Wade and Breden 1986), since they parallel social Lepidoptera in terms of behaviors, colony composition, foraging, and, in the case of the beetles, life history.

The societies considered here exhibit two demographic patterns: multigeneration families (aphids and thrips) and single-generation larval cohorts (caterpillars and sawflies). Note that *multigeneration* refers not to voltinism, but rather to the presence of two or more generations (including juveniles and adults) within a colony. Aphids and thrips share superficial similarities in social form: Both include cecidogenic (gall-making) species or inhabitants of similarly sheltered nest sites, often with both adults and juveniles present and with specialized defender morphs (soldiers). Gregarious caterpillars and sawflies are ecologically and behaviorally similar in that both are holometabolous with eruciform larvae that feed on a wide variety of host plants, and in some cases use silk to construct shelters.

### 10.4.1 Aphid Societies

Aphids are diplodiploid, phloem-feeding insects that usually alternate between primary and secondary host plants. Social aphids tend to be gall-makers and are found in two closely related families, Hormaphididae and Pemphigidae (Hemiptera: Homoptera: Aphididoidea; Aoki 1975, 1977, 1987; see Stern and Foster 1996, 1997 for summaries). Aphids are cyclically parthenogenetic, with

*Table 10.1* Taxonomic Distribution of Insect Herbivore Societies Treated in this Chapter

| Order | Suborder | Superfamily | Family |
|---|---|---|---|
| LEPIDOPTERA | Heteocera | Bombycoidea | Eupterodidae |
| | | | Lasiocampidae |
| | | | Lemoniidae |
| | | | Saturniidae |
| | | Cossoidea | Cossidea |
| | | Gelechioidea | Coleophoridae |
| | | | Ethmiidae |
| | | | Oecophoridae |
| | | Geometroidea | Geometridae |
| | | Hesperioidea | Hesperiidae |
| | | Noctuoidea | Aganaidae |
| | | | Arctiidae |
| | | | Lymantriidae |
| | | | Noctuidae |
| | | | Notodontidae |
| | | | Oenosandridae |
| | | | Thaumetopoeidae |
| | | | Thyretidae |
| | | Pyraloidea | Pyralidae |
| | | Tineoidea | Galacticidae |
| | | Tortricoidea | Tortricidae |
| | | Uranioidea | Uraniidae |
| | | Yponomeutoidea | Heliodinidae |
| | | | Plutellidae |
| | | | Yponomeutidae |
| | | Zygaenoidea | Limacodidae |
| | | | Zygaenidae |
| | Rhopalocera | Papilionoidea | Lycaenidae |
| | | | Nymphalidae |
| | | | Papilionidae |
| | | | Pieridae |
| HYMENOPTERA | Symphyta | Megalodontoidea | Pamphiliidae |
| | | Tenthredinoidea | Argidae |
| | | | Diprionidae |
| | | | Pergidae |
| | | | Tenthredinidae |
| HEMIPTERA | Homoptera | Aphididoidea | Hormaphididae |
| | | | Pemphigidae |
| THYSANOPTERA | | | Phlaeothripidae |

Listed families have at least one reported social group; see Costa and Pierce (1996) for comprehensive treatment of the Lepidoptera.

several asexual generations punctuated by a bout of sexual reproduction, producing a "telescoping" of generations (Moran 1992). This reproductive pattern quickly builds up tremendous population sizes, as a result of which many aphid populations appear aggregated. Such aggregates are distinct from societies, however. Aphid societies are associations of adults and nymphs, often clonal, exhibiting specialized defender or soldier morphs (Aoki 1977, 1979, 1987; Itô 1989). The soldiers, which may or may not be sterile, possess horns or enlarged and spinous forefemorae used to pierce the integument of invertebrate predators. Gall-forming aphids show great host specificity, notably at the level of host family or genus (Wool 1984).

### 10.4.2 Thrips Societies

Thrips differ from aphids in feeding habits (they use piercing–sucking mouthparts to feed on mesophyll, pollen, or fungi), reproduction (sexual reproduction only), genetics (they are haplodiploid), and colony demographic structure (each society may contain one or more reproductive, nonclonal adult). The most complex thrips societies are found in the Phlaeothripidae. The adults of most social species possess enlarged forelegs, often used in male–male or female–female fighting; sexual dimorphism is common (Crespi and Mound 1997). In some species, presence of enlarged forelegs is coincident with reduced wings; such "micropterous" adults function as soldiers that defend the colony (Crespi 1992a, 1992b; Mound and Crespi 1995; Crespi and Mound 1997). Many thrips societies, notably the Australian species, are superficially similar to social aphids in nesting in galls or plant-derived domiciles such as phyllodes or abandoned lepidopteran leaf rolls. Gall thrips are invariably host-specific (Ananthakrishnan 1979), while noncecidogenic species vary from mono- to polyphagous.

### 10.4.3 Caterpillar Societies

Caterpillars are diplodiploid larval Lepidoptera that typically consume leaf or other plant tissue. Sociality is widely distributed among lepidopteran taxa and found in more than 20 families representing 13 ditrysian superfamilies (Costa and Pierce 1997; see Table 10.1). These societies vary considerably in foraging pattern, host specificity, voltinism, behavioral and structural defenses, and nest building, suggesting multiple independent origins of sociality in this order (Costa and Pierce 1997). Caterpillar societies are temporally constrained by the fact that they are composed of immatures that disperse prior to or following pupation. There is no clear pattern of host-plant specialization among social caterpillars; monophagy, oligophagy, and polyphagy are common (Costa and Pierce 1997).

### 10.4.4 Sawfly Societies

Sociality in sawflies is also a phenomenon of the larval stage. Sawflies are found in the hymenopteran suborder Symphyta. Like all Hymenoptera, sawflies are haplodiploid, with a predominantly arrhenotokous (female diploidy, male haploidy) mode of reproduction with scattered instances of thelytoky (parthenogenetic production of diploid females). Social species are found primarily in five families: Pamphiliidae (Megalodontoidea), Tenthredinidae, Diprionidae, Argidae, and Pergidae (Tenthredinoidea). Sawfly larvae are often treated as "caterpillars" because of the morphological and ecological similarities they share with lepidopteran larvae, but the two differ in important genetic and behavioral respects. Sawflies tend to be monophagous or oligophagous, though several are generalists.

## 10.5 Determinants of Genetic Structure at Microgeographic and Colony Scales

The population genetic effects of social interaction are most apparent at the smallest spatial scales. For present purposes, it is useful to focus on the levels of the host plant and the social group, microgeographic and colony structure, respectively. Structure at the colony level ("micro-microgeographic" in scale) often, but not always, corresponds to family structure, or genetic identity by common descent. The genetic structure of social groups is not static, however. Colony and microgeographic genetic structure vary considerably among social species due to varying mating, demographic, and resource-use behaviors, and within species due to local population density, mate competition, and the ability to maintain colony integrity (kin discrimination). For example, the initial makeup or composition of a social group depends on female mating and sperm utilization, and whether the colony is founded as a single- or multiple-family unit. Subsequent to colony founding, composition depends on likelihood of interaction with nonrelatives (itself dependent on resource use and local population density). The following sections will explore determinants of local and colony genetic structure, considering first factors relevant to dispersal and genetic relationships among colonies at microgeographic scales. These factors will then be related to colony founding and development over time.

### 10.5.1 Microgeographic (Host-Plant-Level) Structure: Setting the Stage for Colony Dynamics

Since colonies are spatially nested within host plants, genetic dynamics within and between colonies may be described from the point of view of either spatial scale. Mating and oviposition, which in part dictate intracolony structure, simultaneously influence intercolony (microgeographic) patterns. Life histories of socially interacting species uniquely shape localized levels of inbreeding and gene flow, factors that must be considered jointly with mating and oviposition behaviors.

Microgeographic population processes have been poorly studied in social herbivores, but several general predictions can be made on the basis of life history and behavior, and by extrapolation from solitary species. The following sections will summarize information for each of the four social herbivore groups, shedding light on the ways in which the behavioral ecology and life history of each influences microgeographic genetic patterns. Traits of particular interest are mating and dispersal. All of the insect groups treated here have mobile, winged adults, and gene flow at microgeographic spatial scales is probably substantial (see Peterson and Denno, Chapter 12, this volume). However, other factors such as inbreeding may boost local genetic identity. Microgeographic structure will therefore reflect a balance between dispersal and inbreeding, as well as selection.

### 10.5.1.1 Aphids

Social aphid colonies are usually founded independently by new fundatrices returning to the primary host from the secondary host. In the sexual secondary-host stage, migrants from various primary hosts form mixed aggregations. In addition to directed flight ability, adult aphids are wind dispersed (Taylor 1974; Dixon 1985), and can travel considerable distances (Hardy and Cheng 1986).

Several researchers have analyzed population structure and gene flow in solitary aphids. Loxdale (1990) analyzed gene flow in the cereal aphids *Sitobion avenae, S. fragariae*, and *Rhopalosiphum padi* using allozymes, and found high rates of migration using both $F_{ST}$ (standardized allele frequency variance among populations; Wright 1951) and private alleles (average frequency of alleles found in only a single deme or subpopulation; Slatkin 1985). Wöhrman and Tomiuk (1988) also used allozymes to infer high migration rates and low levels of inbreeding in populations of the rose aphid *Macrosiphum rosae*. Long-distance migration in the corn leaf aphid (*Rhopalosiphum maidis*) was suggested on the basis of seasonal fluctuation in allele frequencies at several allozyme loci by Steiner et al. (1985). Judging from these observations of nonsocial species, dispersal in social aphids may also be high. Dispersal, plus mixing at secondary hosts, probably results in weak or no microgeographic structure in these species.

### 10.5.1.2 Thrips

Thrips, like aphids, are both winged and easily dispersed by air currents in flight. Some species engage in mass flights, and many have been reported to move long distances and achieve great heights (Lewis 1973). Unfortunately, less information is available on thrips microgeographic movement patterns than for even the poorly characterized social aphids. Thrips lack the host alternation that contributes to outcrossing and genetic homogenization of aphid populations. Univoltine species are likely to experience sufficient movement to genetically homogenize populations at microgeographic spatial scales, but populations of multivoltine species may become structured through inbreeding or localized selection (Karban 1989; Crespi and Mound 1997).

### 10.5.1.3 Caterpillars

Lepidopteran population genetics is by far the best studied of the social herbivore taxa discussed here, yet gaps remain even within this intensively scrutinized group. In particular, social species as a group have not been the focus of study, again making it necessary to extrapolate from nonsocial species. Adult Lepidoptera are generally strong fliers, and many studies have established that population structure is weak (Eanes and Koehn 1978; Mitchell 1979; Pashley et al. 1985; Menken et al. 1992).

Peterson and Denno (Chapter 12, this volume) summarize gene flow estimates for 63 lepidopteran species from 15 families (among other insect taxa) and report high rates of gene flow for most, though a few exhibit significant levels of structure. The nymphalid butterflies *Euphydryas chalcedona* and *E. editha*, for example, are known to be relatively poor dispersers (Gilbert and Singer 1973), and this is reflected in relatively high $F_{ST}$ estimates (McKechnie et al. 1975). But population differentiation in *E. editha* may be due less to dispersal behavior than to local adaptations to host plants (Rausher 1982; Thomas and Singer, Chapter 14, this volume), as in other species that are mobile but genetically differentiated along host lines (e.g., *Papilio glaucus*; Hagen 1990).

Costa and Ross (1993, 1994) provide the most comprehensive population genetic treatment of a social lepidopteran, estimating gene flow and genetic structure hierarchically from colony- to macrogeographic spatial scales in eastern tent caterpillar (*Malacosoma americanum*) populations. This lasiocampid moth exhibits high rates of gene flow at all spatial scales above the colony. Strong colony-level structure corresponds to family structure, but multiple colonies occurring on common host plants (microgeographic scale) are unrelated due to high levels of local dispersal.

### 10.5.1.4 Sawflies

Sawfly dispersal is thought to be low, suggesting that populations are genetically structured at microgeographic scales, but this is based on behavioral inference rather than gene flow estimation. For example, Smith (1993) points out that the absence of sawflies in island fauna suggests they are poor long-distance migrants. Benson (1950) suggests that the combination of poor flying ability and relatively large body mass of many sawflies makes for ineffective dispersal either actively by flight or passively by wind. Several authors report differential dispersal abilities between the sexes: Maetô and Yoshida (1988) state that females of the pamphiliid genus *Cephalcia* are poor fliers and, after eclosion, commonly climb trees to lay their eggs, whereas the mobile males of this species fly to the tree crowns where mating occurs. Coppel and Benjamin (1965) state that among diprionid sawflies, males are superior dispersers, perhaps due to their smaller body mass.

Craig and Mopper (1993) consider dispersal in relation to likelihood of sib–sib inbreeding and its possible role in skewing sex ratios. These authors point out that the life history of many species (e.g., diprionids and tenthredinids) serves to undermine sib mating. There are two points at which interfamilial mixing can occur: the larval dispersal phase (typically in the ultimate or penultimate instar, when the group dissolves as the larvae seek to pupate in the soil), and posteclosion of adults, when individuals return to the tree canopy to mate. Local mixing is more likely, because females of many species pheromonally signal for mates (Jewett et al. 1976), and males tend to be quite mobile. Coppel and Benjamin (1965) also mention that the larvae of many species "migrate," presumably in response to local food depletion. The behavioral patterns reported by Craig and Mopper (1993) and Coppel and Benjamin (1965) contribute to localized dispersal and outbreeding but say little about microgeographic patterns of population differentiation. Differentiation is likely to be significant at both microgeographic and macrogeographic spatial scales, judging by the degree of movement reported in the studies cited here.

## 10.5.2 Colony Structure: Spatial and Temporal Patterns

### 10.5.2.1 Colony Founding

Dispersal behavior underlies intra- and intercolony genetic patterns by influencing local colony density and degree of relatedness between individuals within and among colonies. Colonies may be founded either by reproductive adults or by larval cohorts derived from one or more egg mass. Adult-founded colonies may in turn stem from a single inseminated female or male-female pair (haplometrotic founding), or may be founded by a group of coreproducing adults (pleometrotic founding). There are two interrelated factors determining the family structure of a newly founded colony: (1) number and relatedness of founding females, and (2) mating and sperm utilization by founding female(s).

#### 10.5.2.1.1 Number and Relatedness of Foundresses

Colony genetic structure declines in proportion to the number of founding females, but this decline is modified by the kin relationships between them as well as number and relatedness of (and to) their mates. The family relationships of some social insect colonies can be exceedingly complex due to the presence of multiple reproductives related by various degrees (e.g., polygynous fire ant colonies; see Ross 1993), but that of most social herbivore colonies is relatively simple.

*Aphids.* Newly founded aphid colonies are clonal families owing to the parthenogenetic production of offspring by the aphid stem mother (fundatrix). Aphid societies with specialized morphs are gall-makers, and galls are initiated by single females (with a few exceptions; Wool 1984) that often contest for optimal initiation sites on the host plant (Whitham 1979; Aoki and Makino 1982; Stern and Foster 1997; Foster 1996). Most aphids alternate hosts, first establishing a gall on a primary (generally woody) host and after several clonal generations switching to a secondary host, where the nonaggregated aphids feed on the roots. Social structure is largely restricted to the gall-dwelling stage, whereas the non-gall-dwelling clonal populations are more spatially diffuse (Moran 1993; Stern and Foster 1996a, 1996b).

*Thrips.* Mode of colony founding in thrips varies widely. Non-gall-forming social species are often polygynous, cofounded by multiple reproductive females that oviposit into a communal batch (e.g., *Elaphrothrips tuberculatus, Hoplothrips pedicularis,* and *H. karnyi;* Crespi 1986a, 1986b, 1988). Females congregate at male-defended group oviposition sites, and successfully competing males achieve reproductive dominance in the colony. No information exists on the relatedness of associated females or competing males.

Gall-forming thrips societies are founded either by single females (rarely two) or by a single male–female pair (Van Leeuwen 1956; Hill et al. 1982; Crespi 1992b), and so at least upon initiation may be described as monogynous colonies in which relatedness is probably high. Interestingly, as with cecidogenic aphids, females of several gall-forming thrips engage in fighting, apparently to secure superior gall initiation sites (Pelikán 1990; Crespi 1992a, 1992b; Crespi and Mound 1997). Some species that establish colonies within glued phyllodes may also fight for domicile location (e.g., *Panoplothrips* and *Carcinothrips*), whereas other phyllode-dwellers (e.g., *Dunatothrips aneurae*) are founded by either single or multiple females (Crespi and Mound 1997). Crespi and Mound report that colonies established in lepidopteran leaf-ties tend to be founded by multiple females and males, also conforming to a polygynous breeding pattern.

*Caterpillars.* All caterpillar societies are derived from clustered eggs, and adults are almost never present (but see Nafus and Schreiner 1988). Clutches are usually the product of a single female, but there are examples of communal or pooled oviposition in some Lepidoptera, which may occur as a consequence of limited oviposition sites (e.g., *Heliconius* spp.; see Turner 1971; Benson et al. 1976; Mallet and Jackson 1980). The average relatedness of individuals within a pooled egg cluster would, of course, decline in proportion to the number of contributing females, but this can be mitigated by maternal relatedness. No information exists on relatedness between communally ovipositing butterflies, but if limited resource availability is the primary reason for communal oviposition, it is possible that the butterfly populations are also highly patchy, and patchiness, coupled with low dispersal, could lead to local inbreeding and relatedness between females.

Single-maternity clutches are reported in a variety of social species, including eastern (*Malacosoma americanum*) and forest (*M. disstria*) tent caterpillars (Lasiocampidae), fall webworms (*Hyphantria cunea;* Arctiidae), walnut caterpillars (*Datana integerrima;* Notodontidae), and hackberry butterflies (*Asterocampa celtis;* Nymphalidae; Stehr and Cook 1968; Morris 1972; Warren and Tadic 1970; Johnson and Lyon 1988). Females of most social species deposit a single clutch of eggs, but some clutches are loosely clustered or are split into two or more egg masses (e.g., small ermine moths [*Yponomeuta* spp.; Yponomeutidae]; Menken et al. 1992). Colonies from such clutches are likely to merge soon after eclosion, as can those from adjacent egg masses deposited by different females (e.g., Fitzgerald and Willer 1983).

*Sawflies.* Like colonial caterpillars, sawfly societies are founded by cluster-ovipositing females, and adults are rarely present when the larvae emerge (but see Dias 1975, 1976). Simultaneous, pooled oviposition has not been directly observed, but Codella and Raffa (1995) report that the size of *Neodiprion lecontei* egg batches often exceeds the maximum clutch size found in dissections of single females. There are reports of juxtaposed clutches merging upon eclosion, creating a multiple-family supercolony (e.g., *Perga affinis* [Carne 1962], *Themos olfersii* [Dias 1975], and *Neodriprion lecontei* [Codella and Raffa 1995]). Genetic relationships within and between such clutches are unknown.

10.5.2.1.2  Founding Female Mating and Sperm Utilization.

Maternity of eggs in a clutch constitutes only part of the relatedness equation; paternity also influences group relatedness. Frequency of female matings with different (and unrelated) males provides only a crude index of the likely paternity of offspring, since females that mate multiply may selectively utilize the sperm of one or a few mates (Page 1986; Ross 1986). Nonetheless, polyandry is expected to reduce intracolony relatedness.

*Aphids.* Determination of paternity is irrelevant to aphid societies, because fundatrices reproduce parthenogenetically. Aphid societies are extended individuals, composed of up to thousands of identical clones. However, fundatrices produce both male and female clones; in some cases female oocytes are XX, males are XO (Dixon 1985).

*Thrips.* Intracolony relatedness in thrips is influenced by their haplodiploid mode of sex determination, which results in relatedness asymmetries: Thrips females develop from fertilized (diploid) eggs and males from unfertilized (haploid) eggs (arrhenotoky). Females are thus in principle related to each other by 0.75 on average, while they are related to males by 0.25. However, no data are currently available on intracolony relatedness in thrips societies (B. J. Crespi personal communication). On the basis of what is known of colony founding, relatedness is likely to vary widely. Species with single foundresses are likely to exhibit high intracolony relatedness, but little is known about the mating and sperm utilization of females prior to colony founding. Females in the colonial species described by Crespi (1986a, 1986b; 1988) appear to mate primarily with a single dominant male, but subordinate males may achieve up to approximately 20% of the matings. Varadarasan and Ananthakrishnan (1982) suggested that some galling species are singly mated, but mate competition is high in many species, and multiple mating is likely. Crespi (1992b) and Crespi and Mound (1997) indicate that some gall-dwellers spend multiple generations in their gall, and it is possible that some of these species are inbred. One outcome expected of such a breeding system is female-biased sex ratios (Hamilton 1967; Trivers and Hare 1976; Wrensch and Ebbert 1992), and thrips colonies do often exhibit an extreme female-biased sex ratio (Mound 1970, 1971; Raman and Ananthakrishnan 1984; Crespi 1992b; Crespi and Mound 1997).

*Caterpillars.* Mating frequency in Lepidoptera is highly variable, and knowledge of mating behavior is uneven. Single mating may be achieved through post-copulatory competitive measures such as mating plugs (e.g., acraeine and parnas-siine butterflies) or antiaphrodisiac pheromones (heliconiine butterflies; Ackery and Vane-Wright 1984). Spermatophore counts provide a convenient index of copulation frequency, and remarkably high mating frequencies have been inferred in this way. Solitary danaine butterflies, for example, are well known to mate multiply (Pliske 1973; Ehrlich and Ehrlich 1978; Ackery and Vane-Wright 1984), and 14 New World butterfly and hesperiid species analyzed by Pliske (1973) exhibited multiple mating, with a range of 2–15 spermatophores per female.

Mating frequency *per se* may be irrelevant if a "last in, first out" pattern of sperm utilization occurs, ensuring that most fertilizations are secured by a single mate (Parker 1970). It is equally likely, however, that some proportion of nonprimary sperm secures fertilizations. For example, Costa and Ross (1993) found that most fertilizations of eastern tent caterpillars (*M. americanum*) in a northeast Georgia population were attributable to the sperm of a single mate, but a small percentage of siblings manifested different paternity. Since Lepidoptera are diplodiploid, intracolony relatedness is probably on the order of 0.4–0.5 (on average) upon founding.

*Sawflies.* Relatively little is known of sawfly mating behavior, but most species do appear to mate, despite the fact that these haplodiploid insects can reproduce via unfertilized eggs. Lack of mating may be inferred from the presence of all-male colonies, but Craig and Mopper (1993) report that all-male colonies are rare, citing several studies specifically seeking and failing to observe such colonies, and in fact found that female-biased sex ratios are the norm. These authors also summarized life history and behaviors of several gregarious diprionid and ten-thredinid sawflies, and found a high incidence of polygynous and polygamous mating systems. Females of the solitary sawfly *Dineura virididorsata* mate multiply, for example (Walter et al. 1994), whereas in various *Neodiprion* species, males mate multiply while females tend to be monogamous (Benjamin 1955; Woods and Guttman 1987).

Relatedness may be high (on the order of 0.75) in many sawfly colonies because of female-biased sex ratios, since sisters are related by 0.75 on average by virtue of sharing a common haploid father. To the degree that sib-sib inbreeding occurs in sawfly populations (possible in group-pupating species under conditions of low population density; Craig and Mopper 1993), intracolony relatedness may be even greater. Sex-ratio bias may stem from intense mate competition, population genetic structure, or inbreeding (Hamilton 1967; Trivers and Hare 1976; Wrensch and Ebbert 1992), and in some sawfly species may be facultatively induced by host-plant quality (Craig et al. 1992; Mopper and Whitham 1992).

### 10.5.2.2 Colony Development: Temporal Changes in Colony Structure

Defense and foraging behaviors are interrelated factors that influence localized genetic patterns, creating spatial and temporal variance in microgeographic and colony genetic structure by influencing colony makeup and longevity. Colony longevity is especially relevant to temporal patterns; for example, long-lived colonies may cycle through sexual and asexual phases, whereas larval societies experience an annual turnover. Thus, while female number, relatedness, and offspring paternity establish initial colony genetic structure, behavioral traits such as defense and foraging determine how initial conditions are preserved, or change.

Defense is relevant to colony genetic structure, because it may preserve familial (or clonal) colony makeup. Of greatest importance in population genetic terms is defense against conspecifics; invasion or predation by extraspecific organisms may result in colony destruction, but only conspecific mixing can dilute family structure. Foraging pattern refers to mode of host use. Social insects may be patch-restricted, nomadic, or central-place foragers: patch-restricted foragers live in or on their food, feeding statically in their immediate vicinity. Nomadic foragers travel in groups among patches, establishing temporary "bivouacs" at each. Central-place foragers also actively seek food patches but return after each foraging bout to a more or less permanent common nest or resting site (Fitzgerald and Peterson 1988; Fitzgerald 1993, 1995; Costa and Pierce 1997).

Foraging behavior establishes whether colony members contact nonfamily members: patch-restricted foragers are essentially static, unlikely to encounter alien conspecifics unless patches coalesce or food depletion forces the colony to forage elsewhere. Nomadic and central-place foragers, on the other hand, are mobile and more likely to encounter nonfamily members. Family integrity may in principle be retained by kin discrimination (which has not been demonstrated for any social herbivore), or groups may merge. Likelihood or frequency of contact between independent colonies depends jointly on local colony density, resource availability, and foraging behavior.

### 10.5.2.2.1 Aphids

Aphid soldiers generally do not distinguish among invaders, and colonies of some species that experience high local densities may actually merge (Itô 1989). Although aphids do not differentiate kin from nonkin (Itô 1989; Foster 1990; Aoki et al. 1991; Sakata and Itô 1991; Aoki and Kurosu 1992; Carlin et al. 1994), at least one species exhibits "morph-recognition": *Ceratoglyphina bambusae* soldiers attack conspecific nonsoldiers early in the gall cycle, regardless of genetic relationship (Aoki et al. 1991). Generally speaking, aphids appear to be more concerned with invading predators than with conspecifics joining the colony. In fact, several authors report intergall movement in a variety of social aphid species, which are most mobile in the first instar (Aoki 1975; Eastop 1977).

Intergall dispersal has been observed in various pemphigine genera, including young nymphs of *Pachypappa marsupialis* (Aoki 1975) and defender morphs of *Pemphigus obesinymphae* (Moran 1993). Aoki (1979) reported dimorphic "migratory" and "nonmigratory" first-instar *P. marsupialis*; the migratory morphs may disperse to other conspecific galls whether or not those galls are occupied. Setzer (1980) inferred intergall migration in the genus *Pemphigus* from high levels of within-gall allozyme variability. Since galls are induced by single partheno-genetically reproducing female foundresses (the fundatrix), within-gall related-ness is expected to be 1.0. Thus, any intergall migration can dilute intragall genetic relationships, but the degree of dilution depends on migration frequency and interclone relatedness. Even at moderately high levels of invasion, dilution would be minimal if invaders originated in clone populations related to the stem-mother fundatrix.

In light of the prodigious parthenogenic output of fundatrices, it is possible that gall invasion does not affect colony genetic structure to any significant degree. It is also possible that because gall-producing aphids are patch-restricted foragers, opportunities for migration are limited to mobile soldiers that are often sterile (Stern and Foster 1996, 1997) and therefore pose no reproductive threat to the colony. A different gall genetic pattern would arise if the fundatrix were usurped. Several gall aphids are known to engage in gall parasitism, producing no gall of their own but seeking out and displacing congeners from their galls (Dixon 1985). For example, Akimoto (1981) reported that *Eriosoma yangi* females seek out galls produced by other *Eriosoma* species and kill the gall fundatrices. To the extent that this occurs in a conspecific context, galls may become clonal mosaics as the usurper begins reproducing.

*Thrips.* Little is known about genetic structure and intergall behavioral interactions in social thrips. Some insight can be gained from known life-history characteristics such as colony founding and expected gall life span. Nesting strategies vary (founders may induce galls, glue phyllodes, usurp or seek vacant lepidopteran leaf-ties, or opportunistically seek vacant galls or other suitable spaces; Crespi and Mound 1997) and are relevant to host-plant-level genetic patterns through their influence on nest longevity and colony composition and demographics. Domicile sites on leaves are more ephemeral than those on stems, for example, since leaves are periodically shed from the host plant.

It may be more important for thrips soldier morphs to defend the colony from predators and usurpers than from conspecifics, as with aphid defenders. Indeed, conspecifics may pose no threat (conspecific soldiers transplanted into nonnatal galls neither attack gall residents nor are themselves attacked; B.J. Crespi perssonal communication). Thus, colony genetic structure need not be actively preserved at all, or preservation may be an incidental by-product of low local population density (precluding opportunities to mix with nonrelatives) or behavioral constraint (individuals that do not wander or locally disperse do not have the opportunity to encounter other colonies). Unlike some cecidogenic aphids (discussed earlier), gall-dwelling thrips with soldier morphs do not appear to migrate among galls (B.J. Crespi personal communication). Initial conditions of genetic relatedness are likely to be stable in such colonies.

Ephemeral (annual) colonies appear simple in terms of genetic composition, because only a single generation occupies the gall, sometimes without the founding parent(s) (Crespi and Mound 1996). In contrast to this life-history pattern, species-inducing galls on temporally stable parts of the host plant may be multivoltine. The Australian species *Iotatubothrips crozieri*, for example, makes large woody-stem galls on its host plant (Mound and Crespi 1992; Crespi 1992b). Thus, thrips can reproduce within these galls for many generations and is perhaps inbred as a result (Crespi 1992b). Similarly, Mound and Crespi (1994) report that micropterous *Oncothrips tepperi* and *O. habrus* adults reproduce within galls long after the foundress has died, also raising the possibility of inbreeding. Intragall inbreeding over time would lead to a temporal genetic pattern inverse to that expected in gall-aphids with intergall migration: an increase rather than a decline in intracolony genetic identity over time.

*Caterpillars.* The temporal stability of initial genetic structure in caterpillar societies depends on foraging behavior and nutritional needs, both of which influence the likelihood of searching the host plant for optimal feeding sites or abandoning depleted host plants. Nomadic and central-place foragers are likely to encounter conspecifics as they search for food, and colony genetic structure would be preserved only by nestmate or kin discrimination. In the only study addressing discriminatory behavior in caterpillars, Costa and Ross (1993) found that *M. americanum* larvae cannot or do not distinguish kin from nonkin, one effect of which is steady erosion of colony genetic structure as the larvae forage in the vicinity of conspecifics. These caterpillars selectively feed on young, twig-tip foliage. Since this resource is patchily distributed on the host plant, the larvae tend to search increasingly farther from the immediate vicinity of the tent as they deplete nearby young foliage, bringing the larvae into contact with caterpillars from other colonies.

Influencing the likelihood of such interactions is local population density. Multiple family groups deposited on a common host plant are likely to merge in eastern tent caterpillar populations (Fitzgerald and Willer 1983; Costa and Ross 1993), though this depends in part on spatial dispersion of colonies and the size of the tree resource. In the case of eastern tent caterpillars, ovipositing females are apparently unable to assess host-plant size and often leave their egg masses on young saplings with insufficient resources to support the colony through pupation (Fitzgerald 1995). Multiple colonies on a small tree are very likely to encounter one another, whereas the same number distributed on a much larger tree may be able to complete development without foraging very far from the tent (Costa and Ross 1993).

Although one might predict greater temporal stability in patch-restricted foragers capable of completing larval development within their initial food patch, this also depends on local population density. The leaf-and-silk nests of the social geometrid *Hydria prunivorata* are initially constructed by larvae from a single egg mass, but under conditions of heavy infestation larvae from different egg masses may merge to form larger nests (Schultz and Allen 1975).

*Sawflies.* Sawfly colony dynamics conform to the same pattern exhibited by social Lepidoptera. Colonies tend to engage in patch-restricted or nomadic foraging, with *de facto* central-place foraging found among species such as *Perga affinis* that return to rest sites on the bole of the host tree (Carne 1962). Under conditions of high local population density, multiple-family groups of this species appear to form supercolonies. Females of the Australian pergine sawflies insert pod-like clusters of eggs into leaves of their *Eucalyptus* host plant, and larval groups readily merge (Carne 1962; Macdonald and Ohmart 1993). Similarly, *Neodiprion rugifrons* colonies often merge and split as they move about on their jack pine host (Wilkinson et al. 1966). Many diprionid sawflies with gregarious larvae reach outbreak densities (Craig and Mopper 1993), and such species may experience family mixing as a result.

A variety of conditions can maintain initial intracolony relatedness despite intercolony mixing. By split-clutch oviposition, for example, merging colonies may consist of sibs (e.g., *Neodiprion rugifrons*: Wilkinson et al. 1966 and *N. lecontei*: Codella and Raffa 1995). Also, many pergids are known to reproduce through amphitokous parthenogenesis, and in cases where females of such species oviposit multiple clutches (e.g., *Pergagrapta* and *Perga*: Macdonald and Ohmart 1993), merging colonies may be genetically identical.

### 10.5.2.3 Summary and Conclusions

Throughout this chapter, I have emphasized how life history and behavioral ecology jointly shape spatial and temporal genetic patterns in social herbivore populations. In principle, factors maintaining microgeographic and colony genetic structure include (1) low local population density, (2) patch-restricted foraging, (3) kin discrimination, (4) inbreeding/low dispersal, and (5) philopatry. Those tending to undermine structure include (1) multiple mating, (2) group (multifamily) oviposition, (3) nomadic or central-place foraging, and (4) inability to discriminate kin/nestmates. Based on the taxa treated here, it is possible to sketch some generalizations with respect to expected colony and microgeographic genetic patterns in social herbivore populations (Table 10.2).

All groups, with the possible exception of sawflies, are highly dispersed, leading to weak genetic structure above the colony level. Colony-level genetic structure is probably high to moderate for all groups but may change in different ways. Aphid colonies are clonal, and relatedness is probably high despite intercolony migration. Relatedness in thrips societies should vary considerably according to social system (polyandry, polygamy, and monogyny are all represented). Thrips societies are often highly female biased, and haplodiploidy may facilitate high intracolony relatedness. Caterpillar colonies probably experience high initial relatedness, but under high-density conditions, colony fusion could rapidly dilute family structure. Sawflies similarly experience family dilution, but intracolony relatedness is likely to be initially higher in these colonies, again due to haplodiploidy and female-biased sex ratio.

The temporal dimension of colony genetic structure is influenced by foraging and dispersal behavior. Cecidogenic or patch-restricted species are primarily static foragers that experience fewer opportunities for colony mixing than nomadic or central-place foragers. Colony longevity probably plays a role as well; perennial colonies may have inbreeding opportunities unavailable to larval cohorts that annually disperse.

Table 10.2 Life-History, Colony, and Population Genetic Characteristics of Social Insect Herbivores

| Taxon | Colony founding[1] | Demography[2] | Foraging pattern[3] | Specialized defenders | Relatedness[4] Founding | Maturity | Microgeographic dispersal[4] |
|---|---|---|---|---|---|---|---|
| Aphids | M | A + I | PR; Cecidogenic | Yes | H | H | H |
| Thrips | M, PolyG Poly A | A + I | Mostly PR; Cecido. Some N, CP | Yes | M-H | M-H | H |
| Caterpillars | Mostly M; Some PolyA | I | PR, N, CP | No | Mostly H | L-H | H |
| Sawflies | Mostly M; Some PolyA | I | PR, N | No | H | M-H | M-L |

[1] M = Monogynous, PolyA = Polyandrous, PolyG = Polygynous

[2] A = Adults, I = Immatures

[3] PR = Patch-restricted, N = Nomadic, CP = Central-place

[4] H = High, M = Moderate, L = Low

## 10.6 Spatial and Temporal Population Genetic Dynamics: Patterns and Implications

Sociality is a life-history strategy that shapes spatial and temporal population genetic parameters in interesting ways. The spatial dimension of population genetic patterns can be visualized with a topographic "structure landscape" model (Fig. 10.3), representing scalar genetic identity relationships in Euclidean space. In social herbivore populations, at highly localized spatial scales it is appropriate to focus on genetic identity by descent–family structure. The amplitude and contours of the genetic features of a particular population at a particular point in time may be described as *acute* or *diffuse*. An acute social structure (Fig. 10.3A) corresponds to highly localized, discrete family units, as in cecidogenic aphids and thrips upon colony founding, and in early-season caterpillar and sawfly societies founded by singly mated females. A diffuse social structure (Fig. 10.3B) corresponds to colonies founded by multiple females (polygyny) and multiply mated females (polyandry), as well as species experiencing colony mixing by inter-colony migration or fusion later in colony phenology.

The fact that structure landscapes may change over time (peaks may erode) underscores the importance of the *temporal* dimension of genetic patterns in social herbivore populations, though few studies have directly addressed this (but see McCauley et al. 1988; Costa and Ross 1993). Figure 10.4 presents generalized relatedness curves corresponding to social species with alternative life-history and social-behavioral characteristics, including univoltine (Fig. 10.4A-B) and multi-voltine (Fig. 10.4C-E) life cycles, and traits tending to preserve (Fig. 10.4B-C) or undermine (Fig. 10.4A, D-E) colony family structure.

The heuristic patterns diagrammed in Figures 10.3 and 10.4 illustrate the significant role that behavioral ecology plays in shaping small-scale population genetic patterns. This idea is not new, of course, but the suite of behavioral characters constituting sociality and/or aggregation (and thus underlying colony- and microgeographic-scale genetic structure) is commonly overlooked in treatments of phytophagous insect population genetics. Insofar as such behaviors determine the spatial and temporal association of genotypes, they create a population genetic dynamic absent from most solitary insect populations.

This dynamic has practical consequences for studies of evolutionary and ecological genetics, notably in terms of framing and testing models of host-plant use and insect–plant interactions. For example, sampling regimes should be based on knowledge of insect natural history and behavioral ecology, informing how and when to assess populations, depending on the questions of interest. Failure to account for social structure in higher level population genetic surveys may result in sampling error by treating pedigree-linked individuals as independent genetic entities. This is akin to the problem of nonindependence of ramets in plant population genetics—the identity and genetic relationship of individuals composing a putative

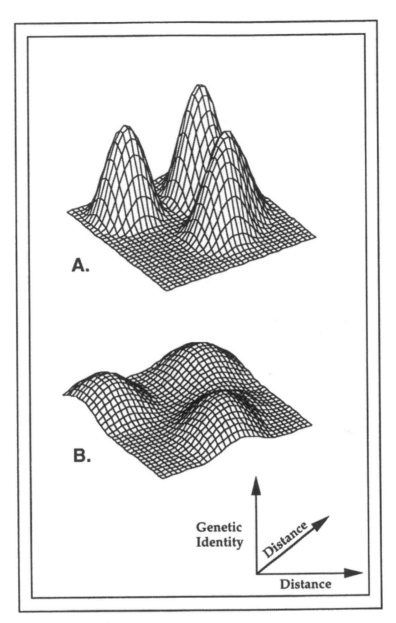

*Figure 10.3* Acute (A) and diffuse (B) genetic structure landscapes. In social insect populations, localized (colony) genetic structure corresponds to family structure. Acute landscapes
are characteristic of well-defined, strongly family-structured populations; diffuse landscapes
are characteristic of mixed-family colonies. Colony structure depends on number of
foundress mates, foundress number, local colony density, and kin discrimination (influencing
genetic admixure between colonies). Prevailing patterns are likely to fluctuate over time;
many species experience an "erosion" from acute to diffuse landscapes as colonies develop.

*Figure 10.4* Generalized intracolony relatedness curves. Tick marks indicate generation midpoints. Patterns over time are broadly determined by life history (uni- vs. multivoltine, annual vs. perennial, immatures only vs. adults plus immatures) and behavior (colony founding, mating, oviposition, kin discrimination). Sources of variance are numerous, including mating frequency, sperm-use patterns, local population density, and frequency of colony usurpation/kleptoparasitism. A. Univoltine colonies, the members of which lack kin-discrimination abilities. Relatedness is initially high, but erodes as groups forage and mix *ad libitum*. B. Univoltine, single-foundress colonies, with mechanisms such as kin discrimination, low colony density, or low dispersal preserving family structure. High relatedness prevails until colony abandonment. C. Multivoltine, single-foundress colonies with family structure preserved. D. Multivoltine colonies founded by a single female or single male–female pair: (i) inbreeding increases intracolony relatedness over time; (ii) relatedness dilution over time due to intercolony migration. E. Multiple-foundress or multiply-mated single foundress colony: (i) mixed-family colony persists as founded; (ii) mixed-family colony experiencing colony fusion.

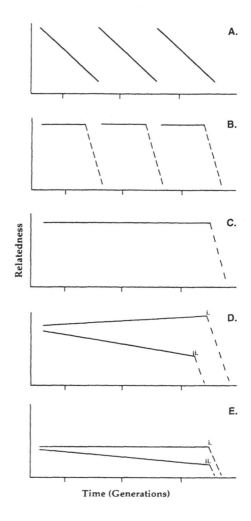

Time (Generations)

population must be known if model assumptions are to be met. In the case of social insect populations, nonindependence of genotypes within families necessitates sampling a single individual per group to avoid obtaining inflated allele and genotypic frequencies for higher level analyses. Socially mediated population dynamics also have relevance to *timing* of population studies. Some life stages are more likely to provide independent population samples than others (e. g., lepidopteran or symphytan adults rather than aggregated larvae), which means that certain times of the year may be more appropriate than others for assessing some population genetic parameters.

Accounting for intrademic or social structure in populations also has important implications for conservation genetics. Sugg et al. (1996) point out that social structure provides an alternative perspective on the causes and consequences of spatial partitioning of genetic variation in populations, and, most importantly, bears on the rate of loss of genetic variation and its implications for population genetic management efforts aimed, for example, at minimizing inbreeding. Such management concerns are perhaps less relevant for social insects than for the social vertebrates discussed by Sugg et al. (1996), but the population genetic and evolutionary effects of sociality are precisely the same: Social behavior can influence genetic diversity through hierarchical partitioning and its effects on local inbreeding and effective population size.

While these examples illustrate how intrademic or social structure should inform *studies* of ecology and evolution, such structure may also bear significantly on the evolutionary process itself. Insofar as sociality sets up a population genetic context not present in populations of solitary insects, it provides an example of how behavioral ecology potentially shapes selection parameters. For example, intrademically structured populations such as those of social organisms may experience group or kin selection, or frequency-dependent selection on traits influenced by group size or behavior. McCauley and Goff (Chapter 9, this volume) show that strong intrademic structure, such as that found in social insect populations, can influence fitness through frequency-dependent selection by favoring the evolution of traits with positive frequency-dependent fitness effects and limiting those with negative frequency-dependent fitness effects.

Strong localized genetic structure may also be the outcome of selection and adaptation to host plants (Mopper 1996). The deme-formation hypothesis (Edmunds and Alstad 1978, 1981), for example, posits that host-specific organisms may experience selection for adaptation to *individual* conspecific host plants (see Hanks and Denno, Chapter 11, this volume; Cobb and Whitham, Chapter 3, this volume for discussion). Such species would most likely be highly sedentary and multivoltine (e.g., coccids), experiencing multiple bouts of reproduction in such a way that host-plant-level "demes" are built. Such "demes" are actually extended family groups, and knowledge of mobility (gene flow) is centrally important to evaluating the intensity of selection should localized adaptation be observed. Mopper et al. (1995) found such fine-scale adaptive patterns in populations of the mobile cosmopterigid leafminer *Stilbosis quadricustatella*, for example, suggesting that host-plant-imposed selection intensity is strong in this insect.

Finally, another intriguing perspective is that strong localized genetic structure may set up its own selective milieu that subsequently influences the amount and distribution of genetic variation in a species over evolutionary time (see Mopper 1996). One example of this is the niche-variation hypothesis (Snyder 1974), which suggests that populations of social insects are buffered by their nest microenvironment, which may explain the low levels of genetic variation observed in many social insects (especially social Hymenoptera; see Shoemaker et al. 1992).

All of these examples illustrate the relevance of jointly addressing both behavioral ecology and population genetics in the study of phytophagous insect populations, recognizing the interrelationship of population genetic patterns and processes, and the behavioral and ecological traits on which they depend. Knowledge of genetic patterns at different spatial scales provides a basis for inferring key behavioral traits such as mating frequency, oviposition pattern, and dispersal, and knowledge of behavioral traits provides a framework in which to construct population genetic hypotheses to better understand the evolutionary significance of observed genetic patterns. The behavioral–population–genetic interface is thus an often neglected spatial scale rich in pattern and process, and social behaviors in particular put a unique spatial and temporal spin on the patterns found at this interface.

## Acknowledgments

I am grateful to Susan Mopper and Sharon Strauss for kindly extending the invitation to contribute to this volume and for their considerable editorial efforts. Many thanks to Bernie Crespi, Merrill Peterson, Leon Raijman, Bill Wcislo, and Dessie Underwood for sharing unpublished or in-press manuscripts, or otherwise sharing ideas and information about their study organisms. Thanks, too, to Mary Sears and Ronnie Broadfoot of the Museum of Comparative Zoology (MCZ)'s Ernst Mayr Library for invaluable assistance tracking down obscure literature, and to David Smith for kindly providing assistance with the sawfly literature. This manuscript has greatly profited from the comments and criticism of Peter Goff, David McCauley, Susan Mopper, Naomi Pierce, Mary Carol Rossiter, Kathrin Sommer, and Sharon Strauss. Finally, I appreciate the support of the National Science Foundation and the Alfred P. Sloan Foundation through their joint Postdoctoral Research Fellowship in Molecular Evolution during the preparation of this manuscript.

## 10.7 References

Ackery, P. R. and R. I. Vane-Wright. 1984. *Milkweed Butterflies: Their Cladistics and Biology.* Comstock Publishing Associates, Ithaca, NY.

Akimoto, S. 1981. Gall formation by *Eriosoma fundatrices* and gall parasitism in *Eriosoma yangi* (Homoptera, Pemphigidae). *Kontyû* 49:426–436.

Ananthakrishnan, T. N. 1979. Biosystematics of Thysanoptera. *Annu. Rev. Entomol.* 24:159–183.

Aoki, S. 1975. Descriptions of the Japanese species of *Pemphigus* and its allied genera (Homoptera: Aphidoidea). *Insecta Matsumurama, New Series* 5:1–61.

Aoki, S. 1977. *Colophina clematis* (Homoptera: Pemphigidae), an aphid species with "soldiers." *Kontyû* 45:276–282.

Aoki, S. 1979. Dimorphic first instar larvae produced by the fundatrix of *Pachypappa marsupialis* (Homoptera, Aphidoidea). *Kontyû* 47:390–398.

Aoki, S. 1987. Evolution of sterile soldiers in aphids. Pages 53–65 *in* Y. Ito, J. L. Brown and J. Kikkawa (Eds.), *Animal Societies: Theories and Facts.* Japanese Scientific Socity Press, Tokyo, Japan.

Aoki, S. and U. Kurosu. 1992. No attack on conspecifics by soldiers of the gall aphid *Ceratoglyphina bambusae* (Homoptera) late in the season. *Japanese J. Entomol 60*:707–713.

Aoki, S., U. Kurosu, and D. Stern. 1991. Aphid soldiers discriminate between soldiers and non-soldiers, rather than between kin and non-kin, in *Ceratoglyphina bambusae. Anim. Behav. 42*:865–866.

Aoki, S. and S. Makino. 1982. Gall usurpation and lethal fighting among fundatrices of the aphid *Epipemphigus niisimae* (Homoptera: Pemphigidae). *Kyontû 50*:365–376.

Avise, J. C. 1994. *Molecular Markers, Natural History, and Evolution.* Chapman & Hall, New York.

Batra, S. W. T. 1966. Nests and social behavior of halictine bees of India (Hymenoptera: Halictidae). *Indian J. Entomol. 28*:375–393.

Benjamin, D. M. 1955. The biology and ecology of the red-headed pine sawfly. USDA Forest Service Technical Bulletin No. 1118.

Benson, R. B. 1950. An introduction to the natural history of British sawflies. *Trans. Soc. Brit. Entomol. 10*:45–142.

Benson, W. W., K. S. Brown, and L. E. Gilbert. 1976. Coevolution of plants and herbivores: Passionflower butterflies. *Evolution 29*:659–680.

Carlin, N. F., D. S. Gladstein, A. J. Berry, and N. E. Pierce. 1994. Absence of kin discrimination behavior in a soldier-producing aphid, *Ceratovacuna japonica* (Hemiptera: Pemphigidae; Cerataphidini). *J. NY Entomol. Soc. 102*:287–298.

Carne, P. B. 1962. The characteristics and behaviour of the saw-fly *Perga affinis affinis* (Hymenoptera). *Aust. J. Zool. 10*:1–38.

Codella, S. G. and K. F. Raffa. 1995. Contributions of female oviposition patterns and larval behavior to group defense in conifer sawflies (Hymenoptera: Diprionidae). *Oecologia 103*:24–33.

Coppel, H. C. and D. M. Benjamin. 1965. Bionomics of the Nearctic pine-feeding diprionids. *Annu. Rev. Entomol. 10*:69–96.

Costa, J. T. and T. D. Fitzgerald. 1996. Developments in social terminology: Semantic battles in a conceptual war. *Trends Ecol. Evol. 11*:285–289.

Costa, J. T. and N. E. Pierce. 1997. Social evolution in the Lepidoptera: Ecological context and communication in larval societies. Pp. 407–442 *in* J. C. Choe and B. J. Crespi (Eds.), *The Evolution of Social Behaviour in Insects and Arachnids.* Cambridge University Press, Cambridge, UK.

Costa, J. T. and K. G. Ross. 1993. Seasonal decline in intracolony genetic relatedness in eastern tent caterpillars: Implications for social evolution. *Behav. Ecol. Sociobiol. 32*:47–54.

Costa, J. T. and K. G. Ross. 1994. Hierarchical genetic structure and gene flow in macrogeographic populations of the eastern tent caterpillar (*Malacosoma americanum*). *Evolution 48*:1158–1167.

Craig, T. M., and S. Mopper. 1993. Sex ratio variation in sawflies. Pp. 61–92 *in* M. Wagner and K. F. Raffa (Eds.), *Sawfly Life History Adaptations to Woody Plants*. Academic Press, San Diego, CA.

Craig, T. P., P. W. Price, and J. K. Itami. 1992. Facultative sex ratio shifts by a herbivorous insect in response to variation in host plant quality. *Oecologia 92*:153–161.

Crespi, B. J. 1986a. Territoriality and fighting in a colonial thrips, *Hoplothrips pedicularis*, and sexual dimorphism in Thysanoptera. *Ecol. Entomol. 11*:119–130.

Crespi, B. J. 1986b. Size assessment and alternative fighting tactics in *Elaphrothrips tuberculatus*, (Insecta: Thysanoptera). *Anim. Behav. 34*:1324–1335.

Crespi, B. J. 1988. Risks and benefits of lethal male fighting in the polygynous, colonial thrips *Hopolothrips karnyi*. *Behav. Ecol. Sociobiol. 22*:293–301.

Crespi, B. J. 1992a. Eusociality in Australian gall thrips. *Nature 359*:724–726.

Crespi, B. J. 1992b. Behavioural ecology of Australian gall thrips (Insecta, Hysanoptera). *J. Nat. Hist. 26*:769–809.

Crespi, B. J. and J. C. Choe (Eds.). 1997. *The Evolution of Social Behaviour in Insects and Arachnids*. Cambridge University Press, Cambridge, UK.

Crespi, B. J., and L. A. Mound. 1997. Ecology and evolution of social behavior among Australian gall thrips and their allies. *In* J.C. Choe and B.J. Crespi (Eds.), *The Evolution of Social Behaviour in Insects and Arachnids*. Cambridge University Press, Cambridge, UK. Pp. 166–180.

Crespi, B. J. and D. Yanega. 1995. The definition of eusociality. *Behav. Ecol. 6*:109–115.

Crozier, R. H., P. Pamilo, and Y. C. Crozier. 1984. Relatedness and microgeographic genetic variation in *Rhytidoponera mayri*, an Australian arid-zone ant. *Behav. Ecol. Sociobiol. 15*:143–150.

Darwin, C. R. 1859. *On the Origin of Species by Means of Natural Selection*. J. Murray, London.

Dias, B. F. de S. 1975. Comportamento pre-social de sinfitas do Brasil Central. I. *Themos olfersii* (Klug) (Hymenoptera: Argidae). *Stud. Entomol. 18*:401–432.

Dias, B. F. de S. 1976. Comportamento pre-social de sinfitas do Brasil Central. II. *Dielocerus diasi* Smith (Hymenoptera: Argidae). *Stud. Entomol. 19*:461–501.

Dixon, A. F. G. 1985. *Aphid Ecology*. Blackie and Son, Ltd., Glasgow, Scotland.

Eanes, W. F. and R. K. Koehn. 1978. An analysis of genetic structure in the monarch butterfly, *Danaus plexippus L. Evolution 32*:784–797.

Eastop, V. F. 1977. Worldwide importance of aphids as virus vectors. Pp. 3–62 *in* K. F. Harris and K. Maramorosch (Eds.), *Aphids as Virus Vectors*. Academic Press, New York.

Edmunds, G. F. and D. N. Alstad. 1978. Coevolution in insect herbivores and conifers. *Science 199*:941–945.

Edmunds, G. F. and D. N. Alstad. 1981. Responses of black pineleaf scales to hostplant variability. Pp. 29–38 *in* R. Denno and H. Dingle (Eds.), *Insect Life History Patterns*. Springer-Verlag, New York.

Ehrlich, A. H. and P. R. Ehrlich. 1978. Reproductive strategies in the butterflies: I. Mating frequency, plugging, and egg number. *J. Kansas Entomol. Soc. 51*:666–697.

Eickwort, G. C. 1981. Presocial insects. Pp. 199–280 *in* H. R. Hermann (Ed.), *Social Insects,* Vol II. Academic Press, New York.

Fisher, R. A. 1930. *The Genetical Theory of Natural Selection.* Claredon Press, Oxford, UK.

Fitzgerald, T. D. 1993. Sociality in caterpillars. Pp. 372–403 *in* N. E. Stamp and T. M. (Eds.), *Casey Caterpillars: Ecological and Evolutionary Constraints on Foraging.* Chapman & Hall, New York.

Fitzgerald, T. D. 1995. *The Tent Caterpillars.* Cornell University Press, Ithaca, NY.

Fitzgerald, T. D. and D. E. Miller. 1983. Tent building behavior of the eastern tent caterpillar *Malacosoma americanum* (Lepidoptera: Lasiocampidae). *Journal of the Kansas Entomological Society 56*:20–31.

Fitzgerald, T. D. and S. C. Peterson. 1983. Elective recruitment communication by the eastern tent caterpillar (*Malacosoma americanum*). *Anim. Behav. 31*:417–442.

Fitzgerald, T. D. and S. C. Peterson. 1988. Cooperative foraging and communication in caterpillars. *BioScience 38*:20–25.

Foster, W. A. 1990. Experimental evidence for effective and altruistic colony defense against natural predators by soldiers of the gall-forming aphid *Pemphigus spyrothecae* (Hemiptera: Pemphigidae). *Behav. Ecol. Sociobiol. 27*:421–430.

Foster, W. A. 1996. Duelling aphids: Intraspecific fighting in *Astegopteryx minuta* (Homoptera: Hormaphididae). *Anim. Behav. 51*:645–655.

Gadagkar, R. 1994. Why the definition of eusociality is not helpful to understand its evolution and what should we do about it. *Oikos 70*:485–487.

Gilbert, L. E. and M. C. Singer. 1973. Dispersal and gene flow in a butterfly species. *Am. Nat. 107*:58–72.

Grafen, A. 1984. Natural selection, kin selection, and group selection. Pp. 62–84 *in* J. R. Krebs and N. B. Davies (Eds.), *Behavioral Ecology: An Evolutionary Approach.* Sinauer Associates, Sunderland, MA.

Hagen, R. H. 1990. Population structure and host use in hybridizing subspecies of *Papilio glaucus* (Lepidoptera: Papilionidae). *Evolution 44*:1914–1930.

Haldane, J. B. S. 1932. The Causes of Evolution. Longmans, Green, New York.

Hamilton, W. D. 1964. The genetical evolution of social behaviour. I and II. *J. Theor. Biol. 7*:1–52.

Hamilton, W. D. 1967. Extraordinary sex ratios. *Science 156*:477–488.

Hamilton, W. D. 1972. Altruism and related phenomena, mainly in social insects. *Annu. Rev. Ecol. Syst. 3*:193–232.

Hardy, A. C. and L. Cheng. 1986. Studies in the distribution of insects by aerial currents: III. Insect drift over the sea. *Ecol. Entomol. 11*:283–290.

Hill, D. S., P. M. Hore, and I. W. B. Thorton. 1982. *Insects of Hong Kong.* Hong Kong University Press, Hong Kong.

Itô, Y. 1989. The evolutionary biology of sterile soldiers in aphids. *Trends Ecol. Evol. 4*:69–73.

Jewett, D. M., F. Matsumura, and H. C. Coppel. 1976. Sex pheromone specificity in the pine sawflies: Interchange of acid moieties in an ester. *Science 192*:51–53.

Johnson, W. T. and H. H. Lyon. 1988. *Insects That Feed on Trees and Shrubs.* Cornell University Press, Ithaca, NY.

Karban, R. 1989. Fine-scale adaptation of herbivorous thrips to individual host plants. *Nature 340*:60–61.

Laidlaw, H. H. Jr. and R. E. Page Jr. 1984. Polyandry in honey bees (*Apis mellifera* L.): Sperm utilization and intracolony genetic relationships. *Genetics 108*:985–997.

Lester, R. J. and R. K. Selander. 1981. Genetic relatedness and the social organization of *Polistes. Am. Nat. 117*:147–166.

Lewis, T. 1973. *Thrips: Their Biology, Ecology, and Economic Importance.* Academic Press, New York.

Loxdale, H. D. 1990. Estimating levels of gene flow between natural populations of cereal aphids (Homoptera: Aphididae). *Bull. Entomol. Res. 80*:331–338.

Macdonald, J. and C. P. Ohmart. 1993. Life history strategies of Australian pergid sawflies and their interactions with hosts. Pp. 485–502 *in* M. Wagner and K. F. Raffa (Eds.), *Sawfly Life History Adaptations to Woody Plants.* Academic Press, San Diego, CA.

Maetô, K. and N. Yoshida. 1988. Characteristics of the oviposition of the red-headed spruce web-spinning sawfly, *Cephalcia isshikii* Takeuchi (Hymenoptera: Pamphiliidae). *Appl. Entomol. Zool. 23*:361–362.

Mallet, J. L. and D. A. Jackson. 1980. The ecology and social behaviour of the Neotropical butterfly *Heliconius xanthocles* Bates in Colombia. *Zool. J. Linn. Soc. 70*:1–13.

Maynard Smith, J. 1976. Group selection. *Quart. Rev. Biol. 51*:277–283.

McCauley, D. E., M. J. Wade, F. J. Breden, and M. Wohltman. 1988. Spatial and temporal variation in group relatedness: Evidence from the imported willow leaf beetle. *Evolution 42*:184–192.

McKechnie, S. W., P. P. Ehrlich, and R. R. White. 1975. Population genetics of *Euphydryas* butterflies: I. Genetic variation and the neutral hypothesis. *Genetics 81*:571–594.

Menken, S. B. J., W. M. Herrebout, and J. T. Wiebes. 1992. Small ermine moths (*Yponomeuta*): Their host relations and evolution. *Annu. Rev. Entomol. 37*:41–66.

Michener, C. D. 1953. Problems in the development of social behavior and communication among insects. *Trans. Kan. Acad. Sci. 56*:1–15.

Michener, C. D. 1958. The evolution of social behavior in bees. *Proc. 10th Int. Cong. Entomol. 2*:441–447.

Michener, C. D. 1969. Comparative social behavior of bees. *Annu. Rev. Entomol. 14*:299–342.

Michod, R. E. 1982. The theory of kin selection. *Annu. Rev. Ecol. Syst. 13*:23–55.

Michod, R. E. and W. W. Anderson. 1979. Measures of genetic relationship and the concept of inclusive fitness. *Am. Nat. 114*:637–647.

Mitchell, E. R. 1979. Migration by *Spodoptera exigua* and *S. frugiperda*, North American style. Pp. 386–393 *in* R. L. Rabb and G. G. Kennedy (Eds.), *Movement of Highly Mobile Insects.* North Carolina State University Graphics, Raleigh, NC.

Mopper, S. 1996. Adaptive genetic structure in phytophagous insect populations. *Trends Ecol. Evol. 11*:235–238.

Mopper, S., M. Beck, D. Simberloff, and P. Stiling. 1995. Local adaptation and agents of selection in a mobile insect. *Evolution 49*:810–815.

Mopper, S. and T. G. Whitham. 1992. The plant stress paradox: Effects on pinyon sawfly fecundity and sex ratios. *Ecology 73*:515–525.

Moran, N. A. 1992. The evolution of aphid life cycles. *Annu. Rev. Entomol. 37*:321–348.

Moran, N. A. 1993. Defenders in the North American aphid *Pemphigus obesinymphae. Insectes Sociaux 40*:391–402.

Morris, R. F. 1972. Fecundity and colony size in natural populations of *Hyphantria cunea. Can. Entomol. 104*:399–409.

Mound, L. A. 1970. Intragall variation in *Brithothrips fuscus* Moulton with notes on other Thysanoptera-induced galls on *Acacia* phyllodes in Australia. *Entomol. Mon. Mag. 105*:159–162.

Mound, L. A. 1971. Gall-forming and allied species (Thysanoptera: Phlaeothripidae) from *Acacia* trees in Australia. *Bull. Brit. Mus. Nat. Hist. (Ent.) 25*:389–466.

Mound, L. A. and B. J. Crespi. 1992. Two new species of Australian gall thrips from woody stem galls on *Casuarina. J. Nat. Hist. 26*:395–406.

Mound, L. A. and B. J. Crespi. 1995. Biosystematics of two new gall-inducing thrips with soldiers (Insecta: Thysanoptera) from *Acacia* trees in Australia. *J. Nat. Hist. 29*:147–157.

Nafus, D. M. and I. H. Schreiner. 1988. Parental care in a tropical nymphalid butterfly *Hypolimnas anomala. Anim. Behav. 36*:1425–1431.

Otte, D. and J. A. Endler (Eds.). 1989. *Speciation and its Consequences.* Sinauer Associates, Sunderland, MA.

Page, R. E. 1986. Sperm utilization in social insects. *Annu. Rev. Ent. 31*:297–320.

Pamilo, P. 1982. Multiple mating in *Formica* ants. *Hereditas 97*:37–45.

Pamilo, P. 1983. Genetic differentiation within subdivided populations of *Formica* ants. *Evolution 37*:1010–1022.

Pamilo, P. 1984. Genotypic correlation and regression in social groups: Multiple alleles, multiple loci and subdivided populations. *Genetics 107*:307–320.

Pamilo, P. 1989. Estimating relatedness in social groups. *Trends Ecol. Evol. 4*:353–355.

Pamilo, P. 1990. Comparison of relatedness estimators. *Evolution 44*:1378–1382.

Pamilo, P. and R. H. Crozier. 1982. Measuring genetic relatedness in natural populations: Methodology. *Theor. Pop. Biol. 21*:171–193.

Parker, G. A. 1970. Sperm competition and its evolutionary consequences in the insects. *Biol. Rev 45*:525–568.

Pashley, D. P., S. J. Johnson, and A. N. Sparks. 1985. Genetic population structure of migratory moths: The fall armyworm (Lepidoptera: Noctuidae). *Ann. Entomol. Soc. Am. 78*:756–762.

Pelikán, J. 1990. Butting in phlaeothripid larvae. Pp. 51–55 *in* J. Holman, J. Pelikán, A. F. G. Dixon, and L. Weisman (Eds.), *Proceedings of the 3rd International Symposium on Thysanoptera.* Kazmierz Dolny, Poland.

Pliske, T. E. 1973. Factors determining mating frequencies in some New World butterflies and skippers. *Ann. Entomol. Soc. Am. 66*:164–169.

Queller, D. C. 1991. Group selection and kin selection. *Trends Ecol. Evol. 6*:64.

Queller, D. C. and K. F. Goodnight. 1989. Estimating relatedness using genetic markers. *Evolution 43*:258–275.

Queller, D. C., J. E. Strassmann, and C. R. Hughes. 1992. Genetic relatedness and population structure in primitively eusocial wasps in the genus *Mischocyttarus* (Hymenoptera: Vespidae). *J. Hymen. Res. 1*:81–89.

Raman, A. and T. N. Ananthakrishnan. 1984. Biology of gall thrips (Thysanoptera: Insecta). Pp. 107–127 *in* T. N. Ananthakrishnan (Ed.), *Biology of Gall Insects.* Oxford and IBH Publishing Co., New Delhi, India.

Rausher, M. D. 1982. Population differentiation in *Euphydryas editha* butterflies: Larval adaptation to different hosts. *Evolution 36*:581–590.

Reilly, L. M. 1987. Measurements of inbreeding and average relatedness in a termite population. *Am. Nat. 130*:339–349.

Ross, K. G. 1986. Kin selection and the problem of sperm utilization in social insects. *Nature 323*:798–800.

Ross, K. G. 1993. The breeding system of the fire ant *Solenopsis invicta*: Effects on colony genetic structure. *Am. Nat. 141*:554–576.

Ross, K. G. and R. W. Matthews. 1989. Population genetic structure and social evolution in the sphecid wasp *Microstigmus comes. Am. Nat. 134*:574–598.

Ross, K. G., E. L. Vargo, and D. J. C. Fletcher. 1988. Colony genetic structure and queen mating frequency in fire ants of the subgenus *Solenopsis* (Hymenoptera: Formicidae). *Biol. J. Lin. Soc. 34*:105–117.

Sakata, K. and Y. Itô. 1991. Life history characteristics and behavior of the bamboo aphid, *Pseudoregma bambucicola* (Hemiptera: Pemphigidae), having sterile soldiers. *Insectes Soc. 38*:317–326.

Schultz, D. E. and D. C. Allen. 1975. Biology and descriptions of the cherry scallop moth *Hydria prunivorata* (Lepidoptera: Geometridae). *Can. Entomol. 107*:99–106.

Schwarz, M. P. 1986. Persistent multi-female nests in an Australian allodapine bee, *Exoneura bicolor* (Hymenoptera: Anthophoridae). *Insectes Soc. 33*:258–277.

Setzer, R. W. 1980. Intergall migration in the aphid genus *Pemphigus. Ann. Entomol. Soc. Am. 73*:327–331.

Sherman, P. W., E. A. Lacey, H. K. Reeve, and L. Keller. 1995. The eusociality continuum. *Behav. Ecol. 6*:102–108.

Shoemaker, D. D., J. T. Costa, and K. G. Ross. 1992. Estimates of heterozygosity in two social insects using a large number of electrophoretic markers. *Heredity 69*:573–582.

Slatkin, M. 1985. Rare alleles as indicators of gene flow. *Evolution 39*:53–65.

Slatkin, M. 1987. Gene flow and the geographic structure of natural populations. *Science 236*:787–792.

Smith, D. R. 1993. Systematics, life history, and distribution of sawflies. Pp. 3–32 *in* M. Wagner and K. F. Raffa (Eds.), *Sawfly Life History Adaptations to Woody Plants*. Academic Press, San Diego, CA.

Snyder, T. P. 1974. Lack of enzyme variability in three bee species. *Evolution 28*:687–688.

Stehr, F. W. and E. F. Cook. 1968. A revision of the genus *Malacosoma* Hübner in North America (Lepidoptera: Lasiocampidae): Systematics, biology, immatures, and parasites. Smithsonian Institution, United States National Museum Bulletin No. 276.

Steiner, W. W. M., D. J. Voegtlin, and M. E. Irwin. 1985. Genetic differentiation and its bearing on migration in North American populations of the corn leaf aphid, *Rhopalosiphum maidis* (Fitch) (Homoptera: Aphididae). *Ann. Entomol. Soc. Am. 78*:518–525.

Stern, D. L. and W. A. Foster. 1996a. The evolution of soldiers in aphids. *Biol. Rev. 71*:27–79.

Stern, D. L. and W. A. Foster. 1997. The evolution of sociality in aphids: A clone-eye's view. *In* J. C. Choe and B. J. Crespi (Eds.), *The Evolution of Social Behaviour in Insects and Arachnids*. Cambridge University Press, Cambridge, Pp. 150–165.

Strassmann, J. E., C. R. Hughes, D. C. Queller, S. Turillazzi, R. Cervo, S. K. Davis, and K. F. Goodnight. 1989. Genetic relatedness in primitively eusocial wasps. *Nature 342*:268–269.

Sugg, D. W., R. K. Chesser, F. S. Dobson, and J. L. Hoogland. 1996. Population genetics meets behavioral ecology. *Trends Ecol. Evol. 11*:338–342.

Taylor, L. R. 1974. Insect migration, flight periodicity, and the boundary layer. *J. Anim. Ecol. 43*:225–238.

Trivers, R. L. and H. Hare. 1976. Haplodiploidy and the evolution of social insects. *Science 191*:249–263.

Turner, J. R. G. 1971. Studies of Müllerian mimicry and its evolution in burnet moths and Heliconiid butterflies. Pages 224–260 *in* E. R. Creed (Ed.), *Ecological Genetics and Evolution*. Blackwell Scientific Publishing, Oxford, UK.

Uyenoyama, M. K. 1984. Inbreeding and the evolution of altruism under kin selection: Effects on relatedness and group structure. *Evolution 38*:778–795.

Uyenoyama, M. K. and M. W. Feldman. 1980. Theories of kin and group selection: A population genetics perspective. *Theor. Pop. Biol. 17*:380–414.

van Leeuwen, W. J. D. 1956. The aetiology of some thrips galls found on Malaysian *Scheffera*. *Acta Bot. Neerlandica 5*:80–89.

Varadarasan, S. and T. N. Ananthakrishnan. 1982. Biological studies on some gall thrips. *Proc. Indian Nat. Acad. Sci. B48 1*:35–43.

Wade, M. J. 1985. Soft selection, hard selection, kin selection, and group selection. *Am. Nat. 125*:61–73.

Wade, M. J. and F. Breden. 1986. Life history of natural populations of the imported willow leaf beetle, *Plagiodera versicolor* (Coleoptera: Chrysomelidae). *Ann. Entomol. Soc. Am. 79*:73–79.

Walter, G. H., K. Ruohomäki, E. Haukioja, and E. Vainio. 1994. Reproductive behaviour of mated and virgin females of a solitary sawfly *Dineura virididorsata*. *Entomol. Exp. Appl. 70*:83–90.

Warren, L. O. and M. Tadic. 1970. The fall webworm, *Hyphantria cunea* (Drury). University of Arkansas Agricultural Experimental Station Bulletin No. 795.

Wcislo, W. T. 1997. Are behavioral classifications blinders to natural variation? *In* B. J. Crespi and J. C. Choe (Eds.), *The Evolution of Social Behaviour in Insects and Arachnids.* Cambridge University Press, Cambridge, UK. Pp. 8–13.

West-Eberhard, M. J. 1975. The evolution of social behavior by kin selection. *Quart. Rev. Biol. 50*:1–33.

Whitham, T. G. 1979. Territorial behavior of *Pemphigus* gall aphids. *Nature 279*:324–325.

Wilkinson, G. S. and G. F. McCracken. 1985. On estimating relatedness using genetic markers. *Evolution 11*:32–39.

Wilkinson, R. C., G. C. Becker, and D. M. Benjamin. 1966. The biology of *Neodiprion rugifrons* (Hymenoptera: Diprionidae), a sawfly infesting jack pine in Wisconsin. *Ann. Entomol. Soc. Am. 59*:786–792.

Williams, G. C. 1966. *Adaptation and Natural Selection.* Princeton University Press, Princeton, NJ.

Williams, G. C. and D. C. Williams. 1957. Natural selection of individually harmful social adaptations among sibs with special reference to social insects. *Evolution 11*:32–39.

Wilson, D. S. 1975. A theory of group selection. *Proc. Natl. Acad. Sci. USA 72*:143–146.

Wilson, D. S. 1977. Structured demes and the evolution of group-advantageous traits. *Am. Nat. 111*:157–185.

Wilson, D. S. 1983. The group selection controversy: History and current status. *Annu. Rev. Ecol. Syst. 14*:159–187.

Wilson, E. O. 1971. *The Insect Societies.* Harvard University Press, Cambridge, MA.

Wöhrman, K. and J. Tomiuk. 1988. Life cycle strategies and genotypic variability in populations of aphids. *J. Genet. 67*:43–52.

Woods, P. E. and S. I. Guttman. 1987. Genetic variation in *Neodiprion* (Hymenoptera: Symphyta: Diprionidae) sawflies and a comment on low levels of genetic diversity within the Hymenoptera. *Ann. Entomol. Soc. Am. 80*:590–599.

Wool, D. 1984. Gall-forming aphids. Pp. 11–58 *in* T. N. Ananthakrishnan (Ed.), *Biology of Gall Insects.* Oxford and IBH Publishing Co., New Delhi, India.

Wrensch, D. L. and M. Ebbert (Eds.). 1992. *Evolution and Diversity of Sex Ratio in Insects and Mites.* Chapman & Hall, New York.

Wright, S. 1951. The genetical structure of populations. *Ann. Eugen. 15*:323–354.

# 11

# Dispersal and Adaptive Deme Formation in Sedentary Coccoid Insects

*Lawrence M. Hanks*
Department Of Entomology, University Of Illinois, Urbana, IL

*Robert F. Denno*
Department of Entomology, University of Maryland, College Park, MD

## 11.1 Introduction

Reproductive isolation of populations may eventually lead to genetic differentiation as gene pools are altered by localized selective factors and drift. The spatial structuring of genetic differentiation depends on the spatial dimensionality of gene flow, which is a reflection of dispersal behavior (e.g., Selander 1970). Edmunds and Alstad (1978) proposed that the sedentary nature of black pineleaf scale *Nuculaspis californica* Coleman (Homoptera: Diaspididae) promoted the formation of demes that were adapted to the unique genotype of individual pine host plants. This specialization to the genotype of one host could reduce fitness on conspecific hosts that differ in genotype, thereby limiting the spread of scale populations among trees and generating a patchy distribution (Edmunds and Alstad 1978). Support for the demic adaptation hypothesis comes from the patchy distribution common to many species of scale insects, whereby heavily infested trees stand among others that appear free of scale (Miller and Kosztarab 1979). However, attempts to document demic adaptation in natural populations of scale insects have yielded conflicting results (Wainhouse and Howell 1983; Unruh and Luck 1987; Cobb and Whitham 1993; Hanks and Denno 1994).

In this chapter, we evaluate the potential for demic adaptation in species of the Homopteran superfamily Coccoidea, which includes armored and soft scale insects, mealybugs, and their allies. Coccoids are sedentary relative to other herbivorous insects and thus may be predisposed to deme formation and local adaptation to individual host trees (Edmunds and Alstad 1978; Miller and Kosztarab 1979; Hanks and Denno 1994). We begin by reviewing the natural history of the coccoids. Subsequently, we consider stage-related dispersal behavior to determine whether coccoid populations are especially likely to be reproductively isolated on individual hosts. Reproductive isolation and the persistence of populations on long-lived perennial hosts set the stage for adaptation to host genotype. Finally, we evaluate the evidence for demic adaptation of coccoids in published field studies to reassess the degree of support for this widely espoused hypothesis.

## 11.2 Natural History of Coccoid Insects

The family Coccoidea comprises about 6,000 described species that have been allocated to as many as 20 families (Miller and Kosztarab 1979). Coccoid families are distinguished by their production of wax fibers that clothe the body or form specialized structures (Table 11.1). Like their closest living relatives the aphids, coccoids are small-bodied insects that feed on plant tissues through needle-like stylets. Eggs are deposited in a single mass under the waxy cover of the female or in a waxy ovisac. The tiny (< 0.5 mm long) mobile first instars, called crawlers, disperse and soon settle to feed on the host. First instars usually remain immobile once they have settled. Female coccoids are neotenic, developing through two to three additional instars before maturing into apterous, reproductively mature nymphs that range in size from 0.5 to 35 mm (Miller and Kosztarab 1979). The morphology of adult females has been used in determining the phylogeny of the Coccoidea and suggests a general decline in mobility from primitive to advanced families (Table 11.1).

Male coccoids are paurometabolous, completing two nymphal instars, a "prepupal" stage, and a "pupal" stage before emerging as adults (Miller and Kosztarab 1979; Koteja 1990). Adult males are morphologically similar across coccoid families, usually being < 1 mm long, winged, and lacking functional mouthparts (Giliomee 1990). Males rely on pheromones to locate females (Beardsley and Gonzalez 1975; Miller and Kosztarab 1979). In some species, or variants within species, males are produced in very small numbers or not at all, and females are reproduced parthenogenetically (Nur 1971; Beardsley and Gonzalez 1975; Miller and Kosztarab 1979).

Host plants of coccoids are diverse, including bryophytes, gymnosperms, and angiosperms, both monocots and dicots (Davidson and Miller 1990). Woody perennials are their most common hosts, and on these hosts coccoid populations may persist for many decades (Davidson and Miller 1990; McClure 1990a). Polyphagy is common (Beardsley and Gonzalez 1975), and host ranges can be extensive, such as the 236 plant genera in 88 families that are hosts of *Aspidiotus nerii* Bouché (Davidson and Miller 1990). Coccoids feed on nearly every plant part, from roots, stems, branches, and leaves to fruit (Beardsley and Gonzalez 1975; Miller and Kosztarab 1979). Feeding stylets may be many times longer than the body and tap into a variety of plant tissues, including parenchyma, mesophyll, cambium, and phloem (e.g., Glass 1944; Agarwal 1960; Hoy 1961; Williams 1970; Walstad et al. 1973; Blackmore 1981; Sadof and Neal 1993; Calatayud et al. 1994).

## 11.3 Dispersal, Behavior, and Capability

The supposition that coccoid insects are predisposed to demic adaptation presumes a sedentary nature (Hanks and Denno 1994). Although some of their life

*Table 11.1*  Characteristics of Adult Females for the Common Families of Coccoidea That Feed on Perennial Plants[a]

| Group | Family | Common name | Characteristics of females |
|---|---|---|---|
| Margarodoidea | Margarodidae | Giant coccids, Ground pearls | Covered with loose wax and most with legs well developed, or nymphs legless in waxy cyst with adult having legs present and reduced (ground pearls). |
| Margarodoidea | Ortheziidae | Ensign coccids | Covered with hard, waxy plates, legs well developed. |
| Lecanoidea | Pseudococcidae | Mealybugs | Covered with mealy wax secretions, legs usually well developed. |
| Lecanoidea | Eriococcidae | Felt scales | Similar to pseudococcids in form, without waxy secretions, but may be covered with wax during oviposition. |
| Lecanoidea | Dactylopiidae | Cochineal insects | Covered with wooly wax or waxy plates, legs reduced. |
| Lecanoidea | Coccidae | Soft scales | Exoskeleton usually without wax, or wax ornate or amorphous; legs present or vestigial. |
| Lecanoidea | Cryptococcidae | Cryptococcids | Covered with waxy filaments, legs vestigial. |
| Lecanoidea | Asterolecaniidae | Pit scales | Covered with tough waxy film or embedded in waxy mass, legs vestigial or absent. |
| Diaspidoidea | Diaspididae | Armored scales | Females legless, under a hard cover of wax and tanned protein. |

[a] From Howell and Williams 1976; Miller and Kosztarab 1979.
Groups are listed in phylogenetic order from primitive to advanced.

stages (nymphs of both sexes and adult females) have little or no mobility, coccoids nevertheless can move very effectively between hosts, as evidenced by the rapid spread of introduced pest species (e.g., Bean and Godwin 1955; Anderson et al. 1976; Hennessey et al. 1990). In this section, we discuss the dispersal behavior of coccoids and the potential for reproductive isolation of populations on individual hosts.

### 11.3.1 First-Instar Crawlers

For coccoid taxa in which adult females are immobile (Table 11.1), colonization of new hosts depends entirely on the first-instar crawlers. Even in more mobile species, crawler dispersal plays the key role in colonization, because later instar females remain rather sedentary (see Second Instar to Adult Females, section 11.3.2). Because of their small energy reserves, crawlers can wander only a brief time before they must settle and feed (Koteja 1990). Estimates of wandering time of coccoid crawlers range from a few hours to more than a week (Table 11.2). Longer wandering times appear to be restricted to the lecanoid families (coccids, dactylopiids, and eriococcids), whereas diaspidid crawlers appear to be consistent in wandering at most two days, but usually less than one day (Table 11.2). Variation in three estimates of wandering time for *Lepidosaphes beckii* (Table 11.2) may reflect differences in environmental factors that affect walking behavior, such as temperature, humidity, density of crawlers, dustiness of the substrate, and host species (e.g., Beardsley and Gonzalez 1975; Willard 1973a, 1973b).

Perhaps due to their greater nutritional requirements, female crawlers may disperse farther than males to fresher or less heavily infested host tissues (e.g., Cumming 1953; Brown 1958; van Halteren 1970; Oetting 1984; Gilreath and Smith 1987; Clark et al. 1989a, 1989b). Nevertheless, crawlers usually settle within 1m of their mother (e.g., Metcalf 1922; Baker 1933; Jones 1935; Bodenheimer and Steinitz 1937; Das et al. 1948; Hill 1952; Carnegie 1957; Gentile and Summers 1958; Basu and Chatterjee 1963; Samarasinghe and Leroux 1966; Patel 1971; Podsiadlo 1976; Tripathi and Tewary 1984; Willink and Moore 1988; Clark et al. 1989a). The small body size of crawlers (< 0.5 mm) and their limited life span without feeding (usually less than one day) prohibits movement between host trees over the ground (e.g., Quayle 1911; Stofberg 1937; Taylor 1935; Gentile and Summers 1958). It was noted early on that colonization of new hosts was effected primarily by aerial dispersal (e.g., Webster 1902; Quayle 1911, 1916; Jones 1935; Bodenheimer and Steinitz 1937). Small body size, flattened body form, and projecting caudal filaments render crawlers buoyant in an airstream (Beardsley and Gonzalez 1975; Greathead 1990), and they are carried downwind like inert particles (e. g, Quayle 1916; Brown 1958; Cornwell 1960; McClure 1977a; Wainhouse 1980; Augustin 1986; Yardeni 1987). Aerial dispersal accounts for the rapid spread of introduced coccoid species (e.g., Bean and Godwin 1955; Anderson et al. 1976; Hennessey et al. 1990).

Table 11.2  Active Wandering Period of Coccoid Crawlers as Indicated by Their Longevity without Feeding (How Long They Can Wander before Settling) and Actual Time Spent Walking (How Long They Typically Wander before Settling).

| Family | Species | Longevity without feeding | Time spent walking | Experiment | Reference |
|---|---|---|---|---|---|
| Pseudococcidae | *Ferrisia virgata* (Ckll.) | . | 2–6 h | n | Awadallah et al. 1979 |
| Eriococcidae | *Eriococcus coriaceus* Maskell | . | 5 m–6 h | n | Patel 1971 |
| Eriococcidae | *Eriococcus orariensis* Hoy | <4 d | . | n | Hoy 1961 |
| Dactylopiidae | *Dactylopius coccus* Costa | <2 d | . | n | Marin and Cisneros 1977 |
| Dactylopiidae | *Dactylopius confusus* (Cockerell) | <10 d | . | n | Gilreath and Smith 1987 |
| Dactylopiidae | *Dactylopius* spp. | <7 d | <3 d | n | Karny 1972 |
| Coccidae | *Pseudaonidia duplex* Ckll. | . | <6 h | y | Bliss et al. 1935 |
| Coccidae | *Pulvinaria vitis* Sign. | <9 d | . | y | Collinge 1911 |
| Coccidae | *Pulvinariella mesembryanthemi* (Vallot) | 4–8 d | . | y | Washburn and Frankie 1981 |
| Coccidae | *Saissetia oleae* (Olivier) | . | <24 h | y | Mendel et al. 1984 |
| Diaspididae | *Aonidiella aurantii* (Maskell) | <18 h | . | y | Willard 1973a |
| Diaspididae | *Aonidiella aurantii* (Maskell) | <6 h | . | n | Nel 1933 |
| Diaspididae | *Aspidiotus destructor* Sign. | <48 h | <12 h | n | Taylor 1935 |
| Diaspididae | *Aulacaspis tegalensis* Zehnt | 24–48 h | . | n | Moutia 1944 |
| Diaspididae | *Carulaspis minima* (Targ.-Tozz.) | <24 h | . | n | Stimmel 1979 |
| Diaspididae | *Chrysomphalus aonidum* (L.) | . | 6–8 h | n | Schweig and Grunberg 1936 |
| Diaspididae | *Diaspis boisduvalii* Signoret | . | <24 h | n | Bohart 1942 |
| Diaspididae | *Diaspis echinocacti* (Bouche) | . | <24 h | y | Oetting 1984 |
| Diaspididae | *Greenaspis decurvata* Green | . | 2–6 h | n | Nandagopal and David 1990 |
| Diaspididae | *Lepidosaphes beckii* (Newm.) | <2 h | 1.1 h | y | Hully 1962 |
| Diaspididae | *Lepidosaphes beckii* (Newm.) | <36 h | <24 h | n | Stofberg 1937 |
| Diaspididae | *Lepidosaphes beckii* (Newm.) | . | <24 h | n | Quayle 1912 |
| Diaspididae | *Lepidosaphes conchiformis* (Gmel.) | . | 24 h | n | Ahmad and Ghani 1971 |

Table 11.2 (continued)

| Family | Species | Longevity with out feeding | Time spent walking | Experiment | Reference |
|--------|---------|---------------------------|--------------------|-----------| ----------|
| Diaspididae | Lepidosaphes ulmi (L.) | . | 0.33 h | n | Samarasinghe and Leroux 1966 |
| Diaspididae | Melanaspis glomerata (Green) | 24 h | . | n | Tripathi and Tewary 1984 |
| Diaspididae | Melanaspis glomerata (Green) | . | 12–36 h | n | Agarwal 1960 |
| Diaspididae | Melanaspis tenebricosa (Comstock) | < 24 h | 1 h | n | Metcalf 1922 |
| Diaspididae | Pseudaulacaspis pentagona Targ. | 16–22 h | . | y | Hanks and Denno 1993a |
| Diaspididae | Quadraspidiotus perniciosus (Comst.) | . | 1–4 h | n | Gentile and Summers 1958 |

The Experiment column indicates whether time estimates are derived from experimental data versus estimation.

Early researchers believed that crawlers actively disperse by launching themselves from their host plant (Balachowsky 1937; Andrewartha and Birch 1954). However, coccoid crawlers appear to show two distinct types of dispersal behavior: "active" and "inactive" (Table 11.3). Newly hatched crawlers of active dispersers do not immediately seek sites for settling, but rather display behaviors that expedite aerial dispersal. For example, newly emerged crawlers of *Antonina graminis, Aulacaspis tegalensis, Melanaspis glomerata, Pulvinariella mesembryanthemi,* and *Dactylopius austrinus* aggregate on the tips of leaves or spines where they are readily dislodged and carried off by wind (Table 11.3). They later settle to feed on the crown and lower nodes of grasses (*A. graminis:* Chada and Wood 1960), under leaf sheaths of sugarcane (*A. tegalensis:* Agarwal 1960; *M. glomerata:* Williams 1970), on maturing iceplant leaves (*P. mesembryanthemi:* Washburn and Frankie 1985), and on cactus cladodes (*D. austrinus:* Moran et al. 1982). Female crawlers of *D. austrinus* also show morphological adaptations for aerial dispersal in the form of abundant, long, waxy filaments that improve their aerial buoyancy (Gunn 1978; Moran et al. 1982). Despite their active dispersal from host plants, few *D. austrinus* crawlers travel farther than 6 m from their natal host (Moran et al. 1982), few dispersing *A. tegalensis* crawlers drift above the tops of sugarcane (Greathead 1972), and few *P. mesembryanthemi* and *D. austrinus* crawlers move more than 1 m above the ground (Washburn and Frankie 1981; Moran et al. 1982). The short distances traveled by these wind-dispersed crawlers suggests that active dispersal primarily results in movement within patches of hosts rather than between patches. However, even local dispersal may be of selective advantage by reducing intraspecific competition and mortality from natural enemies (references in Table 11.3; Washburn et al. 1985).

In contrast to actively dispersing coccoids, inactively dispersing species (Table 11.3) do not readily disperse from their hosts. Crawlers of these species cling to the host and may only take flight when wind speeds are sufficient to tear them free, or after they fall from the host. Hosts of these species are all woody plants (Table 11.3). A disinclination to disperse aerially and a tendency to settle on the natal host have also been observed in other coccoid species that feed on trees (Rabkin and Lejeune 1954; Nielsen and Johnson 1973; Edmunds and Alstad 1981). Emigration and colonization of new hosts by these tree-feeding coccoids are apparently arbitrary and accidental events, as perhaps indicated by the low percentage of crawlers that disperse from trees (usually less than 20%; McClure 1977b; Stephens and Aylor 1978; Hill 1980; Unruh 1985; Wainhouse and Gate 1988).

The behavioral dichotomy between actively and inactively dispersing coccoids appears to reflect the dimensionality of the host plant and associated risks of dispersal. For active dispersers, dispersal involves relatively low risk, because host plants (grasses, sugarcane, iceplant, cacti, ferns) tend to occur in low-profile and tightly grouped, monospecific patches, and crawlers leaving one host will have a reasonable chance of alighting on another. For inactively dispersing species, however, dispersal involves high risks, because host plants (trees and shrubs) have a

Table 11.3  Characteristics of Coccoid Species with Crawlers That Show Active versus Inactive Dispersal Behaviors

| Dispersal behavior | Coccoid family | Species | Host plant | Crawler behavior | References |
|---|---|---|---|---|---|
| Active | Pseudococcidae | *Antonin graminis* Maskell | Rhodesgrass | Ascend host, aggregate on leaf tips | Chada and Wood 1960 |
| Active | Dactylopiidae | *Dactylopius austrinus* De Lotto | Cacti, *Opuntia* spp. | Females ascend hosts and aggregate on tips of spines | Moran and Cobby 1979, Moran et al. 1982 |
| Active | Coccidae | *Pulvinariella mesembryanthemi* Vallot | Iceplant, *Carpobrotus* spp. | Ascend host plant and move to leaf tips, elevate body in response to wind | Washburn and Frankie 1981, 1985; Washburn and Washburn 1984 |
| Active | Diaspididae | *Aulacaspis te galensis* Zhnt. | Sugarcane | Ascend plant, move to leaves, easily dislodged by wind | Williams 1970; Greathead 1972 |
| Active | Diaspididae | *Melanaspis glomerata* Green | Sugarcane | Ascend plant, aggregate on leaf auricles, easilydislodged by wind | Agarwala 1956; Agarwal 1960 |
| Active | Diaspididae | *Pinnaspis aspidistrae* Sign. | Ferns | Easily dislodged from plant, drop off of fronds even in the absence of wind | Werner 1930 |
| Inactive | Margarodidae | *Matsucoccus matsumurae* Kuwana (=*M. resinosae*) | Red pine | Cling to twigs and needles in response to wind | Stephens and Aylor 1978 |

246

Table 11.3 (continued)

| Dispersal behavior | Coccoid family | Species | Host plant | Crawler behavior | References |
|---|---|---|---|---|---|
| Inactive | Pseudococcidae | *Planococcoides njalensis* Laing | Cacao trees | Cling and flatten to host in response to wind | Cornwell 1960 |
| Inactive | Eriococcidae | *Eriococcus orariensis* Hoy | *Leptospermum scoparium* Forst. | Cling to plant, seek shelter in crevices in response to wind | Hoy 1961 |
| Inactive | Coccidae | *Saissetia oleae* Bern. | Olive trees | Cling to host in response to wind | Quayle 1911 |
| Inactive | Diaspididae | *Aonidiella aurantii* Maskell | Lemon trees | Flatten to substrate and remain motionless in response to wind | Willard 1976 |
| Inactive | Diaspididae | *Pseudaulacaspis pentagona* Targ. | White mulberry trees | Cling and flatten to host in response to wind | Hanks and Denno 1993a |

higher profile, may occur in mixed species stands, and are widely separated, such that crawlers dispersing from one host are unlikely to reach a new host.

The horizontal distance that windblown crawlers will be transported depends on their terminal velocity (their maximum rate of fall in still air) and wind speed. Terminal velocities of scale crawlers have been estimated between 0.1 and 0.4 meters/sec (Brown 1958; Willard 1973a; Stephens and Aylor 1978; Wainhouse 1980; Washburn and Washburn 1984; Unruh 1985). Updrafts exceeding this relatively slow wind speed will carry crawlers upward, but the distance they can travel to colonize new hosts will be limited by how long they can survive without feeding (Table 11.2). Given an average longevity of 24 h, and moderate winds with a steady horizontal component of 1 m/s and a vertical component exceeding the terminal velocity, viable crawlers could be transported between trees as far as 86 km. Such long-distance aerial transportation accounts for the spread of coccoid infestations across as much as 14 km of open water (e.g., Brower 1949; Hoy 1961).

Although dispersing crawlers may be carried upward and potentially long distances on the wind, field studies have indicated that the majority move laterally and downward. For example, the percentage of crawlers reaching altitudes above canopy height was only ~ 7% for *Aonidiella aurantii* Maskell (Willard 1976) and ~ 6% for *Planococcoides njalensis* (Laing) (Cornwell 1960). Wainhouse (1980) found that only ~ 0.7% of dispersing *Cryptococcus fagisuga* Lindinger crawlers were carried above the canopy of beech trees where winds were strong enough (> 1 m/s) to carry them away. The remaining ~ 99.3% of dispersers occurred below canopy tops (most less than 3.2 m above the ground) where wind speeds of < 0.7 m/s were too slow to carry them far. The vast majority of crawlers departing from heavily infested trees fall to the ground where they perish (e.g., Hoy 1961; Beardley and Gonzalez 1975; Stephens and Aylor 1978; Wainhouse 1980; Unruh 1985).

The key issue of crawler dispersal in the context of the demic adaptation hypothesis is the rate at which crawlers immigrate into existing populations on novel host trees. Immigration rate will be strongly dependent on the spatial distribution of host trees. We used published field studies of coccoid dispersal to determine how immigration rates will be affected by host distribution, particularly to identify the degree of separation between hosts that is necessary to reduce immigration rates to insignificant levels. Quayle (1916) was the first to study the aerial dispersal behavior of coccoid crawlers by setting sticky traps at a distance from trees infested with *Saissetia oleae* (Olivier). Ten subsequent field studies provided us with data to calculate capture rates of crawlers at different distances from source trees (Fig. 11.1). For those studies in which traps were placed in various orientations around the source tree, or were set up on different dates, we used only the maximum capture rates reported for each trapping distance. Trap capture rates are an overestimate of immigration rate, since not all crawlers arriving at a host tree will succeed in colonizing it (see Wainhouse and Gate 1988).

The rate at which scale crawlers were captured on sticky traps (log transformed) was strongly and negatively correlated with log-distance from source trees (Fig. 11.1). Traps placed immediately adjacent to source trees (points on the y axis

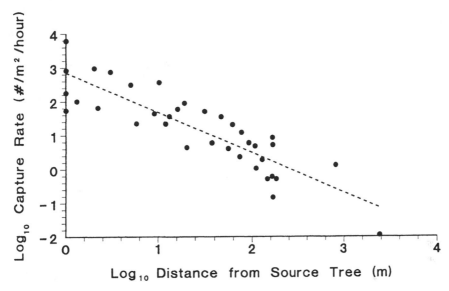

*Figure 11.1* Relationship between rate at which coccoid crawlers were captured on sticky traps and the distance that traps were positioned from source trees (log–log transformed; $Y = -1.18X + 2.84$; $r^2 = 0.77$, $P < 0.001$). Data are derived from the following field studies ($N$ = number of observations): Strickland (1950, $N$ = 2), Rabkin and Lejeune (1954, $N$ = 2), Hoy (1961, N = 1), Timlin (1964, $N$ = 2), Nielsen and Johnson (1973, $N$ = 1), McClure (1977a, $N$ = 8), Stephens and Aylor (1978, $N$ = 11), Wainhouse (1980, $N$ = 5), Willard (1974, $N$ = 4), and Hill (1980, $N$ = 1). These studies were not standardized for a variety of factors that influence crawler dispersal, such as trap height (0–10 m above ground), number of source plants (single trees to orchards), size of source plants (small shrubs to tall trees), density of scales on source plants (usually described as "heavily infested"), wind conditions (usually not measured), and orientation with respect to prevailing wind direction (usually downwind).

in Fig. 11.1) captured crawlers at an average rate of about 1,000 insects/m²/hour, but average capture rate declined sharply to ~ 100 crawlers/m²/hour at 10 m distance and fell to nearly zero crawlers/m²/hour at a distance from source plants greater than 100 m. Thus, considerable numbers of crawlers could immigrate into trees that neighbor other infested trees (less than 10 m apart), but dispersal between trees will be negligible where hosts are separated by more than 100 m.

## 11.3.2 Second Instar to Adult Females

Female diaspidids and some other coccoids (Table 11.1) are entirely sessile and have no alternative but to oviposit on their natal host (Beardley and Gonzalez 1975). Second instars and adults of other species of the more "primitive" groups (margarodoids and lecanoids), have well-developed legs and may be relatively mobile, although they may walk little and are likely to spend their whole lives on their natal host (e.g., Goux 1944; Strickland 1951; Cornwell 1958; Hoy 1961; Manichote and Middlekauff 1967). These primitive coccoids may overwinter, molt, and oviposit in locations different from where they feed (e.g., Hough 1925; Washburn 1965; Unruh 1985; Russell 1987). Nevertheless, movement by walking between host trees is probably minimal except perhaps where branches interdigitate. Females of a few mealybug species are unique in producing lateral wax filaments that render them buoyant in the air, and these individuals may disperse short distances between adjacent hosts (Miller and Denno 1977).

### 11.3.3 Second Instar to Adult Males

Male nymphs of margarodoids and lecanoids may be capable of walking and even leave the host to pupate, but do not travel far (e.g., Bean and Godwin 1955; Unruh and Luck 1987). Diaspidoid males, however, are sessile until they emerge as adults (Beardsley and Gonzalez 1975). Though adult male coccoids usually are alate and capable of flight, males of species in several coccoid families show a consistent disinclination to take flight, instead walking on their natal host in search of mates (Margarodidae: McKenzie 1943, Unruh 1985; Dactylopiidae: Karny 1972; Eriococcidae: Hoy 1961, Patel 1971; Pseudococcidae: Highland 1956, Barrass et al. 1994; Diaspididae: Metcalf 1922, Taylor 1935, Stofberg 1937, Gentile and Summers 1958, Lellakova-Duskova 1963, Tashiro and Moffitt 1968, Stimmel 1979, Alstad et al. 1980, Tripathi and Tewary 1984, Lambdin 1990, Hanks and Denno 1993a, Lambdin et al. 1993). The disinclination to disperse is a reflection of weak flight abilities due to fragility and small body size (Beardsley and Gonzalez 1975).

Flying males may better control their direction by dispersing within the tree canopy where wind speeds are slower (Rice and Moreno 1970), or taking flight during calm weather conditions (Rice and Hoyt 1980; Barrass et al. 1994). For example, in light prevailing winds (~ 0–0.9 m/s) male A. aurantii and A. citrina Coquillett may fly more than 40 m upwind (Rice and Moreno 1970; Moreno et al. 1974). Significant rates of dispersal over short distances are indicated by the capture of large numbers of males with unbaited sticky traps (e.g., McClure 1979; Clark et al. 1989b), female-baited traps (Rice and Hoyt 1980), or synthetic pheromone traps (e.g., Rice and Hoyt 1980; Angerilli and Logan 1986; Walker et al. 1990). The distance males can fly in search of mates is limited by their weak flight abilities and short average life spans (usually < 24 hours; Beardsley and Gonzalez 1975). Wings are probably retained in males for reasons associated with mate location within the canopy of single trees rather than for long-distance dispersal (see Denno 1994).

## 11.4 Dispersal and Deme Formation

Demic adaptation to host-plant individuals would appear to be promoted by the low mobility of all life stages of coccoids and their inefficiency in moving between plants: (1) most crawlers tend to settle on their maternal host tree rather than disperse; (2) second instar to adult females are either incapable of movement or very unlikely to move between hosts; and (3) adult males, though usually alate, nevertheless tend to search for females on the natal host. Thus, each generation is founded primarily by crawlers that are settling on their maternal host, and most matings occur within populations. Gene flow may be restricted at spatial scales even smaller than individual trees, because crawlers disperse only a limited distance from their mother, adult females cannot move or tend not to wander far during their lives, and adult males tend to mate with nearby females. Alstad (Chapter 1, this volume) concludes that these limitations in gene flow account for the genetic structuring of black pineleaf scale populations at spatial scales as small as individual branchlets. This structuring may result from genetic drift: however, it could arise by adaptation and, if so, reflects within-tree variation in selective factors (Alstad, Chapter 1, this volume). In the latter case, each population on an individual tree may comprise an aggregation of independently evolving subpopulations. Nevertheless, if scales on the same host tree experience a more similar selective environment than do scales on different trees, the population as a whole may be adapted to the unique characteristics of host-tree individuals.

Although the sedentary nature of coccoids may result in reproductive isolation and could favor local adaptation to individual trees, the potential for deme formation will depend on the proximity of host trees. Because crawlers are dispersed on the wind, and adult males have little control over their long-range course, rates of immigration into populations will be highest when host trees are adjacent but low when they are widely separated. Immigration rates will also be affected by the presence of nonhost tree species that present obstacles to dispersers as well as reduce wind speed, limiting the distance that crawlers are carried.

The immigration of small numbers of coccoids into populations may foster demic adaptation to host trees by providing genetic variation that is grist for the natural selection mill, but high rates of gene flow could hinder adaptation (Slatkin 1985, 1987). Adaptation should be less likely to occur where immigration rates are high (the host tree stands within 10 m of other infested hosts) and selection pressures exerted by the host plant are weak. The opportunity for demic adaptation should be greatest where immigration rates are low (the host-tree stands > 100 m from other infested hosts) and selection pressures are great. Strong selection pressures could promote demic adaptation in spite of high rates of gene flow (Ehrlich and Raven 1969; Mopper 1996).

## 11.5 Testing the Demic Adaptation Hypothesis

The demic adaptation hypothesis (Edmunds and Alstad 1978) has been tested in four later independent field studies of the coccoid insects *Cryptococcus fagisuga* (Wainhouse and Howell 1983), *Matsucoccus acalyptus* Herbert (Unruh and Luck

1987; Cobb and Whitham 1993), and *Pseudaulacaspis pentagona* Targ. (Hanks and Denno 1994). These studies used similar reciprocal transfer designs in which scales from each infested study tree were transferred to other infested trees, as well as back-transferred to their original host or clone. Survivorship should be higher on natal hosts than on novel hosts if scales are indeed adapted to their host-plant individual.

These four transfer studies support the hypothesis that the opportunity for demic adaptation depends on restriction of immigration rates by spatial isolation of host trees. In four transfer experiments, the lack of evidence for adaptation may be attributed to dispersal between host trees:

1. Wainhouse and Howell (1983) transferred eggs of *C. fagisuga* within a seed orchard of 22–year-old beech trees (*Fagus sylvatica* L; also see Wainhouse and Deeble 1980). Trees were of two clone lines that tended to support high scale densities. Scale eggs were collected from a single tree of each clone line and were transferred to five or six trees of each clone line. Similar survivorship across clone lines may have been due to dispersal between the two source trees which stood only ~ 6 m apart. *C. fagisuga* is unisexual, and dispersal of males is therefore not an issue, but trapping studies showed that an abundance of crawlers dispersed from heavily infested trees and traveled an average distance of 10 m (Wainhouse 1980).

2. Unruh and Luck (1987) conducted transfer experiments in a pinyon pine forest in southern California, transferring *M. acalyptus* crawlers between and among nine pine trees in one study, and another six trees in a second study. Host trees were 40–100 years old. In sticky trap studies, Unruh (1985) discovered that only a very small percentage (~ 2%) of *M. acalyptus* crawlers dispersed from trees, and most fell to the ground within 6 m of their natal host. He concluded that dispersal of crawlers between trees was probably minimal. Although adult males tended to walk rather than fly in their search for mates, dispersal rates of adult males were nevertheless sufficient to hinder genetic differentiation of populations on individual trees (Unruh 1985). Panmixis across populations on different trees was further indicated by allozyme variation (Unruh 1985).

3. Cobb and Whitham (1993) transferred *M. acalyptus* between and among 15 infested *Pinus edulis* hosts (averaging 31.8 years in age) and also detected no evidence of demic adaptation to individual trees. Adaptation to host tree was also further refuted by increasing mortality rates within incipient (newly founded) populations over a six-year period, and similar mortality rates in incipient and established (presumably adapted) populations.

Cobb and Whitham (Chapter 3, this volume) proposed that adaptation of *M. acalyptus* to host tree individuals was confounded by (1) temporal variation in selection pressures related to scale population density and host-plant resistance, (2) within-tree variation in resistance traits, (3) gene flow in the form of adult males dispersing between male-biased and female-biased populations, (4) emigration of crawlers from high-density populations, and (5) their immigration into trees whose populations had been decimated by natural catastrophes. Gene flow would have been facilitated by the close proximity of study trees (all stood within 1 hectare), as suggested by the work of Unruh (1985). Thus, the possibility remains that adapted demes of *M. acalyptus* could arise where host trees are more widely separated.

4. Hanks and Denno (1994) conducted transfers of *P. pentagona* between and among paired white mulberry, *Morus alba* L., trees (10–20 years old) in an urban landscape. Crawlers and adult males of this species are sedentary and tend not to disperse from their natal hosts (Hanks and Denno 1993a). Nevertheless, the potential for dispersal between study trees was maximized because of their close proximity (< 5 m apart), and dispersal may have inhibited demic adaptation.

The two transfer experiments involving *C. fagisuga* and *P. pentagona* supported the demic adaptation hypothesis by demonstrating significantly higher scale survivorship on natal compared to novel hosts:

1. Wainhouse and Howell (1983) transferred *C. fagisuga* eggs between and among five heavily infested beech trees, two of which occurred in the same plantation but were about 30 m apart, and three trees occurred in different, widely separated plantations. Immigration of crawlers into trees was minimized, because study trees were relatively isolated from other heavily infested trees (D. Wainhouse personal communication). Wainhouse (1980) showed that few aerially dispersing *C. fagisuga* crawlers are likely to travel farther than 25 m.

2. Hanks and Denno (1994) transferred *P. pentagona* between pairs of mulberry host trees, where each study tree was separated from the nearest infested tree by at least 300 m. Extreme isolation of host trees assured very low rates of immigration by crawlers and adult males.

In these studies, trees were relatively isolated, minimizing dispersal between hosts, and this reproductive isolation may have fostered local adaptation.

## 11.6 Discussion

Reproductive isolation of coccoid populations on individual trees seems likely to result from their sedentary nature. Gene flow between populations in the form of dispersing crawlers and adult males may impede genetic differentiation. However, because dispersal is primarily passive and is strongly influenced by arbitrary wind currents, gene flow into populations will be significant only where trees stand in close proximity (< 10 m apart), but negligible when trees are more widely separated (> 100 m apart). The critical role of host-tree distribution in reproductive isolation is supported by field experiments that showed evidence of adaptation only where host trees were isolated from other infested trees, but not where they stood in close proximity (Wainhouse and Howell 1983; Hanks and Denno 1994).

Because reproductive isolation of coccoid populations depends on spatial separation of host trees, deme formation seems an unlikely cause of their patchy distribution within stands of closely situated trees. Dispersal between adjacent trees should preclude specialization to individual trees and maintain a generalist ability to colonize new hosts (Slatkin 1985, 1987). Local, patchy distributions of coccoids more likely reflect small-scale spatial variation in ecological factors that influence population densities such as host-plant quality (e.g, McClure 1990b; Hanks and Denno 1993b), host-plant resistance (e.g., Ghose 1983; McClure and Hare 1984; Wainhouse et al. 1988), natural enemies (e.g., Hanks and Denno 1993b), and the abundance of ants that discourage natural enemies (e.g, Cornwell 1957; Buckley 1987; Hanks and Sadof 1990).

The role of demic adaptation in colonization and development of populations is called into question by outbreaks and the spread of populations across hosts where natural controls are eliminated. For example, coccoids may reach high densities where natural enemies are discouraged by drifting road dust (Bartlett 1951) or insecticides (Edmunds 1973; Luck and Dahlsten 1975). Similarly, coccoid pests introduced into new regions without their natural enemies rapidly spread through stands of hosts (e.g., Bean and Godwin 1955; Anderson et al. 1976; Houston et al. 1979; Hennessey et al. 1990). The ease with which unrestrained coccoids colonize novel hosts and erupt to high densities suggests that, at least for some species, adaptation to host-plant genotype is neither necessary for colonization nor of general significance in the dynamics of populations on individual hosts. Nevertheless, genetic differentiation of populations on isolated hosts may play an important role in coccoid evolution and speciation.

## Acknowledgments

We thank John Davidson, Susan Mopper, Doug Miller, Merrill Peterson, Rick Redak, and Sharon Strauss for their comments on an earlier draft of this chapter.

## 11.7 References

Agarwal, R. A. 1960. The sugarcane scale, *Melanaspis (Targionia) glomerata* (Green), its biology and control. *Indian Sugar* 10:523–544.

Agarwala, S. B. D. 1956. *Melanaspis glomerata* (Green)—a new coccid pest of sugarcane. *Proc. Bihar Acad. Agr. Sci.* 5:24–31.

Ahmad, R. and M. A. Ghani. 1971. Laboratory studies on the biology of *Lepidosaphes conchiformis* (Gmel.) (Hem., Diaspididae) and of its parasite *Aphytis maculicornis* (Masi) (Hym., Aphelinidae). *Bull. Entomol. Res.* 61:69–74.

Alstad, D. N., G. F. Edmunds Jr., and S. C. Johnson. 1980. Host adaptation, sex ratio, and flight activity in male black pineleaf scale. *Annu. Entomol. Soc. Am.* 73:665–667.

Anderson, J. F., R. P. Ford, J. D. Kegg, and J. H. Risley. 1976. The red pine scale in North America. Connecticut Agricultural Experimental Station Bulletin 765.

Andrewartha, H. G. and L. C. Birch. 1954. *The Distribution and Abundance of Animals.* The University Chicago Press, Chicago, IL.

Angerilli, N. P. D. and D. M. Logan. 1986. The use of pheromones and barrier traps to monitor San Jose scale (Homoptera: Diaspididae) phenology in the Okanagan Valley of British Columbia. *Can. Entomol. 118*:767–774.

Augustin, S. 1986. Etude de la dispersion des larves de premier stade de *Cryptococcus fagisuga* Lind. (Hom., Coccoidea) dans une hêtraie. *J. Appl. Entomol. 102*:178–194.

Awadallah, K. T., E. D. Ammar, M. F. S. Tawfik, and A. Rashad. 1979. Life-history of the white mealybug *Ferrisia virgata* (Ckll.). *Dtsch. Entomol. Z. 26*:101–110.

Baker, H. 1933. The obscure scale on the pecan and its control. United States Department of Agriculture Circular 295.

Balachowsky, A. 1937. Les cochenilles de France, d'Europe, du nord de l'Afrique et du bassin Mediterranean. Caracteres generaux des cochenilles. Parts I–III. Hermann et Cie., Paris.

Barrass, I. C., P. Jerie, and S. A. Ward. 1994. Aerial dispersal of first- and second-instar longtailed mealybug, *Pseudococcus longispinus* (Targioni Tozzetti) (Pseudococcidae: Hemiptera). *Aust. J. Exp. Agr. 34*:1205–1208.

Bartlett, B. R. 1951. The action of certain "inert" dust materials on parasitic hymenoptera. *J. Econ. Entomol. 44*:891–896.

Basu, A. C. and P. B. Chatterjee. 1963. Study on the behaviour and control of *Ferrisiana virgata* (Ckll.)-a new mealy bug pest of betel vine, *Piper betle* Linn. in West Bengal, India. *Indian Agricult. 7*:112–117.

Bean, J. L. and P. A. Godwin. 1955. Description and bionomics of a new red pine scale, *Matsucoccus resinosae. For. Sci. 1*:164–176.

Beardsley, J. W. Jr. and R. H. Gonzalez. 1975. The biology and ecology of armored scales. *Annu. Rev. Entomol. 20*:47–73.

Blackmore, S. 1981. Penetration of the host plant tissues by the stylets of the coccoid *Icerya seychellarum* (Coccoidea: Margarodidae) on Aldabra Atoll. *Atoll Res. Bull. 225*:33–38.

Bliss, C. I., A. W. Cressman, and B. M. Broadbent. 1935. Productivity of the camphor scale and the biology of its eggs and crawler stages. *J. Agr. Res. 50*:243–266.

Bodenheimer, F. S. and H. Steinitz. 1937. Studies in the life history of the citrus mussel scale (*Lepidosaphes pinnaeformis* Bché) in Palestine. *Hadar 10*:153–159.

Bohart, R. M. 1942. Life history of *Diaspis boisduvalii* and its control on cattleya with calcium cyanide. *J. Econ. Entomol. 35*:365–368.

Brower, A. E. 1949. The beech scale and beech bark disease in Acadia national park. *J. Econ. Entomol. 42*:226–228

Brown, C. E. 1958. Dispersal of the pine needle scale, *Phenacaspis pinifoliae* (Fitch), (Diaspididae: Homoptera). *Can. Entomol. 90*:685–690.

Buckley, R. C. 1987. Interactions involving plants, Homoptera, and ants. *Annu. Rev. Entomol. 18*:111–135.

Calatayud, P. A., Y. Rahbé, W. F. Tjallingii, M. Tertuliano, and B. Le Rü. 1994. Electrically recorded feeding behaviour of cassava mealybug on host and non-host plants. *Entomol. Exp. Appl. 72*:219–232.

Carnegie, A. J. M. 1957. Observations on the behavior of crawlers of *Lepidosaphes beckii* Newm. (Homoptera: Diaspididae). *J. Entomol. Soc. S. Africa 20*:164–169.

Chada, J. L. and E. A. Wood Jr. 1960. Biology and control of the rhodesgrass scale. United States Department of Agriculture (USDA) Technical Bulletin Agricultural Research Service #1221.

Clark, S. R., G. L. DeBarr, and C. W. Berisford. 1989a. Life history of the woolly pine scale *Pseudophilippia quaintancii* Cockerell (Homoptera: Coccidae) in loblolly pine seed orchards. *J. Entomol. Sci. 24*:365–372.

Clark, S. R., G. L. DeBarr, and C. W. Berisford. 1989b. The life history of *Toumeyella pini* (King) (Homoptera: Coccidae) in loblolly pine seed orchards in Georgia. *Can. Entomol. 121*:853–860.

Cobb, N. S. and T. G. Whitham. 1993. Herbivore deme formation on individual trees: A test case. *Oecologia 94*:496–502.

Collinge, W. E. 1911. On the locomotion and length of life of the young of *Pulvinaria vitis* var. *ribesiae*, Sign. *J. Econ. Bio. 6*:139–142.

Cornwell, P. B. 1957. An investigation into the effect of cultural conditions on populations of the vectors of virus diseases of cacao in Ghana with an evaluation of seasonal population trends. *Bull. Entomol. Res. 48*:375–396.

Cornwell, P. B. 1958. Movements of the vectors of virus diseases of cacao in Ghana: I. Canopy movements in and between trees. *Bull. Entomol. Res. 49*:613–630.

Cornwell, P. B. 1960. Movements of the vectors of virus diseases on cacao in Ghana: II. Wind movements and aerial dispersal. *Bull. Entomol. Res. 51*:175–201.

Cumming, M. E. P. 1953. Notes on the life history and seasonal development of the pine needle scale, *Phenacaspis pinifoliae* (Fitch) (Diaspididae: Homoptera). *Can. Entomol. 85*:347–352.

Das, G. M., T. D. Mukherjee, and N. S. Gupta. 1948. Biology of the common mealybug, *Ferrisia virgata* Ckll. (Coccidae), a pest on jute, *Corchorus olitorius* L., in Bengal. *Proc. Zool. Soc. Bengal 1*:109–114.

Davidson, J. A. and D. R. Miller. 1990. Ornamental plants. Pp. 603–632 *in* D. Rosen (Ed.), *The Armored Scale Insects, Their Biology, Natural Enemies, and Control*, Vol. 4B. Elsevier Science Publishing, Amsterdam, The Netherlands.

Denno, R. F. 1994. Life history variation in planthoppers. Pp. 163–215 *in* R. F. Denno and J. Perfect (Eds.), *Planthoppers: Their Ecology and Management*. Chapman & Hall, New York.

Edmunds, G. F. Jr. 1973. The ecology of black pineleaf scale (Homoptera: Diaspididae). *Environ. Entomol. 2*:765–777.

Edmunds, G. F. Jr. and D. N. Alstad. 1978. Coevolution in insect herbivores and conifers. *Science 199*:941–945.

Edmunds, G. F. Jr. and D. N. Alstad. 1981. Responses of black pineleaf scales to host plant variability. Pp. 29–38 *in* R. F. Denno and H. Dingle (Eds.), *Insect Life History Patterns*. Springer-Verlag, New York.

Ehrlich, P. R. and P. H. Raven. 1969. Differentiation of populations. *Science* *165*:1228–1232.

Gentile, A. G. and F. M. Summers. 1958. The biology of San Jose scale on peaches with special reference to the behavior of males and juveniles. *Hilgardia 27*:269–285.

Ghose, S. K. 1983. Biology of parthenogenetic race of *Dysmicoccus brevipes* (Cockerell) (Pseudococcidae, Hemiptera). *Indian J. Agric. Sci. 53*:939–942.

Giliomee, J. H. 1990. The adult male. Pp. 21–27 *in* D. Rosen (Ed.), *The Armored Scale Insects, Their Biology, Natural Enemies, and Control,* Vol. 4A. Elsevier Science Publishing, Amsterdam, The Netherlands.

Gilreath, M. E. and J. W. Smith Jr. 1987. Bionomics of *Dactylopius confusus* (Homoptera: Dactylopiidae). *Annu. Entomol. Soc. Am. 80*:768–774.

Glass, E. H. 1944. Feeding habits of two mealybugs, *Pseudococcus comstocki* (Kuw.) and *Phenacoccus colemani* (Ehrh.). Technical Bulletin, Virginia Polytechnic Institute Agricultural Experiment Station #95.

Goux, L. 1944. Contribution to the study of the biology of *Eriococcus* (Hemiptera: Coccidae) diet, alimentary specificity and biological races. *Bull. Mus. Hist. Nat. Marseille T.* 4(3–4):135–152.

Greathead, D. J. 1972. Dispersal of the sugar-cane scale *Aulacaspis tegalensis* (Zhnt.) (Hemiptera: Diaspididae) by air currents. *Bull. Entomol. Res. 61*:547–558.

Greathead, D. J. 1990. Crawler behavior and dispersal. Pp. 305–308 *in* D. Rosen (Ed.), *The Armored Scale Insects, Their Biology, Natural Enemies, and Control,* Vol. 4A. Elsevier Science Publishing, Amsterdam, The Netherlands.

Gunn, B. H. 1978. Sexual dimorphism in the first instar of the cochineal insect *Dactylopius austrinus* De Lotto (Homoptera: Dactylopiidae). *J. Entomol. Soc. S. Africa 41*:333–338.

Hanks, L. M. and R. F. Denno. 1993a. The white peach scale, *Pseudaulacaspis pentagona* (Targioni-Tozzetti) (Homoptera: Diaspididae): Life history in Maryland, host plants, and natural enemies. *Proc. Entomol. Soc. Washington 95*:79–98.

Hanks, L. M. and R. F. Denno. 1993b. Natural enemies and plant water relations influence the distribution of an armored scale insect. *Ecology 74*:1081–1091.

Hanks, L. M. and R. F. Denno. 1994. Local adaptation in the armored scale insect *Pseudaulacaspis pentagona* (Homoptera: Diaspididae). *Ecology 75*:2301–2310.

Hanks, L. M. and C. S. Sadof. 1990. The effect of ants on nymphal survivorship of *Coccus viridis* (Homoptera: Coccidae). *Biotropica 22*:210–213.

Hennessey, R. D., P. Neuenschwander, and T. Muaka. 1990. Spread and current distribution of the cassava mealybug, *Phenacoccus manihoti* (Homoptera: Pseudococcidae), in Zaire. *Trop. Pest Manag. 36*:103–107.

Highland, H. A. 1956. The biology of *Ferrisiana virgata*, a pest of azaleas. *J. Econ. Entomol. 49*:276–277.

Hill, C. H. 1952. The biology and control of the scurfy scale on apples in Virginia. Virginia Agr. Exp. Stn. Bull. 119.

Hill, M. G. 1980. Wind dispersal of the coccid *Icerya seychellarum* (Homoptera: Margarodidae) on Aldabra Atoll. *J. Anim. Ecol. 49*:939–957.

Hough, W. S. 1925. Biology and control of Comstock's mealy bug on the umbrella catalpa. Virginia Agric. Exp. Sta. Tech. Bull. 29.

Houston, D. R., E. J. Parker, and D. Lonsdale. 1979. Beech bark disease: Patterns of spread and development of the initiating agent *Cryptococcus fagisuga. Can. J. For. Res. 9*:336–344.

Howell, J. O. and M. L. Williams. 1976. An annotated key to the families of scale insects (Homoptera: Coccoidea) of America, north of Mexico, based on characteristics of adult females. *Ann. Entomol. Soc. Am. 69*:181–189.

Hoy, J. M. 1961. *Eriococcus orariensis* Hoy and other Coccoidea (Homoptera) associated with *Leptospermum* Forst species in New Zealand. *New Zealand Dept. Sci. Ind. Res. Bull. 141*:1–70.

Hully, P. E. 1962. On the behavior of crawlers of the citrus mussel scale, *Lepidosaphes beckii* (Newm.) (Homoptera: Diaspididae). *Entomol. Soc. S. Africa 25*:56–72.

Jones, E. P. 1935. III. The bionomics and ecology of red scale, *Aonidiella aurantii* Mask., in southern Rhodesia. *Mazoe Citrus Exp. Stn. 5*:11–52.

Karny, M. 1972. Comparative studies on three *Dactylopius* species (Homoptera: Dactylopiidae) attacking introduced opuntias in South Africa. Antomoloy Memoir, Dept. of Agricultural Technical Services, Republic of South Africa #26.

Koteja, J. 1990. Life history. Pp. 243–254 *in* D. Rosen (Ed.), *The Armored Scale Insects, their Biology, Natural Enemies, and Control,* Vol. 4A. Elsevier Science Publishing, Amsterdam, The Netherlands.

Lambdin, P. L. 1990. Development of the black willow scale, *Chionaspis salicisnigrae* (Homoptera: Diaspididae), in Tennessee. *Entomol. News 101*:288–292.

Lambdin, P. L., D. Paulsen, and J. D. Simpson. 1993. Development and behavior of the walnut scale on flowering dogwood in Tennessee. *Tenn. Farm Home Sci. 166*:26–29.

Lellakova-Duskova, F. 1963. Morphology, metamorphosis and the life cycle of the scale insect *Quadraspidiotus gigas* (Thiem et Gerneck) (Homoptera: Coccoidea, Diaspididae). *Act. Entomol. Mus. Nat. Pragae 35*:611–648.

Luck, R. F. and D. L. Dahlsten. 1975. Natural decline of a pine needle scale (*Chionaspis pinifoliae* [Fitch]), outbreak at South Lake Tahoe, California following cessation of adult mosquito control with malathion. *Ecology 56*:893–904.

Manichote, P. and W. W. Middlekauff. 1967. Life history studies of the cactus mealybug, *Spilococcus cactearum* McKenzie. *Hilgardia 37*:639–660.

Marin, R. and Cisneros, F. 1977. Biologia y morfologia de la cochinilla del carmin, *Dactylopius coccus* Costa (Homoptera: Dactylopiidae). *Rev. Peru. Entomol. 20*:115–120.

McClure, M. S. 1977a. Dispersal of the scale *Fiorinia externa* (Homoptera: Diaspididae) and effects of edaphic factors on its establishment on hemlock. Environ. Entomol. 6:539–544.

McClure, M. S. 1977b. Population dynamics of the red pine scale, *Matsucoccus resinosae* (Homoptera: Margarodidae): The influence of resinosis. *Environ. Entomol. 6*:789–795.

McClure, M. S. 1979. Spatial and seasonal distribution of disseminating stages of *Fiorinia externa* (Homoptera: Diaspididae) and natural enemies in a hemlock forest. *Environ. Entomol. 8*:869–873.

McClure, M. S. 1990a. Habitats and hosts. Pp. 285–288 *in* D. Rosen (Ed.), *The Armored Scale Insects, Their Biology, Natural Enemies, and Control,* Vol. 4A. Elsevier Science Publishing, Amsterdam, The Netherlands.

McClure, M. S. 1990b. Influence of environmental factors. Pp. 319–330 *in* D. Rosen (Ed.), *The Armored Scale Insects, Their Biology, Natural Enemies, and Control,* Vol. 4A. Elsevier Science Publishing, Amsterdam, The Netherlands.

McClure, M. S. and J. D. Hare. 1984. Foliar terpenoids in *Tsuga* species and the fecundity of scale insects. *Oecologia (Berlin) 63*:185–193.

McKenzie, H. L. 1943. The seasonal history of *Matsucoccus vexillorum* Morrison (Homoptera: Coccoidea, Margarodidae). *Microentomology 8*:42–52.

Mendel, Z., H. Podoler, and D. Rosen. 1984. Population dynamics of the Mediterranean black scale, *Saissetia olea* (Olivier), on citrus in Israel: 5. The crawlers. *J. Entomol. Soc. S. Africa 47*:23–34.

Metcalf, Z. P. 1922. The gloomy scale. North Carolina Agr. Exp. Sta. Tech. Bull 21.

Miller, D. R., and R. F. Denno. 1977. A new genus and species of mealybug with a consideration of morphological convergence in three species (Homoptera: Pseudococcidae). *Syst. Entomol. 2*:111–157.

Miller, D. R. and M. Kosztarab. 1979. Recent advances in the study of scale insects. *Ann. Rev. Entomol. 24*:1–27.

Mopper, S. 1996. Adaptive genetic structure in phytophagous insect populations. *Trends Ecol. Evol. 11*:235–238.

Moran, V. C. and B. S. Cobby. 1979. On the life history and fecundity of the cochineal insect, *Dactylopius austrinus* De Lotto (Homoptera: Dactylopiidae), a biological control agent for the cactus *Opuntia aurantiaca. Bull. Entomol. Res. 69*:629–636.

Moran, V. C., B. H. Gunn, and G. H. Walter. 1982. Wind dispersal and settling of first-instar crawlers of the cochineal insect *Dactylopius austrinus* (Homoptera: Coccoidea, Dactylopiidae). *Ecol. Entomol. 7*:409–419.

Moreno, D. S., G. E. Carman, J. Fargerlund, and J. G. Shaw. 1974. Flight and dispersal of the adult male yellow scale. *Ann. Entomol. Soc. Am. 67*:15–20.

Moutia, L. A. 1944. The sugarcane scale, *Aulacaspis tegalensis*, Zehnt. *Bull. Entomol. Res. 35*:69–77.

Nandagopal, V. and H. David. 1990. Biology of a leaf scale insect, *Greenaspis decurvata* Green (Homoptera: Diaspididae) in sugarcane. *Entomology 15*:63–68.

Nel, R. G. 1933. A comparison of *Aonidiella aurantii* and *Aonidiella citrina*, including a study of the internal anatomy of the latter. *Hilgardia 7*:417–466.

Nielsen, D. G. and N. E. Johnson. 1973. Contribution to the life history and dynamics of the pine needle scale, *Phenacaspis pinifoliae*, in central New York. *Ann. Entomol. Soc. Am. 66*:34–43.

Nur, U. 1971. Parthenogenesis in coccids (Homoptera). Am. Zool. 11: 301–308.

Oetting, R. D. 1984. Biology of the cactus scale, *Diaspis echinocacti* (Bouché) (Homoptera: Diaspididae). *Ann. Entomol. Soc. Am.* 77:88–92.

Patel, J. D. 1971. Morphology of the gum tree scale *Eriococcus coriaceus* Maskell (Homoptera: Eriococcidae), with notes on its life history and habits near Adelaide, South Australia. *J. Aust. Entomol. Soc.* 10:43–56.

Podsiadlo, E. 1976. Dispersal of individuals of *Asterodiaspis variolosa* (Ratzeburg) (Homoptera: Coccoidea, Asterolecaniidae) on host plants from their hatching place. *Ekologia Polska* 24:69–76.

Quayle, H. J. 1911. Locomotion of certain young scale insects. *J. Econ. Entomol.* 4:301–306.

Quayle, H. J. 1912. The purple scale. University of California Agricultural Experiment Station Bulletin 226.

Quayle, H. J. 1916. Dispersion of scale insects by the wind. *J. Econ. Entomol.* 9:486–493.

Rabkin, F. B. and R. R. Lejeune. 1954. Some aspects of the biology of the pine tortoise scale, *Toumeyella numismaticum* (Pettit and McDaniel) (Homoptera: Coccidae). *Can. Entomol.* 86:570–575.

Rice, R. E. and S. C. Hoyt. 1980. Response of San Jose scale to natural and synthetic pheromones. *Environ. Entomol.* 9:190–194.

Rice, R. E. and D. S. Moreno. 1970. Flight of male California red scale. *Ann. Entomol. Soc. Am.* 63:91–96.

Russell, L. M. 1987. Habits and biology of the beech mealybug, *Peliococcus serratus* (Ferris) (Coccoidea), Pseudococcidae). *Proc. Entomol. Soc. Wash.* 89:359–362.

Sadof, C. S. and J. J. Neal. 1993. Use of host plant resources by the euonymus scale, *Unaspis euonymi* (Homoptera: Diaspididae). *Ann. Entomol. Soc. Am.* 86:614–620.

Samarasinghe, S. and E. J. Leroux. 1966. The biology and dynamics of the oystershell scale, *Lepidosaphes ulmi* (L.) (Homoptera: Coccidae), on apple in Quebec. *Ann. Entomol. Soc. Quebec* 11:206–292.

Schweig, C. and A. Grunberg. 1936. The problem of black scale (*Chrysomphalus ficus,* Ashm.) In Palestine. *Bull. Entomol. Res.* 27:677–713.

Selander, R. K. 1970. Behavior and genetic variation in natural populations. Am. Zool. 10:53–66.

Slatkin, M. 1985. Gene flow in natural populations. *Annu. Rev. Ecol. Syst.* 16:393–430.

Slatkin, M. 1987. Gene flow and the geographic structure of natural populations. *Science* 236:787–792.

Stephens, G. R. and D. E. Aylor. 1978. Aerial dispersal of red pine scale, *Matsucoccus resinosae* (Homoptera: Margarodidae). *Environ. Entomol.* 7:556–563.

Stimmel, J. F. 1979. Seasonal history and distribution of *Carulaspis minima* (Targ.-Tozz.) In Pennsylvania (Homoptera: Diaspidae). *Proc. Entomol. Soc. Wash.* 81:222–229.

Stofberg, F. J. 1937. Biology of citrus mussel scale, *Lepidosaphes pinnaeformis* (Bouché) Kirk. Science Bulletin, Dept. of Agriculture and Forestry, Union of South Africa #165.

Strickland, A. H. 1950. The dispersal of Psuedococcidae (Hemiptera-Homoptera) by air currents in the Gold Coast. *Proc. Roy. Entomol. Soc. London (A) 25*:1–9.

Strickland, A. H. 1951. The entomology of swollen shoot of cacao: I. The insects species involved, with notes on their biology. *Bull. Entomol. Res. 41*:725–748.

Tashiro, H. and C. Moffitt. 1968. Reproduction in the California red scale, *Aonidiella aurantii*: II. Mating behavior and postinsemination female changes. *Ann. Entomol. Soc. Am. 61*:1014–1020.

Taylor, T. H. C. 1935. The campaign against *Aspidiotus destructor,* Sign. Fiji. *Bull. Entomol. Res. 26*:1–102.

Timlin, J. S. 1964. The biology, bionomics, and control of *Parlatoria pittospori* Mask. (Hemiptera: Diaspididae): A pest on apples in New Zealand. *New Zealand J. Agric. Res.* 7:536–550.

Tripathi, S. R. and P. Tewary. 1984. Biology of the sugarcane scale insect *Melanaspis glomerata* (Green) (Diaspidae: Coccoidea). *J. Adv. Zool. 5*:68–78.

Unruh, T. R. 1985. Insect–plant interactions of the pinyon pine scale and single leaf pinyon in southern California. Ph.D. dissertation, University of California, Riverside, CA.

Unruh, T. R. and R. F. Luck. 1987. Deme formation in scale insects: A test with the pinyon needle scale and a review of other evidence. *Ecol. Entomol. 12*:439–449.

van Halteren, P. 1970. Note on the biology of the scale insect *Aulacaspis mangiferae* Newst. (Diaspididae: Hemiptera) on mango. *Ghana J. Agric. Sci. 3*:83–85.

Wainhouse, D. 1980. Dispersal of first instar larvae of the felted beech scale, *Cryptococcus fagisuga. J. Appl. Ecol. 17*:523–532.

Wainhouse, D. and R. Deeble. 1980. Variation in susceptibility of beech (*Fagus* spp.) to beech scale (*Cryptococcus fagisuga*). *Annu. Sci. Forest. 37*:279–289.

Wainhouse, D. and I. M. Gate. 1988. The beech scale. Pp. 67–85 *in* A. A. Berryman (Ed.), *Dynamics of Forest Insect Populations.* Plenum Press, New York.

Wainhouse, D. I. M. Gate, and D. Lonsdale. 1988. Beech resistance to the beech scale: A variety of defenses. Pp. 277–293 *in* W. J. Mattson, J. Levieux, and C. Bernard-Dagan (Eds.), *Mechanisms of Woody Plant Defenses against Insects: Search for a Pattern.* Springer-Verlag, New York.

Wainhouse, D. and R. S. Howell. 1983. Intraspecific variation in beech scale populations and in susceptibility of their host *Fagus sylvatica. Ecol. Entomol. 8*:351–359.

Walker, G. P., D. C. G. Aitken, N. V. O'Connell, and D. Smith. 1990. Using phenology to time insecticide applications for control of California red scale (Homoptera: Diaspididae) on citrus. *J. Econ. Entomol. 83*:189–196.

Walstad, J. D., D. G. Nielsen, and N. E. Johnson. 1973. Effect of the pine needle scale on photosynthesis of Scots pine. *For. Sci. 19*:109–111.

Washburn, J. O. and G. W. Frankie. 1981. Dispersal of a scale insect, *Pulvinariella mesembryanthemi* (Homoptera: Coccidae) on iceplant in California. *Environ. Entomol.* 10:724–727.

Washburn, J. O. and G. W. Frankie. 1985. Biological studies of iceplant scales, *Pulvinariella mesembryanthemi* and *Pulvinaria delottoi* (Homoptera: Coccidae), in California. *Hilgardia 53*:1–27.

Washburn, J. O., G. W. Frankie, and J. K. Grace. 1985. Effects of density on survival, development, and fecundity of the soft scale, *Pulvinariella mesembryanthemi* (Homoptera: Coccidae), and its host plant. Environ. *Entomol. 14*:755–761.

Washburn, J. O. and L. Washburn. 1984. Active aerial dispersal of minute wingless arthropods: Exploitation of boundary-layer velocity gradients. *Science 223*:1088–1089.

Washburn, R. I. 1965. Description and bionomics of a new species of *Puto* from Utah (Homoptera: Coccoidea: Pseudococcidae). *Ann. Entomol. Soc. Am. 58*:293–297.

Webster, F. M. 1902. Winds and storms as agents in the diffusion of insects. *Am. Nat. 36*:795–801.

Werner, W. H. R. 1930. Observations on the life-history and control of the fern scale, *Hemichionaspis aspidistrae* Sign. *Pap. Mich. Acad. Sci. Arts Lett. 13*:517–541.

Willard, J. R. 1973a. Survival of crawlers of California red scale, *Aonidiella aurantii* (Mask.) (Homoptera: Diaspididae). *Aust. J. Zool. 21*:567–573.

Willard, J. R. 1973b. Wandering times of the crawlers of California red scale, *Aonidiella aurantii* (Mask.) (Homoptera: Diaspididae) on citrus. *Aust. J. Zool. 21*:217–229.

Willard, J. R. 1974. Horizontal and vertical dispersal of California red scale, *Aonidiella aurantii* (Mask.), (Homoptera: Diaspididae) in the field. *Aust. J. Zool. 22*:531–548.

Willard, J. R. 1976. Dispersal of California red scale (*Aonidiella aurantii* [Maskell]) (Homoptera: Diaspididae) in relation to weather variables. *J. Aust. Entomol. Soc. 15*:395–404.

Williams, J. R. 1970. Studies on the biology, ecology and economic importance of the sugarcane scale insect, *Aulacaspis tegalensis* (Zhnt.) (Diaspididae), in Mauritius. *Bull. Entomol. Res. 60*:61–95.

Willink, E. and D. Moore. 1988. Aspects of the biology of *Rastrococcus invadens* Williams (Hemiptera: Pseudococcidae), a pest of fruit crops in West Africa, and one of its primary parasitoids, *Gyranusoidea tebygi* Noyes (Hymenoptera: Encyrtidae). *Bull. Entomol. Res. 78*:709–715.

Yardeni, A. 1987. Evaluation of wind-dispersed soft-scale crawlers (Homoptera: Coccidae), in the infestation of a citrus grove in Israel. *Israel J. Entomol. 21*:25–31.

# 12

# Life-History Strategies and the Genetic Structure of Phytophagous Insect Populations

*Merrill A. Peterson and Robert F. Denno*
Department of Entomology, University of Maryland, College Park, MD

## 12.1 Introduction

### 12.1.1 Gene Flow and the Spatial Scale of Local Adaptation

Spatial variation in selection creates the potential for local adaptation, but the realization of this potential is governed by the balance between selection and the countering effects of both genetic drift and gene flow (Slatkin 1973, 1987; Endler 1977). Strong selection can generally overcome the effects of all but the most extreme levels of genetic drift (Wright 1931; Fisher 1958), but moderate gene flow from nearby populations in which alternate traits are favored can theoretically prevent the evolution of locally adapted demes, even under fairly strong selective regimes (Slatkin 1973, 1985a; May et al. 1975). Furthermore, the spatial scale at which local adaptations develop is influenced by the spatial scale of gene flow. For species with broadscale gene flow, adaptation may occur at a regional scale, whereas for species with limited gene flow, it may occur over much smaller spatial scales (Slatkin 1973; Endler 1979; Hanks and Denno 1994; Thomas and Singer, Chapter 14, this volume). Thus, to thoroughly understand both the conditions that favor the evolution of local adaptations in phytophagous insects and the scale at which those adaptations occur, it is essential to elucidate the factors that influence gene flow among populations.

### 12.1.2 Factors Potentially Influencing Gene Flow among Insect Populations

#### 12.1.2.1 Dispersal Ability

Various authors have suggested a number of factors that influence how gene flow homogenizes gene frequencies among insect populations (e.g., Zera 1981; King 1987; Wade and McCauley 1988; Whitlock 1992). The most intuitively obvious is dispersal behavior. Simply stated, a common way for genes to move among populations, and indeed the only way for it to occur in insects, is for individuals bearing those genes to disperse among the populations (Slatkin 1985a). Thus, the *dispersal–gene flow hypothesis* states that there should be a positive correlation between the extent of dispersal and levels of gene flow.

### 12.1.2.2 Habitat Persistence

A second factor that may influence the degree to which populations exchange individuals, and potentially genes, is habitat persistence. Habitat persistence could influence gene flow by its effect on the evolution of dispersal behavior, which in turn may influence the movement of genes among populations (Roderick 1996). In particular, it has been argued that selection favoring dispersal is much stronger in temporary habitats than in persistent ones (Southwood 1962, 1977; Roff 1986, 1990; Denno et al. 1991; Gandon et al., Chapter 13, this volume), suggesting that gene flow may be greater in species that occupy temporary habitats. A second effect of decreased habitat persistence may be an increase in the frequency of recolonization and extinction events, which can theoretically either augment or diminish the genetic differentiation of populations (Slatkin 1985a; Wade and McCauley 1988; Whitlock and McCauley 1990; Hastings and Harrison 1994). These effects may be most pronounced in species characterized by a metapopulation structure (Hastings and Harrison 1994; Gandon et al. 1996; Chapter 13, this volume). Thus, the hypothesis hereafter referred to as the *habitat persistence–gene flow hypothesis* is that gene flow is reduced among populations occupying persistent habitats, compared to levels of gene flow among populations in temporary habitats.

### 12.1.2.3 Habitat Patchiness

In addition to dispersal ability and habitat persistence, habitat patchiness may also influence levels of gene flow among insect populations (Caccone and Sbordoni 1987; King 1987; McCauley 1991; Descimon and Napolitano 1993; Hastings and Harrison 1994; Britten et al. 1995; Britten and Rust 1996; Roderick 1996). Comparisons of genetic variation in metapopulations of species differing in dispersal suggest that habitat patchiness may have its greatest effects on the genetic structure of sedentary species (Hastings and Harrison 1994). For sedentary species, barriers of unsuitable habitat may restrict movement among habitat patches, leading to reduced gene flow among populations occupying those patches (Caccone and Sbordoni 1987; Descimon and Napolitano 1993). In contiguous habitats, on the other hand, gene flow over tremendous distances may be possible in these species due to the lack of barriers. In addition to the direct isolating effect of habitat patchiness on population genetic structure, at a small spatial scale, an added isolating effect of habitat patchiness may be selection against dispersal (Roff 1990; Denno et al., 1996). Thus, the *habitat patchiness–gene flow hypothesis* predicts that populations in patchy environments will experience reduced gene flow relative to those in contiguous habitats.

### 12.1.2.4 Population Age

The genetic structure of populations is largely determined by the equilibrium between (1) stochastic forces such as genetic drift and founder effects, which tend to cause populations to diverge over time; (2) natural selection, a force that may either increase or diminish population genetic differentiation; and (3) gene flow, a force that homogenizes gene pools in different populations (Slatkin 1985a). Under equilibrium conditions, the degree to which populations are genetically differentiated at selectively neutral alleles provides a fairly accurate indication of the historical average gene flow among those populations, but this is not the case under nonequilibrium conditions (Slatkin 1985a, 1993; Wade and McCauley 1988; Whitlock and McCauley 1990; Nürnberger and Harrison 1995). For example, nonequilibrium population structure may exist if extinction–recolonization events occur frequently, such that there is never enough time for a balance to be reached between gene flow and stochastic forces (Wade and McCauley 1988; Whitlock and McCauley 1990). In such a case, apparent levels of gene flow would be either much lower or much higher than true levels of gene flow, depending on the relative strengths of gene flow and stochastic forces during the colonization events (Wade and McCauley 1988; Whitlock and McCauley 1990). This theory predicts that colonization of extirpated habitats by a small number of individuals relative to those individuals moving among extant populations will increase the genetic variance among populations, reducing apparent gene flow. On the other hand, if the number of colonists is relatively large compared to the number of individuals moving among extant populations, the genetic variance among populations will be reduced, increasing apparent gene flow (Wade and McCauley 1988; Whitlock and McCauley 1990). It is our opinion that conditions in which colonization promotes greater population differentiation are far more representative of typical insect populations, as evidenced by the increased population structure in introduced populations of both walnut husk flies (Berlocher 1984) and cynipid gallwasps (Stone and Sunnucks 1993). Thus, our *population age–gene flow hypothesis* is that apparent gene flow is decreased among young populations compared to old populations. It is important to note that this hypothesis is not entirely independent from the habitat persistence–gene flow hypothesis, simply because species occupying ephemeral habitats are more likely to experience frequent extinctions and colonizations, and thus be typified by young populations, compared to species that occupy highly persistent habitats. However, since species occupying relatively persistent habitats can also undergo frequent extinction–colonization events (e.g., Harrison et al. 1988; Thomas and Harrison 1992; Hanski et al. 1995), we treat this as a separate hypothesis.

### 12.1.3 Measuring Gene Flow among Populations

#### 12.1.3.1 Direct versus Indirect Measures of Gene Flow

The two methods by which researchers typically estimate gene flow provide what are commonly called "direct" and "indirect" estimates (Slatkin 1985a, 1987; Roderick 1996). Direct estimates are the result of combining observations of the distances moved by dispersing individuals and the reproductive success of those individuals, whereas indirect estimates are obtained by comparing the frequencies of putatively neutral alleles in different populations and estimating gene flow using a variety of different analyses (e.g., Wright 1951; Nei 1973; Sokal and Wartenberg 1983; Weir and Cockerham 1984; Slatkin 1985b). Both methods have positive and negative aspects. Direct methods may provide a more accurate picture of current levels of genetic exchange than indirect methods, which provide a better estimate of historical patterns of gene flow, assuming that the populations are at an equilibrium between gene flow and genetic drift (Slatkin 1985a, 1987; Roderick 1996). On the other hand, direct methods may miss evolutionarily important but rare long-distance dispersal events and require an extensive amount of long-term fieldwork. Indirect methods are much better in this regard, in that the genetic evidence of rare dispersal events is captured in the allele-frequency patterns in different populations, and the necessary surveys of allelic variation are relatively easy to perform (Slatkin 1985a, 1987). Most studies of gene flow to date have relied on indirect estimates to test each of the four hypotheses outlined earlier. Nonetheless, as Porter and Geiger (1995) discuss, these indirect estimates may be useful only over relatively small spatial scales, since at a regional scale, the rate of approach to an equilibrium between gene flow and genetic drift may be exceedingly slow.

### 12.1.3.2 The Need for Spatially Explicit Analyses

Because gene flow theoretically declines with geographic distance (Wright 1943), any tests of hypotheses regarding factors that influence gene flow should involve spatially explicit analyses. Simply comparing levels of gene flow in different species is a woefully inadequate approach if the species were not all studied over a similar spatial scale. For example, Goulson (1993), upon reviewing the literature on lepidopteran population genetic structure, claimed that the weak genetic subdivision of populations of the sedentary butterfly, *Maniola jurtina*, matched that of highly vagile lepidopteran species. In part because of this inconsistency, Goulson concluded that selection, rather than gene flow, is a better explanation for the genetic homogeneity of populations. However, because the geographic scale over which Goulson sampled *M. jurtina* populations was much smaller than that sampled for vagile species that also exhibited little population structure, this conclusion may be ill-founded. Indeed, many studies have shown that gene flow over small spatial scales is fairly rampant, even in sedentary species (Sillén-Tullberg 1983; McCauley 1991; Michalakis et al. 1993; Peterson 1995). A much more appropriate comparison would have been to determine if over the same geographic scale, *M. jurtina* exhibits a level of population genetic structure similar to highly mobile lepidopterans. Many studies to date have tested the hypotheses outlined herein by comparing gene-flow levels within different species or groups of populations, but very few (e.g., King 1987; Stone and Sunnucks 1993; Britten and Rust 1996) have used spatially explicit analyses in making these comparisons. Clearly, for each of the hypotheses to be adequately tested, we need to pay closer attention to the scale of distance separating populations.

## 12.1.4 Tests of the Hypotheses

### 12.1.4.1 Dispersal–Gene Flow Hypothesis

To date, the dispersal–gene flow hypothesis has been more frequently tested than any of the other hypotheses presented herein. In the first direct test of this hypothesis with insects, Zera (1981) compared the heterogeneity of allele frequencies among populations of a wingless waterstrider with that found among populations of a wing-polymorphic waterstrider sampled over a similar area. His discovery of greater heterogeneity among populations of the wingless species was consistent with the predictions of the dispersal–gene flow hypothesis.

In sharp contrast to the waterstrider study, Liebherr (1988) found that dispersal ability was a poor predictor of gene flow among populations of five ground beetle species. Because altitudinal distribution was a much better predictor of gene flow, he argued that the relatively great persistence and/or fragmentation of upland habitats probably influenced gene flow among populations more than dispersal ability. Although all three of these studies explicitly examined the association between vagility and gene flow, the small number of species in each made it impossible to statistically test whether dispersal ability had an effect on levels of gene flow. At the extreme, in the waterstrider study (Zera 1981), dispersal capability was not replicated; there was only one species within each category.

The authors of several reviews of insect population genetic structure (Eanes and Koehn 1978; Pashley 1985; Pashley et al. 1985; McCauley 1987; McCauley and Eanes 1987; Daly 1989) have also concluded that variation among species in levels of gene flow is consistent with differences in their dispersal ability, but in all cases, the authors have reached this conclusion without statistical analysis. Furthermore, with the exception of McCauley (1987) and McCauley and Eanes (1987), authors of these reveiws have not taken into account the discrepant geographic scales over which different studies were performed. Thus, although many authors have concluded that gene flow is greater in mobile insects compared to their sedentary counterparts, these conclusions are invariably flawed due to a lack of statistical rigor and/or the failure to perform spatially explicit analyses.

### 12.1.4.2 Habitat Persistence–Gene Flow Hypothesis

Liebherr's (1988) discovery of an altitudinal increase in population genetic structure (assayed using allozymes) in ground beetle species remains the best published test of the hypothesis that gene flow is reduced in persistent habitats, in spite of the fact that habitat persistence was confounded by habitat patchiness. Liebherr argued that the greater long-term persistence of high-elevation habitats during climatic shifts had made the upland species less prone to local extinction and colonization events (see also Roff 1990), thus minimizing gene flow in these species. As with the tests of the dispersal–gene flow hypothesis and the habitat patchiness–gene flow hypothesis, this test was compromised, because the small number of species made it impossible to statistically analyze the importance of habitat persistence. Furthermore, without a direct assessment of habitat persistence, the validity of the assumption that habitat persistence increases with elevation remains unclear. One might argue that the pattern Liebherr observed was due to greater among-population differences in selective regimes at high elevations compared to low elevations. Although it has been demonstrated that some allozyme loci are indeed under selective constraints (e.g., Watt 1977; Watt et al. 1985), it is unlikely that selection would explain the observation that numerous, independent allozyme loci provide similar estimates of population genetic structure (Slatkin 1987).

In a recently completed study of population genetic structure in the saltmarsh planthopper, *Prokelisia marginata*, we have provided strong evidence that regional, intraspecific variation in population genetic structure is shaped by regional variation in habitat persistence. These planthoppers are wing dimorphic and, in most populations, both long-winged, flight-capable adults and short-winged, flightless adults occur (Wilson 1982; Denno et al. 1987; Denno et al. 1996). From region to region, the proportion of long-winged individuals varies with the persistence of preferred habitats, as measured by the ability of planthoppers to remain in those habitats throughout the year. Specifically, in regions where preferred habitats do not remain suitable year-round, the proportion of long-winged individuals is much higher than in regions where those habitats can be occupied all year (Denno et al. 1996), as predicted by theory (Southwood 1962, 1977; Roff 1986, 1990). Estimates of population genetic structure based on allozyme variation within and among populations indicate that in regions typified by persistent habitats and reduced levels of long-wingedness, populations are more differentiated than in regions where habitats do not persist year-round and long-winged individuals predominate (Peterson and Denno in press).

### 12.1.4.3 Habitat Patchiness–Gene Flow Hypothesis

One of the first tests of the hypothesis that insect gene flow is reduced in patchy, compared to contiguous, habitats was performed by Caccone and Sbordoni (1987), who surveyed population genetic structure in several species of cave crickets. Their documentation that species in continuous caves were less genetically structured than species restricted to fragmented caves supports the habitat patchiness–gene flow hypothesis. In a similar study, King (1987) examined allozyme variation among populations of the melyrid beetle, *Collops georgianus*, a species restricted to rock outcrops. She found that at a large spatial scale, the contiguity of outcrops explained a significant amount of the apparent gene flow occurring between resident beetle populations. Similarly, Britten and Rust (1996) found that, after controlling for distance effects, gene flow was greater among populations of the dune-obligate beetle, *Eusattus muricatus*, if the populations were part of the same pluvial lake basin than if they occupied different basins. Descimon and Napolitano (1993) also showed that gene flow was much greater among montane populations of the butterfly, *Parnassius mnemosyne*, in the contiguous core of its distribution than among peripheral, isolated populations, independent of distance. In the checkerspot butterfly, *Euphydryas editha*, gene flow among populations, estimated using allozymes, is greater in the central Rocky Mountains than in the Great Basin (Britten et al. 1995). Britten and colleagues attributed this difference to greater levels of habitat patchiness in the Great Basin. From this collection of examples, it would appear that the influence of habitat patchiness on gene flow among insect populations has been well documented. However, only the studies of King (1987) and Britten and Rust (1996) statistically assessed the relationship between habitat patchiness and gene flow, making them the only rigorous tests of the habitat patchiness–gene flow hypothesis.

### 12.1.4.4 Population Age–Gene Flow Hypotheses

To date, nearly all tests of the population age–gene flow hypothesis have demonstrated that apparent gene flow is indeed weaker among young populations than among old populations. Two noteworthy studies have provided convincing evidence that this pattern is due to the effects of extinction and colonization events (Whitlock 1992; Nürnberger and Harrison 1995). By using existing theory to interpret his observations of population size, migration, extinction, and colonization in the forked fungus beetle, *Bolitotherus cornutus*, Whitlock (1992) predicted that younger populations would be more genetically subdivided than older populations. This prediction was borne out by a survey of allozyme variation. Nürnberger and Harrison (1995) showed that the population structure of the whirligig beetle, *Dineutus assimilis*, is shaped by strong gene flow and genetic drift. In this highly mobile species, gene flow homogenizes mitochondrial DNA (mtDNA) haplotype frequencies among populations over a large spatial scale. At a local scale, however, frequent extinction–colonization events have resulted in much greater population genetic subdivision than would be predicted from its strong dispersal ability. Thus, in both this species and forked fungus beetles, at presumably neutral loci, the extinction and establishment of local populations has led to a population structure shaped by genetic drift, masking the extent to which gene flow actually occurs among populations.

In addition to the extinction–recolonization studies, strong support of the hypothesis that apparent gene flow is relatively low among young populations has also come from studies that compare the genetic structure of native and introduced populations. One such example is the work of Berlocher (1984), who compared the genetic structure of native, midwestern U.S. populations of the tephritid fly, *Rhagoletis completa*, with that found among introduced populations in California and Oregon. Berlocher found that populations in the introduced range were much more differentiated than populations in the native range, consistent with the hypothesis. In a similar example, Stone and Sunnucks (1993) showed that populations of the cynipid gallwasp, *Andricus quercuscalicis*, in recently invaded portions of central and western Europe, were much more genetically subdivided than native populations in southern Europe. In both studies, the authors argued that bottlenecks and founder events had influenced patterns of genetic variation, although Stone and Sunnucks also suggested that the increased patchiness of the gallwasp's host plants in the invaded region had contributed to reduced levels of gene flow. In each system, there has apparently been insufficient time for the homogenizing force of gene flow to diminish the diversifying effects of founder events and genetic drift. Unfortunately, in these studies, as well as those of Whitlock (1992) and Nürnberger and Harrison (1995), it was impossible to statistically determine the effects of population age, since there were no replicated sets of young and old populations.

### 12.1.4.5 Our Tests of the Hypotheses

In this chapter, we conduct an extensive review of the literature on the population structure of phytophagous insects in order to provide spatially explicit statistical tests of each of the hypotheses regarding factors that might influence gene flow in these species. We restricted our analysis to phytophagous insects in part because they are the focus of this book, but also because it is relatively easy to define life-history traits for phytophagous insects as a group, compared to defining those traits for phytophagous, predaceous, parasitic, and detritivorous insects.

The first test we conduct is a direct examination of the dispersal–gene flow hypothesis that gene flow is positively correlated with dispersal ability. In addition, we indirectly test each of the remaining three hypotheses. As a test of the habitat persistence–gene flow hypothesis, we compare levels of gene flow in species that feed on herbaceous and woody plants. It has been documented that these growth forms differ in persistence times (Brown 1986; Roff 1990). Our prediction is that due to the longer persistence of stands of woody plants, their resident insect populations should experience less gene flow than the insects that exploit relatively less persistent herbaceous plant patches. The third and fourth analyses we conduct are indirect tests of the habitat patchiness–gene flow hypothesis. In particular, we hypothesize that due to the relatively patchy nature of their host plants, populations of specialist herbivores exchange genes less frequently than populations of generalists. Similarly, we hypothesize that gene flow among populations of agricultural pests is more pronounced than among populations occupying natural habitats, since agricultural settings typically provide vast expanses of host plants. Finally, to test the population age–gene flow hypothesis, we test the prediction that apparent gene flow is diminished among populations of introduced species compared to their native counterparts.

## 12.2 Methods

### 12.2.1 The Data Set

We surveyed the literature on patterns of allozyme variation among phytophagous insect populations so that we could compare gene-flow estimates among species that differ in a variety of life-history characteristics. We restricted this survey to allozyme studies rather than including studies of chromosomal and mtDNA variation among populations, because estimates of gene flow from allozyme and DNA data can differ markedly (e.g., Loxdale and Brookes 1988 vs. Martinez et al. 1992; Latorre et al. 1992; Haag et al. 1993; Mitton 1994; Baruffi et al. 1995). This difference can be attributed to a number of reasons: (1) The ability to assess more variation at some DNA markers potentially allows a more accurate estimate of among-population differences than can be resolved by allozymes; (2) the maternal inheritance of mtDNA means that evolutionarily important gene flow resulting from male dispersal cannot be measured; (3) because mtDNA is haploid, it has a smaller effective population size than nuclear markers, and is thus more sensitive to genetic drift and founder events; and (4) estimates of population structure from mtDNA are single-locus estimates, and are thus more likely to be biased than allozyme-based estimates involving multiple loci (Mitton 1994; Roderick 1996).

We restricted our review to sexual, nonsocial, terrestrial phytophagous insects in order to minimize variation in gene flow due to factors other than those which we wanted to explicitly analyze. In this review we obtained data from 230 surveys of allozyme variation in 151 species of phytophagous insects from 35 families in 7 orders (Table 12.1). For a study to be included in our review, it had to meet several criteria. First, either an estimate of gene flow or allele frequencies and sample sizes were provided from which we could calculate a gene-flow estimate for each species. Second, distances among sampled populations were either indicated or could be obtained from a description of locations. Because precise locations of populations could not be determined from the information in many studies, we estimated the average pairwise distance separating all populations in each study and assigned these to three distance categories: zero-50 km, 50–500 km, > 500 km. This allowed us to include many more studies in our review than we could have had we included only those studies for which we could treat distance as a continuous variable. The final criterion for inclusion in our review was that information on at least one of the following characteristics of each species could be obtained: dispersal ability, growth form of the host plant, host range, whether or not the species was native, and whether or not the species was an agricultural pest. For those studies in which allozyme variation was assessed both within and among host races (McPheronn 1990; McPheron et al. 1988a; Guttman and Weigt 1989; Guttman et al. 1989; Feder et al. 1990a; Waring et al. 1990; Roininen et al. 1993; Herbst and Heitland 1994), we only included within-host race comparisons.

To assign each species to a category of dispersal ability (zero-1 km, 1–25 km, > 25 km), we relied on data from mark-release-recapture studies (most cases), studies of the invasion of new habitats (either natural range expansions or invasion by introduced species), or observations of the appearance of individuals outside of breeding grounds. Our choice of dispersal categories discriminates among very sedentary, moderately mobile, and highly vagile species. It is probable

Table 12.1  Ecological and Population Genetic Data for Phytophagous Insect Species Included in Our Review

| Order: Family: Species | # Loci | Nm | Dist. (km) | Disp. (km) | H.F. | H.R. | Res. | Hab. | Source for Gene Flow Estimate | Additional Sources, if Needed |
|---|---|---|---|---|---|---|---|---|---|---|
| ORTHOPTERA | | | | | | | | | | |
| Acrididae | | | | | | | | | | |
| Chorthippus albomarginatus[1] | 3 | 15.4 | 0–50 | | H | O | N | N | Gill 1981a | Marshall & Haes 1990[4,6,7]; R. Chapman, pers. comm.[5] |
| C. brunneus[1] | 4 | 15.4 | 50–500 | 0–1 | H | O | N | N | Gill 1981b | Richards & Waloff 1954[3]; Marshall & Haes 1990[4,6,7]; R. Chapman, pers. comm.[5] |
| C. brunneus[1] | 3 | 24.8 | 50–500 | 0–1 | H | O | N | N | Gill 1981a | Richards & Waloff 1954[3]; Marshall & Haes 1990[4,6,7]; R. Chapman, pers. comm[5] |
| MEAN C. brunneus | | 19.0 | 50–500 | 0–1 | H | O | N | N | | |
| C. parallelus[1] | 4 | 10.2 | 50–500 | | H | O | N | N | Gill 1981a | Marshall & Haes 1990[4,6,7]; Bernays & Chapman 1970[5] |

Table 12.1  (continued)

| Order: Family: Species | # Loci | Nm | Dist. (km) | Disp. (km) | H.F. | H.R. | Res. | Hab. | Source for Gene Flow Estimate | Additional Sources, if Needed |
|---|---|---|---|---|---|---|---|---|---|---|
| *Melanoplus sanguinipes*[1] | 3 | 31.0 | 50–500 | >25 | H | P | N | A | | Chapco & Bidochka 1986, Johnson 1969[3]; Vickery & McE. Kevan 1983[4]; Mulkern et al. 1969[5] |
| *Myrmeleotettix maculatus*[1] | 2 | 6.2 | 0–50 | | H | O | N | N | Gill, 1981a | Marshall & Haes 1990[4,6,7]; R. Chapman, pers. comm.[5] |
| *Omocestus viridulus*[1] | 2 | 8.1 | 0–50 | | H | O | N | N | Gill, 1981a | Marshall & Haes 1990[4,6,7]; R. Chapman, pers. comm.[5] |
| Tettigoniidae | | | | | | | | | | |
| *Ephippiger ephippiger*[1] | 4 | 0.8 | 50–500 | 0–1 | W | P | N | N | Oudman et al. 1990 | Duijm & Oudman 1983[4,5,7] |
| Gryllidae | | | | | | | | | | |
| *Gryllus pennsylvanicus*[1] | 7 | 1.9 | 50–500 | | H | P | N | A | Harrison 1979 | |
| *G. rubens*[1] | 5 | 124.8 | 50–500 | | H | P | N | A | Harrison 1979 | |
| *G. veletis*[1] | 6 | 11.7 | >500 | | H | P | N | A | Harrison 1979 | |

*Table 12.1* (continued)

| Order: Family: Species | # Loci | Nm | Dist. (km) | Disp. (km) | H.F. | H.R. | Res. | Hab. | Source for Gene Flow Estimate | Additional Sources, if Needed |
|---|---|---|---|---|---|---|---|---|---|---|
| **HETEROPTERA** | | | | | | | | | | |
| Lygaeidae | | | | | | | | | | |
| Lygaeus equestris | 2 | 24.8 | 0–50 | 1–25 | H | O | N | N | Sillén-Tullberg 1983 | Solbreck 1971[3,7] |
| L. equestris | 2 | 12.9 | 50–500 | 1–25 | H | O | N | N | Sillén-Tullberg 1983 | Solbreck 1971[3,7] |
| **HOMOPTERA** | | | | | | | | | | |
| Membracidae | | | | | | | | | | |
| Enchenopa "binotata"— Juglans race | 4 | 6.7 | 0–50 | 0–1 | W | M | N | N | Guttman et al. 1989 | |
| E. "binotata"—Juglans race | 12 | 0.6 | >500 | 0–1 | W | M | N | A | Guttman & Weigt 1989 | |
| E. "binotata"—Celastrus race | 12 | 2.5 | 0–50 | 0–1 | W | M | N | N | Guttman & Weigt 1989 | |
| E. "binotata"—Cercis race | 4 | 27.5 | 0–50 | 0–1 | W | M | N | N | Guttman et al. 1989 | |
| E. "binotata"—Cercis race | 12 | 1.2 | >500 | 0–1 | W | M | N | N | Guttman & Weigt 1989 | |
| E. "binotata"—Ptelea race | 4 | 19.0 | 0–50 | 0–1 | W | M | N | N | Guttman et al. 1989 | |
| E. "binotata"—Ptelea race | 12 | 0.4 | 50–500 | 0–1 | W | M | N | N | Guttman & Weigt 1989 | |
| E. "binotata"—Robinia race | 12 | 2.2 | 50–500 | 0–1 | W | M | N | N | Guttman & Weigt 1989 | |
| E. "binotata"—Viburnum race | 12 | 1.4 | >500 | 0–1 | W | M | N | N | Guttman & Weigt 1989 | |
| Cercopidae | | | | | | | | | | |
| Philaenus loukasi[1] | 7 | 11.7 | 50–500 | | H | M | N | N | Loukas & Drosopoulos 1992 | |
| P. signatus[1] | 3 | undef. | >500 | | H | M | N | N | Loukas & Drosopoulos 1992 | |

Table 12.1 (continued)

| Order: Family: Species | # Loci | Nm | Dist. (km) | Disp. (km) | H.F. | H.R. | Res. | Hab. | Source for Gene Flow Estimate | Additional Sources, if Needed |
|---|---|---|---|---|---|---|---|---|---|---|
| P. spumarius[1] | 7 | 1.0 | 0–50 | 0–1 | H,W | P | N | N | Saura et al. 1973 | Weaver & King 1954[3] |
| P. spumarius[1] | 4 | 49.8 | 50–500 | 0–1 | H,W | P | N | A | Loukas & Drosopoulos 1992 | Weaver & King 1954[3] |
| Cicadellidae | | | | | | | | | | |
| Nephotettix malayanus[1] | 2 | 6.3 | 50–500 | | H | M | N | A | Tan et al. 1994 | |
| N. nigropictus[1] | 4 | 11.7 | 50–500 | | H | O | N | A | Tan et al. 1994 | |
| N. virescens[1] | 4 | undef. | 50–500 | 1–25 | H | M | N | A | Tan et al. 1994 | Widiarta et al. 1990[3] |
| Delphacidae | | | | | | | | | | |
| Nilaparvata lugens | 4 | 2.6 | > 500 | > 25 | H | M | N | A | Demayo et al. 1990 | Kisimoto 1981[5] |
| N. lugens | 1 | 2.3 | > 500 | > 25 | H | M | N | A | Hoshizaki 1994 | Kisimoto 1981[5] |
| N. lugens | 1 | 17.6 | > 500 | > 25 | H | M | N | A | Hoshizaki 1994 | Kisimoto 1981[5] |
| MEAN N. lugens | | 3.5 | > 500 | > 25 | H | M | N | A | | |
| Prokelisia dolus[1] | 6 | 7.3 | > 500 | | H | M | N | N | Peterson & Denno in press | |
| P. dolus[1] | 6 | 8.7 | > 500 | > 25 | H | M | N | N | Peterson & Denno in press | |
| MEAN P. dolus | | 7.9 | > 500 | | H | M | N | N | | |
| P. marginata[1] | 7 | 10.6 | > 500 | > 25 | H | M | N | N | Peterson & Denno in press | Sparks et al. 1986[3] |
| P. marginata[1] | 7 | 16.4 | > 500 | > 25 | H | M | N | N | Peterson & Denno in press | Sparks et al. 1986[3] |
| MEAN P. marginata | | 12.9 | > 500 | > 25 | H | M | N | N | | |

Table 12.1   (continued)

| Order: Family: Species | # Loci | Nm | Dist. (km) | Disp. (km) | H.F. | H.R. | Res. | Hab. | Source for Gene Flow Estimate | Additional Sources, if Needed |
|---|---|---|---|---|---|---|---|---|---|---|
| Tumidigena minuta[1] | 3 | 16.4 | 0–50 | 0–1 | H | M | N | N | Peterson & Denno in press | |
| Psyllidae | | | | | | | | | | |
| Cacopsylla pyricola | 8 | 3.2 | 50–500 | > 25 | W | M | I | A | Unruh 1990 | |
| Aphididae | | | | | | | | | | |
| Macrosiphum rosae | 2 | 1.5 | 0–50 | | H,W | P | N | A | Wöhrmann & Tomiuk 1990 | Tomiuk et al. 1990[4,5], Blackman & Eastop 1984[6] |
| M. rosae | 2 | 0.7 | > 500 | | H,W | P | N | A | Wöhrmann & Tomiuk 1990 | Tomiuk et al. 1990[4,5], Blackman & Eastop 1984 |
| M. rosae | 2 | 14.5 | > 500 | | H,W | P | N | A | Wöhrmann & Tomiuk 1988 | Tomiuk et al. 1990[4,5], Blackman & Eastop 1984[6] |
| MEAN M. rosae | | 1.5 | > 500 | | H,W | P | N | A | | |
| Rhopalosiphum padi[1] | 2 | 8.1 | 50–500 | > 25 | H,W | P | N | A | Loxdale & Brookes 1988 | Blackman & Eastop 1984[6] |
| Schoutedenia lutea | 3 | 0.8 | 0–50 | | W | M | N | N | Tomiuk et al. 1991 | Hales & Carver 1976[4,6] |
| Sitobion avenae | 14 | 1.6 | > 500 | > 25 | H | O | N | A | Loxdale et al. 1984; Loxdale 1990 | Kieckhefer & Gellner 1988[4,5], Blackman & Eastop 1984[6] |

276

*Table 12.1  (continued)*

| Order: Family: Species | # Loci | Nm | Dist. (km) | Disp. (km) | H.F. | H.R. | Res. | Hab. | Source for Gene Flow Estimate | Additional Sources, if Needed |
|---|---|---|---|---|---|---|---|---|---|---|
| **Diaspididae** | | | | | | | | | | |
| *Nuculaspis californica*[2] | 3 | 3.4 | 0–50 | 0–1 | W | O | N | N | Alstad & Corbin 1990 | Edmunds 1973[5] |
| *N. californica*[2] | 3 | 3.7 | 0–50 | 0–1 | W | O | N | N | Alstad & Corbin 1990 | Edmunds 1973[5] |
| *N. californica*[2] | 3 | 5.4 | 0–50 | 0–1 | W | O | N | N | Alstad & Corbin 1990 | Edmunds 1973[5] |
| MEAN *N. californica* | | 4.0 | 0–50 | 0–1 | W | O | N | N | | |
| **COLEOPTERA** | | | | | | | | | | |
| **Scarabaeidae** | | | | | | | | | | |
| *Prodontria modesta* | 7 | 0.5 | 0–50 | | | | N | N | Emerson & Wallis 1994 | |
| **Cerambycidae** | | | | | | | | | | |
| *Tetraopes tetraophthalmus* | 3 | 13.6 | 0–50 | 1–25 | H | M | N | N | McCauley 1991 | Lawrence 1988[3] |
| *T. tetraophthalmus* | 3 | 62.3 | 0–50 | 1–25 | H | M | N | N | McCauley 1991 | Lawrence 1988[3] |
| MEAN *T. tetraophthalmus* | | 22.5 | 0–50 | 1–25 | H | M | N | N | | |
| *T. tetraophthalmus* | 3 | 1.5 | > 500 | 1–25 | H | M | N | N | McCauley & Eanes 1987 | Lawrence 1988[3] |
| **Chrysomelidae** | | | | | | | | | | |
| *Aulacophora lewisii*[1] | 7 | 9.8 | 0–50 | | H | P | N | A | Yong 1993 | |
| *Chrysolina aurichalcea* | 7 | 1.3 | > 500 | 0–1 | H | O | N | N | Sakanoue & Fujiyama 1987 | Suzuki 1978[3]; Matsuda et al. 1982[4] |
| *Chrysomela aeneicollis*[2] | 5 | 3.5 | 0–50 | | W | M | N | N | Rank 1992 | |

Table 12.1 (continued)

| Order: Family: Species | # Loci | Nm | Dist. (km) | Disp. (km) | H.F. | H.R. | Res. | Hab. | Source for Gene Flow Estimate | Additional Sources, if Needed |
|---|---|---|---|---|---|---|---|---|---|---|
| C. aeneicollis[2] | 5 | 12.5 | 0–50 | | W | M | N | N | Rank 1992 | |
| C. aeneicollis[2] | 5 | 5.6 | 0–50 | | W | M | N | N | Rank 1992 | |
| MEAN C. aeneicollis[2] | | 5.5 | 0–50 | | W | M | N | N | | |
| Diabrotica barberi | 11 | 12.4 | 0–50 | | H | O | N | A | Krafsur et al. 1993 | |
| D. barberi[1] | 6 | 2.3 | > 500 | | H | O | N | A | McDonald et al. 1985 | |
| Gonioctena viminalis | 10 | 0.2 | 50–500 | | W | M | N | N | Knoll et al. in press | P. Mardulyn, pers. comm.[4,5,6,7] |
| G. olivacea | 15 | 1.5 | > 500 | | W | M | N | N | Knoll et al. in press | P. Mardulyn, pers. comm.[4,5,6,7] |
| G. quinquepunctata | 10 | 0.7 | > 500 | | W | M | N | N | Knoll et al. in press | P. Mardulyn, pers. comm.[4,5,6,7] |
| G. pallida | 9 | 2.3 | 50–500 | | W | P | N | N | Knoll et al. in press | P. Mardulyn, pers. comm.[4,5,6,7] |
| Leptinotarsa decemlineata | 7 | 3.4 | > 500 | > 25 | H | O | I | A | Jacobson & Hsiao 1983 | Johnson 1969[3]; Hsiao 1978[4] |
| Ophraella communa | 10 | 0.5 | > 500 | | H | O | N | N | Futuyma & McCafferty 1990 | |
| O. notulata | 4 | 0.5 | > 500 | | H | M | N | N | Futuyma & McCafferty 1990 | |

Table 12.1 (continued)

| Order: Family: Species | # Loci | Nm | Dist. (km) | Disp. (km) | H.F. | H.R. | Res. | Hab. | Source for Gene Flow Estimate | Additional Sources, if Needed |
|---|---|---|---|---|---|---|---|---|---|---|
| O. pilosa | 9 | 0.4 | > 500 | | H | O | N | N | Futuyma & McCafferty 1990 | |
| Oreina bifrons[1] | 4 | 24.8 | 50–500 | | H | O | N | N | Rowell-Rahier & Pasteels 1994 | M. Rowell-Rahier pers. comm.[6,7] |
| O. cacaliae | 6 | 0.8 | 50–500 | 0–1 | H | O | N | N | Rowell-Rahier 1992 | M. Rowell-Rahier pers. comm.[6,7] |
| O. gloriosa[1] | 3 | 1.6 | 0–50 | 0–1 | H | M | N | N | Eggenberger & Rowell-Rahier 1991 | |
| O. speciosa[1] | 5 | 1.0 | 50–500 | | H | O | N | N | Rowell-Rahier & Pasteels 1994 | M. Rowell-Rahier pers. comm.[6,7] |
| O. speciosissima | 6 | 4.7 | 50–500 | 0–1 | H | O | N | N | Rowell-Rahier 1992 | M. Rowell-Rahier pers. comm.[6,7] |
| Plagiodera versicolora | 3 | 4.1 | > 500 | | W | M | I | N | McCauley et al. 1988 | |
| Curculionidae | | | | | | | | | | |
| Anthonomus grandis[1] | 9 | 2.3 | 50–500 | > 25 | H | O | I | A | Bartlett 1981 | Cross et al. 1975[5] |
| A. grandis[1] | 8 | 2.5 | > 500 | > 25 | H | O | I | A | Terranova et al. 1990 | Cross et al. 1974[5] |
| Hypera postica[1] | 3 | 3.8 | > 500 | | H | O | I | A | Sell et al. 1978 | Essig & Michelbacker 1933[4,5] |

Table 12.1 (continued)

| Order: Family: Species | # Loci | Nm | Dist. (km) | Disp. (km) | H.F. | H.R. | Res. | Hab. | Source for Gene Flow Estimate | Additional Sources, if Needed |
|---|---|---|---|---|---|---|---|---|---|---|
| H. postica[1] | 11 | 9.0 | > 500 | | H | O | I | A | Hsiao & Stutz 1985 | |
| MEAN H. postica | | 5.4 | > 500 | | H | O | I | A | | |
| Larinus cynarae | 4 | 16.1 | 0–50 | 0–1 | H | O | N | N | Michalakis et al. 1993 | |
| L. cynarae | 4 | 18.5 | 50–500 | 0–1 | H | O | N | N | Michalakis et al. 1993 | |
| Pissodes strobi | 11 | 2.3 | > 500 | 1–25 | W | O | N | N | Phillips & Lanier 1985 | |
| Polydrosus mollis[1] | 6 | 1.5 | 50–500 | | W | P | N | N | Lokki et al. 1976 | |
| Scolytidae | | | | | | | | | | |
| Dendroctunus frontalis[1] | 1 | 249.8 | 0–50 | 1–25 | W | M | N | N | Florence et al. 1982 | Turchin & Thoeny 1993[3]; Payne 1981[5] |
| D. frontalis[1] | 2 | 3.8 | > 500 | 1–25 | W | M | N | N | Anderson et al. 1979 | Turchin & Thoeny 1993[3]; Payne 1981[5] |
| D. frontalis[1] | 5 | 5.2 | > 500 | 1–25 | W | M | N | N | Namkoong et al. 1979 | Turchin & Thoeny 1993[3]; Payne 1981[5] |
| MEAN D. frontalis | | 4.4 | > 500 | 1–25 | W | M | N | N | | |
| D. jeffreyi[1] | 6 | 7.6 | 0–50 | | W | M | N | N | Higby & Stock 1982 | |
| D. ponderosae[1] | 5 | 22.5 | 0–50 | | W | M | N | N | Sturgeon & Mitton 1986 | |

Table 12.1  (continued)

| Order: Family: Species | # Loci | Nm | Dist. (km) | Disp. (km) | H.F. | H.R. | Res. | Hab. | Source for Gene Flow Estimate | Additional Sources, if Needed |
|---|---|---|---|---|---|---|---|---|---|---|
| D. ponderosae[1] | 6 | 14.5 | 50–500 | | W | M | N | N | Higby & Stock 1982 | |
| D. ponderosae[1] | 3 | 0.7 | 50–500 | | W | M | N | N | Stock & Guenther 1979 | |
| D. ponderosae[1] | 4 | 13.6 | 650–500 | | W | M | N | N | Stock & Amman 1980 | |
| MEAN D. ponderosae | | 2.3 | 50–500 | | W | M | N | N | | |
| D. ponderosae[1] | 5 | 7.1 | >500 | | W | M | N | N | Stock et al. 1984 | |
| D. pseudotsugae[1] | 7 | 6.5 | >500 | | W | O | N | N | Stock et al. 1979 | Bedard 1950[5] |
| D. terebrans[1] | 2 | 8.1 | >500 | | W | M | N | N | Anderson et al. 1983 | |
| Ips calligraphus[1] | 4 | 27.5 | >500 | | W | M | N | N | Anderson et al. 1983 | Wood & Stark 1968[4,5] |
| LEPIDOPTERA | | | | | | | | | | |
| Pyralidae | | | | | | | | | | |
| Diatraea grandiosella | 4 | 0.7 | >500 | >25 | H | O | I | A | McCauley et al. 1990 | Chippendale 1979[3,4,5] |
| Homeosoma electellum[1] | 2 | 7.6 | >500 | >25 | H | P | N | A | Beregovoy & Gill 1986 | Teetes & Randolph 1969[5]; Arthur & Bauer 1981[6,7] |
| Ostrinia nubilalis[1,8] | 5 | 249.8 | 50–500 | >25 | H | P | I | A | Harrison & Vawter 1977 | Brindley & Dicke 1963[3]; Covell 1984[5] |

Table 12.1 (continued)

| Order: Family: Species | # Loci | Nm | Dist. (km) | Disp. (km) | H.F. | H.R. | Res. | Hab. | Source for Gene Flow Estimate | Additional Sources, if Needed |
|---|---|---|---|---|---|---|---|---|---|---|
| Olethreutidae | | | | | | | | | | |
| Rhyacionia frustrana[1] | 5 | 11.1 | 0–50 | | W | M | N | N | Namkoong et al. 1982 | |
| R. frustrana[1] | 5 | 13.6 | >500 | | W | M | N | N | Namkoong et al. 1982 | |
| Tortricidae | | | | | | | | | | |
| Choristoneura fumiferana[1] | 3 | 83.1 | 50–500 | >25 | W | O | N | N | May et al. 1977 | Covell 1984[5] |
| C. occidentalis[1] | 13 | 3.1 | 50–500 | >25 | W | O | N | N | Willhite & Stock 1983 | |
| Cydia pomonella | 4 | 3.5 | >500 | 1–25 | W | P | I | A | Pashley 1985 | Covell 1984[4,5] |
| Plutellidae | | | | | | | | | | |
| Plutella xylostella | 4 | 6.3 | 50–500 | >25 | H | O | I | A | Caprio & Tabashnik 1992a | |
| Yponomeutidae | | | | | | | | | | |
| Yponomeuta cagnagellus | 4 | 22.5 | 0–50 | >25 | W | M | N | N | Menken et al. 1980 | |
| Y. cagnagellus | 5 | 11.7 | 50–500 | >25 | W | M | N | N | Menken et al. 1980 | |
| Y. cagnagellus | 5 | 5.7 | 50–500 | >25 | W | M | N | N | Menken et al. 1980 | |
| Y. cagnagellus | 5 | 12.9 | 50–500 | >25 | W | M | N | N | Menken et al. 1980 | |
| MEAN Y. cagnagellus | | 8.9 | 50–500 | >25 | W | M | N | N | | |

*Table 12.1*  (continued)

| Order: Family: Species | # Loci | Nm | Dist. (km) | Disp. (km) | H.F. | H.R. | Res. | Hab. | Source for Gene Flow Estimate | Additional Sources, if Needed |
|---|---|---|---|---|---|---|---|---|---|---|
| Y. "padellus"—Crataegus race | 5 | 20.6 | 50–500 | 0–1 | W | M | N | N | Menken 1981 | Brookes & Butlin 1994[3] |
| Y. "padellus"—Crataegus race | 5 | 35.5 | 50–500 | 0–1 | W | M | N | N | Menken 1981 | Brookes & Butlin 1994[3] |
| MEAN Y. "padellus"—Crataegus race | | 26.1 | 50–500 | 0–1 | W | M | N | N | | |
| Nepticulidae | | | | | | | | | | |
| Stigmella roborella | 8 | 11.6 | 50–500 | | W | M | N | N | Cronau & Menken 1990 | Emmett 1976[4,5] |
| Geometridae | | | | | | | | | | |
| Thera obeliscata[1] | 2 | undef. | 50–500 | | W | O | N | N | Jelnes 1975a | Björkman & Larsson 1991[5] |
| T. variata[1] | 1 | 4.5 | 50–500 | | W | O | N | N | Jelnes 1975a | Björkman & Larsson 1991[5] |
| Noctuidae | | | | | | | | | | |
| Agrotis ipsilon | 5 | 35.5 | 50–500 | > 25 | H | P | N | A | Buès et al. 1994 | Showers et al 1989[3]; Covell 1984[4] |
| Alabama argillacea | 13 | 35.5 | > 500 | > 25 | H | O | N | A | Pashley 1985 | |
| Anticarsia gemmatalis | 19 | 11.7 | > 500 | > 25 | H | O | N | A | Pashley 1985; Pashley & Johnson 1986 | |
| Helicoverpa armigera | 8 | 10.6 | > 500 | > 25 | H | P | N | A | Daly & Gregg 1985; Daly 1989 | |

Table 12.1 (continued)

| Order: Family: Species | # Loci | Nm | Dist. (km) | Disp. (km) | H.F. | H.R. | Res. | Hab. | Source for Gene Flow Estimate | Additional Sources, if Needed |
|---|---|---|---|---|---|---|---|---|---|---|
| H. punctigera | 5 | 27.0 | > 500 | > 25 | H | P | N | A | Daly & Gregg 1985; Daly 1989 | |
| H. virescens | 13 | 124.8 | 0–50 | > 25 | H | P | N | A | Korman et al. 1993 | Covell 1984[5] |
| H. virescens | 13 | 249.8 | 50–500 | > 25 | H | P | N | A | Korman et al. 1993 | Covell 1984[5] |
| H. virescens[1] | 10 | 5.0 | > 500 | > 25 | H | P | N | A | Sluss & Graham 1979 | Covell 1984[5] |
| Spodoptera exempta[1] | 3 | 249.8 | > 500 | > 25 | H | O | N | A | den Boer 1978 | Rainey 1979[3] |
| S. frugiperda | 12 | 2.7 | > 500 | > 25 | H | P | N | A | Pashley et al. 1985 | Covell 1984[5] |
| Xesita adela[1] | 4 | 4.5 | 50–500 | | H,W | P | N | A | Hudson & Lefkovitch 1982 | Covell 1984[4,5,7] |
| X. dolosa[1] | 5 | 1.5 | 50–500 | | H,W | P | N | A | Hudson & Lefkovitch 1982 | Covell 1984[4,5,7] |
| Lymantriidae | | | | | | | | | | |
| Lymantria dispar[1] | 8 | 0.1 | > 500 | 1–25 | W | P | N | N | George 1984 | Taylor & Reling 1986[3]; Covell 1984[4] |
| L. dispar[1] | 10 | 1.3 | > 500 | 1–25 | W | P | N | N | Harrison et al. 1983 | Taylor & Reling 1986[3]; Covell 1984[4] |
| MEAN L. dispar | | 0.3 | > 500 | 1–25 | W | P | N | N | | |

Table 12.1 (continued)

| Order: Family: Species | # Loci | Nm | Dist. (km) | Disp. (km) | H.F. | H.R. | Res. | Hab. | Source for Gene Flow Estimate | Additional Sources, if Needed |
|---|---|---|---|---|---|---|---|---|---|---|
| Lasiocampidae | | | | | | | | | | |
| *Malacosoma americanum* | 7 | 3.4 | 0–50 | | W | O | N | N | Costa & Ross 1993 | Covell 1984[5] |
| *M. americanum* | 7 | 5.1 | 50–500 | | W | O | N | N | Costa & Ross 1993 | Covell 1984[5] |
| *M. americanum* | 11 | 5.9 | 50–500 | | W | O | N | N | Costa & Ross 1994 | Covell 1984[5] |
| MEAN *M. americanum* | | 5.4 | 50–500 | | W | O | N | N | | |
| *M. americanum* | 11 | 17.7 | > 500 | | W | O | N | N | Costa & Ross 1994 | Covell 1984[5] |
| Saturniidae | | | | | | | | | | |
| *Hemileuca oliviae*[1] | 6 | 6.2 | 50–500 | 0–1 | H | O | N | N | Dubach et al. 1988 | Bellows et al. 1984[3]; Watts & Everett 1976[5] |
| *H. oliviae*[1] | 5 | 6.9 | > 500 | 0–1 | H | O | N | N | Dubach et al. 1988 | Bellows et al. 1984[3]; Watts & Everett 1976[5] |
| Lycaenidae | | | | | | | | | | |
| *Aricia agestis*[1] | 3 | 6.3 | 50–500 | | H | O | N | N | Jelnes 1975b | Jarvis 1958[7] |
| *A. artaxerxes*[1] | 4 | 1.2 | 50–500 | | H | M | N | N | Jelnes 1975c | Jarvis 1958[7] |
| *Euphilotes enoptes*[1] | 6 | 7.3 | 0–50 | 1–25 | W | M | N | N | Peterson 1995 | Peterson 1997 |
| *E. enoptes* | 6 | 5.3 | 50–500 | 1–25 | W | M | N | N | Peterson 1996 | Peterson 1997 |
| *Hemiargus isola*[1] | 1 | 124.8 | 50–500 | > 25 | H,W | O | N | A | Burns & Johnson 1971 | Scott 1986[3,4,5,7] |

Table 12.1 (continued)

| Order: Family: Species | # Loci | Nm | Dist. (km) | Disp. (km) | H.F. | H.R. | Res. | Hab. | Source for Gene Flow Estimate | Additional Sources, if Needed |
|---|---|---|---|---|---|---|---|---|---|---|
| Pieridae | | | | | | | | | | |
| Anthocharis sara[1] | 5 | 2.3 | 50–500 | | H | O | N | N | Geiger & Shapiro 1986 | Scott 1986[5] |
| A. (sara) stella[1] | 6 | 49.8 | 0–50 | | H | O | N | N | Geiger & Shapiro 1986 | Scott 1986[5] |
| Colias meadii[1] | 1 | 0.8 | 0–50 | 1–25 | H | O | N | N | Johnson 1975 | Watt et al. 1977[3] |
| C. meadii[1] | 1 | 1.2 | 0–50 | 1–25 | H | O | N | N | Johnson 1976 | Watt et al. 1977[3] |
| C. meadii[1] | 1 | 2.9 | 0–50 | 1–25 | H | O | N | N | Johnson 1976 | Watt et al. 1977[3] |
| MEAN C. meadii | | 1.3 | 0–50 | 1–25 | H | O | N | N | | |
| Pieris marginalis | 12 | 1.0 | > 500 | 1–25 | H | O | N | N | Geiger & Shapiro 1992 | Chew 1981[3] |
| P. napi | 9 | 47.8 | 50–500 | 0–1 | H | O | N | N | Porter & Geiger 1995 | Ohsaki 1980[3] |
| P. napi | 9 | 2.1 | > 500 | 0–1 | H | O | N | N | Porter & Geiger 1995 | Ohsaki 1980[3] |
| P. napi | 9 | 2.6 | > 500 | 0–1 | H | O | N | N | Porter & Geiger 1995 | Ohsaki 1980[3] |
| P. napi[1] | 7 | 2.6 | > 500 | 0–1 | H | O | N | N | Geiger & Scholl 1985 | Ohsaki 1980[3] |
| P. napi | 9 | 9.5 | > 500 | 0–1 | H | O | N | N | Porter & Geiger 1995 | Ohsaki 1980[3] |
| MEAN P. napi | | 3.0 | > 500 | 0–1 | H | O | N | N | | |
| P. rapae | 4 | 17.6 | > 500 | 1–25 | H | P | I | A | Vawter, in Eanes & Koehn 1978 | Jones et al. 1980[3]; Scott 1986[5] |

Table 12.1  (continued)

| Order: Family: Species | # Loci | Nm | Dist. (km) | Disp. (km) | H.F. | H.R. | Res. | Hab. | Source for Gene Flow Estimate | Additional Sources, if Needed |
|---|---|---|---|---|---|---|---|---|---|---|
| Pontia occidentalis[1] | 5 | 41.4 | 50–500 | > 25 | H | O | N | N | Shapiro & Geiger 1986 | Shapiro 1977[3]; Scott 1986[5] |
| P. protodice[1] | 4 | 24.8 | > 500 | 1–25 | H | P | N | N | Shapiro & Geiger 1986 | Shapiro 1982[3]; Scott 1986[5] |
| Papilionidae | | | | | | | | | | |
| Papilio (glaucus) canadensis[1] | 9 | 9.8 | > 500 | | W | P | N | N | Hagen 1990 | Scott 1986[5] |
| P. glaucus[1] | 11 | 12.3 | 50–500 | 1–25 | W | P | N | N | Hagen 1990 | Scott 1986[5]; Fales 1959[3] |
| P. glaucus | 9 | 100.0 | > 500 | 1–25 | W | P | N | N | Bossart & Scriber 1995 | Scott 1986[5]; Fales 1959[3] |
| P. machaon[1] | 8 | 2.7 | > 500 | | H | P | N | N | Sperling 1987 | |
| P. zelicaon[1] | 3 | 2.7 | 50–500 | 1–25 | H,W | P | N | A | Tong & Shapiro 1989 | Shields 1967[3] |
| P. zelicaon[1] | 7 | 7.3 | > 500 | 1–25 | H | O | N | N | Sperling 1987 | Shields 1967[3] |
| Parnassius mnemosyne[1] | 9 | 1.1 | 50–500 | 1–25 | H | M | N | N | Napolitano et al. 1988 | |
| P. mnemosyne | 9 | 1.3 | 50–500 | 1–25 | H | M | N | N | Napolitano & Descimon 1994 | |
| MEAN P. mnemosyne | | 1.2 | 50–500 | 1–25 | H | M | N | N | | |
| Nymphalidae | | | | | | | | | | |
| Boloria improba | 20 | 0.3 | > 500 | 0–1 | W | M | N | N | Britten & Brussard 1992 | |
| B. titania | 18 | 0.3 | > 500 | | H,W | P | N | N | Britten & Brussard 1992 | Scott 1986[4,5] |

*Table 12.1* (continued)

| Order: Family: Species | # Loci | Nm | Dist. (km) | Disp. (km) | H.F. | H.R. | Res. | Hab. | Source for Gene Flow Estimate | Additional Sources, if Needed |
|---|---|---|---|---|---|---|---|---|---|---|
| *Chlosyne palla*[1] | 4 | 124.8 | 0–50 | 1–25 | H | P | N | N | Schrier et al. 1976 | |
| *Coenonympha tullia*[1] | 9 | 2.6 | 50–500 | 1–25 | H | O | I | N | Wiernasz 1989 | Scott 1986[5] |
| *C. tullia*[1] | 10 | 4.6 | 50–500 | 1–25 | H | O | N | N | Porter & Mattoon 1989 | Scott 1986[5] |
| *C. tullia* | 12 | 4.7 | 50–500 | 1–25 | H | O | N | N | Porter & Geiger 1988 | Scott 1986[5] |
| MEAN *C. tullia* | | 3.7 | 50–500 | 1–25 | H | O | I,N | N | | |
| *Danaus plexippus* | 6 | 27.5 | > 500 | > 25 | H | M | N | N | Eanes & Koehn 1978 | |
| *Dryas iulia* | 2 | 108.4 | 50–500 | | H,W | M | N | N | Haag et al. 1993 | Scott 1986[5] |
| *Erebia embla* | 5 | 10.2 | 0–50 | | H | P | N | N | Douwes & Stille 1988 | Hendriksen 1982[5] |
| *Euphydryas chalcedona*[1] | 7 | 3.2 | 50–500 | 1–25 | H,W | P | N | N | McKechnie et al. 1975 | Brown & Ehrlich 1980[3]; Scott 1986[5] |
| *E. editha* | 13 | 0.9 | 50–500 | 1–25 | H | P | N | N | Britten et al. 1995 | Scott 1986[5] |
| *E. editha* | 13 | 5.1 | 50–500 | 1–25 | H | P | N | N | Britten et al. 1995 | Scott 1986[5] |
| *E. editha* | 7 | 7.8 | 50–500 | 1–25 | H | P | N | N | McKechnie et al. 1975 in Slatkin 1987 | Scott 1986[5] |
| *E. editha* | 13 | 16.8 | 50–500 | 1–25 | H | P | N | N | Britten et al. 1995 | Scott 1986[5] |

Table 12.1  (continued)

| Order: Family: Species | # Loci | Nm | Dist. (km) | Disp. (km) | H.F. | H.R. | Res. | Hab. | Source for Gene Flow Estimate | Additional Sources, if Needed |
|---|---|---|---|---|---|---|---|---|---|---|
| MEAN E. editha | | 3.1 | 50–500 | 1–25 | H | P | N | N | | |
| E. editha | 14 | 53.1 | > 500 | 1–25 | H | P | N | N | Baughman et al. 1990 | Scott 1986[5] |
| E. editha[1] | 6 | 0.4 | > 500 | 1–25 | H | P | N | N | Ehrlich & White 1980 | Scott 1986[5] |
| MEAN E. editha | | 1.0 | > 500 | 1–25 | H | P | N | N | | |
| E. phaeton[1] | 3 | 20.6 | 0–50 | 0–1 | H | P | N | N | Brussard & Vawter 1975 | Peterson & Denno in press; Scott 1986[5] |
| Limenitis lorquini | 11 | 4.9 | 50–500 | 1–25 | W | P | N | N | Porter 1990 | Scott 1986[5] |
| L. weidemeyerii | 8 | 5.8 | 50–500 | 1–25 | W | P | N | N | Rosenberg 1989 | Scott 1986[5] |
| L. weidemeyerii | 11 | 1.6 | > 500 | 1–25 | W | P | N | N | Porter 1990 | Scott 1986[5] |
| Maniola jurtina[1] | 1 | 3.1 | 0–50 | 0–1 | H | O | N | N | Handford 1973a | Brakefield 1982[3,4,5] |
| M. jurtina[1] | 1 | 2.1 | 0–50 | 0–1 | H | O | N | N | Handford 1973b | Brakefield 1982[3,4,5] |
| MEAN M. jurtina | | 2.5 | 0–50 | 0–1 | H | O | N | N | | |
| M. jurtina | 4 | 16.4 | 50–500 | 0–1 | H | O | N | N | Goulson 1993 | Brakefield 1982[3,4,5] |
| Oeneis chryxus | 9 | 2.8 | 50–500 | | H | P | N | N | Porter & Shapiro 1989 | Scott 1986[5] |
| Phyciodes tharos[1] | 5 | 49.8 | > 500 | | H | O | N | N | Vawter & Brussard 1975 | Scott 1986[5] |
| Speyeria coronis[1] | 6 | 1.6 | 50–500 | | H | M | N | N | Brittnacher et al. 1978 | |
| S. nokomis | 4 | 24.8 | 0–50 | | H | M | N | N | Britten et al. 1994 | |
| S. nokomis | 3 | 11.1 | 0–50 | | H | M | N | N | Britten et al. 1994 | |

Table 12.1  (continued)

| Order: Family: Species | # Loci | Nm | Dist. (km) | Disp. (km) | H.F. | H.R. | Res. | Hab. | Source for Gene Flow Estimate | Additional Sources, if Needed |
|---|---|---|---|---|---|---|---|---|---|---|
| MEAN S. nokomis | | 15.4 | 0–50 | | H | M | N | N | Britten et al. 1994 | |
| S. nokomis | 6 | 0.9 | 50–500 | | H | M | N | N | Britnacher et al. 1978 | |
| S. zerene[1] | 6 | 2.9 | 50–500 | | H | M | N | N | | |
| **DIPTERA** | | | | | | | | | | |
| Tephritidae | | | | | | | | | | |
| Anastrepha distincta | 18 | 25.5 | > 500 | | H,W | P | N | A | Steck 1991 | Norrbom & Kim 1988[4,5,7] |
| A. fraterculus | 21 | 11.4 | > 500 | | H,W | P | N | A | Steck 1991 | Norrbom & Kim 1988[4,5,7] |
| A. obliqua | 20 | 21.3 | > 500 | | H,W | P | N | A | Steck 1991 | Norrbom & Kim 1988[4,5,7] |
| A. striata | 14 | 1.0 | > 500 | | H,W | P | N | A | Steck 1991 | Norrbom & Kim 1988[4,5,7] |
| Ceratitis capitata | 22 | 1.8 | > 500 | 1–25 | H,W | P | I | A | Gasperi et al. 1991 | Johnson 1969[3] |
| Dacus neohumeralis[1] | 2 | undef. | > 500 | | W | P | N | A | McKechnie 1975 | Drew 1989[4,5,6,7] |
| D. oleae[1] | 2 | 27.5 | > 500 | 1–25 | W | M | I | A | Tsakas & Zouros 1980 | Economopoulos et al. 1978[3], Munroe 1984[5,6,7] |
| D. tryoni[1] | 2 | 13.5 | > 500 | 1–25 | H,W | P | N | A | McKechnie 1975 | Prokopy & Fletcher 1987[4,5,7], Drew 1989[6] |

*Table 12.1*  (*continued*)

| Order: Family: Species | # Loci | Nm | Dist. (km) | Disp. (km) | H.F. | H.R. | Res. | Hab. | Source for Gene Flow Estimate | Additional Sources, if Needed |
|---|---|---|---|---|---|---|---|---|---|---|
| *D. umbrosus*[1] | 2 | 1.8 | 50–500 | 0–1 | H,W | P | N | N | Yong 1988 | Tan & Serit 1988[3]; Drew 1989[4,5,6,7] |
| *Eurosta "solidaginis"*— *S. altissima* race[1] | 5 | 2.8 | > 500 | | H | M | N | N | Waring et al. 1990 | |
| *E. "solidaginis"*—*S. gigantea* race[1] | 3 | 7.3 | > 500 | | H | M | N | N | Waring et al. 1990 | |
| *Oxyna parietina* | 8 | 12.3 | 50–500 | | H | M | N | N | Eber et al. 1992 | |
| *Rhagoletis basiola*[1] | 8 | 5.1 | > 500 | | W | M | N | A | Berlocher & Bush 1982 | |
| *R. cingulata*[1] | 5 | 12.3 | > 500 | 0–1 | W | M | N | N | Berlocher & Bush 1982 | Johnson 1969[3] |
| *R. completa* | 4 | 4.4 | 50–500 | 1–25 | W | M | I | A | Berlocher 1984 | |
| *R. completa*[1] | 5 | 11.7 | > 500 | 1–25 | W | M | N | A | Berlocher & Bush 1982 | |
| *R. completa* | 4 | 36.8 | > 500 | 1–25 | W | M | N | A | Berlocher 1984 | |
| MEAN *R. completa* | | 17.6 | > 500 | 1–25 | W | M | N | A | | |
| *R. fausta*[1] | 7 | 24.8 | > 500 | | W | M | N | N | Berlocher & Bush 1982 | |
| *R. juglandis*[1] | 5 | 62.3 | 50–500 | | W | M | N | A | Berlocher & Bush 1982 | |
| *R. mendax*[1] | 10 | 6.7 | > 500 | | W | O | N | A | Berlocher & Bush 1982 | |
| *R. mendax* | 13 | 16.4 | > 500 | | W | O | N | A | Berlocher 1995 | |
| MEAN *R. mendax* | | 9.6 | > 500 | | W | O | N | A | | |

*Table 12.1*  (*continued*)

| Order: Family: Species | # Loci | Nm | Dist. (km) | Disp. (km) | H.F. | H.R. | Res. | Hab. | Source for Gene Flow Estimate | Additional Sources, if Needed |
|---|---|---|---|---|---|---|---|---|---|---|
| R. *"pomonella"—Crataegus* race | 6 | 7.5 | 0–50 | 1–25 | W | M | N | N | Feder et al. 1990a | |
| R. *"pomonella"—Crataegus* race | 10 | 41.4 | 0–50 | 1–25 | W | M | N | N | McPheron et al. 1988a | |
| R. *"pomonella"—Crataegus* race | 5 | 27.5 | 0–50 | 1–25 | W | M | N | N | McPheron et al. 1988b | |
| R. *"pomonella"—Crataegus* race | 5 | 49.8 | 0–50 | 1–25 | W | M | N | N | McPheron et al. 1988b | |
| MEAN R. *"pomonella"— Crataegus* race | | 19.0 | 0–50 | 1–25 | W | M | N | N | | |
| R. *"pomonella"—Crataegus* race | 6 | 3.7 | 50–500 | 1–25 | W | M | N | N | Feder & Bush 1989 | |
| R. *"pomonella"¹—Crataegus* race | 8 | 6.0 | > 500 | 1–25 | W | M | N | N | Berlocher & Bush 1982 | |
| R. *"pomonella"—Crataegus* race | 13 | 5.3 | > 500 | 1–25 | W | M | N | N | Feder et al. 1990b | |
| R. *"pomonella"—Crataegus* race | 10 | 7.1 | > 500 | 1–25 | W | M | N | N | McPheron 1990 | |
| MEAN R. *"pomonella"— Crataegus* race | | 6.1 | > 500 | 1–25 | W | M | N | N | | |
| R. *"pomonella"—Malus* race | 10 | 49.8 | 0–50 | 1–25 | W | M | N | A | McPheron et al. 1988a | |
| R. *"pomonella"—Malus* race | 6 | 35.0 | 0–50 | 1–25 | W | M | N | A | Feder et al. 1990a | |

Table 12.1  (continued)

| Order: Family: Species | # Loci | Nm | Dist. (km) | Disp. (km) | H.F. | H.R. | Res. | Hab. | Source for Gene Flow Estimate | Additional Sources, if Needed |
|---|---|---|---|---|---|---|---|---|---|---|
| MEAN R. "pomonella"—Malus race | | 41.4 | 0–50 | 1–25 | W | M | N | N | | |
| R. "pomonella"—Malus race | 6 | 19.8 | 50–500 | 1–25 | W | M | N | A | Feder & Bush 1989 | |
| R. "pomonella"—Malus race | 10 | 10.2 | >500 | 1–25 | W | M | N | A | McPheron 1990 | |
| R. "pomonella"—Malus race | 13 | 21.2 | >500 | 1–25 | W | M | N | A | Feder et al. 1990b | |
| MEAN R. "pomonella"—Malus race | | 13.6 | >500 | 1–25 | W | M | N | N | | |
| Tephritis bardanae[1] | 6 | 3.2 | 50–500 | | H | M | N | N | Eber et al. 1991 | |
| **HYMENOPTERA** | | | | | | | | | | |
| Argidae | | | | | | | | | | |
| Schizocerella pilicornis leaf mining race[1] | 3 | 16.4 | >500 | | H | M | I | A | Gorske & Sell 1976 | Gorske et al. 1977[5]; D. Smith, pers. comm.[6] |
| Tenthredinidae | | | | | | | | | | |
| Euura "atra"-S. alba race[1] | 4 | 35.5 | 50–500 | | W | M | N | N | Roininen et al. 1993 | |
| Platycampus "luridiventris"—A. incana race | 8 | 3.6 | >500 | | W | M | N | N | Herbst & Heitland 1994 | |
| P. "luridiventris"—A. glutinosa race | 8 | 2.5 | 50–500 | | W | M | N | N | Herbst & Heitland 1994 | Wong et al. 1977[6] |

Table 12.1 (continued)

| Order: Family: Species | # Loci | Nm | Dist. (km) | Disp. (km) | H.F. | H.R. | Res. | Hab. | Source for Gene Flow Estimate | Additional Sources, if Needed |
|---|---|---|---|---|---|---|---|---|---|---|
| Cynipidae | | | | | | | | | | |
| Andricus quercuscalicis | 4 | 1.3 | 50–500 | > 25 | W | M | I | N | Stone & Sunnucks 1993 | |
| A. quercuscalicis | 4 | 9.8 | 50–500 | > 25 | W | M | N | N | Stone & Sunnucks 1993 | |
| MEAN A. quercuscalicis | | 2.6 | 50–500 | > 25 | W | M | I,N | N | | |
| A. quercuscalicis | 4 | 1.0 | > 500 | > 25 | W | M | I | N | Stone & Sunnucks 1993 | |

[1] Gene-flow estimate based on our calculation of $F_{st}$

[2] Gene-flow estimates from averaged $F_{st}$ values

[3] Source for dispersal ability

[4] Source for host-plant growth form

[5] Source for host range

[6] Source for residence status

[7] Source for principle habitat

[8] Gene-flow estimate from 4:96 pheromone trap samples

For each species, the following data are listed: (1) number of loci (# Loci) from which gene flow estimates were obtained; (2) gene flow estimates (Nm); (3) the average pairwise distance (Dist.) among populations surveyed for genetic variation (three categories: 0–50 km, 50–500 km, and > 500 km); (4) dispersal ability (Disp.; three categories: 0–1 km, 1–25 km, and > 25 km); (5) host-plant growth forms (H.F.), either herbaceous (H), woody (W), or both (H,W); (6) host range (H.R.), categorized as monophagous (M), oligophagous (O), and polyphagous (P), (7) residence status (Res.), listed as native (N) and introduced (I); and (8) habitat of occurrence (Hab.), categorized as natural (N) and agricultural (A). See methods for criteria by which we assigned species to these categories. For each species, the source for the gene flow estimate and additional sources (if needed) for ecological data are listed. Mean gene flow estimates for species–host races represented at a distance category more than once were obtained from averaged $F_{ST}$ values.

that the mark-release-recapture studies underestimate the scale of evolutionarily important dispersal, because such studies generally miss rare, long-distance dispersal events (Slatkin 1985a). Similarly, observations of individuals outside of breeding grounds probably underestimate dispersal, since such stragglers are noticed only when they are found in inhospitable habitats, but not when they have successfully traversed such habitats. On the other hand, estimates based on observed range expansions may slightly overestimate typical levels of dispersal, particularly if the invasion of a new habitat is facilitated by human transport. Although there is undoubtedly some error in our assignment of species to different dispersal categories, we feel that the species we have designated as highly vagile are indeed much more mobile than species we categorized as poor dispersers.

Placing species into different categories of diet breadth proved much easier than categorizing them based on dispersal ability. For this review, we designated a species as "monophagous" if it fed exclusively on the members of one plant genus, "oligophagous" if it was restricted to feeding within one plant family, and "polyphagous" if it attacked more than one plant family. The few omnivorous species (*Gryllus* field crickets) were categorized as polyphagous. Our categorization of species' host ranges was based on their host ranges in the regions in which their population genetic structure was assessed.

In gathering data on host range, we also noted whether the plants fed upon by each species were herbaceous (grasses and forbs) or woody (shrubs and trees). Some species fed primarily on plants of one growth form but occasionally utilized the other growth form. We assigned such species to the category representing their typical host plants. If a species normally fed on both herbaceous and woody plants, we designated it as feeding on both. On the rare occasions (e.g., the butterfly *Papilio zelicaon*) in which a species fed on plants of one growth form in one region and both growth forms in another, we assigned that species to the appropriate host-plant growth form for the region from which its allozyme variation was assessed.

We also sorted the species by residence status, designating each as either introduced or native, depending on where the allozyme study was performed relative to the native range. For our statistical analyses, species that were studied throughout their native and introduced ranges were categorized as "native" if more than half of the populations were from the native range. Conversely, species for which more than half of the sampled populations were from a recently colonized range were categorized as "introduced." Similarly, species that had undergone recent natural expansions from their native range were assigned to the category of native species if more than half of the sampled populations were from the native range, but were called "introduced" if more than half were the result of the recent range expansion.

The final category into which we placed all species was their habitat of occurrence in the region in which their population structure was assessed. Species were scored as being restricted to either natural or agricultural habitats. The great majority of species in the "natural" habitat category occurred in nonmanaged habitats, but a few conifer-feeding species (the scale, *Nuculaspis californica*, *Pissodes* weevils, scolytid beetles, and the moths, *Thera* spp., *Rhyacionia frustrana*, and *Choristoneura* spp.) fed in natural as well as silvicultural habitats. In addition to those species occurring almost exlusively on agricultural crops, species exploiting both natural and agricultural habitats were included in the "agricultural" habitat category. Our inclusion of conifer-feeding species in the "natural" category is justified, because forestry practices mimic the natural dispersions of their host plants much better than do most cropping schemes (including those used for most fruit trees). However, since our categorization of these species may bias the results in important ways, we also repeated the analyses with the conifer-feeding insects scored as occurring in agricultural habitats.

We used indirect estimates of gene flow for each species included in this review. For nearly all species, we based these estimates on Wright's (1951) $F_{ST}$ or one of two estimators of this statistic, $G_{ST}$ (Nei 1973) and theta (Weir and Cockerham 1984). Although these statistics all describe the amount of genetic variation in a sample that is attributable to differences among subpopulations, their relative suitability for estimating population structure has been a subject of recent debate (Slatkin and Barton 1989; Cockerham and Weir 1993). Nonetheless, they all provide estimates of gene flow (*Nm*) by the equation (after Wright 1951), $F_{ST} = 1/(1 + 4Nm)$.

Approximately one-half of the studies we examined provided an estimate of either gene flow (using Wright's formula) or one of the above three statistics from which we could estimate gene flow. If we were given a choice of estimators of population genetic structure, we first chose $F_{ST}$, if available, followed by $G_{ST}$, and then theta. For a small number of studies, the only estimate of gene flow was based on the private alleles method of Slatkin (1985b), a method that has been shown to give comparable results to $F_{ST}$ and its related statistics (Slatkin and Barton 1989). For the remaining half of the studies, we used the program BIOSYS-1 (Swofford and Selander 1981) to calculate $F_{ST}$, from which we estimated gene flow using Wright's (1951) formula. Estimation of this statistic using BIOSYS-1 requires data on allele frequencies and sample sizes, so we could only perform this analysis if both were available. In performing these analyses, we included only those loci in which the frequency of the commonest allele was not greater than 95% in all populations. In the event that not all populations were scored for variation at all loci (a situation that prevented the program from calculating $F_{ST}$), we eliminated either loci or populations from the data set to allow the execution of the analysis. All such data removals were done to minimize data loss. For several species, allele frequencies were so similar among populations that $F_{ST}$ took a value of zero, yielding an undefined estimate of gene flow (because obtaining an estimate of *Nm* in this case requires division by zero). We did not include these species in subsequent analyses, unless the estimate of $F_{ST}$ for these species could be averaged with others in determining family means (see Analyses, section 12.2.3).

## 12.2.2 *Phylogenetic Independence and the Comparative Approach*

The main strength of the comparative method, such as we employ in this chapter, is that it involves comparing evolutionary trends across a broad range of taxa, thereby allowing an assessment of the generality of those trends (Pagel and Harvey 1988; Harvey and Pagel 1991). For example, if we were to show that gene flow was correlated with dispersal ability across such divergent taxa as grasshoppers, aphids, beetles, butterflies, and flies, we would be able to conclude with reasonable conviction that for insects as a whole, variation in dispersal ability is an important factor influencing population genetic differentiation. However, a potential weakness of this approach is that species do not always represent evolutionarily independent origins of a given trait. In such a case, it is inappropriate to treat them as independent for statistical purposes (Ridley 1983). To illustrate by example, if sedentary dispersal strategies evolved only once in each insect family, it would be a clear violation of statistical assumptions to treat two closely related, sedentary chrysomelid beetles as independent data points in the assessment of the relationship between dispersal ability and gene flow. How then can one be confident in conclusions based on such comparative approaches?

Although many of the general conclusions about evolutionary pattern and process have come from studies using the comparative method, researchers have only recently begun to be concerned with the rigor of this approach (Pagel and Harvey 1988). Fortunately, these researchers have developed a host of analyses that differ both in their applicability to real-world data, and in the degree to which they limit phylogenetic nonindependence (see Pagel and Harvey 1988; Harvey and Pagel 1991 for thorough reviews of these approaches). Probably the most commonly used of these methods, and certainly the most readily applicable to our data set, is the analysis of higher nodes (e.g., Harvey and Zammuto 1985; Krebs et al. 1989). This method involves averaging data across species within a higher node (genus, family, etc.), under the logic that evolutionary independence increases at higher and higher nodes. One of the biggest problems with this approach is that information on variation within the higher nodes (e.g., among species within a family) is lost, limiting degrees of freedom for statistical analyses (Harvey and Pagel 1991). Nonetheless, this approach is attractive, because it does not require a known phylogeny for the group under consideration (one is not available for the phytophagous insects surveyed herein), yet does allow the use of all of the data in a cross-species survey.

For these reasons, we chose to use the higher nodes approach in our survey of the ecological correlates of gene flow in phytophagous insects, using family means to assess the relationship between these correlates and gene flow. Our decision to average data within families is justified by the fact that the ecological traits we examined varied in many of the 35 families in our survey (Table 12.1). For example, there were among-species differences in dispersal ability in 10 of the 15 families represented by more than one species for which dispersal ability was known. Similarly, 15 of the 20 families represented by more than one species exhibited among-species variation in host range. Indeed, all of the ecological traits in our review varied among some pairs of congeners (e.g., compare the butterflies, *Pieris napi* and *P. rapae* in Table 12.1). These observations suggest that these ecological traits are evolutionarily labile, at least to the point that they have likely evolved independently within each family. Thus, it is reasonable to treat members of different families that share a trait (e.g., poor dispersal ability) as evolutionarily independent. Furthermore, since there were 35 families in our survey, we could be assured of a reasonably large data set, even after averaging data within each family.

## 12.2.3 Analyses

To determine the effects of dispersal ability, host plant growth form, host range, residence status, and habitat on family mean levels of gene flow, we analyzed the effect of each factor using analysis of variance (ANOVA; SYSTAT 1992). To make the analyses spatially explicit, we included the average pairwise distance separating populations (zero–50 km, 50–500 km, > 500 km) as a factor in each analysis, both as a main effect and in interaction with each of the other factors. Thus, our analyses allowed us to examine how each factor contributed to overall levels of gene flow among populations separated by comparable distances, as well as to determine from the interaction how each influenced the pattern of isolation by distance (Wright 1943).

To obtain family averages for these analyses, we averaged $F_{ST}$ estimates for each set of species that shared the ecological trait of concern (e.g., poor dispersal ability) and were studied over a similar spatial scale, and then obtained a gene flow estimate for this averaged $F_{ST}$ using Wright's (1951) equation. It should be noted that for species–host races represented at a distance category by more than one study, we determined an average level of gene flow for that species using averaged $F_{ST}$ values. These within-species averages were then used in computing family averages. Thus, each species-host race was represented only once in the family average at a given spatial scale. Following Slatkin (1993), we used log-transformed gene-flow estimates in all analyses. Ideally, we would have been able to include all main effects and their interactions in a single ANOVA. Unfortunately, this was computationally impossible, because many of the possible combinations of the different factors were not represented in the data set, so we resorted to conducting separate analyses for each factor.

Because the higher nodes approach reduces degrees of freedom, potentially compromising our ability to detect a significant relationship between an ecological trait and population structure, we repeated the analyses using species as independent data points. This allowed us to see if the failure to detect an effect was due to a lack of degrees of freedom. In addition, we repeated all analyses after removing single-locus studies (< 5% of the studies) from the data set, because the observation that different loci can give drastically different estimates of population structure (e.g., Slatkin 1987; Rank 1992) suggests that single-locus studies may provide biased estimates of gene flow. We also conducted the analyses after removing introduced species, incase the inclusion of these species biased the results in any way. Because the results of all of these additional analyses were qualitatively and statistically similar to the analysis of family averages using the entire data set, we report herein only the results for the analyses of family averages.

To determine which of the ecological factors were autocorrelated, we conducted G-tests of independence (Sokal and Rohlf 1981) for all pairwise combinations of dispersal ability, host-plant growth form, host range, residence status, and habitat. For these tests, we applied the simple correction of $G$ for an $R \times C$ table, and we reduced the tablewide type-I error rate by using a sequential Bonferroni test (Rice 1989).

## 12.3 Results

### 12.3.1 Dispersal Ability

After removing the effect of reduced gene flow with increasing distance ($F_{2,58} = 8.300, p = 0.001$), it was clear that highly vagile species exhibited greater levels of gene flow across all distance categories than did less mobile species ($F_{2,58} = 3.519, p = 0.036$) (Fig. 12.1A). The nonsignificant interaction between dispersal ability and distance ($F_{4,58} = 1.228, p = 0.309$) indicated that vagile species exhibited the same pattern of isolation by distance as did their sedentary counterparts. Examples of sedentary species that displayed low levels of gene flow included the boreal relict butterfly, *Boloria improba* (Britten and Brussard 1992), the wingless katydid, *Ephippiger ephippiger* (Oudman et al. 1990), and the black pineleaf scale, *Nuculaspis californica* (Alstad and Corbin 1990). Vagile species exhibiting pronounced gene flow included the migratory grasshopper, *Melanoplus sanguinipes* (Chapco and Bidochka 1986), the saltmarsh planthopper, *Prokelisia marginata* (Peterson and Denno in press), the spruce budworm moth, *Choristoneura fumiferana* (May et al. 1977), the tobacco budworm, *Helicoverpa virescens* (Korman et al. 1993), the monarch butterfly, *Danaus plexippus* (Eanes and Koehn 1978), and the fruitfly, *Anastrepha distincta* (Steck 1991).

### 12.3.2 Host-Plant Growth Form

The analysis of the effects of host-plant growth form on gene flow revealed that, after accounting for distance effects ($F_{2,82} = 4.603, p = 0.013$), species utilizing plants of different growth forms indeed differed in levels of gene flow ($F_{2,82} = 4.111, p = 0.020$). Examination of Figure 12.1B reveals that this effect is due to greater levels of gene flow at a given distance in species attacking herbaceous hosts, compared to their counterparts utilizing woody hosts. The relationship between gene flow and geographic distance did not vary with host-plant growth form (interaction term: $F_{4,82} = 1.927, p = 0.114$). Insects attacking herbaceous plants and exhibiting high levels of gene flow included the cotton leafworm moth, *Alabama argillacea* (Pashley 1985), and the checkerspot butterfly, *Chlosyne palla*

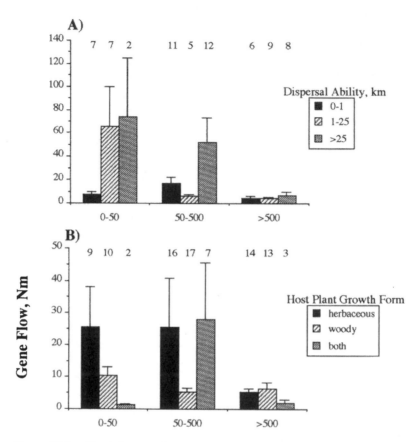

*Figure 12.1.* The influence of selected life-history traits (dispersal ability, host-plant growth form, and host range) on the population genetic structure of phytophagous insects. (A) In addition to the reduction in gene flow (mean ± SE) associated with increasing distance ($F_{2,58}$ = 8.300, $p$ = 0.001), levels of gene flow at a given distance were influenced by dispersal ability ($F_{2,58}$ = 3.519, $p$ = 0.036). The pattern of reduction in gene flow with distance did not vary with mobility ($F_{4,58}$ = 1.228, $p$ = 0.309). (B) Furthermore, host plant growth form ($F_{2,82}$ = 4.603, $p$ = 0.013) had a significant influence on gene flow (mean ± SE), after accounting for the isolating effect of geographic distance ($F_{2,82}$ = 4.603, $p$ = 0.013), but the interaction between geographic distance and host plant growth form ($F_{4,82}$ = 1.927, $p$ = 0.114) did not.

(Schreier et al. 1976). Low levels of gene flow were evident in many species exploiting woody host plants, including the white pine weevil, *Pissodes strobi* (Phillips and Lanier 1985), the mountain pine beetle, *Dendroctonus ponderosae* (Stock and Guenther 1979), the lycaenid butterfly, *Euphilotes enoptes* (Peterson 1996), the eastern tent caterpillar, *Malacosoma americanum* (Costa and Ross 1993), and the gallwasp, *Andricus quercuscalicis* (Stone and Sunnucks 1993).

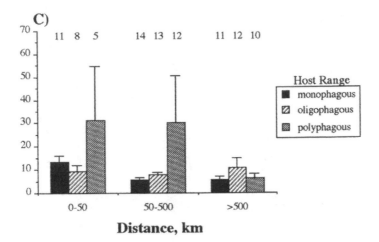

*Figure 12.1. (continued)* (C) However, gene flow (mean ± SE) was not influenced by host range ($F_{2,87} = 0.556, p = 0.576$) or the interaction between geographic distance and host range ($F_{4,87} = 0.674, p = 0.612$). In this analysis, geographic distance ($F_{2,87} = 2.473, p = 0.090$) also did not have a significant influence on gene flow. In all three panels, sample sizes (number of families from which the means were obtained) are indicated above each bar.

### 12.3.3 Host Range

For insects with different host ranges, neither geographic distance ($F_{2,87} = 2.473$, $p = 0.090$) and host range ($F_{2,87} = 0.556, p = 0.576$), nor the interaction between host range and distance ($F_{4,87} = 0.674, p = 0.612$) influenced levels of gene flow among populations (Fig. 12.1C). The nontransformed data shown in Figure 12.1C gave the illusion of a trend for greater gene flow in polyphagous species. This trend, which was due to a few families represented by single species exhibiting extraordinarily high levels of gene flow, disappeared with log-transformation. Furthermore, treating species as evolutionarily (and statistically) independent in a second analysis did not produce a different result, indicating that the failure to detect an effect of host range on gene flow was not due to a lack of statistical power resulting from reduced degrees of freedom.

### 12.3.4 Principle Habitat

In the comparison of species occupying natural and agricultural habitats, geographic distance again had a significant isolating effect ($F_{2,80} = 4.393, p = 0.015$; Fig. 12.2A). After taking this effect into account, agricultural pests exhibited greater levels of gene flow across all distances than did their counterparts in natural habitats ($F_{1,80} = 5.888, p = 0.017$). The nonsignificant distance by habitat interaction ($F_{2,80} = 0.699, p = 0.500$) revealed that isolation by distance was similar

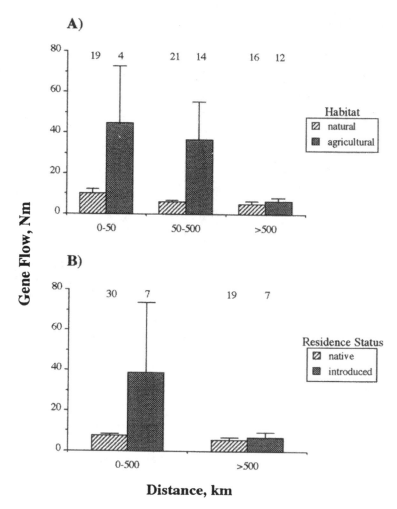

*Figure 12.2.*   The influence of habitat of occurrence (natural vs. agricultural) and residence status on the population genetic structure of phytophagous insects. (A) Geographic distance significantly reduced gene flow (mean ± SE) among populations ($F_{2.80}$ = 4.393, $p$ = 0.015), and gene flow was greater among agricultural pests than species in natural habitats ($F_{1.80}$ = 5.888, $p$ = 0.017). The pattern of isolation by distance (interaction term) was the same for herbivores in both habitat categories ($F_{2.80}$ = 0.699, $p$ = 0.500). (B) Neither geographic distance ($F_{1.222}$ = 0.742, $p$ = 0.390), nor residence status ($F_{1.222}$ = 0.415, $p$ = 0.520), nor their interaction ($F_{1.222}$ = 0.044, $p$ = 0.834) had any effect on gene flow (mean ± SE). Sample sizes (number of families from which the means were obtained) in both panels are indicated above each bar.

in the two types of species. Typifying agricultural species that exhibit pronounced gene flow were the brown planthopper, *Nilaparvata lugens* (Hoshizaki 1994); the meadow spittlebug, *Philaenus spumarius* (Loukas and Drosopoulos 1992); the European corn borer, *Ostrinia nubilalis* (Harrison and Vawter 1977); the cotton leafworm, *Alabama argillacea* (Pashley 1985); the olive fruit fly, *Dacus oleae* (Tsakas and Zouros 1980); and the apple maggot fly, *Rhagoletis pomonella* (Feder et al. 1990b). The milkweed cerambycid beetle, *Tetraopes tetraophthalmus* (McCauley and Eanes 1987); Old World populations of the gypsy moth, *Lymantria dispar* (Harrison et al. 1983); the checkerspot butterfly, *Euphydryas editha* (Ehrlich and White 1980); the meadow brown butterfly, *Maniola jurtina* (Handford 1973a, 1973b); and the goldenrod ball-gall fly, *Eurosta solidaginis* (Waring et al. 1990) were representative of species occurring in natural habitats and showing relatively less gene flow. Inclusion of conifer-feeders in the category of agricultural pests did not qualitatively change the results.

### 12.3.5 Residence Status

Introduced species and native species in our survey did not differ in levels of gene flow ($F_{1,59} = 0.002$, $p = 0.966$; Fig. 12.2B). This analysis also failed to demonstrate an effect of geographic distance ($F_{1,59} = 1.772$, $p = 0.188$), or an interaction between distance and residence status ($F_{1,59} = 0.008$, $p = 0.931$), on levels of gene flow among populations. The lack of an effect of residence status on gene flow may indicate that these species truly do not differ in levels of gene flow. Alternatively, the relatively small number of introduced species in our data set ($N = 18$) may have limited our ability to detect an effect of residence status.

### 12.3.6 Autocorrelation of Factors

Numerous ecological traits were autocorrelated in our survey, as revealed by the G-tests of independence (Table 12.2). Dispersal ability was more limited in native

*Table 12.2* Tests of Statistical Nonindependence for All Pairwise Combinations of Ecological Factors Included in Our Analyses

|  | Growth Form | Host Range | Residence | Habitat |
|---|---|---|---|---|
| Dispersal ability | $G_{corr.,4d.f.}$ = 9.520 | $G_{corr.,4d.f.}$ = 9.801 | *$G_{corr.,2d.f.}$ = 13.705 | *$G_{corr.,2d.f.}$ = 33.278 |
| Growth form |  | *$G_{corr.,4d.f.}$ = 74.408 | $G_{corr.,2d.f.}$ = 0.894 | *$G_{corr.,2d.f.}$ = 23.456 |
| Host range |  |  | $G_{corr.,2d.f.}$ = 0.748 | *$G_{corr.,2d.f.}$ = 22.287 |
| Residence |  |  |  | *$G_{corr.,1d.f.}$ = 16.671 |

G-values indicated with an asterisk indicate that a pair of factors is statistically nonindependent at $p < 0.05$, according to the sequential Bonferroni method described by Rice (1989).

than introduced species. Similarly, agricultural pests were substantially more vagile than species in natural habitats. Agricultural pests were also more polyphagous, fed more frequently on herbaceous plants, and were more likely to be introduced than their counterparts in natural habitats. Finally, monophagous species were restricted more to woody host plants compared to their oligophagous and polyphagous counterparts, which primarily attacked herbaceous hosts.

## 12.4 Discussion

The results of our survey reveal that a wide variety of factors can influence both overall levels of gene flow and patterns of isolation by distance among phytophagous insect populations (see Roderick 1996 for a further discussion of factors that may influence the genetic structure of insect populations). Our spatially explicit analyses demonstrated that geographic distance does isolate populations from each other, as evidenced by a significant distance effect on gene flow in the majority of our analyses. This result makes it clearly imperative that any future comparisons of population structure across taxa take distance into account, and every effort should be made to compare species at similar spatial scales.

The central focus of this chapter, however, is an examination of life-history and habitat factors that influence gene flow among populations of phytophagous insects. In clear support of the dispersal-gene flow hypothesis, our analysis revealed that across all distances, gene flow increased with mobility (Fig. 12.1A). Furthermore, the analysis revealed that gene flow among populations of mobile species (both species moving 1–25 km and species moving > 25 km) was reduced by distance, but this was not the case for sedentary species (dispersing < 1 km). We do not mean to suggest that isolation by distance does not occur in sedentary species. Rather, it occurs at a spatial scale finer than the data allowed us to examine. Thus, we suggest that the failure of our analysis to detect isolation by distance in sedentary species may be evidence that the homogenizing effects of gene flow have been overwhelmed by the effects of genetic drift (or selection, if allozyme variants are not selectively neutral) over all but the smallest distances (e.g., Whitlock 1992; Nürnberger and Harrison 1995). For mobile species, gene flow apparently has overcome the effects of drift and produced a clear pattern of isolation by distance.

To test the habitat persistence–gene flow hypothesis, we compared species that utilize arguably persistent, woody host plants with species that exploit arguably ephemeral, herbaceous host plants. Indeed, when the analysis was restricted to studies of variation at more than one allozyme locus, it revealed that gene flow is more pronounced in insects on herbaceous plants than in species utilizing woody host plants (Fig. 12.1B). Because dispersal ability was not autocorrelated with host-plant growth form, this effect cannot be attributed to differences in vagility. Thus, it appears that the most plausible explanation for this pattern is that herbaceous plant patches are less persistent than patches of woody plants (Brown 1986;

Roff 1990), so the resident populations of herbivores on herbaceous plants are more prone to extinction and recolonization events, resulting in higher levels of gene flow in insects on these relatively ephemeral plants (Slatkin 1985a; Wade and McCauley 1988; Whitlock and McCauley 1990).

Our two tests of the effect of habitat patchiness on gene flow provided conflicting evidence of whether gene flow is augmented in contiguous compared to patchy habitats. We predicted that polyphagous insects would show greater levels of gene flow than species with more specialized diets, after correcting for distance, because suitable habitats are probably more continuous for generalists than specialists (Futuyma and Moreno 1988). Such an increase in genetic subdivision is one mechanism by which many authors have suggested that specialization could promote speciation (Stanley 1979; Price 1980; Vrba 1984; Futuyma and Moreno 1988). Our finding of no effect of host range on either overall levels of gene flow or patterns of isolation by distance (Fig. 12.1C) clearly does not support this hypothesis.

Our second test of the habitat patchiness–gene flow hypothesis involved a comparison of gene flow and patterns of isolation by distance in species in natural and agricultural habitats. As with the test using host range, we hypothesized that gene flow would be greater among populations of agricultural pests, due to the contiguous dispersion of their agricultural crops (most of which were large monocultures). Although the pattern of isolation by distance was similar in the two categories of species, the greater overall levels of gene flow in agricultural pest species compared to species in natural habitats supported this hypothesis (Fig. 12.2A). This pattern was evident as well when the analysis was restricted to native species, indicating that residence status was not confounding the effect of habitat type. The autocorrelation of habitat type with dispersal ability suggests that one explanation for the effect of habitat type on gene flow is not habitat contiguity, but rather the greater mobility of agricultural pests compared to species in natural habitats. Another explanation for this result is that tilling of agricultural habitats renders these habitats less persistent than natural habitats, facilitating gene flow among populations of crop pests.

Although several studies have provided convincing support of the hypothesis that apparent gene flow increases with population age (Berlocher 1984; Whitlock 1992; Stone and Sunnucks 1993; Nürnberger and Harrison 1995), we did not find such an effect in our comparison of native and introduced species (Fig. 12.2B). It is possible that the effect of population age was not strong enough to overcome the small sample size of introduced species in our analysis. Clearly, to address this hypothesis adequately, more studies of the population structure of introduced species must be conducted.

*12.4.1 Life History and Local Adaptation*

Although it has never been demonstrated before with statistical rigor, our finding that gene flow in insects is enhanced by mobility comes as no surprise. This result suggests that, all else being equal, local adaptation at small spatial scales is more likely in sedentary species than in vagile species. However, the observation that both vagile and sedentary insects can display local adaptation at a small spatial scale (Mopper et al. 1995; Mopper 1996) makes it clear that levels of gene flow are not the only determinant of local adaptation. It is likely that the selective forces operating on those vagile species that exhibit local adaptations (e.g., leaf-mining moths and gall-forming midges, reviewed in Mopper 1996) have been strong, since they countered substantial levels of gene flow. In more sedentary species (e.g., scale insects and thrips, reviewed in Mopper 1996), selection need not have been as great to produce locally adapted demes. Gandon et al. (1996; Chapter 13, this volume) have suggested an alternate explanation for the observation that populations of mobile insects can be locally adapted to their host populations. They argue that if coevolutionary processes allow host populations to evolve resistance to local insect populations, counteradaptation by the insects may be promoted by gene flow among the insect populations, as long as this gene flow leads to the immigration of potentially useful genotypes.

Because gene flow is greater among populations of species that utilize herbaceous host plants compared to those that exploit woody plants, it appears that host-plant growth form is also an important factor influencing local adaptation. Clearly, the increased longevity of woody plants makes them ideally suited for the formation of demes adapted to individual plants (Edmunds and Alstad 1978), and it is undoubtedly for this reason that most efforts to demonstrate fine-scale adaptation (to individual host plants) by phytophagous insects have focused on species that utilize long-lived host plants (Mopper 1996). In addition, the diminished levels of gene flow in species utilizing woody hosts may indicate that local adaptation above the scale of individual plants is more likely in these species than in those that feed on herbaceous plants.

Local adaptation is also more likely in species occurring in natural habitats, compared to insect pests of agricultural crops, due to the high levels of gene flow that typify crop pests. Thus, for most agricultural pest insects, spatial variation in pesticide use, plant genotypes, and biocontrol agents are only likely to result in locally adapted insect populations if selective differences are extreme (Comins 1977). It may be encouraging to agricultural interests that high levels of gene flow from neighboring populations experiencing different management regimes may either prevent or delay the evolution of local adaptations for countering those tactics (Caprio and Tabashnik 1992b). However, once local adaptation has occurred, gene flow will promote the rapid spread of locally adaptive traits, quickly rendering the control tactic useless over large areas (Daly and Gregg 1989; Caprio and Tabashnik 1992b; Korman et al. 1993).

The degree to which populations can evolve local adaptations is determined by the balance between the deterministic force of natural selection and countering forces such as genetic drift and gene flow (Slatkin 1973, 1987; Endler 1977). Our extensive survey revealed that several life-history traits, including dispersal ability, host-plant growth form, and the exploitation of natural versus agricultural habitats may all play roles in determining the genetic structure of phytophagous insect populations. Because of their influence on population genetic structure, these life-history traits are likely to influence, in turn, the spatial scale at which local adaptations can evolve in herbivorous insects.

## Acknowledgments

We are indebted to S. Mopper and S. Strauss for including our work in this volume. In addition, we thank C. Björkman, R. Chapman, F. Chew, P. Mardulyn, C. Mitter, A. Norrbom, N. Ohsaki, M. Rowell-Rahier, and D. Smith for contributing and/or verifying ecological data, and C. Mitter, S. Mopper, and M. Rowell-Rahier for providing copies of their in-press manuscripts. We also thank S. Mopper, B. Shaffer, S. Strauss, and C. Yoon for helpful comments on an earlier draft of this chapter. C. Mitter provided computer facilities, for which we are grateful. Many of the ideas and analyses in this chapter were improved as a result of stimulating discussions with a number of colleagues, including A. Bohonak, A. de Queiroz, B. Farrell, S. Kelley, A. McCune, and C. Mitter. This research was supported in part by National Science Foundation Grants DEB-9209693 to R. F. Denno and J. R. Ott and DEB-9527846 to R. F. Denno, and Maryland Agricultural Experiment Station Competitive Grant ENTO-95-10 to M. A. Peterson and R. F. Denno.

## 12.5 References

Alstad, D. N. and K. W. Corbin. 1990. Scale insect allozyme differentiation within and between host trees. *Evol. Ecol. 4:*43–56.

Anderson, W. W., C. W. Berisford, and R. H. Kimmich. 1979. Genetic differences among five populations of the southern pine beetle. *Ann. Entomol. Soc. Am. 72:*323–327.

Anderson, W. W., C. W. Berisford, R. J. Turnbow, and C. J. Brown. 1983. Genetic differences among populations of the black turpentine beetle, *Dendroctonus terebrans*, and an engraver beetle, *Ips calligraphis* (Coleoptera: Scolytidae). *Ann. Entomol. Soc. Am. 76:*896–902.

Arthur, A. P. and D. J. Bauer. 1981. Evidence of northerly dispersal of the sunflower moth by warm winds. *Environ. Entomol. 10:*528–533.

Bartlett, A. C. 1981. Isozyme polymorphisms in boll weevils and thurberia weevils from Arizona. *Ann. Entomol. Soc. Am. 74:*359–362.

Baruffi, L., G. Damiani, C. R. Guglielmino, C. Bandis, A. R. Malacrida, and G. Gasperi. 1995. Polymorphism within and between populations of *Ceratitis capitata*: Comparison between RAPD and multilocus enzyme electrophoresis data. *Heredity 74:*425–437.

Baughman, J. F., P. F. Brussard, P. R. Ehrlich, and D. D. Murphy. 1990. History, selection, drift, and gene flow: Complex differentiation in checkerspot butterflies. *Can. J. Zool. 68:*1967–1975.

Bedard, W. D. 1950. The Douglas-fir beetle. U.S. Department of Agriculture Circular No. 817.

Bellows, T. S., J. C. Owens, and E. W. Huddleston. 1984. Flight activity and dispersal of range caterpillar moths, *Hemileuca oliviae* (Lepidoptera: Saturniidae). *Can. Entomol.* 116:247–252.

Beregovoy, V. H. and D. S. Gill. 1986. Isozyme polymorphism in the sunflower moth. *J. Heredity* 77:101–105.

Berlocher, S. H. 1984. Genetic changes coinciding with the colonization of California by the walnut husk fly, *Rhagoletis completa. Evolution* 38:906–918.

Berlocher, S. H. 1995. Population structure of *Rhagoletis mendax*, the blueberry maggot. *Heredity* 74:542–555.

Berlocher, S. H. and G. L. Bush. 1982. An electrophoretic analysis of *Rhagoletis* (Diptera: Tephritidae) phylogeny. *Syst. Zool.* 31:136–155.

Bernays, E. A. and R. F. Chapman. 1970. Food selection by *Chorthippus parallelus* (Zetterstedt) (Orthoptera: Acrididae) in the field. *J. Anim. Ecol.* 39:383–394.

Björkman, C. and S. Larsson. 1991. Host-plant specialization in needle-eating insects of Sweden. Pp. 1–49 *in* Y. N. Baranchikov, W. J. Mattson, F. P. Hain, and T. L. Payne (Eds.), *Forest Insect Guilds: Patterns of Interaction with Host Trees*. U.S. Department of Agriculture Forest Service general and technical reports NE-153.

Blackman, R. L. and V. F. Eastop. 1984. *Aphids on the Worlds Crops*. John Wiley, New York.

Bossart, J. L. and J. M. Scriber. 1995. Maintenance of ecologically significant genetic variation in the tiger swallowtail butterfly through differential selection and gene flow. *Evolution* 49:1163–1171.

Brakefield, P. M. 1982. Ecological studies on the butterfly *Maniola jurtina* in Britain: I. Adult behaviour, microdistribution and dispersal. J. Anim. Ecol. 51:713–726.

Brindley, T. A. and F. F. Dicke. 1963. Significant developments in European corn borer research. *Annu. Rev. Entomol.* 8:155–176.

Britten, H. B. and P. F. Brussard. 1992. Genetic divergence and the Pleistocene history of the alpine butterflies *Boloria improba* (Nymphalidae) and the endangered *Boloria acrocnema* (Nymphalidae) in western North America. *Can. J. Zool.* 70:539–548.

Britten, H. B., P F. Brussard, D. D. Murphy, and G. T. Austin. 1994. Colony isolation and isozyme variability of the western seep fritillary, *Speyeria nokomis apacheana* (Nymphalidae) in the western Great Basin. *Great Basin Naturalist* 54:97–105.

Britten, H. B., P. F. Brussard, D. D. Murphy, and P. R. Ehrlich. 1995. A test for isolation-by-distance in central Rocky Mountain and Great Basin populations of Edith's checkerspot butterfly (*Euphydryas editha*). *J. Hered.* 86:204–210.

Britten, H. B. and R. W. Rust. 1996. Population structure of a sand dune–obligate beetle, *Eusattus muricatus*, and its implications for dune management. *Conserv. Biol.* 10:647–652.

Brittnacher, J. G., S. R. Sims, and F. J. Ayala. 1978. Genetic differentiation between species of the genus *Speyeria* (Lepidoptera: Nymphalidae). *Evolution* 32:199–210.

Brookes, M. I. and R. K. Butlin. 1994. Population structure in the small ermine moth *Yponomeuta padellus*: An estimate of male dispersal. *Ecol. Entomol.* 19:97–107.

Brown, I. L. and P. R. Ehrlich. 1980. Population biology of the checkerspot butterfly, *Euphydryas chalcedona*: Structure of the Jasper Ridge colony in California. *Oecologia* 47:239–251.

Brown, V. K. 1986. Life cycle strategies and plant succession. Pp. 105–124 *in* F. Taylor and R. Karbarn (Eds.), *The Evolution of Insect Cycles.* Springer-Verlag, New York.

Brussard, P. F. and A. T. Vawter. 1975. Population structure, gene flow and natural selection in populations of *Euphydryas phaeton. Heredity 34*:407–415.

Buès, R., J. Freuler, J. F. Toubon, S. Gerber, and S. Poitout. 1994. Stabilité du polymorphisme enzymatique dans les populations d'un Lépidoptère migrant, *Agrotis ipsilon. Entomol. Exp. Appl. 73*:187–191.

Burns, J. M. and F. M. Johnson. 1971. Esterase polymorphism in the butterfly *Hemiargus isola*: Stability in a variable environment. *Proc. Natl. Acad. Sci. USA 68*:34–37.

Caccone, A. and V. Sbordoni. 1987. Molecular evolutionary divergence among North American cave crickets: I. Allozyme variation. *Evolution 41*:1198–1214.

Caprio, M. A. and B. E. Tabashnik. 1992a. Allozymes used to estimate gene flow among populations of diamondback moth (Lepidoptera: Plutellidae) in Hawaii. *Environ. Entomol. 21*:808–816.

Caprio, M. A. and B. E. Tabashnik. 1992b. Gene flow accelerates local adaptation among finite populations: Simulating the evolution of insecticide resistance. *J. Econ. Entomol. 85*:611–620.

Chapco, W. and M. J. Bidochka. 1986. Genetic variation in prairie populations of *Melanoplus sanguinipes*, the migratory grasshopper. *Heredity 56*:397–408.

Chew, F. 1981. Coexistence and local extinction in two pierid butterflies. *Am. Nat. 118*:655–672.

Chippendale, G. M. 1979. The southwestern corn borer, *Diatraea grandiosella*: Case history of an invading insect. *Missouri Agric. Exp. Sta. Res. Bull. 1031*:1–52.

Cockerham, C. C. and B. S. Weir. 1993. Estimation of gene flow from *F*-statistics. *Evolution 47*:855–863.

Comins, H. N. 1977. The development of insecticide resistance in the presence of migration. *J. Theor. Biol. 64*:177–197.

Costa, J. T. III and K. G. Ross. 1993. Seasonal decline in intracolony genetic relatedness in eastern tent caterpillars: Implications for social evolution. *Behav. Ecol. Sociobiol. 32*:47–54.

Costa, J. T. III and K. G. Ross. 1994. Hierarchical genetic structure and gene flow in macrogeographic populations of the eastern tent caterpillar (*Malacosoma americanum*). *Evolution 48*:1158–1167.

Covell, C. V. Jr. 1984. *A Field Guide to the Moths of Eastern North America.* Houghton Mifflin Co., Boston, MA.

Cronau, J. P. and S. B. J. Menken. 1990. Biochemical systematics of the leaf mining moth family Nepticulidae (Lepidoptera): II. Allozymic variability in the *Stigmella ruficapitella* group. *Neth. J. Zool. 40*:499–512.

Cross, W. H., M. J. Lukefahr, P. A. Fryxell, and H. R. Burke. 1975. Host plants of the boll weevil (*Anthonomus grandis*). *Environ. Entomol. 4*:19–26.

Daly, J. C. 1989. The use of electrophoretic data in a study of gene flow in the pest species *Heliothis armigera* (Hübner) and *H. punctigera* Wallengren (Lepidoptera: Noctuidae). Pp. 115–141 *in* H. D. Loxdale and J. den Hollander (Eds.), *Electrophoretic Studies on Agricultural Pests.* Clarendon Press, Oxford, UK.

Daly, J. C. and P. Gregg. 1985. Genetic variation in *Heliothis* in Australia: Species identification and gene flow in the two pest species *H. armigera* (Hübner) and *H. punctigera* Wallengren (Lepidoptera: Noctuidae). *Bull. Ent. Res. 75*:169–184.

Demayo, C. G., R. C. Saxena, and A. A. Barrion. 1990. Allozyme variation in local populations of the brown planthopper, *Nilaparvata lugens* (Stal) in the Philippines. *Philipp. Ent. 8*:737–748.

den Boer, M. H. 1978. Isoenzymes and migration in the African armyworm *Spodoptera exempta* (Lepidoptera: Noctuidae). *J. Zool. 185*:539–553.

Denno, R. F., G. K. Roderick, K. L. Olmstead, and H. G. Döbel. 1991. Density-related migration in planthoppers (Homoptera: Delphacidae): The role of habitat persistence. *Am. Nat. 138*:1513–1541.

Denno, R. F., G. K. Roderick, M. A. Peterson, A. F. Huberty, H. G. Döbel, M. D. Eubanks, J. E. Losey, and G. A. Langellotto. 1996 Habitat persistence underlies intraspecific variation in the dispersal strategies of planthoppers. *Ecol. Monogr. 66*:389–408.

Denno, R. F., M. E. Schauff, S. W. Wilson, and K. L. Olmstead. 1987. Practical diagnosis and natural history of two sibling salt marsh–inhabiting planthoppers in the genus *Prokelisia* (Homoptera: Delphacidae). *Proc. Entomol. Soc. Wash. 89*:687–700.

Descimon, H. and M. Napolitano. 1993. Enzyme polymorphism, wing pattern variability, and geographical isolation in an endangered butterfly species. *Biol. Conserv. 66*:117–123.

Douwes, P. and B. Stille. 1988. Selective versus stochastic processes in the genetic differentiation of populations of the butterfly *Erebia embla* (Thnbg) (Lepidoptera, Satyridae). *Hereditas 109*:37–43.

Drew, R. A. I. 1989. The tropical fruitflies (Diptera: Tephritidae: Dacineae) of the Australasian and Oceanic Regions. *Mem. Queensland Mus. 26*:1–521.

Dubach, J. M., D. B. Richman, and R. B. Turner. 1988. Genetic and morphological variation among geographical populations of the range caterpillar, *Hemileuca oliviae* (Lepidoptera: Saturniidae). *Ann. Entomol. Soc. Am. 81*:132–137.

Duijm, M. and L. Oudman. 1983. Interspecific mating in *Ephippiger* (Orthoptera: Tettigonioidea). *Tijdschrift voor Entomologie 126*:97–108.

Eanes, W. F. and R. K. Koehn. 1978. An analysis of genetic structure in the monarch butterfly, *Danaus plexippus* L. *Evolution 32*:784–797.

Eber, S., R. Brandl, and S. Vidal. 1992. Genetic and morphological variation among populations of *Oxyna parietina* (Diptera: Tephritidae) across a European transect. *Can. J. Zool. 70*:1120–1128.

Eber, S., P. Sturm, and R. Brandl. 1991. Genetic and morphological variation among biotypes of *Tephritis bardanae*. *Biochem. Syst. Ecol. 19*:549–557.

Economopoulos, A. P., G. E. Haniotakis, J. Mathioudis, N. Missis, and P. Kinigakis. 1978. Long-distance flight of wild and artificially reared *Dacus oleae* (Gmelin) (Diptera: Tephritidae). *Z. Ang. Ent. 87*:101–108.

Edmunds, G. F. Jr. 1973. Ecology of black pineleaf scale (Homoptera: Diaspididae). *Environ. Entomol. 2*:765–777.

Edmunds, G. F. Jr. and D. N. Alstad. 1978. Coevolution in insect herbivores and conifers. *Science 199*:941–945.

Eggenberger, F. and M. Rowell-Rahier. 1991. Chemical defence and genetic variation: Interpopulational study of *Oreina gloriosa* (Coleoptera: Chrysomelidae). *Naturwissenschaften 78*:317–320.

Ehrlich, P. R. and R. R. White. 1980. Colorado checkerspot butterflies: Isolation, neutrality, and the biospecies. *Am. Nat. 115*:328–341.

Emerson, B. C., and G. P. Wallis. 1994. Species status and population genetic structure of the flightless chafer beetles *Prodontria modesta* and *P. bicolorata* (Coleoptera; Scarabaeidae) from South Island, New Zealand. *Molec. Ecol. 3*:339–345.

Emmett, A. M. 1976. Nepticulidae. Pp. 171–267 in J. Heath (Ed.), *The Moths and Butterflies of Great Britain and Ireland,* Vol. 1. Curwen Press Ltd., London.

Endler, J. A. 1977. *Geographic Variation, Speciation, and Clines.* Princeton University Press, Princeton, NJ.

Endler, J. A. 1979. Gene flow and life history patterns. *Genetics 93*:263–284.

Essig, E. O. and A. E. Michelbacher. 1933. The alfalfa weevil. California Agricultural Experiment Station Bulletin 567.

Fales, J. H. 1959. A field study of the flight behavior of the tiger swallowtail butterfly. *Ann. Entomol. Soc. Am. 52*:486–487.

Feder, J. L. and G. L. Bush. 1989. Gene frequency clines for host races of *Rhagoletis pomonella* in the midwestern United States. *Heredity 63*:245–266.

Feder, J. L., C. A. Chilcote, and G. L. Bush. 1990a. Regional, local and microgeographic allele frequency variation between apple and hawthorne populations of *Rhagoletis pomonella* in western Michigan. *Evolution 44*:595–608.

Feder, J. L., C. A. Chilcote, and G. L. Bush. 1990b. The geographic pattern of genetic differentiation between host associated populations of *Rhagoletis pomonella* (Diptera: Tephritidae) in the eastern United States and Canada. *Evolution 44*:570–594.

Fisher, R. A. 1958. *The Genetical Theory of Natural Selection,* 2nd ed. Dover Press, New York.

Florence, L. Z., P. C. Johnson, and J. E. Coster. 1982. Behavioral and genetic diversity during dispersal: Analysis of a polymorphic esterase locus in southern pine beetle, *Dendroctonus frontalis. Environ. Entomol. 11*:1014–1018.

Futuyma, D. J. and S. S. McCafferty. 1990. Phylogeny and the evolution of host plant associations in the leaf beetle genus *Ophraella* (Coleoptera: Chrysomelidae). *Evolution 44*:1885–1913.

Futuyma, D. J. and G. Moreno. 1988. The evolution of ecological specialization. *Annu. Rev. Ecol. Syst. 19*:207–233.

Gandon, S., Y. Capowiez, Y. Dubois, Y. Michalakis, and I. Olivieri. 1996 Local adaptation and gene-for-gene coevolution in a metapopulation model. *Proc. R. Soc. Lond. B., 263*:1003–1009.

Gasperi, G., C. R. Guglielmino, A. R. Malacrida, and R. Milani. 1991. Genetic variability and gene flow in geographical populations of *Ceratitis capitata* (Wied.)(medfly). *Heredity 67*:347–356.

Geiger, H. J. and A. Scholl. 1985. Systematics and evolution of holarctic Pierinae (Lepidoptera): An enzyme electrophoretic approach. *Experientia 41*:24–29.

Geiger, H. and A. M. Shapiro. 1986. Electrophoretic evidence for speciation within the nominal species *Anthocharis sara* Lucas (Pieridae). *J. Res. Lepid. 25*:15–24.

Geiger, H. and A. M. Shapiro. 1992. Genetics, systematics and evolution of holarctic *Pieris napi* species group populations (Lepidoptera, Pieridae). *Zeitschrift für Zoologische Systematik und Evolutionsforschung 30*:100–122.

George, C. 1984. Allozyme variation in natural populations of *Lymantria dispar* (Lepidoptera). *Génét. Sél. Evol. 16*:1–14.

Gill, P. 1981a. Allozyme variation in sympatric populations of British grasshoppers—evidence of natural selection. *Biol. J. Linn. Soc. 16*:83–91.

Gill, P. 1981b. Enzyme variation in the grasshopper *Chorthippus brunneus* (Thunberg). *Biol. J. Linn. Soc. 15*:247–258.

Gorske, S. F., H. J. Hopen, and R. Randell. 1977. Bionomics of the purslane sawfly, *Schizocerella pilicornis. Ann. Entomol. Soc. Am. 70*:104–106.

Gorske, S. F. and D. K. Sell. 1976. Genetic differences among purslane sawfly biotypes. *J. Hered. 67*:271–274.

Goulson, D. 1993. Allozyme variation in the butterfly, *Maniola jurtina* (Lepidoptera: Satyrinae) (L.): Evidence for selection. *Heredity 71*:386–393.

Guttman, S. I. and L. A. Weigt. 1989. Macrogeographic genetic variation in the *Enchenopa binotata* complex (Homoptera: Membracidae). *Ann. Entomol. Soc. Am. 82*:156–165.

Guttman, S. I., T. Wilson, and L. A. Weigt. 1989. Microgeographic genetic variation in the *Enchenopa binotata* complex (Homoptera: Membracidae). *Ann. Entomol. Soc. Am. 82*:225–231.

Haag, K.L., A. M. Araújo, and A. Zaha. 1993. Genetic structure of natural populations of *Dryas iulia* (Lepidoptera: Nymphalidae) revealed by enzyme polymorphism and mitochondrial DNA (mtDNA) restriction fragment length polymorphism (RFLP). *Biochem. Genet. 31*:449–460.

Hagen, R. H. 1990. Population structure and host use in hybridizing subspecies of *Papilio glaucus* (Lepidoptera: Papilionidae). *Evolution 44*:1914–1930.

Hales, D. F. and M. Carver. 1976. A study of *Schoutedenia lutea* (van der Goot, 1917) (Homoptera: Aphididae). *Aust. Zool. 19*:85–94.

Handford, P. T. 1973a. Patterns of variation in a number of genetic systems in *Maniola jurtina*: The Isles of Scilly. *Proc. R. Soc. Lond. B. 183*:285–300.

Handford, P. T. 1973b. Patterns of variation in a number of genetic systems in *Maniola jurtina*: The boundary region. *Proc. R. Soc. Lond. B. 183*:265–284.

Hanks, L. M. and R. F. Denno. 1994. Local adaptation in the armored scale insect *Pseudalacaspis pentagona* (Homoptera: Diaspididae). *Ecology 75*:2301–2310.

Hanski, I., J. Pöyry, T. Pakkala, and M. Kuussaari. 1995. Multiple equilibria in metapopulation dynamics. *Nature 377*:618–621.

Harrison, R. G. 1979. Speciation in North American field crickets: Evidence from electrophoretic comparisons. *Evolution 33*:1009–1023.

Harrison, R. G. and A. T. Vawter. 1977. Allozyme differentiation between pheromone strains of the European corn borer, *Ostrinia nubilalis*. *Ann. Entomol. Soc. Am. 70*:717–720.

Harrison, R. G., S. F. Wintermeyer, and T. M. Odell. 1983. Patterns of genetic variation within and among gypsy moth, *Lymantria dispar* (Lepidoptera: Lymantriidae), populations. *Ann. Entomol. Soc. Am. 76*:652–656.

Harrison, S., D. D. Murphy, and P. R. Ehrlich. 1988. Distribution of the bay checkerspot butterfly, *Euphydryas editha bayensis*: Evidence for a metapopulation model. *Am. Nat. 132*:360–382.

Harvey, P. H. and M. D. Pagel. 1991. *The Comparative Method in Evolutionary Biology*. Oxford University Press, Oxford, UK.

Harvey, P. H. and R. M. Zammuto. 1985. Patterns of mortality and age at first reproduction in natural populations of mammals. *Nature 315:* 319–320.

Hastings, A. and S. Harrison. 1994. Metapopulation dynamics and genetics. *Ann. Rev. Ecol. Syst. 25*:167–188.

Hendriksen, H. J. 1982. *The Butterflies of Scandinavia in Nature*. Skandinavisk Bogforlag, Odense, Denmark.

Herbst, J. and W. Heitland. 1994. Genetic differentiation among populations of the sawfly-species *Platycampus luridiventris* associated with different alder species (Hymenoptera: Tentredinidae). *Entomol. Gener. 19*:39–48.

Higby, P. K. and M. W. Stock. 1982. Genetic relationships between two sibling species of bark beetles (Coleoptera: Scolytidae), Jeffrey pine beetle and mountain pine beetle, in northern California. *Ann. Entomol. Soc. Am. 75*:668–674.

Hoshizaki, S. 1994. Detection of isozyme polymorphism and estimation of geographic variation in the brown planthopper, *Nilaparvata lugens* (Homoptera: Delphacidae). *Bull. Ent. Res. 84*:503–508.

Hsiao, T. H. 1978. Host plant adaptations among geographic populations of the Colorado potato beetle. *Entomol. Exp. Appl. 24*:437–447.

Hsiao, T. H. and J. M. Stutz. 1985. Discrimination of alfalfa weevil strains by allozyme analysis. *Entomol. Exp. Appl. 37*:113–121.

Hudson, A. and L. P. Lefkovitch. 1982. Allozymic variation in four Ontario populations of *Xestia adela* and *Xestia dolosa* and in a British population of *Xestia c-nigrum* (Lepidoptera: Noctuidae). *Ann. Entomol. Soc. Am. 75*:250–256.

Jacobson, J. W., and T. H. Hsiao. 1983. Isozyme variation between geographic populations of the Colorado potato beetle, *Leptinotarsa decemlineata* (Coleoptera: Chrysomelidae). *Ann. Entomol. Soc. Am. 76*:162–166.

Jarvis, F. V. L. 1958. Biological notes on *Aricia agestis* (Schiff.) in Britain, part I. *Entomol. Res. 70*:141–148.

Jelnes, J. E. 1975a. Electrophoretic studies on two sibling species *Thera variata* and *Thera obeliscata* (Lepidoptera: Geometridae) with special reference to phosphoglucomutase and phosphoglucose isomerase. *Hereditas 79*:67–72.

Jelnes, J. E. 1975b. Isoenzyme heterogeneity in Danish populations of *Aricia agestis* (Lepidoptera: Rhopalocera). *Hereditas 79*:53–60.

Jelnes, J. E. 1975c. Isoenzyme heterogeneity in Danish populations of *Aricia artaxerxes* (Lepidoptera: Rhopalocera). *Hereditas 79*:47–52.

Johnson, C. G. 1969. *Migration and Dispersal of Insects by Flight.* Methuen, London.

Johnson, G. B. 1976. Polymorphism and predictability at the α-glycerophosphate dehydrogenase locus in *Colias* butterflies: Gradients in allele frequency within single populations. *Biochem. Genet. 14*:403–426.

Jones, R. E., N. Gilbert, M. Guppy, and V. Nealis. 1980. Long-distance movement of *Pieris rapae. J. Anim. Ecol. 49*:629–642.

Kieckhefer, R. W. and J. L. Gellner. 1988. Influence of plant growth stage on cereal aphid reproduction. *Crop Sci. 28:* 688–690.

King, P. S. 1987. Macro- and microgeographic structure of a spatially subdivided beetle species in nature. *Evolution 41*:401–416.

Kisimoto, R. 1981. Development, behaviour, population dynamics and control of the planthopper, *Nilaparvata lugens* Stal. *Rev. Plant Prot. Res. 14*:26–58.

Knoll, S., M. Rowell-Rahier, P. Mardulyn, and J. M. Pasteels. 1996. Spatial genetic structure of leaf beetle species with special emphasis on alpine populations. *In* Cox, M. L. and Jolivet, P. H. A. (Eds), Chrysomelidae Biology, The Classification, Phylogeny and Genetics. SPB Academic Publishing, Amsterdam, The Netherlands. 1:379–388.

Korman, A. K., J. Mallet, J. L. Goodenough, J. B. Graves, J. L. Hayes, D. E. Hendricks, R. Luttrell, S. D. Pair, and M. Wall. 1993. Population structure of *Heliothis virescens* (Lepidoptera: Noctuidae): An estimate of gene flow. *Ann. Entomol. Soc. Am. 86*:182–188.

Krafsur, E. S., P. Nariboli, and J. J. Tollefson. 1993. Gene diversity in natural *Diabrotica berberi* Smith and Lawrence populations (Coleoptera: Chrysomelidae). *Ann. Entomol. Soc. Am. 86*:490–496.

Krebs, J. R., D. F. Sherry, S. D. Healy, V. H. Perry, and A. L. Vaccarino. 1989. Hippocampal specialization in food-storing birds. *Proc. Natl. Acad. Sci. USA 86*:1388–1392.

Latorre, A., C. Hernández, D. Martínez, J. A. Castro, M. Ramón, and A. Moya. 1992. Population structure and mitochondrial DNA gene flow in Old World populations of *Drosophila subobscura. Heredity 68*:15–24.

Lawrence, W. S. 1988. Movement ecology of the red milkweed beetle in relation to population size and structure. *J. Anim. Ecol. 57*:21–35.

Liebherr, J. K. 1986. Comparison of genetic variation in two carabid beetles (Coleoptera) of differing vagility. *Ann. Entomol. Soc. Am. 79*:424–433.

Liebherr, J. K. 1988. Gene flow in ground beetles (Coleoptera: Carabidae) of differing habitat preference and flight-wing development. *Evolution 42*:129–137.

Lokki, J., A. Saura, P. Lankinen, and E. Suomalainen. 1976. Genetic polymorphism and evolution in parthenogenetic animals: VI. Diploid and triploid *Polydrosus mollis* (Coleoptera: Curculionidae). *Hereditas 82*:209–216.

Loukas, M. and S. Drosopoulos. 1992. Population genetics of the spittlebug genus *Philaenus* (Homoptera: Cercopidae) in Greece. *Biol. J. Linn. Soc. 46*:403–413.

Loxdale, H. 1990. Estimating levels of gene flow between natural populations of cereal aphids (Homoptera: Aphididae). *Bull. Ent. Res. 80*:331–338.

Loxdale, H. D. and C. P. Brookes. 1988. Electrophoretic study of enzymes from cereal aphid populations: V. Spatial and temporal genetic similarity of holocyclic populations of the bird-cherry oat aphid, *Rhopalosiphum padi* (L.) (Hemiptera: Aphididae), in Britain. *Bull. Ent. Res. 78*:241–249.

Loxdale, H. D., I. J. Tarr, C. P. Weber, C. P. Brookes, P. G. N. Digby, and P. Castañera. 1985. Electrophoretic study of enzymes from cereal aphid populations: III. Spatial and temporal genetic variation of populations of *Sitobion avenae* (F.) (Hemiptera: Aphididae). *Bull. Ent. Res. 75*:121–141.

Marshall, J. A. and E. C. M. Haes. 1990. *Grasshoppers and Allied Insects of Great Britain and Ireland.* Harley Books, Colchester, UK.

Martinez, D., A. Moya, A. Latorre, and A. Fereres. 1992. Mitochondrial DNA variation in *Rhopalosiphum padi* (Homoptera: Aphididae) populations from four Spanish localities. *Ann. Entomol. Soc. Am. 85*:241–246.

Matsuda, K., S. Watanabe, and T. Sugiyama. 1982. Feeding response of *Chrysolina aurichalcea* (Mannerheim) to polyacetylenes (Coleoptera: Chrysomelidae). *Tohoku J. Agric. Res. 33*:51–54.

May, B., D. E. Leonard, and R. D. Vadas. 1977. Electrophoretic variation and sex linkage in spruce budworm. *J. Hered. 68*:355–359.

May, R. M., J. A. Endler, and R. E. McMurtrie. 1975. Gene frequency clines in the presence of selection opposed by gene flow. *Am. Nat. 109*:659–676.

McCauley, D. E. 1987. Population genetic consequences of local colonization: Evidence from the milkweed beetle *Tetraopes tetraophthalmus.* Fla. *Entomol. 70*:21–30.

McCauley, D. E. 1991. The effect of host plant patch size variation on the population structure of a specialist herbivore insect, *Tetraopes tetraophthalmus. Evolution 45*:1675–1684.

McCauley, D. E., F. J. Breden, G M. Chippendale, and J. A. Mihm. 1990. Genetic differentiation of populations of the southwestern corn borer (Lepidoptera: Pyralidae) from the United States and Mexico. *Ann. Entomol. Soc. Am. 83*:586–590.

McCauley, D. E., and W. F. Eanes. 1987. Hierarchical population structure analysis of the milkweed beetle, *Tetraopes tetraophthalmus* (Forster). *Heredity 58*:193–201.

McCauley, D. E., M. J. Wade, F. J. Breden, and M. Wohltman. 1988. Spatial and temporal variation in group relatedness: Evidence from the imported willow leaf beetle. *Evolution 42*:184–192.

McDonald, I. C., J. L. Krysan, and O. A. Johnson. 1985. Genetic variation within and among geographic populations of *Diabrotica barberi* (Coleoptera: Chrysomelidae). *Ann. Entomol. Soc. Am. 78*:271–278.

McKechnie, S. W. 1975. Enzyme polymorphism and species discrimination in fruit flies of the genus *Dacus* (Tephritidae). *Aust. J. Biol. Sci. 28*:405–411.

McKechnie, S. W., P. R. Ehrlich. and R. R. White. 1975. Population genetics of *Euphydryas* butterflies. I. Genetic variation and the neutrality hypothesis. *Genetics 81*:571–594.

McMullen, L. H. and S. F. Condrashoff. 1973. Notes on dispersal, longevity and overwintering of adult *Pissodes strobi* (Peck) (Coleoptera: Curculionidae) on Vancouver Island. *J. Entomol. Soc. B.C. 70*:22–26.

McPheron, B. A. 1990. Genetic structure of apple maggot fly (Diptera: Tephritidae) populations. *Ann. Entomol. Soc. Am. 83*:568–577.

McPheron, B.A., D. C. Smith, and S. H. Berlocher. 1988a. Genetic differences between host races of *Rhagoletis pomonella. Nature 336*:64–66.

McPheron, B.A., D. C. Smith, and S. H. Berlocher. 1988b. Microgeographic genetic variation in the apple maggot *Rhagoletis pomonella. Genetics 119*:445–451.

Menken, S. B. J. 1981. Host races and sympatric speciation in small ermine moths, Yponomeutidae. *Entomol. Exp. Appl. 30*:280–292.

Menken, S. B. J., J. T. Wiebes, and W. M. Herrebout. 1980. Allozymes and the population structure of *Yponomeuta cagnagellus* (Hübner) (Lepidoptera). *Neth. J. Zool. 30*:228–242.

Michalakis, Y., A.W. Sheppard, V. Noël, and I. Olivieri. 1993. Population structure of a herbivorous insect and its host plant on a microgeographic scale. *Evolution 47*:1611–1616.

Mitton, J. B. 1994. Molecular approaches to population biology. *Annu. Rev. Ecol. Syst. 25*:45–69.

Mopper, S. 1996. Adaptive genetic structure in phytophagous insect populations. *Trends Ecol. Evol. 11*:235–238.

Mopper, S., M. Beck, D. Simberloff, and P. Stiling. 1995. Local adaptation and agents of selection in a mobile insect. *Evolution 49*:810–815.

Mulkern, G. B., K. P. Pruess, H. Knutson, A. F. Hagen, J. B. Campbell, and J. D. Lambley. 1969. Food habits and preferences of grassland grasshoppers of the north central Great Plains. North Dakota Agricultural Experiment Station Bulletin 481.

Munroe, H. K. 1984. A taxonomic treatise on the Dacidae (Tephritoidea: Diptera) of Africa. Entomol. Memoir of the Department of Agriculture and Water Supply of the Republic of South Africa No. 61.

Namkoong, G., J. A. Richmond, J. H. Roberds, L. B. Nunnally, B. C. McClain, and J. L. Tyson. 1982. Population genetic structure of Nantucket pine tip moth. *Theor. Appl. Genet. 63*:1–7.

Namkoong, G., J. H. Roberds, L. B. Nunnally, and H. A. Thomas. 1979. Isozyme variations in populations of southern pine beetles. *For. Sci. 25*:197–203.

Napolitano, M. and H. Descimon. 1994. Genetic structure of French populations of the mountain butterfly *Parnassius mnemosyne* L. (Lepidoptera: Papilionidae). *Biol. J. Linn. Soc. 53*:325–341.

Napolitano, M., H. Geiger, and H. Descimon. 1988. Structure démographique et génétique de quatre populations provençales de *Parnassius mnemosyne* (L.) (Lepidoptera: Papilionidae): Isolement et polymorphisme dans des populations "menacées." *Génét. Sél. Evol. 20*:51–62.

Nei, M. 1973. Analysis of gene diversity in subdivided populations. *Proc. Natl. Acad. Sci. USA 70*:3321–3323.

Norrbom, A. L. and K. C. Kim. 1988. A list of the reported host plants of the species of *Anastrepha* (Diptera: Tephritidae). Department of Agriculture Animal & Plant Health Inspection Service, plant protection and quarantine 81–52.

Nürnberger, B., and R. G. Harrison. 1995. Spatial population structure in the whirligig beetle *Dineutus assimilis*: Evolutionary inferences based on mitochondrial DNA and field data. *Evolution 49*:266–275.

Ohsaki, N. 1980. Comparative population studies of three *Pieris* butterflies, *P. rapae*, *P. melete*, and *P. napi*, living in the same area: II. Utilization of patchy habitats by adults through migratory and non-migratory movements. *Res. Popul. Ecol. 22*:163–183.

Oudman, L., M. Duijm, and W. Landman. 1990. Morphological and allozymic variation in the *Ephippiger ephippiger* complex (Orthoptera: Tettigonioidea). *Neth. J. Zool. 40*:454–483.

Pagel, M. D. and P. H. Harvey. 1988. Recent developments in the analysis of comparative data. *Q. Rev. Biol. 63*:413–440.

Pashley, D. P. 1985. The use of population genetics in migration studies: A comparison of three noctuid species. Pp. 305–324 *in* D. R. MacKenzie, C. S. Barfield, G. G. Kennedy, R. D. Berger, and D. J. Taranto (Eds.), *The Movement and Dispersal of Agriculturally Important Biotic Agents.* Claitor's Publishing Division, Baton Rouge, LA.

Pashley, D. P. and S. J. Johnson. 1986. Genetic population structure of migratory moths: The velvetbean caterpillar (Lepidoptera: Noctuidae). *Ann. Entomol. Soc. Am. 79*:26–30.

Pashley, D. P., S. J. Johnson, and A. N. Sparks. 1985. Genetic population structure of migratory moths: The fall armyworm (Lepidoptera: Noctuidae). *Ann. Entomol. Soc. Am. 78*:756–762.

Payne, T. L. 1981. Life history and habits. Pp. 7–28 *in* R.C. Thatcher, J. L. Searcy, J. E. Coster, and G. D. Hertel (Eds.), The Southern Pine Beetle. USDA Forest Service Science and Education Administration Technical Bulletin No.1631.

Peterson, M. A. 1995. Phenological isolation, gene flow and developmental differences among low- and high-elevation populations of *Euphilotes enoptes* (Lepidoptera: Lycaenidae). *Evolution 49*:446–455.

Peterson, M. A. 1997. Host plant phenology and dispersal by a montane butterfly: causes and consequences of uphill movement. *Ecology, 78*:167–180.

Peterson, M. A. 1996. Long-distance gene flow in the sedentary butterfly, *Euphilotes enoptes* (Lepidoptera: Lycaenidae). *Evolution, 50*:1990–1999.

Peterson, M. A. and R. F. Denno. The influence of intraspecific variation in dispersal strategies on the genetic structure of planthopper populations. *Evolution,* in press.

Phillips, T. W. and G. N. Lanier. 1985. Genetic divergence among populations of the white pine weevil, *Pissodes strobi* (Coleoptera: Curculionidae). *Ann. Entomol. Soc. Am. 78*:744–750.

Porter, A. H. 1990. Testing nominal species boundaires using gene flow statistics: The taxonomy of two hybridizing admiral butterflies (*Limenitis*: Nymphalidae). *Syst. Zool. 39*:131–147.

Porter, A. H. and H. Geiger. 1988. Genetic and phenotypic population structure of the *Coenonympha tullia* complex (Lepidoptera: Nymphalidae: Satyrinae) in California: No evidence for species boundaries. *Can. J. Zool. 66*:2751–2765.

Porter, A. H. and H. Geiger. 1995. Limitations to the inferences of gene flow at regional geographic scales—an example from the *Pieris napi* group (Lepidoptera: Pieridae) in Europe. *Biol. J. Linn. Soc. 54*:329–348.

Porter, A. H. and S. O. Mattoon. 1989. A new subspecies of *Coenonympha tullia* (Müller) (Nymphalidae: Satyrinae) confined to the coastal dunes of northern California. *J. Lepid. Soc. 43*:229–238.

Porter, A. H. and A. M. Shapiro. 1989. Genetics and biogeography of the *Oeneis chryxus* complex (Satyrinae) in California. *J. Res. Lepid. 28*:263–276.

Price, P. W. 1980. *The Evolutionary Biology of Parasites.* Princeton University Press, Princeton, NJ.

Prokopy, R. J. and B. S. Fletcher. 1987. The role of adult learning in the acceptance of host fruit for egglaying by the Queensland fruit fly, *Dacus tryoni. Entomol. Exp. Appl. 45*:259–263.

Rainey, R. C. 1979. Dispersal and redistribution of some Orthoptera and Lepidoptera by flight. *Mitt. Schweiz. Entomol. Ges. 52*:125–132.

Rank, N. E. 1992. A hierarchical analysis of genetic differentiation in a montane leaf beetle *Chrysomela aeneicollis* (Coleoptera: Chrysomelidae). *Evolution 46*:1097–1111.

Rice, W. R. 1989. Analyzing tables of statistical tests. *Evolution 43*:223–225.

Richards, O. W. and N. Waloff. 1954. Studies on the biology and population dynamics of British grasshoppers. *Anti-Locust Bull. 17*:1–182.

Ridley, M. 1983. The explanation of organic diversity: The comparative method and adaptions for mating. University Press, Oxford, UK.

Roderick, G. K. 1996. Geographic structure of insect populations: Gene flow, phylogeography, and their uses. *Annu. Rev. Entomol. 41*:325–352.

Roff, D. A. 1986. The evolution of wing dimorphism in insects. *Evolution 40*:1009–1020.

Roff, D. A. 1990. The evolution of flightlessness in insects: An evaluation of the hypotheses. *Ecol. Monogr. 60*:389–421.

Roininen, H., J. Vuorinen, J. Tahvanainen, and R. Julkunen-Tiitto. 1993. Host preference and allozyme differentiation in shoot galling sawfly, *Euura atra. Evolution 47*:300–308.

Rosenberg, R. 1989. Genetic differentiation among populations of Weidemeyer's admiral butterfly. *Can. J. Zool. 67*:2294–2300.

Rowell-Rahier, M. 1992. Genetic structure of leaf-beetles populations: Microgeographic and sexual differentiation in *Oreina cacaliae* and *O. speciosissima. Entomol. Exp. Appl. 65*:247–257.

Rowell-Rahier, M. and J. M. Pasteels. 1994. A comparison between allozyme data and phenotypic distances from defensive secretion in *Oreina* leaf beetles (Chrysomelidae). *J. Evol. Biol. 7*:489–500.

Sakanoue, S. and S. Fujiyama. 1987. Allozymic variation among geographic populations of *Chrysolina aurichalcea* (Coleoptera, Chrysomelidae). *Kontyû, 55*:437–449.

Saura, A., O. Halkka, and J. Lokki. 1973. Enzyme gene heterozygosity in small island populations of *Philaenus spumarius* (L.) (Homoptera). *Genetica 44*:459–473.

Schrier, R. D., M. J. Cullenward, P. R. Ehrlich, and R. R. White. 1976. The structure and genetics of a montane population of the checkerspot butterfly, *Chlosyne palla. Oecologia 25*:279–289.

Scott, J. A. 1986. *The Butterflies of North America.* Stanford University Press, Stanford, CA.

Sell, D. K., E. J. Armbrust, and G. S. Whitt. 1978. Genetic differences between eastern and western populations of the alfalfa weevil. *J. Hered. 69:*37–50.

Shapiro, A. M. 1977. Apparent long-distance dispersal by *Pieris occidentalis* (Pieridae). *J. Lepid. Soc. 31:*202–203.

Shapiro, A. M. 1982. A new elevational record for *Pieris protodice* in California (Lepidoptera: Pieridae). *Pan-Pac. Entomol. 58:*162.

Shapiro, A. M. and H. Geiger. 1986. Electrophoretic confirmation of the species status of *Pontia protodice* and *P. occidentalis* (Pieridae). *J. Res. Lepid. 25:*39–47.

Shields, O. 1967. Hilltopping: An ecological study of summit congregation behavior of butterflies on a southern California hill. *J. Res. Lepid. 6:*69–178.

Showers, W. B., R. D. Smelser, A. J. Keaster, F. Whitford, J. F. Robinson, J. D. Lopez, and S. E. Taylor. 1989. Recapture of marked black cutworm (Lepidoptera: Noctuidae) males after long-range transport. *Environ. Entomol. 18:*447–458.

Sillén-Tullberg, B. 1983. An electrophoretic study of allozyme frequencies in local populations of *Lygaeus equestris* (Heteroptera: Lygaeidae). *Hereditas 99:*153–156.

Slatkin, M. 1973. Gene flow and selection in a cline. *Genetics 75:*733–756.

Slatkin, M. 1985a. Gene flow in natural populations. *Annu. Rev. Ecol. Syst. 16:*393–430.

Slatkin, M. 1985b. Rare alleles as indicators of gene flow. *Evolution 39:*53–65.

Slatkin, M. 1987. Gene flow and the geographic structure of natural populations. *Science 236:*787–792.

Slatkin, M. 1993. Isolation by distance in equilibrium and non-equilibrium populations. *Evolution 47:*264–279.

Slatkin, M. and N. H. Barton. 1989. A comparison of three indirect methods for estimating average levels of gene flow. *Evolution 43:*1349–1368.

Sluss, T. P. and H. M. Graham. 1979. Allozyme variation in natural populations of *Heliothis virescens*. *Ann. Entomol. Soc. Am. 72:*317–322.

Sokal, R. R. and F. J. Rohlf. 1981. *Biometry,* 2nd ed. W. H. Freeman, New York.

Sokal, R. R. and D.E . Wartenberg. 1983. A test of spatial autocorrelation analysis using an isolation-by-distance model. *Genetics 105:*219–237.

Solbreck, C. 1971. Displacement of marked *Lygaeus equestris* (L.) (Hemiptera: Lygaeidae) during pre- and posthibernation migrations. *Acta Entomol. Fenn. 28:*74–83.

Southwood, T. R. E. 1962. Migration of terrestrial arthropods in relation to habitat. *Biol. Rev. 37:*171–214.

Southwood, T. R. E. 1977. Habitat, the templet for ecological strategies. *J. Anim. Ecol. 46:*337–365.

Sparks, A. N., R. D. Jackson, J. E. Carpenter, and R. A. Muller. 1986. Insects captured in light traps in the Gulf of Mexico. *Ann. Entomol. Soc. Am. 79:*132–139.

Sperling, F. A. H. 1987. Evolution of the *Papilio machaon* species group in western Canada (Lepidoptera: Papilionidae). *Quaest. Ent. 23:*198–315.

Stanley, S. M. 1979. *Macroevolution: Pattern and Process.* W. H. Freeman, San Francisco, CA.

Steck, G. J. 1991. Biochemical systematics and population genetic studies of *Anastrepha fraterculus* and related species (Diptera: Tephritidae). *Ann. Entomol. Soc. Am. 84:*10–28.

Stock, M. W. and G. D. Amman. 1980. Genetic differentiation among mountain pine beetle populations from lodgepole pine and ponderosa pine in northeast Utah. *Ann. Entomol. Soc. Am. 73*:472–478.

Stock, M. W., G. D. Amman, and P. K. Higby. 1984. Genetic variation among mountain pine beetle (*Dendroctonus ponderosae*) (Coleoptera: Scolytidae) populations from seven western states. *Ann. Entomol. Soc. Am. 77*:760–764.

Stock, M. W. and J. D. Guenther. 1979. Isozyme variation among mountain pine beetle (*Dendroctonus ponderosae*) populations in the Pacific Northwest. *Environ. Entomol. 8*:889–893.

Stock, M. W., G. B. Pitman, and J. D. Guenther. 1979. Genetic differences between Douglas-fir beetles (*Dendroctonus pseudotsugae*) from Idaho and central Oregon. *Ann. Entomol. Soc. Am. 72*:394–397.

Stone, G. W. and P. Sunnucks. 1993. Genetic consequences of an invasion through a patchy environment—the cynipid gallwasp *Andricus quercuscalicis* (Hymenoptera: Cynipidae). *Molec. Ecol. 2*:251–268.

Sturgeon, K. B. and J. B. Mitton. 1986. Allozyme and morphological differentiation of mountain pine beetles *Dendroctonus ponderosae* Hopkins (Coleoptera: Scolytidae) associated with host tree. *Evolution 40*:290–302.

Suzuki, K. 1978. Discovery of a flying population in *Chrysolina aurichalcea* (Mannerheim) (Coleoptera: Chrysomelidae). *Kontyû 46*:549–551.

Swofford, D. L. and R. B. Selander. 1981. *BIOSYS-1, release 1.7, a computer program for the analysis of allelic variation in genetics*. Department Genetics and Development, University of Illinois, Urbana.

SYSTAT. 1992. *SYSTAT: Statistics*, Version 5.2 ed. Systat, Inc., Evanston, Ill.

Tan, K. H. and M. Serit. 1988. Movements and population density comparisons of native male adult *Dacus dorsalis* and *Dacus umbrosus* (Diptera: Tephritidae) among three ecosystems. *J. Plant Prot. Trop. 5*:17–21.

Tan, S. G., M. Y. Omar, K. W. Mahani, M. Rahani, and O. S. Selvaraj. 1994. Biochemical genetic studies on wild populations of three species of green leafhoppers, *Nephotettix*, from peninsular Malaysia. *Biochem. Genet. 32*:415–422.

Taylor, R. A. J. and D. Reling. 1986. Density/height profile and long-range dispersal of first-instar gypsy moth (Lepidoptera: Lymantriidae). *Environ. Entomol. 15*:431–435.

Teetes, G. L. and N. M. Randolph. 1969. Some new host plants of the sunflower moth in Texas. *J. Econ. Entomol. 62*:264–265.

Terranova, A. C., R. G. Jones, and A. C. Bartlett. 1990. The southeastern boll weevil: An allozyme characterization of its population structure. *Southwest. Entomol. 15*:481–496.

Thomas, C. D. and S. Harrison. 1992. Spatial dynamics of a patchily distributed butterfly species. *J. Anim. Ecol. 61*:437–446.

Tomiuk, J., D. F. Hales, K. Wöhrmann, and D. Morris. 1991. Genotypic variation and structure in Australian populations of the aphid *Schoutedenia lutea*. *Hereditas 115*:17–23.

Tomiuk, J., K. Wöhrmann, I. Böhm, and J. Stamp. 1990. Variability of quantitative characters and enzyme loci in rose aphid populations. *Entomologist 109*:84–92.

Tong, M. L. and A. M. Shapiro. 1989. Genetic differentiation among California populations of the anise swallowtail butterfly, *Papilio zelicaon* Lucas. *J. Lepid. Soc. 43*:217–228.

Tsakas, S. C. and E. Zouros. 1980. Genetic differences among natural and laboratory-reared populations of the olive-fruit fly *Dacus oleae* (Diptera: Tephritidae). *Entomol. Exp. Appl. 28*:268–276.

Turchin, P. and W. T. Thoeny. 1993. Quantifying dispersal of southern pine beetles with mark-recapture experiments and a diffusion model. *Ecol. Applic. 3*:187–198.

Unruh, T. R. 1990. Genetic structure among 18 West Coast pear psylla populations: Implications for the evolution of resistance. *Am. Entomol. 36*:37–43.

Vawter, A. T. and P. F. Brussard. 1975. Genetic stability of populations of *Phyciodes tharos* (Nymphalidae: Melitaeinae). *J. Lepid. Soc. 29*:15–23.

Vickery, V. R. and D. K. McE. Kevan. 1983. A monograph of the orthopteroid insects of Canada and adjacent regions. *Lyman Entomological Museum and Research Laboratory Memoir 13*:681–1462.

Vrba, E. S. 1984. Evolutionary pattern and process in the sister-group Alcelaphini-Aepycerotini (Mammalia: Bovidae). Pp. 62–79 *in* N. Eldredge and S. M. Stanley (Eds.), *Living Fossils*. Springer-Verlag, New York.

Wade, M. J. and D. E. McCauley. 1988. Extinction and recolonization: Their effects on the genetic differentiation of local populations. *Evolution 42*:995–1005.

Waring, G. L., W. G. Abrahamson, and D. J. Howard. 1990. Genetic differentiation among host-associated populations of the gallmaker *Eurosta solidaginis* (Diptera: Tephritidae). *Evolution 44*:1648–1655.

Watt, W. B. 1977. Adaptation at specific loci: I. Natural selection on phosphoglucose isomerase of *Colias* butterflies: Biochemical and population aspects. *Genetics 87*:177–194.

Watt, W. B., P. A. Carter, and S. M. Blower. 1985. Adaptation at specific loci: IV. Differential mating success among glycolytic allozyme genotypes of *Colias* butterflies. *Genetics 109*:157–175.

Watt, W. B., F. S. Chew, L. R. G. Snyder, A. G. Watt, and D. E. Rothschild. 1977. Population structure of pierid butterflies: I. Numbers and movements of some montane *Colias* species. *Oecologia 27*:1–22.

Watts, J. G. and T. D. Everett. 1976. Biology and behaviour of the range caterpillar. New Mexico Agricultural Experimental Station Bulletin. 646.

Weaver, C.R. and D. R. King. 1954. Meadow spittlebug. *Ohio Agric. Exp. Sta. Res. Bull. 741*:1–99.

Weir, B. S. and C. C. Cockerham. 1984. Estimating F-statistics for the analysis of population structure. *Evolution 38*:1358–1370.

Whitlock, M. C. 1992. Nonequilibrium population structure in forked fungus beetles: Extinction, colonization, and the genetic variation among populations. *Am. Nat. 139*:952–970.

Whitlock, M. C. and D. E. McCauley. 1990. Some population genetic consequences of colony formation and extinction: Genetic correlations within founding groups. *Evolution 44*:1717–1724.

Widiarta, I. N., Y. Susuki, H. Sawada, and F. Nakasuji. 1990. Population dynamics of the green leafhopper, *Nephotettix virescens* Distant (Hemiptera: Cicadellidae) in synchronized and staggered transplanting areas of paddy fields in Indonesia. *Res. Pop. Ecol.* 32:319–328.

Wiernasz, D. C. 1989. Ecological and genetic correlates of range expansion in *Coenonympha tullia*. *Biol. J. Linn. Soc. 38*:197–214.

Willhite, E. A. and M. W. Stock. 1983. Genetic variation among western spruce budworm (*Choristoneura occidentalis*) (Lepidoptera: Tortricidae) outbreaks in Idaho and Montana. *Can. Entomol. 115*:41–54.

Williams, C. F. and R. P. Guries. 1994. Genetic consequences of seed dispersal in three sympatric forest herbs: I. Hierarchical population genetic structure. *Evolution* 48:791–805.

Wilson, S. W. 1982. The planthopper genus *Prokelisia* in the United States (Homoptera: Fulgoroidea: Delphacidae). *J. Kans. Entomol. Soc. 55*:532–546.

Wöhrmann, K. and J. Tomiuk. 1988. Life cycle strategies and genotypic variability in populations of aphids. *J. Genet. 67*:43–52.

Wöhrmann, K. and J. Tomiuk. 1990. The population biological consequences of a mosaic-like population structure in *Macrosiphum rosae*. Pp. 51–67 *in* R. K. Campbell and R.D. Eikenberry (Eds.), *Aphid–Plant Genotype Interactions*. Elsevier Press, Amsterdam, The Netherlands.

Wong, H. R., J. C. E. Melvin, and J. A. Drouin. 1977. Damage by a willow shoot-boring sawfly in Alberta. *Tree Planters' Notes 27*:18–20.

Wood, D. L. and R. W. Stark. 1968. The life history of *Ips calligraphis* (Coleoptera: Scolytidae) with notes on its biology in California. *Can. Entomol. 100*:145–151.

Wright, S. 1931. Evolution in Mendelian populations. *Genetics 16*:97–159.

Wright, S. 1943. Isolation by distance. *Genetics 28*:139–156.

Wright, S. 1951. The genetical structure of populations. *Ann. Eugen. 15*:323–354.

Yong, H. S. 1988. Allozyme variation in the Artocarpus fruit fly *Dacus umbrosus* (Insecta: Tephritidae) from peninsular Malaysia. *Comp. Biochem. Physiol. 91B*:85–89.

Yong, H. S. 1993. Biochemical genetic differentiation between two *Aulacophora* leaf beetles (Insecta: Coleoptera: Chrysomelidae) from peninsular Malaysia. *Comp. Biochem. Physiol. 106B*:317–319.

Zera, A.J. 1981. Genetic structure of two species of waterstriders (Gerridae: Hemiptera) with differing degrees of winglessness. *Evolution 35*:218–225.

# PART IV
## Local Adaptation, Host Race Formation, and Speciation

# 13

## Differential Adaptation in Spacially Heterogeneous Environments and Host–Parasite Coevolution

*Sylvain Gandon*
Laboratorire d'Ecologie, Universite Pierre et Marie Curie, Paris, France

*Dieter Ebert*
Zoologishes Institut, Universitat Basel, Basel, Switzerland

*Isabelle Olivieri*
Lab Genetique et Environnement, Institut des Sciences de l'Evolution, Universite Montpellier II, Montpellier, France

*Yannis Michalakis*
Laboratorire d'Ecologie, Universite Pierre et Marie Curie, Paris, France

### 13.1 Introduction

The terms *adaptive deme formation* and *local adaptation* have been used in the plant–herbivore and host–parasite literature, respectively, to designate one of the following two situations. The first one is when the mean fitness of a population (or deme) is on average larger in the environment this population originated from than in other environments. The second situation is when the mean fitness of a population on its natal environment is on average larger than the mean fitness of populations issued from other environments. We will use the term *local adaptation* to designate the situation when both conditions are satisfied, though this is not always the case. This definition emphasizes the potential differential response of populations with respect to their natal versus nonnatal environments, a phenomenon that should not be restricted to biotic interactions only.

With this definition, the concept of local adaptation implicitly assumes that the relative fitnesses vary in space. This condition is likely to be satisfied in most, if not all, natural situations, because the local environment of each population is composed of biotic and abiotic factors that typically vary in time and space. In this chapter, we attempt to clarify the concept of local adaptation in the context of biotic interactions and, in particular, host–parasite systems. We use the term *host–parasite* in a broad sense, including all interactions involving reciprocal selection and some specificity. A parasite represents all small organisms with a parasitic lifestyle, such as viruses, bacteria, protozoa, fungi, helminths, or small herbivores.

Local adaptation can be relatively easily studied within host–parasite systems, because the local environment of parasites is usually well defined. Each individual host could be seen as an ephemeral island for the parasite. Moreover, if one assumes sufficient genetic and phenotypic diversity in the host population, each individual host may represent a different type of habitat. In this case, as an analogy of the source–sink concept (Pulliam 1988; Dias 1996), sensitive hosts could be considered as sources, whereas resistant hosts could be considered as sinks for the parasites. However, it should also be noted that adaptation to abiotic environments may be conceptually different from adaptation to biotic environments. The difference arises from the fact that the biotic environment (the host) might evolve in response to the adaptation of the parasite, if the parasite affects host's fitness. Such a coevolutionary process can greatly affect the predictions concerning adaptation to local environments.

Host–parasite interactions may be formalized using population genetics theory. Population differentiation and local adaptation result mainly from a balance between natural selection and gene flow. We first present the factors and the mechanisms that promote or prevent local adaptation of parasites and then use this theoretical background to propose an experimental design in order to test the local adaptation hypothesis.

## 13.2 Local Adaptation and Dispersal

### 13.2.1 Evolution of Dispersal

Dispersal is a life-history trait that has profound demographic and genetic effects on populations. The evolution of dispersal has been studied theoretically by many authors (Hamilton and May 1977; Comins et al. 1980; Levin et al. 1984; Frank 1986; Johnson and Gaines 1990; Denno 1994; Olivieri et al. 1995). Some processes favor an increase in dispersal rates, whereas others act against dispersal. The balance between these opposing forces drives the evolution of dispersal.

In this chapter, we will assume that dispersal reflects gene flow between different habitats. It is known, however, that many herbivorous insects and some parasites do not disperse randomly but preferentially settle on a specific type of habitat (e.g., Thomas and Singer, Chapter 14, this volume; Bernays and Chapman 1994; de Meeûs et al. 1994). We briefly discuss the consequences of such habitat selection mechanisms in a later section.

### 13.2.2 What Selects against Dispersal?

Two factors select against dispersal. First are costs that are associated with dispersal itself. Dispersing individuals might incur a cost due to either increased mortality or costs during the settling period in the novel environment. Second are costs due to the spatial structure of the environment. In a spatially heterogeneous environment, dispersal will often lead individuals to unsuitable environments. If there is sufficient genetic variability for local adaptation, dispersal is selected against because of negative associations between genes coding for local adaptation and genes increasing the dispersal rate (Balkau and Feldman 1973). Such associations arise because genes increasing the dispersal rate have a higher probability to settle on different environments and hence be, on average, selected against. But even in the absence of genetic variability for local adaptation, dispersal will be selected against in the presence of spatial heterogeneity. This would happen because passive diffusion moves individuals from favorable to less favorable habitats more often than the reverse, since favorable habitats tend to have more individuals (Hastings 1983; Holt 1985).

### 13.2.3 What Selects for Dispersal?

When the environment is variable, some level of dispersal will be selected for (Levin et al. 1984). An extreme case of temporal variability is the local extinction of populations. Indeed, when extinctions occur, dispersal will be selected for, because each particular deme will eventually become extinct, and only offspring that have emigrated will be able to reproduce (Olivieri et al. 1995). Such temporal variability is likely to occur frequently in many environments either because of environmental stochasticity (abiotic and biotic) or demographic stochasticity. Furthermore, temporal variability will result in spatial variability if the different populations are not perfectly synchronized. For instance, if variability is due to temporal changes in population size, then unless all populations change size at the same time, one would observe populations of various sizes across space at a given time. Finally, the degree of relatedness in each population is also involved in the evolution of dispersal. Dispersal can be adaptive if it reduces competition between close relatives (Frank 1986), and the evolution of dispersal can also be seen as a mechanism for the avoidance of inbreeding depression (Shields 1982).

### 13.2.4 Local Adaptation and Dispersal

Gene flow is often regarded as a constraining force in evolution, because in a spatially heterogeneous environment, it counteracts selective forces that lead to local adaptation (Slatkin 1987). Therefore, everything else being equal, one should logically expect a negative correlation between dispersal and local adaptation (points 1 and 2 in Fig. 13.1). This prediction has been tested in many host–parasite systems by relating the dispersal ability of the parasite to the presence or absence of local adaptation (Mopper 1996). Contrary to this prediction, parasite mobility does not seem to be strongly related to local adaptation. In some host–parasite systems, no local adaptation was found for sessile parasites (point 3 in Fig. 13.1), whereas in other cases, mobile parasites exhibited local adaptation (point 4 in Fig. 13.1). In the following, we present some arguments that could contribute to the explanation of these findings.

### 13.2.5 Conventional Wisdom: Negative Correlation between Dispersal and Local Adaptation

This negative correlation can be well explained by the operation of a migration–selection balance in a spatially heterogeneous environment. Natural selection leads

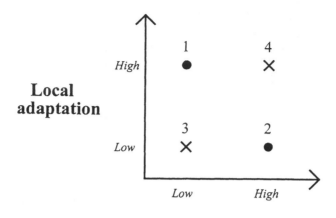

*Figure 13.1* Local adaptation and dispersal. Points 1 and 2 represent conventional wisdom: negative correlation between local adaptation and level of dispersal. Points 3 and 4 represent paradoxical results: positive correlation between local adaptation and level of dispersal as explained in the text.

to adaptation to local environmental conditions. When dispersal is low relative to selection, gene flow cannot overcome the effect of selection, and local adaptation occurs (point 1 in Fig. 13.1). Conversely, for larger dispersal rates, immigrants from other populations will frequently introduce genes leading to adaptation to different conditions and will prevent local adaptation (point 2 in Fig. 13.1). In this latter case, because dispersal rates are also likely to evolve, one might ask why the parasite dispersal rate is so large if it prevents local adaptation. This may be explained by the metapopulation dynamics of parasite populations. In particular, some dispersal is adaptive when population extinctions occur (Olivieri et al. 1995). In host–parasite systems, each parasite population is mortal, because each individual host can be seen as an ephemeral habitat. Larger extinction rates can be induced by many factors (e.g., the death of an infected host, the extinction of a host population, or extinction of the parasite population due to its own natural enemies) and, therefore, select for dispersal to colonize new hosts. Peterson and Denno (Chapter 12, this volume) illustrated this point by looking at the effect of host-plant growth form (host persistence–gene flow hypothesis, see their Fig. 9.2) on gene flow. As mentioned before, within-host competition between close relatives (Frank 1986), or avoidance of inbreeding depression (Shields 1982) could also lead to higher migration rates. These mechanisms can be easily understood if the intrademic (within individual host) genetic structure of the parasite is taken into account (Frank 1994; McCauley and Goff, Chapter 9, this volume).

## 13.2.6 Positive Correlation between Dispersal and Local Adaptation

### 13.2.6.1 Spatial Heterogeneity as a Prerequisite

Certain transplant experiments with sessile parasites (Rice 1983; Unruh and Luck 1987; Cobb and Whitham 1993) did not reveal local adaptation (point 3 in Fig. 13.1). One possible explanation of such results is the lack of spatial heterogeneity of the environment (the host) at the scale studied. This argument may also explain the lack of local adaptation despite large parasite dispersal rates (point 2 in Fig. 13.1). Indeed, if the environment is not heterogeneous at the spatial scale of the study, transplant experiments will fail to detect local adaptation, even if the parasite has large dispersal rates. This point raises a methodological problem, since, in most studies on local adaptation, the degree of heterogeneity of the environment is only assessed by the relative performances of the parasites. We will further discuss this in section 13.3 (see also Boecklen and Mopper, Chapter 4, this volume; Mopper, Chapter 7, this volume).

### 13.2.6.2 Mobility versus Dispersal

Mobility, or the relative ability of an organism to travel in space, has sometimes been used as a predictor for dispersal rates. However, the apparent paradox of local adaptation despite a potentially high dispersal rate (point 4 in Fig. 13.1) may be explained by the fact that mobility does not always reflect dispersal. Dispersal could be low despite very high mobility either because animals return to breed to their natal sites, or because animals actively choose to breed on specific habitats (and avoid breeding on others). Indeed, if a parasite actively chooses on which host to oviposit, there can be very little gene flow between parasites of different kinds of hosts, despite potentially high mobility capacities. In this context, focusing on parasite mobility rather than quantifying gene flow can be misleading (Thomas and Singer, Chapter 14, this volume).

### 13.2.6.3 Coevolution and Dispersal

If the environment is temporally variable, a genotype selected for at one point in time may be selected against at a different time in the same site. Such temporal variability can be induced by a coevolutionary process between the parasite and other species (e.g., the host or natural enemies of the parasite). Gandon et al. (1996a) studied theoretically the effects of both host and parasite dispersal rates on local adaptation for a horizontally transmitted parasite that has the same generation time as its host. They used a mathematical model based on Lotka–Volterra equations, and on the assumption that each host genotype can resist one parasite genotype and is susceptible to attack by all other parasites. Therefore, no host genotype is able to resist to all parasite genotypes (i.e., the matching allele model as defined by Frank 1993).

Within this context, the degree of adaptation of hosts to a given parasite population can be defined as the probability of resistance of hosts to this parasite population. Similarly, the degree of adaptation of parasites to a given host population can be defined as the probability of susceptibility of these hosts to the parasites. Their results indicate that local adaptation of either the host or parasite is very sensitive to the ratio of the host and parasite dispersal rates. When both the host and parasite dispersed at high or equal rates, no local adaptation occurred, because hosts resisted sympatric and allopatric parasites equally (Fig. 13.2, white area). When the host dispersal rate was very low, the parasite exhibited local adaptation as long as the dispersal rate of the parasite was different from zero (Fig. 13.2, vertically shaded area). On the other hand, when the parasite dispersal rate was very low, the host exhibited local adaptation (Fig. 13.2, horizontally shaded area). To illustrate these results, consider the biologically trivial case where hosts do not disperse, while parasites do disperse at some rate. In this case, hosts cannot escape the parasitic attack, because they do not disperse: A given host genotype is unable to escape attack by parasites able to overcome its resistance mechanisms. On the other hand, because parasites do disperse, each parasite genotype has a chance to encounter a host population composed of many hosts susceptible to it. If host genotypes are not uniformly distributed, different host populations will constitute different selective environments for the parasites, and as a consequence, parasites well adapted to their local host population will on average be less adapted to nonlocal host populations. Local adaptation of parasites will arise. By reversing these arguments, one might see why hosts exhibit local adaptation when the host dispersal rate is larger than the parasite dispersal rate.

The previous paragraph referred only to the differential response of host or parasite populations to sympatric and allopatric populations of the species with which they coevolve. Dispersal rates also affect the degree to which each host or parasite population is adapted to its local environment (i.e., the local coevolving population). Gandon et al. (1996a) showed that the proportion of local hosts that parasites were able to attack was maximized by minimizing host and parasite dispersal rates. The previous paragraph explained why very low host dispersal rates are beneficial to the parasites: Hosts are unable to escape parasitic attack in a particular site by dispersing to a site with fewer parasites able to attack them. Low parasite dispersal rates are beneficial to the parasite, because large parasite dispersal rates tend to homogenize parasite populations that can no longer take advantage of the heterogeneous distribution of hosts. In other words, large parasite dispersal rates decrease the degree of adaptation of the parasites to the local hosts by introducing locally maladapted genotypes to each parasite population. This mechanism is exactly the same as that operating in selection–migration balance in population genetics or in source–sink dynamics (Pulliam 1988; Dias 1996). Similar arguments apply to host dispersal rates and adaptation to local environment.

The results of Gandon et al. (1996a) emphasize the importance of the host dispersal rate in determining the level of temporal and spatial variability of the parasitic environment through the distribution of host resistance genes in the metapopulation (Fig. 13.3). For intermediate host dispersal rates, the temporal variability

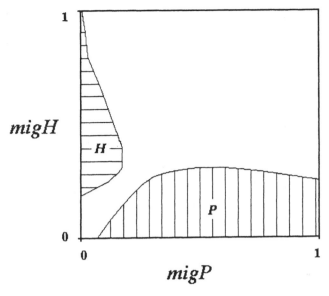

*Figure 13.2* Local adaptation and dispersal. Schematic representation of the effect of both the host and the parasite dispersal rates (*migH* and *migP*, respectively; logarithmic scale) on the difference between resistance to sympatric and allopatric parasites. In horizontal shading, the hosts are more resistant to sympatric parasites than allopatric parasites (*H*). In the unshaded area, there are few differences in resistance to sympatric versus allopatric parasites. In vertical shading, parasites are locally adapted (*P*); hosts are less resistant to sympatric than to allopatric parasites (modified from Gandon et al. 1996a).

of the environment is maximized and dispersal can be adaptive, because it enables the parasite to have the right genes at the right time and place (Gandon et al. 1996b). Therefore, if the environment is variable in time, this mechanism leads to the prediction that there should be a positive correlation between dispersal rate and local adaptation (points 3 and 4 in Fig. 13.1).

### 13.2.6.4 Natural Enemies of Parasites

Parasites' natural enemies (parasites, parasitoids, predators) can also affect the level of local adaptation. For example, Mopper et al. (1995) found that although leaf miners (*Stilbosis quadricustatella*) were locally adapted to phenotypes of individual host trees (*Quercus geminata*), natural enemies induced higher levels of mortality on natal hosts than on novel hosts. These results can be interpreted in different ways. First, if one does not assume any specificity in the interaction between the parasite and its natural enemies, this differential mortality rate could

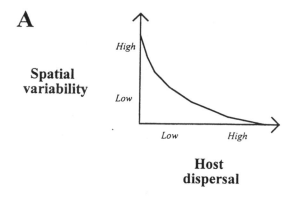

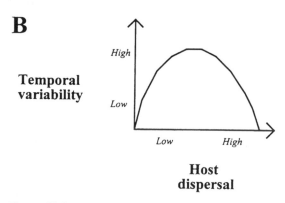

*Figure 13.3* (A) Spatial and (B) temporal variability of the environment in relation to the level of host dispersal. The larger the host dispersal rate, the smaller the differentiation among host populations and, therefore, the smaller the spatial heterogeneity of the environment of the parasites (A). In the absence of dispersal, temporal variability is reduced because, at the scale of the population, there is no novelty in the genes that are involved in the coevolutionary interaction. When host dispersal rate is very high, temporal variability is also reduced, because all the genes are in all the populations. The diversity at the scale of the population is very high, and migration always introduces genes that are already present locally. Therefore, the temporal variability of the environment is maximized for intermediate levels of host dispersal (B).

well be explained by the fact that locally adapted parasites are more apparent to natural enemies. Indeed, locally adapted parasites may be bigger, have larger populations, or produce more chemical compounds that could increase their conspicuousness to natural enemies (Mopper et al. 1995; Mopper 1996).

Second, if one assumes some form of specificity in the interaction between the parasite and its natural enemies, there would be two interactions at the same time: one between the host and the parasite, and another between the parasite and its natural enemies. These two processes would impose opposing forces of selection on the parasite's life history. In the biological system studied by Mopper et al. (1995), the host does not represent a highly variable environment in time, because the host life span is very long compared to that of the parasite, and the host–parasite interaction should select against parasite dispersal. Conversely, if one assumes that natural enemies represent a temporally highly variable environment (shorter generation time and/or higher migration rates than the parasites themselves), the coevolution between parasites and natural enemies would select for parasite dispersal. Therefore, the selected level of dispersal would be a balance between these two antagonistic selective forces. More generally, a positive correlation between dispersal and the degree of adaptation to the local environment would arise whenever offspring fitness is on average worse in the parental environment than in other environments. The mechanisms mentioned earlier illustrate ways in which this situation might arise.

## 13.3 Recombination and Local Adaptation

### 13.3.1 Recombination and the Red Queen Hypothesis

It has been proposed that antagonistic coevolutionary interactions between hosts and their harmful parasites may represent a short-term advantage sufficient to compensate the twofold cost of sexual reproduction (Jaenike 1978; Hamilton 1980; Hamilton et al. 1990; Bell 1982; Ebert and Hamilton 1996). The twofold cost of sexual reproduction, otherwise termed the "twofold disadvantage of producing males" (Maynard Smith 1978), arises from the fact that the proportion of parthenogenetic females within a population increases twice as fast as that of sexual females (Maynard Smith 1978). Assuming a sex ratio of 1:1, this difference is due to the fact that in sexual lineages, only half of the offspring are females, the other half being males, whereas in parthenogenetic lineages all the offspring are female (extended discussions on this topic can be found in Maynard Smith 1978 and Bell 1982). One of the mechanisms proposed to overcome this twofold cost of sexual reproduction, and of interest to us here, involves some sort of frequency-dependent selection (induced by the host–parasite coevolution), which confers higher fitness to rare host genotypes. This mechanism favors sexual individuals, because they can produce genetically variable progeny. In the absence of such spatial and temporal genetic variability (e.g., monocultures), host populations are more vulnerable to rapidly evolving parasites (Brown 1994). This mechanism was called the *Red Queen hypothesis*, referring to a term first coined in evolutionary biology by van Valen (1973), who was inspired by the character in L. Carroll's *Through the Looking Glass*. According to van Valen's hypothesis, each evolutionary advance of any species results in the deterioration of the environment of all other species; hence, the environment of all species changes continually, and only species that can evolve fast enough survive. Bell (1982) used the term to refer to the parasite hypothesis of the maintenance of sexual reproduction.

The major assumptions of the Red Queen hypothesis for the maintenance of sex are that (1) there is genetic variation in the parasites for virulence and/or infectivity; (2) there is genetic variation in the host population for resistance to specific strains of parasites; (3) infection by parasites reduces the fitness of individual hosts; and (4) parasites constantly adapt to host genotypes. The first three points are necessary for coevolution to occur (genetic variation and antagonistic selection). The fourth point, pertaining to local adaptation, is the subject of this chapter.

Given these assumptions, the prediction is that asexual hosts should be more prone to infection by parasites than sexual hosts. In other words, the Red Queen hypothesis predicts that parasites should be more locally adapted on asexual individual hosts than on sexual ones. This prediction is supported by several field studies on very different types of organisms (Lively 1987, 1992; Lively et al. 1990; Morritz et al. 1991; Burt and Bell 1991; see Ladle 1992 for a review), which revealed a strong correlation between asexual reproduction and some parasite fitness traits (e.g., parasite load, prevalence). A large monoculture of short-living hosts might represent the same temporal stability as a single long-living host (e.g., a tree). Heavy parasite loads on asexual host lines can therefore be conceptually interpreted in the same way as local adaptation of parasites to long-living, sexually reproducing hosts.

### 13.3.2 Recombination and Dispersal

The Red Queen hypothesis postulates that recombination allows the host to prevent local adaptation of the parasites (Hamilton 1980). Gandon et al. (1996a) showed that large host dispersal rates could be another way to counteract adaptation of the parasites. We now discuss the interaction of these two traits.

Ladle et al. (1993) used Hamilton's model (Hamilton et al. 1990) to study the evolution of sexual reproduction in a metapopulation. This allowed the authors to study the effect of both the host and the parasite dispersal rates on the evolution of sex. They found that host recombination was selected only for intermediate or large dispersal rates of both the host and the parasite (cf. Fig. 13.4). These results suggest that sex could be unnecessary to release hosts from pathogens when at least one of the dispersal rates, that of parasites or hosts or both, is small. Their results could be explained in the light of two other studies. First, Gandon et al. (1996a) demonstrated that the host was locally adapted when the hosts migrate

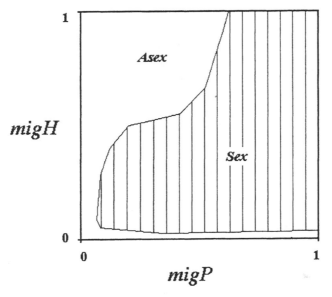

*Figure 13.4* Recombination and dispersal. Schematic representation of the effect of both the host and the parasite dispersal rates (*migH* and *migP,* respectively; linear scale) on the selection of sexual recombination in the host metapopulation. The vertical shading represents the area where sexual reproduction succeeds (*Sex*) over asexual reproduction (modified from Ladle et al. 1993). In the unshaded area, the asexual strain succeeds (*Asex*). We included in the shaded area the cases where both the host and the parasite have large dispersal rates because of the results obtained by Ladle et al. 1993, page 157.

more than the parasites. Although the two models are not identical, this result suggests that sex may not be selected for when host dispersal rates are larger than parasite dispersal rates, because dispersal provides the host population with sufficient genetic diversity. Indeed, when the parasites are not locally adapted, recombination is unnecessary to counteract the effect of parasites (fourth assumption of the Red Queen hypothesis), because their average deleterious impact is very low. Second, when the host dispersal rate is very small for small to moderate population sizes, one would expect the genetic diversity of the host to be very low at the population level (Judson 1995). This could greatly affect the efficiency of recombination to produce genetically variable progeny (second assumption of the Red Queen hypothesis) and, therefore, to prevent parasites from being locally adapted.

There are different ways to prevent parasite local adaptation: sexual reproduction and dispersal. Furthermore, the study of the interaction between dispersal and recombination leads to interesting predictions. When the host dispersal rate is very large, sexual reproduction is unnecessary to allow the host to become locally adapted, and, because of the twofold cost of producing males, recombination is not selected for. Conversely, when the host dispersal rate is too small, recombination is inefficient in preventing parasites from becoming locally adapted (cf. Figs. 13.2 and 13.4). One has to keep in mind that these results are not sufficient to make long-term predictions about patterns of local adaptation in host–parasite systems. Other life-history traits such as host and parasite migration rates, and parasite recombination rate, that are not studied by these models are also likely to evolve and consequently affect the predictions.

## 13.4 Local Adaptation and Experimental Design

### 13.4.1 Transplant Experiments

The most common way to test for the local adaptation phenomenon is to experimentally manipulate parasites and compare their performances on native and novel hosts. For herbivorous insects, a novel environment can be an individual plant from the same population, a different population, or even a separate host species (Mopper et al. 1995). This type of experiment allows the researcher to test whether parasites are locally adapted, examine the spatial scale in which local adaptation occurs and, provided many populations are sampled, determine whether local adaptation is correlated with distance (Parker 1985; Ebert 1994; Lively and Jokela 1996). As noted before, however, a lack of differential responses of parasites could simply mean that hosts are not heterogeneous at the spatial scale examined.

There are two ways to circumvent this problem. One is to compare patterns of local adaptation at multiple scales. Another is to first measure differentiation among host populations. Ideally, differentiation among host populations should be evaluated on traits relevant to the host–parasite interaction. But, because such traits are often difficult to measure or identify, neutral markers may be employed to quantify genetic differentiation of host populations. It should be clear, though, that differentiation measured by neutral markers may not reflect the potential differentiation of traits relevant to the interaction, especially for sexually reproducing species.

Moreover, hosts should be chosen with care. In particular, the presence or absence of sexual reproduction on the differential success of parasites could greatly affect the results (Burt and Bell 1991). Furthermore, age differences between parasitized hosts could bias the experiment, and if the age of an individual host is correlated with the age of its parasite population, parasites may be more adapted on older hosts (Burt and Bell 1991; Cobb and Whitham, Chapter 3, this volume). This pattern could arise because, under the previous assumption, the age of a host would be correlated with the period during which selection acts among parasites within a single host. Younger hosts could be colonized by parasites that are not necessarily adapted to them, but within-host competition in the parasite population could eliminate such locally maladapted parasites on older hosts. Moreover, the level of resistance of an individual host may vary with its age. It can either decrease because of host senescence (Miller 1996) or increase (as in acquired immunity). For example, while studying developmental changes in resistance to herbivory, Kearsley and Whitham (1989) found that a single plant could change very rapidly in its resistance traits. Therefore, host individuals should derive from the same reproductive regime (sexual or asexual) and be similar in age.

Furthermore, as noted by Karban (1989), potential conditioning effects during parasite development on individual hosts may bias the interpretation of transplant experiments: what looks like genetic differentiation and local adaptation could emerge because of such maternal effects and not due to genetic adaptation. To avoid this, it has been suggested that parasites and hosts used in transplant experiments should be kept separated for at least one generation (Karban 1989; Ebert 1994).

There is evidence from some host–parasite systems that there is a significant effect of the infectious dose (the number of parasites inoculated) on the outcome of the inoculation. Ebert (1995) studied the interaction between *Daphnia magna* and a microsporidian parasite, and found that high inoculation doses increase the sporeload, but that the relation between infectious dose and sporeload is not always monotonic. Hochberg (1991) found that higher doses of the granulosis virus of *Pieris brassicae* resulted in substantial reductions in production of pathogen progeny because of intrahost competition. Therefore, the number of infections initiated per host should be carefully controlled (M. Hochberg personal communication).

### 13.4.2 What Should Be Measured?

Because the aim of local adaptation experiments is to study the adaptive ability of the parasites, it is essential to measure parasite traits that affect parasites' fitness. In particular, measuring only host damage (parasite virulence/levels of herbivory) may be inadequate, because such damage may not reflect parasite fitness (Levin and Svanborg Eden 1990). A frequently used variable is the parasite infectivity (Parker 1985; Lively 1989; Lively and Jokela 1996), which measures the proportion of successful attacks on a given host population. A parasite would be locally adapted if its infectivity is higher on the host population from which it derived than on a remote host population (Parker 1985; Lively 1989; Lively and Jokela 1996). Infectivity could also be used at the individual-host scale by comparing the proportion of successful attack on the ancient (natal) host compared to novel hosts (e.g., Edmunds and Alstad 1978; Mopper et al. 1995; see also Alstad, Chapter 1, this volume). Infectivity alone, however, might reveal local adaptation only at large spatial scales, whereas quantitative measures of parasite fitness might reveal local adaptation at finer scales. This was most likely the case in the interaction of a microsporidian and its *Daphnia* host populations (Ebert 1994). Because host resistance can also be quantitative, parasite fitness could also be measured by variables such as parasite mortality rate (Mopper et al. 1995) or reproductive output (Ebert 1995).

It is well known that the fitness of a given organism is dependent on many biotic and abiotic factors (Hunter and Price 1992), which could affect the level of local adaptation of an organism. Examining the sources of insect mortality in *Stilbosis quadricustatella* populations, Mopper et al. (1995) found that the host plant (*Quercus geminata*) was not the sole source of insect mortality. Leaf-miner natural enemies could significantly affect fitness and possibly counteract insect local adaptation to host-plant traits. Therefore, when transplant experiments are conducted in the wild, selective pressures independent from the host–parasite interaction should be included.

Such selective pressures, however, will affect local adaptation only if there is a significant interaction with the host plant. For instance, if the rate of parasitoid attack on a given insect species is the same on all host plants the insect parasitizes, it is unlikely that parasitoid attack may explain differential responses of insects to host plants, even if parasitoid-induced mortality is very large. Stiling and Rossi report such a case in Chapter 2 of this volume. These authors transplanted clones of the plant *Borrichia frutescens* among four islands to test whether populations of the gall midge *Asphondylia borrichiae* were adapted to specific host clones. They found that midges were locally adapted, for which host-related differences in midge fecundity was the most likely explanation. Even though parasitism by four wasp species accounted for approximately 40% of midge mortality, parasitoid attack rates were equivalent on natal and transplanted host clones.

## 13.5 Conclusion

Conventional wisdom predicts that gene flow should prevent local adaptation. However, a major assumption of this prediction is the absence of temporal variability in the environment. Several factors might cause the environment to be variable over time. Depending on the factors acting and the potential of organisms to respond to them, we distinguish three forms of temporal variability. The first is when the habitat quality is temporally variable in a way that the organism cannot adapt to the new environment. Such variability could be induced by environmental or demographic stochasticity, an extreme case being the extinctions of populations. Theoretical studies have shown that in such temporally variable environments, some level of dispersal is adaptive (McPeek and Holt 1992; Olivieri et al. 1995).

A second form of variability occurs when habitat quality is temporally variable such that the organism living in it may track the new environment (i.e., may become adapted to it). For example, such variability can be induced by abiotic factors such as temperature, nutrients, or humidity variations, which may vary over time in a given location. In this case, a given organism may become adapted to the local environment by incorporating mutations advantageous in the new environmental conditions (for an experimental test of such adaptive evolution see

Bennet et al. 1992). The incorporation of such beneficial mutations in a temporally variable environment can be enhanced if the organism evolves in one (or a combination) of the three following ways: increasing its recombination rate (Maynard Smith 1980), its mutation rate (Ishii et al. 1989), or its migration rate.

The third type of temporal variability occurs when there is reciprocal selection between the habitat and the organism (coevolution). Such variability typically occurs in host–parasite systems when there is a certain level of specificity involved in the interaction, and when host and parasite generation times do not differ by too many orders of magnitude. This might also enhance the evolution of traits that facilitate local adaptation. Hamilton showed that in such a heterogeneous environment, sexual recombination can evolve (Hamilton 1980; Hamilton et al. 1990). We believe this argument could also be applied for dispersal abilities. Indeed, Gandon et al. (1996a) found that migration is necessary for local adaptation, and Ladle et al. (1993) found a significant interaction between migration and host recombination. All these results suggest that migration and sexual reproduction are important life-history traits that allow a given organism to catch the Red Queen (to be locally adapted in a temporally variable environment) by possessing good genes at the right time and place (Gandon et al. 1996b).

All three types of temporal variability are involved in host–parasite interactions and could lead to counterintuitive predictions concerning the correlation between the level of local adaptation and the level of gene flow. For example, parasite dispersal may be adaptive if the hosts are ephemeral (the death of an infected host leads to the extinction of a parasite population). If temporal variability is induced by variable, antagonistic selection pressures between the parasite and its biotic environment (including the host and the natural enemies), dispersal might also be adaptive because, contrary to the conventional wisdom, it may lead to local adaptation. However, dispersal may also lead to less suitable hosts and have costs in terms of reduced fitness. Therefore, selection for the optimal dispersal rate is constrained by colonization of new hosts, encountering unsuitable hosts, and avoidance of natural enemies.

## Acknowledgments

We would like to thank M. Hochberg, J.-Y. Lemel, O. Ronce, and M. Slatkin for helpful comments. S. Otto provided interesting insights. P. Leberg and S. Strauss helped to significantly clarify the manuscript. We particularly thank Susan Mopper for numerous discussions, going over the manuscript many times and, in general, for the patience and effort she put to ameliorate the text. This project was funded by a French Ministère de l'Environnement grant to Y. Michalakis (DGAD-SRAE-94205) and by a Swiss National Grant (3100-43093.95) to D. Ebert. I. Olivieri acknowledges financial support from the Bureau de Ressources Génétiques. This is publication number 96-145 of the Institut des Sciences de l'Evolution, Université Montpellier 2.

## 13.6 References

Balkau, B. J. and M. W. Feldman. 1973. Selection for migration modification. *Genetics* 74:171–174.

Bell, G. 1982. *The Masterpiece of Nature: The Evolution and Genetics of Sexuality.* University of California Press, Berkeley, CA.

Bennett, A. F., R. A. Lenski, and J. E. Mittler. 1992. Evolutionary adaptation to temperature: I. Fitness responses of *Escherichia coli* to changes in its thermal environment. *Evolution 46*:16–30.

Bernays, E. A. and R. F. Chapman. 1994. *Host-Plant Selection by Phytophagous Insects.* Chapman & Hall, New York.

Brown, J. K. M. 1994. Chance and selection in the evolution of barley mildew. *Trends Microbiol. 2*:470–475.

Burt, A. and G. Bell. 1991. Seed reproduction is associated with transient escape from parasite damage in American beech. *Oikos 61*:145–148.

Cobb, N. S. and T. G. Whitham. 1993. Herbivore deme formation on individual trees: A test case. *Oecologia 94*:496–502.

Comins, H. N., W. D. Hamilton, and R. M. May. 1980. Evolutionary stable dispersal strategies. *J. Theor. Biol. 82*:205–230.

de Meeûs, T., M. E. Hochberg, and F. Renaud. 1994. Maintenance of two genetic entities by habitat selection. *Evol. Ecol. 8*:1–8.

Denno, R. F. 1994. The evolution of dispersal polymorphisms in insects: The influence of habitat, host plants and mates. *Res. Pop. Ecol. 36*:127–135.

Dias, P. C. 1996. Sources and sinks in population biology. *Trends Ecol. Evol. 11*:326–330.

Ebert, D. 1994. Virulence and local adaptation of a horizontally transmitted parasite. *Science 265*:1084–1086.

Ebert, D. 1995. The ecological interactions between a microsporidian parasite and its host *Daphnia* magna. *J. Anim. Ecol. 64*:361–369.

Ebert, D. and W. D. Hamilton. 1996. Sex against virulence: The coevolution of parasitic diseases. *Trends in Ecology and Evolution 11*:79–82.

Edmunds, G. F. J. and D. N. Alstad. 1978. Coevolution in insects herbivore and conifers. *Science 199*:941–945.

Frank, S. A. 1986. Dispersal polymorphisms in subdivided populations. *J. Theor. Biol. 122*:303–309.

Frank, S. A. 1993. Evolution of host–parasite diversity. *Evolution 47*:1721–1732.

Frank, S. A. 1994. Kin selection and virulence in the evolution of protocells and parasites. *Proc. R. Soc. Lond. B. 258*:153–161.

Gandon, S., Y. Capowiez, Y. Dubois, Y. Michalakis, and I. Olivieri. 1996a. Local adaptation and gene-for-gene coevolution in a metapopulation model. *Proc. Roy. Soc. Lond. B., 263*:1003–1009.

Gandon, S., Y. Michalakis, and D. Ebert. 1996b. Metapopulations and local adaptation in host-parasite interactions. *Trends Ecol. Evol., 11*:431.

Hamilton, W. D. 1980. Sex versus non-sex versus parasite. *Oikos 35*:282–290.

Hamilton, W. D., R. Axelrod, and R. Tanese. 1990. Sexual reproduction as an adaptation to resist parasites (a review). *Proc. Natl. Acad. Sci. USA 87*:3566–3573.

Hamilton, W. D. and R. M. May. 1977. Dispersal in stable habitats. *Nature 269*:578–581.

Hastings, A. 1983. Can spatial selection alone lead to selection for dispersal? *Theor. Pop. Biol. 24*:244–251.

Hochberg, M. E. 1991. Intra-host interactions between a braconid endoparasitoid, *Apanteles glomeratus*, and a baculovirus for larvae of *Pieris brassicae*. *J. Anim. Ecol. 60*:51–63.

Holt, R. D. 1985. Population dynamics in two-patch environments: Some anomalous consequences of an optimal habitat distribution. *Theor. Pop. Biol. 28*:181–208.

Hunter, M. D. and P. W. Price. 1992. Playing chutes and ladders: Heterogeneity and the relative roles of bottom-up and top-down forces in natural communities. *Ecology 73*(3): 724–732.

Ishii, K., H. Matsuda, Y. Iwasa, and A. Sasaki. 1989. Evolutionary stable mutation rate in a periodically changing environment. *Genetics 121*:163–174.

Jaenike, J. 1978. An hypothesis to account for the maintenance of sex within populations. *Evol. Theory 3*:191–194.

Johnson, M. L. and M. S. Gaines. 1990. Evolution of dispersal: Theoretical models and empirical tests using birds and mammals. *Annu. Rev. Ecol. Syst. 21*:449–480.

Judson, O. P. 1995. Preserving genes: A model of the maintenance of genetic variation in a metapopulation under frequency-dependent selection. *Genet. Res. Camb. 65*:175–191.

Karban, R. 1989. Fine-scale adaptation of herbivorous thrips to individual host plants. *Nature 340*:60–61.

Kearsley, M. C. and T. G. Whitham. 1989. Developmental changes in resistance to herbivory: Implications for individuals and populations. *Ecology 70*:422–434.

Ladle, R. J. 1992. Parasites and sex: Catching the Red Queen. *Trends Ecol. Evol. 7*:405–408.

Ladle, R. J., R. A. Johnstone, and O. P. Judson. 1993. Coevolutionary dynamics of sex in a metapopulation: Escaping the Red Queen. *Proc. Roy. Soc. Lond. B. 253*:155–160.

Levin, S. A., D. Cohen, and A. Hastings. 1984. Dispersal strategies in patchy environments. *Theor. Pop. Biol. 26*:165–191.

Levin, B. R. and C. Svandborg Eden. 1990. Selection and evolution of virulence in bacteria: An ecumenical excursion and modest suggestion. *Parasitology 100*:S103–S115.

Lively, C. M. 1987. Evidence from a New Zealand snail for the maintenance of sex by parasitism. *Nature 328*:519–521.

Lively, C. M. 1989. Adaptation by a parasitic trematode to local populations of its snail host. *Evolution 43* (8):1663–1671.

Lively, C. M. 1992. Parthenogenesis in a freshwater snail: Reproductive assurance versus parasitic release. *Evolution 46* (4):907–913.

Lively, C. M., C. Craddock, and R. C. Vrijenhoek. 1990. Red Queen hypothesis supported by parasitism in sexual and clonal fish. *Nature 344*:864–866.

Lively, C. M. and J. Jokela. 1996. Clinal variation for local adaptation in a host–parasite interaction. *Proc. R. Soc. Lond. B. 263*:891–897.

Maynard Smith, J. 1978. *The Evolution of Sex.* Cambridge Univ. Press, Cambridge.

Maynard Smith, J. 1980. Selection for recombination in a polygenic model. *Genet. Res. Camb. 35*:269–277.

McPeek, M. A. and R. D. Holt. 1992. The evolution of dispersal in spatially and temporally varying environments. *Am. Nat. 140*:1010–1027.

Miller, R. A. 1996. The aging immune system: Primer and prospectus. *Science 273*:70–74.

Mopper, S. 1996. Adaptive genetic structure in phytophagous insect populations. *Trends Ecol. Evol. 11*:235–238.

Mopper, S., M. Beck, D. Simberloff, and P. Stiling. 1995. Local adaptation and agents of selection in a mobile insects. *Evolution 49* (5):810–815.

Morritz, C., H. McCallum, S. Donnelan, and J. D. Roberts. 1991. Parasite loads in parthenogenetic and sexual lizards (*Heteronotia binoei*): Support for the Red Queen hypothesis. *Proc. R. Soc. Lond. B. 244*:145–149.

Olivieri, I., Y. Michalakis, and P.-H. Gouyon. 1995. Metapopulation genetics and the evolution of dispersal. *Am. Nat. 146*:202–228.

Parker, M. A. 1985. Local population differentiation for compatibility in an annual legume and its host-specific fungal-pathogen. *Evolution 39*(4):713–723.

Pulliam, H. R. 1988. Sources, sinks, and population regulation. Am. Nat. 132:652–661.

Pulliam, H. R. and B. J. Danielson. 1991. Sources, sinks, and habitat selection: A landscape perspective on population dynamics. *Am. Nat. 137*:S50–S66.

Rice, W. R. 1983. Sexual reproduction: An adaptation reducing parent–offspring contagion. *Evolution 37*:1317–1320.

Shields, W. M. 1982. *Philopatry, Inbreeding, and the Evolution of Sex.* State University New York Press, NY.

Slatkin, M. 1987. Gene flow and the geographic structure of natural populations. *Science 236*:787–792.

Unruh, T. R. and R. F. Luck. 1987. Deme formation in scale insects: A test with the pinyon needle scale and a review of other evidence. *Ecol. Entomol. 12*:439–449.

van Valen, L. 1973. A new evolutionary law. *Evol. Theory 1*:1–30.

# 14

# Scale-Dependent Evolution of Specialization in a Checkerspot Butterfly: From Individuals to Metapopulations and Ecotypes

*Chris D. Thomas*
Department of Biology, University of Leeds, Leeds, UK

*Michael C. Singer*
Department of Biology, University of Texas, Austin, TX

## 14.1 Introduction

The population size of insects associated with any resource patch is determined by local birth and death in that patch and by migration into and out of the patch. When resource patches are small and close together, individuals move readily between patches, so very high emigration and immigration rates dominate patterns of local distribution (Kareiva 1983; Harrison 1991). These emigration and immigration rates will be determined principally by behavioral responses to patch attributes. Resource patches may be separate fallen fruits or fungi for *Drosophila* flies in a small wood (Shorrocks et al. 1990), host-plant individuals of one or more species for the butterfly *Euphydryas editha* in a meadow (Mackay 1985; Parmesan 1991), *Quercus* trees in scrub woodland for the leaf miner *Stilbosis quadricustatella* (Mopper et al. 1995), or widely separated patches of host plant in a landscape for more mobile species, such as patches of *Asclepias* for migrant *Danaus plexippus* butterflies (Malcolm and Zalucki 1993). If we are interested in the evolution of resource-use patterns in such a system, the focus is likely to be on patch-choice behaviors that lead to oviposition or feeding and the consequences of these behaviors for individual fitness.

Resource patches are usually aggregated in landscapes, and these aggregations can often be delineated as habitat patches. Some individuals may remain within one habitat patch throughout their life, while others migrate between patches; networks of local populations in habitat patches form metapopulations (Gilpin and Hanski 1991). In the context of this chapter, "local populations within metapopulations" can be regarded as virtually synonymous with "sub-populations in spatially structured populations." When habitat patches are small and close together, the distribution of insects among patches is again determined principally by patch choice decisions affecting migration among patches. As habitat patch size and spacing increase, one moves from situations in which individuals flow in and out of patches at will, through metapopulations where patch-specific birth, death, emigration, and immigration may be equally important to insect distributions, to separate populations where local birth and death processes are of primary importance (cf. Lima and Zollner 1996; Singer and Thomas 1996). When resource patches of different types are nested within habitat patches and resource choice affects fitness, behavioral choices at the smaller scale will impact birth and death processes at larger scales.

In this chapter, we ask how scale affects patterns of specialization observed in the diet of a herbivorous insect, the butterfly *Euphydryas editha*. The chapter is divided into four main areas. First, we describe patterns of oviposition behavior and host specialization by *E. editha* at different spatial scales. We outline *E. editha*'s host-use patterns at the level of (1) resource patches (plant individuals or patches) within habitats, (2) habitat patches within metapopulations, and (3) populations or metapopulations within the species. Second, we consider various factors that may affect diet breadth within populations. Third, we outline a verbal model of diet evolution in *E. editha*, and fourth, we ask why *E. editha* shows more local differentiation than do swallowtail butterflies, but less than some other insects.

## 14.2 Life Cycle of *Euphydryas editha*

*E. editha* populations have one generation per year, with adults flying in spring or summer. Females lay batches of eggs on or near individual host plants, and the larvae then feed gregariously on the plant that their mother selected. Larvae remain as groups for the first two instars, and the groups begin to fragment in the third instar. After three (sometimes four) feeding instars, larvae enter diapause. Onset of diapause coincides with summer drought and host-plant senescence over much of the range and with early fall in high elevation and northern populations. Larvae become active again in spring but then forage individually. Thus, postdiapause larvae contribute to their own choice of host plant, albeit within a fairly localized area. Postdiapause (and third-instar) larval foraging behavior can be important, but it will not be discussed here. On completing development in spring, the larvae pupate and give rise to the next spring–summer adult generation.

Because larvae are restricted to feeding on the host-plant individual or species chosen by their mother for the first two or three instars, most of the evolutionary consequences of oviposition choice are manifested before diapause. If we are interested in measuring fitness consequences of different oviposition behaviors, it is appropriate to measure offspring performance before diapause.

In most populations, females lay all available eggs in one oviposition event per day (Moore 1987; K. Agnew personal communication). This is likely to facilitate local specialization within a habitat patch, because each female should have time to assess the quality of many resource patches, and so be able to find and select the resource that confers highest offspring fitness (Rausher 1983). In contrast, time-limited insects may be faced with a choice of laying some eggs on hosts (resource patches) with relatively poor offspring survival, or leaving those eggs unlaid (Courtney 1982a, 1982b). However, not all batch-layers show local specialization, for example, where females are either too mobile to allow differentiation or, conversely, so immobile that host choice is left to the larvae (Gibbs 1962; Futuyma et al. 1984).

## 14.3 Individuals and Resource Patches

### 14.3.1 Specialists, Generalists, and Resource Use

Differentiation between populations is highly dependent on levels of gene flow (Peterson and Denno, Chapter 12, this volume). In *E. editha*, mark–release–recapture studies show high proportions of individuals remaining within 100–200 m of a release point, but occasional individuals may move 1–10 km and perhaps beyond (Ehrlich et al. 1975; Thomas and Singer 1987; Thomas et al. 1996; Harrison et al. 1988; Harrison 1989). Given this information on migratory capacity, we first consider areas of relatively uniform habitat with linear dimensions < 1 km. For ovipositing female *E. editha* searching at this scale, resource patches are either individual plants or groups of plants growing within the habitat. At Rabbit Outcrop, which is part of the Generals' Highway metapopulation (below), female butterflies were followed in an area of light coniferous woodland, where they were searching for perennial *Pedicularis semibarbata* (Scrophulariaceae) plants. These butterflies identified *P. semibarbata* rosettes from the air, landed on them disproportionately, and laid egg batches on them, avoiding the annual *Collinsia torreyi* (Scrophulariaceae) plants that grew in the same area (Mackay 1985; Moore 1989; Parmesan et al. 1995; Singer and Thomas 1996). Oviposition on *C. torreyi* would not be adaptive, because the plant senesces at Rabbit Outcrop before the larvae are large enough to enter diapause, even though *C. torreyi* is the sole host of some other *E. editha* populations, where the plant's phenology is more favorable.

The role of female behavior in determining local variation in resource use was also seen after the females had landed on the *P. semibarbata* plants. A female that has a batch of eggs ready to lay and lands (or is placed) on a plant shows a stereotyped behavioral sequence: (1) She taps with her foretarsi, apparently tasting the plant (oviposition behavior can be elicited by filter paper soaked in host plant extracts). She also bends the antennae towards the plant, sometimes touching it. If she likes what she tastes/smells, she (2) curls the abdomen, (3) extrudes the ovipositor, and (4) seeks a suitable location on the plant. Because step 4 takes perhaps 10–30 seconds, we can remove her at this stage, before the first egg is laid, put her in a cage to "cool off" for five minutes, and record that the plant has been accepted. A female that is replaced on a plant it has previously accepted will almost certainly accept it again (Singer 1982), but when a female is placed on a different plant species or individual, she may either accept or reject the second plant. If she rejects the second plant, having accepted the first, then we say that the female prefers the first plant over the second. This is her "postalighting oviposition preference" (Singer 1982). Individual and population-level variation in postalighting oviposition preference has a genetic basis (Singer et al. 1988, 1991, 1993; Singer and Parmesan 1993; Singer and Thomas 1996).

When Ng (1988) used this postalighting preference test at Rabbit Outcrop, he was able to classify females into two categories: those that accepted both *P. semibarbata* plants they were offered (generalists), and those that accepted one *P. semibarbata* plant but rejected the other (specialists). Ng then split groups of eggs from specialist and generalist females, and placed half of each group on *P. semibarbata* plants that had been favored by the specialists and half on plants rejected by the specialists. The result was a preference–performance correlation. Offspring of specialist females survived better on the plants that their mothers chose than on the rejected plants. Meanwhile, the offspring of the generalist females performed equally well on both categories of plant and significantly better than offspring of specialists on the plants that specialist mothers had rejected. Ng's pairs of accepted and rejected *P. semibarbata* plants were separated by 0.2–10 m. We know that these preference trials on individual *P. semibarbata* plants reflected the behavior of undisturbed butterflies (Rausher et al. 1981), and that plants acceptable to both categories of females received more naturally laid eggs (Ng 1988). Therefore, we can say that resource-use patterns at this scale were determined principally by behavior, in the form of oviposition preference.

Three to six species of plants in *E. editha*'s potential host range (Scrophulariaceae plus *Plantago*) grow in most locations where the butterfly is found. The postalighting choice protocol, described earlier, allowed us to obtain a rank order of preference for these potential hosts in 12 separate populations. Each female was repeatedly placed on each plant species, in rotation, and the order in which plants became acceptable was recorded. Plant species that were top-ranked in these preference trials received most eggs per plant in the field (Tables 14.1 and 14.2).

Two of the 12 populations we studied, Rabbit and Schneider, were engaged in rapid evolution of host preference (Singer et al. 1993). We describe Rabbit and Schneider in some detail later, and the rest of this section deals with the remaining 10 populations. These 10 populations showed no change in preference when tested in different years (Singer et al. 1994). In each of the 10 populations, the rank order of plant species in the preference hierarchy was concordant with the rank order in survival of larvae that were experimentally established on all potential hosts (Fig. 14.1; Singer et al. 1994). At the population level, females made disproportionate use of the resource types that conferred the highest fitness.

In 6 of the 10 populations, some females (specialists) preferred one top-ranking plant species, but other females (generalists) in the same population ranked the two "top" plant species equal (no preference in Table 14.1; ignore Rabbit and Schneider). Even among specialists, there is considerable variation in level of be-

*Table 14.1*   Oviposition Preferences of *E. editha* from 12 Populations[a]

| Population | Preferred host | Plant 1 | No pref. | Plant 2 | Second preference |
|---|---|---|---|---|---|
| | | for Number of individuals preferring | | | |

| Population | Preferred host | Plant 1 | No pref. | Plant 2 | Second preference |
|---|---|---|---|---|---|
| **Monophagous populations** | | | | | |
| Indian Flat | *Collinsia tinctoria* | 34 | 0 | 0 | *Castilleja densiflorus* |
| Gardisky | *Castilleja nana* | 8 | 2 | 0 | *Penstemon eterodoxus* |
| Pozo | *Pedicularis densiflorus* | 32 | 0 | 0 | *Castilleja foliolosa* |
| Franklin Point | *Pedicularis densiflorus* | 23 | 2 | 0 | *Castilleja foliolosa* |
| **Oligophagous populations** | | | | | |
| Del Puerto | *Pediculads densiflorus* | 6 | 11 | 0 | *Castilleja foliolosa* |
| Frenchman | *Penstemon rydbergii* | 0 | 14 | 0 | *Castilleja nana* |
| Piute Mtn | *Castilleja martinii* | 0 | 23 | 0 | *Pedicularis semibarbata* |
| Sonora | *Castilleja pilosa* | 13 | 9 | 0 | *Collinsia parviflora* |
| Tamarack | *Collinsia torreyi* | 5 | 8 | 0 | *Veronica serpyllifolia* |
| Tuolumne | *Castilleja lemmoni* | 8 | 16 | 1 | *Pedicularis atollens* |
| Rabbit (1987) | *Pedicularis semibarbata* | 17 | 30 | 15 | *Castilleja disticha* |
| Schneider 1986 | *Collinsia parviflora* | 16 | 11 | 8 | *Plantago lanceolata* |

[a] From Singer et al. 1994

havioral specialization (specificity), with some females accepting the second-ranking plant species a few minutes after the top-ranked plant became acceptable (weak specificity), and other females not accepting the second-ranked plant for days (strong specificity). Therefore, the specialist–generalist dichotomy described here is really a simplification of more complex continuous variation in specificity. Every specialist female (except for one at Tuolomne Meadow) agreed with every other specialist female within the same population on the order in which to rank every potential host-plant species growing locally; this agreement extended well beyond the two top-ranked plants.

Table 14.2  Host Plant Use in Relation to Rank Oviposition Preference for Scrophulariaceae and Plantago Species Present at 12 E. editha Populations[a]

| Population | Preferred species | Second preference | Third preference | Fourth preference |
|---|---|---|---|---|
| Monophagous populations | | | | |
| Indian Flat | *Collinsia tinctoria | Castilleja densitiorus | Castilleja affinis | Keckiella breviflorus |
| Gardisky | *Castilleja nana | Penstemon heterodoxus | | |
| Pozo | *Pedicularis densiflorus | Castilleja foliolosa | Collinsia heterophylla | |
| Franklin Point | *Pedicularis densiflorus | Castilleja foliolosa | Pantago lanceolata | Collinsia greenel |
| Oligophagous populations | | | | |
| Del Puerto | *Pedicularis densiflorus | *Castilleja foliolosa | Castilleja affinis | |
| Frenchman | *Penstemon rydbergii | *Castilleja nana | *Collinsia parviflora | |
| Piute Mtn | *Castilleja martinii | *Pedicularis semibarbata | Collinsia torreyi | |
| Sonora | *Castilleja pilosa | *Collinsia parviflora | *Penstemon rydbergii | |
| Tamarack | *Collinsia torreyi | *Veronica serpyllifolia | *Mimulus breweri | Pedicularis semibarbata |
| Tuolumne | *Collinsia lemmoni | *Pedicularis atollens | *Penstemon heterodoxus | Mimulus primuloides |
| Rabbit | *Pedicularis semibarbata | *Castilleja disticha | *Collinsia torreyi | |
| Schneider (1982) | *Collinsia parviflora | *Pantago lanceolata | *Penstemon rydbergii | Castilleja applegatei |
| Schneider (1990) | *Pantago lanceolata | *Collinsia parviflora | *Penstemon rydbergii | Castilleja applegatei |

* Indicates species that received eggs in the field

[a] From Singer et al. 1994

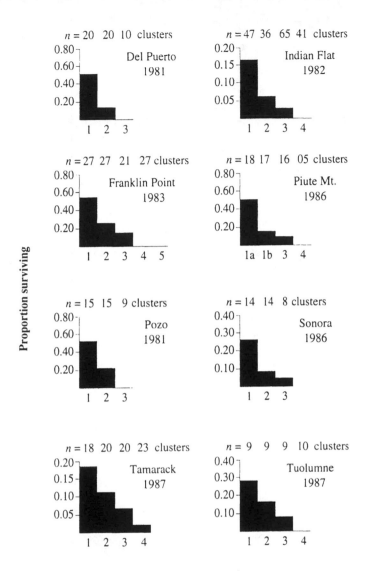

**Rank order of host preference**

*Figure 14.1* Survival of *E. editha* early stages in eight populations, in relation to the order in which females ranked host-plant species in behavioral preference tests. Note the perfect concordance between preference and survival ranks (overall $p = 10^{-6}$). The identities of the ordered plant species are given in Table 14.2. From Singer et al. (1994).

The remaining four populations were variants on the same theme. At French-man and Piute, all females were generalists and did not discriminate between the top two plant species (Table 14.1). But the ranking of these two plants over other potential hosts was consistent, and females agreed on the order in which to accept other, lower ranked potential hosts. In two other populations, Indian Flat and Pozo, all females were specialists, as defined, but specialist females still showed considerable differences in specificity. Therefore, in the entire set of 10 popula-tions, we found a consistent pattern of behavioral variation along a specialist–generalist axis, with the plants favored by the specialists providing higher aver-age offspring survival in the field (Fig. 14.1). Of 215 females preference-tested in the 10 populations, 129 were specialists preferring plant A over B, 85 were gen-eralists (A = B), and only one female (at Tuolomne Meadow) showed the oppo-site rank preference, preferring B over A (Table 14.1; continue to ignore Rabbit and Schneider).

Behavioral specificity in these populations determined population diet breadth. Four of the populations examined were naturally monophagous, and the remain-der were oligophagous (Tables 14.1 and 14.2). In the monophagous populations, only 4 out of 101 females tested were generalists; in these populations, all 100 + natural egg batches that have been found were laid on the top-ranked plant. In contrast, 81 out of 114 females tested were generalists in the six naturally oligophagous populations, in which more than one plant species received eggs. Differences in proportions of specialists and generalists in different locations cor-responded to large-scale differences in diet breadth.

### 14.3.2 Several Specialists

Not surprisingly, *E. editha* has gained a reputation for containing populations where more than one type of specialist is present (Thompson 1994), because our papers have concentrated on two such populations, Rabbit and Schneider. How-ever, as described earlier, most *E. editha* populations are not nearly so variable. Variation is usually along a specialist to generalist axis, rather than a specialist-on-A to generalist to specialist-on-B axis. This pattern of preference variation re-sembles that described for the swallowtail butterfly *Papilio zelicaon* (Thompson 1993; Wehling and Thompson unpublished data).

At Schneider's Meadow, on the east side of the Sierra Nevada, *E. editha* lay their eggs on *Collinsia parviflora* (an ancestral host) and on the introduced Euro-pean weed *Plantago lanceolata*, which is a novel host that has been incorporated into the diet in the last 100 years (Thomas et al. 1987). Females differ from one an-other in the rank order of these plants in their postalighting oviposition preference. Some prefer *P. lanceolata*, some prefer *C. parviflora*, and some do not discrimi-nate between the plants, so there are two types of specialist as well as generalists (Table 14.1); differences in preference were heritable (Singer et al. 1988, 1993).

At Schneider's Meadow, *C. parviflora* and *P. lanceolata* grow near to each other, so they represent two types of resource patch within one area. At this scale, we might expect females to be able to choose the plants they prefer. When females caught in the act of oviposition in the field were tested for preference in captivity, females that had been found ovipositing on *P. lanceolata* usually either preferred this plant or were generalists, whereas females caught ovipositing on *C. parviflora* either preferred *C. parviflora* or were generalists (Singer et al. 1989). Thus, different resource-use patterns by different females in the population were determined by heritable differences in adult behavior (Singer et al. 1988, 1989). Moreover, the offspring of *P. lanceolata–* and *C. parviflora–*preferring females differed in growth rates on the two plants. The more strongly a female preferred *C. parviflora*, the faster her offspring grew on *C. parviflora* (Singer et al. 1988). Differences in female behaviors were correlated with differences in larval performance (a preference–performance correlation), even though overall survival rates were much higher on *P. lanceolata* than on *C. parviflora*.

Annual *C. parviflora* plants in the Schneider area senesce and become inedible to caterpillars in early/midsummer, giving rise to high levels of prediapause larval mortality. The perennial *P. lanceolata* plants survive into summer much better and support higher larval survival (Fig. 14.2; Singer et al. 1994). Under strong natural selection for increased use of the introduced *P. lanceolata*, changes in behavior and use of the two types of resource patch occurred between 1982 and 1990 (Singer et al. 1993). In 1982, < 10% of females were *P. lanceolata* specialists, but this percentage increased significantly to over 50% by 1990 (Fig. 14.3). The proportion of generalists also increased. This shift in behavior generated a shift in resource use; sample censuses revealed 28 egg clusters/larval webs on *C. parviflora* and 4 on *P. lanceolata* in 1982, versus 6 on *C. parviflora* and 25 on *P. lanceolata* in 1990 (see Singer et al. 1993, for intervening years). Environmental perturbation (introduction of *P. lanceolata* and irrigation) resulted in a period of

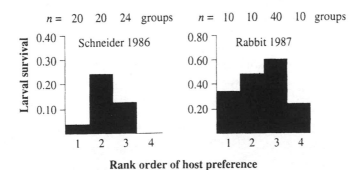

*Figure 14.2* Survival of *E. editha* early stages in two populations where oviposition preferences and diets were observed to evolve. Note the mismatch between preference and survival ranks. Host-plant names are given in Table 14.2. From Singer et al. (1994).

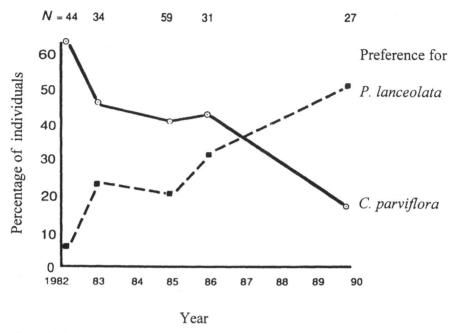

*Figure 14.3* Evolution of an increase in preference for *Plantago lanceolata* and decline in preference for *Collinsia parviflora* by female *E. editha* at Schneider's Meadow during a period when offspring survival was greater on *P. lanceolata* than on *C. parviflora*. From Singer et al. (1993).

rapid evolution at Schneider. During this period, a wide variety of different specialists have been observed in the population (Table 14.3). We do not know whether this diversity of preference rank was initiated by selection acting on variation that was already present at low frequency (remember the eccentric female at Tuolomne; Table 14.1), whether mutations arose after the arrival of *P. lanceolata*, or whether low frequency gene flow from other ecotypes was responsible for initially destabilizing the existing preference structure. Since about 20% of females in nearby undisturbed *C. parviflora*–feeding populations were generalists (they accepted *P. lanceolata* immediately, even though they had not been exposed to it evolutionarily), a few egg batches would probably have been laid on *P. lanceolata* as soon as it was introduced, so selection on preference could have started immediately (Thomas et al. 1987).

### 14.3.3 Host-Associated Fitness Variation within Populations

At Schneider, the main difference in fitness between larvae on *C. parviflora* and *P. lanceolata* was due to differences in plant phenology. In this and other populations of *E. editha*, host-associated variation in fitness is generated by differences in (1) plant nutritional quality (Ng 1988, shown indirectly by elimination of other

*Table 14.3* Pairwise comparisons of plants at Schneider's Meadow[a] and Rabbit.[b] Rank variation in preference was detected for five of the six pairwise comparisons at Schneider, and for four of the six comparisons at Rabbit. All plants listed received eggs in the wild, in the year preference data were obtained, except for *Castilleja applegatei* at Schneider.

| | Number of individuals preferring | | | |
|---|---|---|---|---|
| | Plant 1 | No pref. | Plant 2 | |
| Schneider 1986 | | | | |
| *Collinsia parviflora* | 16 | 11 | 8 | *Plantago lanceolata* |
| *Collinsia parviflora* | 7 | 0 | 1 | *Penstemon rydbergii* |
| *Collinsia parviflora* | 10 | 0 | 0 | *Castilleja applegatei* |
| *Plantago lanceolata* | 7 | 0 | 2 | *Penstemon rydbergii* |
| *Plantago lanceolata* | 8 | 1 | 1 | *Castilleja applegatei* |
| *Penstemon rydbergii* | 3 | 0 | 3 | *Castilleja applegatei* |
| Rabbit 1987 | | | | |
| *Pedicularis semibarbata* | 17 | 30 | 15 | *Castilleja disticha* |
| *Pedicularis semibarbata* | 111 | 51 | 20 | *Collinsia torreyi* |
| *Pedicularis semibarbata* | 19 | 4 | 0 | *Mimulus whitneyi* |
| *Castilleja disticha* | 32 | 6 | 2 | *Collinsia torreyi* |
| *Castilleja disticha* | 19 | 8 | 0 | *Mimulus whitneyi* |
| *Collinsia torreyi* | 6 | 13 | 8 | *Mimulus whitneyi* |

[a] Data from Singer et al., 1989

[b] Data from Singer et al. 1994 and unpublished

factors), (2) competition for food (Moore 1989; Thomas et al. 1990; D. Boughton unpublished data), and (3) variation in mortality from natural enemies associated with different plants (Moore 1989). Regardless of the mechanistic cause of host-associated fitness patterns, it is the patterns of fitness variation themselves and their consistency in time and space that are important for the evolution of host specialization.

At the spatial scale of resource patches within habitats, female behaviors appear to be major determinants of the distribution of eggs. Behaviors generate different rates of alighting on different resource patch types, and differential oviposition once a female has landed. Differences in behavior result in varying diet breadths of different populations. Where females occur together in one habitat but differ from one another in behavior, these behavioral differences generate divergent resource-use patterns.

## 14.4 Local Populations in Metapopulations

We now move to the scale of habitat patches, within which resource patches can be found. In most of the locations described earlier, a number of rather similar habitat patches occur in the same area, and the local populations that inhabit them are likely to be connected by some level of dispersal. When habitat patches are similar to one another in resource abundance and quality, then evolution of resource-use patterns in each patch will be similar to that of the metapopulation as a whole, and a metapopulation perspective may not fundamentally alter our evolutionary understanding of the system. However, habitat patches are never quite identical to one another and, in some circumstances, more than one habitat type may be occupied within a region, with the potential for different selection pressures in different habitats.

A situation like this occurs in two habitat types along the Generals' Highway in the Sierra Nevada in California, where there are areas of light woodland associated with natural granite outcrops (Rabbit Outcrop, described earlier) and also clearings in previously dense forest, made by clear-cut logging in the late 1960s and early 1970s (Fig. 14.4). Four host plant species (all Scrophulariaceae) have received eggs in this region since 1980, but for simplicity we restrict our account to the two major hosts. On the outcrops, such as Rabbit Outcrop, selection favors oviposition on *Pedicularis semibarbata* and disfavors use of *Collinsia torreyi*, which dries up and becomes inedible before the larvae can reach diapause (Singer and Thomas 1996). Before logging, the metapopulation was restricted to outcrops, where eggs were laid on *P. semibarbata* and on the minor host *Castilleja disticha*. In contrast, *P. semibarbata* has been exterminated from logged clearings, whereas *C. torreyi* is present and thriving in this habitat. In logged clearings, the soil is deeper (it was physically disturbed by logging) and the brushwood was burned when the tree trunks were removed. Clearing *C. torreyi* is physically larger and is phenologically more favorable than outcrop *C. torreyi*, such that *E. editha* larvae can reach diapause successfully on it. Therefore, within-habitat selection favored oviposition on *P. semibarbata* in outcrops and on *C. torreyi* in clearings. The butterflies behaved appropriately during the 1980s, ovipositing on *P. semibarbata* in outcrops and on *C. torreyi* in clearings. Survival was higher on clearing *C. torreyi* than on outcrop *P. semibarbata* (Fig. 14.2; Moore 1989; Singer and Thomas 1996), so between-habitat natural selection favored use of clearings.

When habitat types are distinct, yet close enough together for moderate levels of migration between them, the distribution of adults between habitat types is likely to be generated by some combination of behavioral choice, passive movement, and population dynamics in each habitat. A mark–release–recapture program in 1984 revealed that 70% (for clearings) and 85% (for outcrops) of recaptured butterflies stayed within the habitat type where they were marked (Thomas et al. 1996). Distributional patterns of adult preferences, adult abundances, and egg densities showed exactly the pattern we would expect if behavior and population dynamics were both important. A diversity of postalighting oviposition preferences was observed in each habitat in the disturbed Generals' Highway metapopulation. When females were caught in the act of ovipositing in habitat patches 100–500 m apart (Singer 1983; Thomas and Singer unpublished data), females found ovipositing on *P. semibarbata* on an outcrop usually preferred this

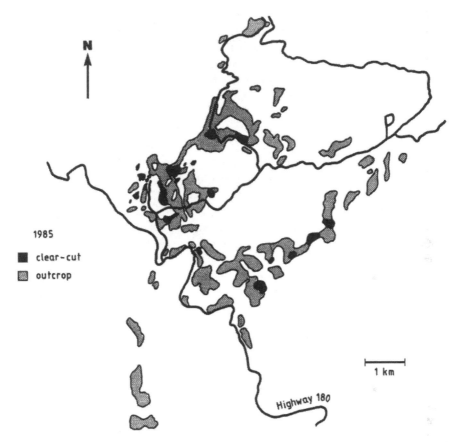

*Figure 14.4* Map of the Generals' Highway metapopulation showing the distribution of logged clearings, where *Collinsia torreyi* was the principal host, and light woodland on natural granite outcrops, where *Pedicularis semibarbata* was the principal host. From Thomas et al. (1996).

plant or were generalists (using the postalighting test), whereas females found ovipositing on *C. torreyi* in the neighboring clearing were usually generalists or preferred *C. torreyi*. When eggs from the two adjacent habitats were reared in a common environment on *Collinsia*, preferences of the offspring from the two areas differed significantly (Singer and Thomas 1996). Thus, female postalighting preferences differed genetically between adjacent habitats.

The importance of spatial scale became increasingly clear as we witnessed evolution in the two habitats during the 1980s. Although selection pressures were different in the two habitats, evolution in the two patches was linked by the exchange of individuals. Preferences in both patches evolved toward increasing preference for *C. torreyi*, but preferences evolved just as fast in this direction on the outcrop, where the change was not adaptive, as in the clearing, where it was (Singer and Thomas 1996). Thus, responses to selection in the clearing drove preferences in the outcrop away from their adaptive optimum. Had we been studying *E. editha* in outcrops alone, we would have been at a loss to understand why their behaviors were evolving in the wrong direction; in the clearing, we might have wondered why the rate of evolution was relatively slow.

Given the 15–30% exchange rates between habitats, why did females from the two habitats differ in behaviors at all? The difference arose because postalighting oviposition preference was linked to, or a component of, habitat-selection behavior. Females tended to remain in habitats that contained host plants they preferred (based on the postalighting preference test) and to leave those that did not contain the preferred host (Thomas and Singer 1987; Singer and Thomas 1996). This assortative behavior generated a consistent pattern of greater preference for *C. torreyi* in the clearing than in the outcrop, during a period when mean preference changed within both habitats.

The population dynamics in the two habitats were also different and partly responsible for the between-habitat difference in selection pressures. During the 1980s, the clearings acted as population sources (birth > death) and the outcrops consumed individuals (death > birth). Although the clearing habitat received fewer eggs, it consistently generated more adults; mark–release–recapture revealed twice the rate of transfer from clearing to outcrop (30%) as vice versa (15%; Thomas et al. 1996). There was a net flow of individuals away from clearings that raised insect densities on nearby outcrops, and densities were so high on outcrops next to clearings that there was severe competition (Thomas et al. 1996). Thus, the population dynamics and related movement reinforced the between-habitat natural selection in favor of using *C. torreyi* in clearings and against *P. semibarbata* on outcrops. Preference for the ancestral host *P. semibarbata* increased with isolation from clearing population sources, suggesting that the balance of gene flow (over 0.5 to 10 km) versus selection was responsible for metapopulation-wide patterns of oviposition preference on outcrops. We could not test whether preferences in clearings were affected by isolation from outcrops, because none of the clearings was isolated from outcrops. However, there were higher proportions of generalists and females that preferred *C. torreyi* in large clearings, presumably because large patches were less influenced by gene flow from outcrops (Singer and Thomas 1996). This increase in preference for *C. torreyi* contributed to greater larval densities in large clearings (Fig. 14.5). Population-dynamic success and evolution of local specialization went hand in hand in large clearings. Spatial-population dynamics and behavioral evolution were inextricably linked and dependent on the size and spacing of habitat patches within the metapopulation.

Situations like those in the Generals' Highway region, with different selection pressures in different habitats, seem predisposed to maintain behavioral polymorphisms. However, in other regions where *E. editha* occurs, there is usually a mix-

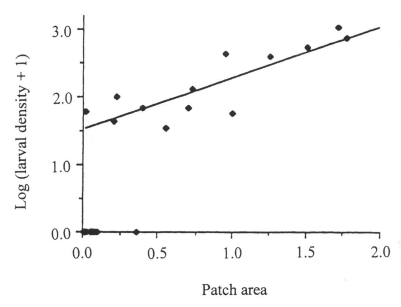

*Figure 14.5*   Relationship between clearing patch area (ha) and *E. editha* larval density (webs/ha). Excluding patches with zero density; $Log_{10}$ Density + 1 = 1.53 + 0.77 Area, $n = 13$, $t = 5.52$, $p < 0.001$, $r = 0.857$. Greatest use was made of large clearings, where adaptations to clearing habitat could be maintained in the face of extensive exchanges between clearings and outcrops.

ture of habitats or microhabitat types available containing different Scrophulariaceae and/or *Plantago* species. Despite the potential for maintained polymorphisms with different genotypes at a selective advantage in different habitats, we typically observe a specialist–generalist axis of variation (Table 14.1) and usually find that metapopulations are restricted to just one general type of habitat patch in a region. Based on experience rather than any theoretical expectation, we think that long-term maintenance of two types of oviposition specialist in the Generals' Highway metapopulation is unlikely, although it is certainly possible. However, the outcome is uncertain. Following a severe summer frost that eliminated all clearing populations (*C. torreyi* plants died and the caterpillars starved) but did not harm those in outcrops, the direction of between-habitat selection was reversed in 1992, and preference evolution also reversed (Thomas et al. 1996; Singer and Thomas 1996). After this brief reversal of selection and evolution, *C. torreyi* in the clearing habitat has once again (1993–1995) become more suitable than *P. semibarbata* in the outcrop habitat (D. Boughton unpublished data), so the whole evolutionary experiment of the 1980s looks set to be repeated.

In *E. editha,* responses to selection for use of different resource patches in different habitat patches are not entirely independent at a scale of habitat patches a few hundred meters to a few kilometers apart. Patterns of resource use at this scale are determined by both behavior (choice of resource patches within habitat patches, choice of habitat patches) and population dynamics (different mortality rates in different habitat patches, passive movement). The juxtaposition of habitat patches in metapopulations certainly influences the rate of evolution and, possibly, the final outcome.

## 14.5 Divergence between Populations

Groups of habitat patches are normally aggregated in space: wet meadows in valleys, rocky outcrops along ridges, serpentine outcrops, and so on. *E. editha* metapopulations usually occupy groups of such habitat patches, often over some kilometers, among which the insects normally show no obvious differentiation (except at Generals' Highway). If we now move up a scale again to habitats that are 5–50 km apart, we may either find very similar populations feeding on the same host plant(s) in similar environments, or very different populations feeding on different host plant(s), usually in different environments. This spatial scale of 5–50 km is very important. Occasional gene flow may take place (see Harrison et al. 1988; Harrison 1989), through a network of habitat patches, if not directly (Peterson and Denno, Chapter 12, this volume), but levels of exchange are likely to be so low that gene flow will not seriously disrupt local adaptations. We give two examples.

First, we compared the evolving *E. editha* population at Schneider's Meadow, where *Plantago lanceolata* had been incorporated into the diet, with three other populations where *P. lanceolata* had not been introduced, and where the diet was still restricted to the traditional host plant, *Collinsia parviflora*. We assumed that the Schneider's Meadow population was similar to these other three, prior to the introduction of *P. lanceolata* at Schneider. The other three localities were about 3.5 km, 45 km, and 50 km from the nearest locations where *P. lanceolata* was known to have been incorporated into the diet. In 1985 and 1986, behavioral preferences were significantly different between Schneider and the other three localities. Schneider females were relatively accepting of *P. lanceolata*, with *C. parviflora* and *P. lanceolata* specialists both present, whereas females from the other populations were mostly *C. parviflora* specialists, with some generalists (Thomas et al. 1987). The Schneider population continued to evolve toward *P. lanceolata* preference over the next five years (Fig. 14.3, Singer et al. 1993). Under selection for different patterns of resource use in separate populations, divergence was possible and presumably increased during the 1980s. Contrast this with the situation of adjacent habitat patches at Rabbit Outcrop, where selection did not result in increasing divergence over a few hundred meters.

Second, we considered an area at low to midelevation on the western slopes of the southern half of the Sierra Nevada in California (Fig. 14.6). At relatively low elevation, in the foothills, *E. editha* lay their eggs on *Collinsia tinctoria* (northern populations) or *Collinsia heterophylla* (southern populations), mostly in open or scrub woodland. These two *Collinsia* species are closely related and quite similar

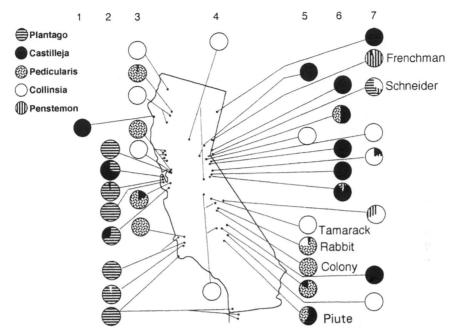

*Figure 14.6* Geographic distribution of oviposition by *Euphydryas editha* in California. Each pie chart shows the proportion of egg batches laid on each host genus for each population. Plant genera are shown for simplicity, but all populations shown as ovipositing on one host genus actually lay their eggs on a single plant species within that genus. Numbers indicate groupings in the coastal ranges (1, sea cliff; 2, coastal grasslands; 3, chaparral in inner coast ranges) and Sierra Nevada (4–6). Category 4 populations are at low elevation (< 1500 m) in the western foothills, and oviposit on *Collinsia tinctoria* or *C. heterophylla*. Category 5 populations are at midelevation (1,800–2,500 m) on the western slopes, predominantly ovipositing on *Collinsia torreyi* to the north of Kings Canyon (Tamarack northward), and on *Pedicularis semibarbata* and *Castilleja* species to the south of Kings Canyon (Rabbit southward). Category 6 populations occur at higher elevations (> 2,500 m), and Category 7 populations occur at 1,200–2,500 m on the eastern slopes of the mountains. Many other populations have also been recorded, but sample sizes are small, so pie charts are not shown. The records that we have not plotted reinforce the pattern of similarity in diet for populations that are close together and occur in similar environments. Map modified from Singer (1995).

to one another. At midelevation, *E. editha* lays eggs on (1) *Collinsia torreyi* (quite different in growth form from low elevation *C. tinctoria* and *C. heterophylla*) to the north of Kings Canyon (Thomas et al. 1990), (2) on *P. semibarbata*, *C. torreyi*, *Castilleja disticha*, and *Mimulus whitneyi* immediately to the south of the Canyon, at Generals' Highway, and (3) on *P. semibarbata* and *Castilleja* species to the south of Generals' Highway.

Low- and midelevation sites can be geographically quite close together and occasionally within 5 km of one another. For the populations to remain behaviorally distinct suggests that selection is adequate to counter low levels of gene flow between low- and midelevation metapopulations. Where environments are very different and there is partial temporal separation of the populations (low vs. midelevation), this differentiation is achieved over geographic distances of 4–10 km.

Where environments are more similar, diets sometimes change over distances as short as 10–30 km, as has happened at Generals' Highway. However, it is more usual to see a single diet over longer distances. South of Generals' Highway, *Pedicularis/Castilleja* metapopulations stretch several hundred kms in the same habitat type, with one isolated example about 400 km to the north (Fig. 14.6). Given the potential for selection to result in changes in preferences and diets over much shorter distances, we presume that this similarity in diet is maintained because this diet is adaptive for *E. editha* in this climatic/vegetation zone, in the absence of major human habitat disruption (logging has also taken place at Piute Mountain but is more restricted and has apparently not destabilized the existing diet).

Neighboring *E. editha* populations that are at least 5–10 km apart are apparently free to evolve separate patterns of resource use if selection pressures differ, because gene flow does not disrupt local adaptation. The result of this freedom is substantial variation in host preference over scales of 5–1,500 km, with resulting variation in diet at this scale (Fig. 14.6; e.g., Singer 1971, 1995; White and Singer 1974; Thomas et al. 1987; cf. Fox and Morrow 1981; Thompson 1994). Some of the geographic variation of diet is simply due to plant availability: different combinations of plant species are available in different vegetation types (e.g., Shapiro 1995). However, most of the diet variation shown in Figure 14.6 is due to geographic variation in oviposition preference (Singer et al. 1994) and, less frequently, to the interaction of insect preference with geographic variation in host resistance (Singer and Parmesan 1993).

The habitat types occupied by *E. editha* in California include desert margins, undercliffs, Serpentine grassland, open forests, damp meadows, and stabilized scree on some of the highest mountains in the Sierra Nevada (3,500 m). Despite this ecological flexibility and the range of specific adaptations observed in different places (Ehrlich et al. 1975; Singer et al. 1995), the butterfly is surprisingly localized over most of this range, occurring as apparently distinct populations and metapopulations in certain habitats. If it can occur in such a wide range of environments, why does it not use more habitats and host plants locally? The fact that it does not suggests that local specialization is adaptive, and that genotypes adapted to other local host plants and nearby habitats cannot normally invade.

## 14.6 Host Races or Ecotypes?

In Figure 14.6, we can recognize groups of populations with similar patterns of resource use and occurring in relatively similar environments. Such groups are often referred to as "host races" (see Feder et al., Chapter 16, this volume). It is tempting to describe the *E. editha* populations that feed on *C. tinctoria/heterophylla* at low elevation in the western foothills of the Sierra Nevada as one host race, and to describe the midelevation *P. semibarbata/Castilleja* feeders as another. However, before we do this, we must address a pair of important questions. Are host races single phylogenetic subdivisions of the species that may eventually diverge to the point of speciation? Or do these groups have multiple evolutionary origins, suggesting (1) that (convergent) local-diet evolution is relatively easy, and (2) that differences in diet between populations do not necessarily represent fundamental phylogenetic subdivisions within the species. Both mitochondrial DNA (mtDNA) and allozyme evidence suggest that the host races examined so far are not fundamental subdivisions of *E. editha* (Radtkey and Singer 1995). Populations that shared diets were not more closely related than expected from the geographic distance between them. This means either that there were several independent evolutionary colonizations of each host genus or, if *E. editha* evolved each diet only once, ongoing or subsequent gene flow between habitats has prevented divergence in other characters. If use of the term *host race* implies a fundamental phylogenetic division within a phytophagous insect species, it would be inappropriate to describe groups of *E. editha* populations as host races, even when they share a diet. We prefer to call similar populations *ecotypes*, recognizing that similarities may be due either to common ancestry or to convergent evolution, or both.

## 14.7 The Hierarchy of Diet Pattern

We have presented a hierarchy of diet patterns in *E. editha*. Variation in the use of resource patches within habitats is generated by individual behavior. Differences in resources used in different habitat patches within metapopulations are determined by the interaction of natural selection within habitat patches, behavioral responses to resource and habitat patches, and population-dynamic consequences of oviposition on different resource types in each habitat. In separate populations, local differences are free to evolve in response to local conditions. Despite this hierarchy, it is important to recognize that all female *E. editha* oviposit on individual resource patches (individual plants or groups of plants), and fitness differences among females depend on their behavioral responses to different resource types. The resource patch is the basic unit. Resource patches occur within habitat patches, and networks of populated habitat patches make up metapopulations. Therefore, the fundamental questions relate to resource patches. How is variation in use of resource types maintained within a population? How and why do diets change? And at what spatial scale can evolved differences in diet establish, and why? These questions are tackled in the remaining sections.

## 14.8 Maintenance of Behavioral Variation

In this section, we consider how specialist–generalist variation in oviposition preference might be maintained, because it was a major determinant of diet pattern for the 10 relatively stable populations examined. Within-population variation in oviposition preference along a specialist–generalist axis has also been observed in other butterflies and has been especially well documented in swallowtails, where it appears to have a genetic basis (Wiklund 1975, 1981; Thompson 1988a, 1988b, 1993). In *E. editha*, we have direct evidence that specialist–generalist variation in preference has a genetic basis only at Rabbit Outcrop, but we have only looked for it in this one population (Singer and Thomas 1996). Presuming a genetic basis, several major hypotheses can be put forward to explain why this variation could be maintained, although more than one explanation may operate within any population. Swallowtail butterflies, *Papilio machaon,* oviposit singly on a range of umbellifers. Wiklund (1975) gave individual females a choice of potential umbellifer host plants. He found that some females laid most eggs on the top-ranked plant species (specialists), and others distributed their eggs more evenly between plant species (generalists, although the largest number of eggs was still deposited on the top-ranked plant). As in *E. editha*, the overall ranking matched larval survival on the plants rather well. Thompson has obtained comparable results for North American swallowtails and found a genetic basis for some of the variation (Thompson 1988a, 1988b, 1993). Wiklund (1975) argued that spatial variation in the distribution of resources was responsible for the maintenance of this variation; females should lay eggs on the top-ranked plant in areas where it was available (specialists favored) but not delay oviposition too much in habitats where the best plant was not present (generalists favored). There would be extensive gene flow between habitats, because swallowtails are quite mobile, so spatial variation in selection pressures would result in variation in specificity in butterflies caught within any area. Evolution of specialization would be predicted in the absence of gene flow between habitats, which is exactly what has happened in extensive fenland areas of East Anglia in England. In East Anglia, swallowtails inhabit only one habitat type, where they lay on just one umbellifer species (Dempster 1994).

We list five hypotheses to account for the existence of within- and between-population variation in oviposition specificity.

1. *Rarity of suitable host species.* In insects such as *E. editha* that become less discriminating as they search, a rare plant that confers high fitness should be accepted for oviposition if it is encountered (there is no cost to this), but it should not be preferred so strongly that oviposition is substantially delayed if the plant cannot be found (this would result in a reduction of fecundity). This may be part of the explanation for the preponderance of generalist insects at Sonora and Piute (Table 14.1), where the top-ranked plant species is much rarer and more localized than the second-ranked host. As Wiklund (1975) suggested for swallowtails, spatial variation in the abundance of a rare but highly suitable plant may maintain specialist-generalist variation.

2. *Density-dependent selection.* All of the experiments to assess the suitability of different plant species and individuals for prediapause survival have involved removing any competitors and placing a single batch of eggs on a plant. In the

wild, favored plant individuals often receive two or more batches of eggs, and intraspecific competition takes place. In standard foraging theory, we might expect individual females to assess conspecific density on each plant, and they might then lay eggs roughly equally on "good plants with many eggs" and "poorer plants without eggs," thus achieving an ideal-free distribution of equal offspring survival on all plants at natural densities. However, egg-avoidance behavior does not appear to have evolved in any *E. editha* population we have studied (L. Ramakrishna unpublished data). Therefore, the butterflies may achieve a distributional pattern of eggs on plants that approaches an ideal-free distribution through evolutionary density-dependence rather than through instantaneous behavioral modification. In years of high density, specialists would suffer most competition and would be at a disadvantage, but in years of low density, specialists would be favored over generalists. Variation in the extent to which parasitoids concentrate their attacks on relatively high local densities of larvae (mostly the offspring of specialists), depending on overall larval and parasitoid densities, could also favor specialists in some years and generalists in others.

3. *Temporal variation in plant quality.* This could take several forms. (1) Environmental stochasticity could change the ranking of offspring survival on the top two plant species in some years. In a "bad year," survival might be higher on the plant that is normally second best, and generalist females would be favored in that year. (2) Environmental stochasticity could simply change the quantitative difference in survival on the top two plants. Any time delay associated with finding plant A, if it is rarer than plant B, might be worthwhile in years in which survival on A was much greater than on B (be a specialist), but not in years when survival was only slightly higher on A than on B (be a generalist). (3) Environmental stochasticity may result in variation in the density of plants that are suitable for oviposition in a given year. A butterfly that chooses to lay eggs on particular plant phenotypes within a favored plant species may find that many host individuals are acceptable in a normal year, but few are available in a drought year. Thus, environmental stochasticity would generate temporal variation in intraspecific competition on the top-ranked plant species (Ehrlich et al. 1980; Singer and Ehrlich 1979). If the second-ranked plant was not affected in the same way by the same environmental variation, selection for specialists and generalists could alternate.

4. *Transient effects of diet evolution.* As we argue later, changes in diet in *E. editha* are associated with the generation of transient genetic diversity of preference, followed by specialization on a new diet. The presence of generalists in *E. editha* populations that are currently monophagous may be because weak selection has not yet removed all generalists from the population, following some previous diet change. This explanation may be appropriate for monophagous populations that contain just a few generalists (Franklin Point) or oligophagous populations where < 5% of eggs are laid on secondary hosts (Tamarack). However, it is unlikely that populations in which all individuals are generalists (Piute) are evolving toward monophagy.

5. *Behavioral variation may be selectively neutral.* In populations in which all individuals are specialists and, provided that every specialist can find the preferred host species, every individual is phenotypically identical in terms of oviposition. Insects with high or moderate specificity may all pick the same favored host category. Therefore, some of the variation in specificity that is expressed in our preference-testing protocol may not be expressed in the field, and therefore is not subject to natural selection. Nonadaptive variation in specificity may either accumulate through mutation and occasional immigration, or may remain semipermanently as a legacy of prior diet changes.

This listing of possible ways in which specialist–generalist variation could be maintained is by no means exhaustive. Given the apparently widespread nature of specialist–generalist variation within populations of plant-feeding insects, exploration of its maintenance requires urgent attention.

### 14.8.1 Preference–Performance Correlations

*Preference–performance correlations* exist if individual females with different preferences choose to oviposit on different resources on which their own offspring will perform (survive or grow) best (Via 1986). We restrict use of this term to situations in which both preference and performance are variable. We have found preference–performance correlations in the only two experiments designed to search for them in *E. editha* (Ng 1988; Singer et al. 1988), so they may be frequent in this insect. Elsewhere (Singer et al. 1994), we have suggested the terms *preference–performance concordance* or *adaptive host choice* to describe the results of population-level studies in which the rank order in which plants are preferred is compared with average offspring growth and survival on the same plants, as illustrated in Figure 14.1.

Preference–performance correlations have the potential to maintain relatively broad patterns of resource use within a single population, because different individuals specialize on different resources. Preference–performance correlations in combination with frequency-dependent selection may be another explanation for the maintenance of both specialist and generalist phenotypes within a population. Suppose, as described for Rabbit Outcrop, that behaviorally specialist females have offspring that survive particularly well on the top-ranked hosts, and generalist females have relatively generalist larval offspring. Intraspecific competition could lead to specialists outcompeting generalists in one environment (top-ranked plants) and generalists outcompeting specialists in another (second-ranked plants), even if the top-ranked plants would be best for all larvae in a competition-free environment. Frequency-dependent selection could then maintain specialist–generalist variation, provided that a second type of specialist could not establish.

What are the underlying mechanisms generating preference–performance correlations? In *E. editha* we perceive four possibilities, but critical experiments to distinguish between them have not been carried out. First, pleiotropy could be responsible. The same genes might affect adult-female oviposition preference and larval growth rate (e.g., genes that affect sensitivity of taste receptors on female foretarsi *and* on larval mouthparts could, respectively, affect female oviposition preference and larval ingestion rate, and thereby growth). Second, the same result could be obtained if different loci controlling larval growth rate and adult preference were linked. We have no evidence for or against this. Third, the result could be obtained, even if there was no linkage, pleiotropy, or assortative mating (to be discussed), but provided that there was strong within-generation selection. Suppose that there is heritable variation in female behavior and, completely independent of this, there is heritable variation in larval survival. If females that prefer plant A lay eggs on plant A, their *surviving* offspring will tend to be the ones that also have genes for high survival on A. Similarly, B-preferring females will have surviving offspring that tend to have genes for high survival on B. Therefore, the emerging adult population in the next generation has a preponderance of adult females that prefer A and carry genes for survival on A, and females that prefer B and carry genes to survive on B, whereas A-preferers that carry genes for survival on B and B-preferers that carry A survival genes will be relatively rare. Even if mating is random, females will, on average, tend to produce offspring that perform relatively well of the host plant they prefer. These alternative possible explanations are hard to explore in *E. editha,* because this species is difficult to rear in captivity and not amenable to detailed genetic analyses (common-environment rearings and parent–offspring correlations are about the best we can do).

A fourth mechanism generating preference–performance correlations may be assortative mating, which could be achieved by temporal, spatial, or behavioral separation. In *E. editha*, it is only likely to occur when different resource types are sufficiently spatially separated that two sets of larvae (which are less mobile than the adults) inhabit different microclimates and feed on different Scrophulariaceae after diapause; the two sets of offspring may then show slightly different emergence dates and thus show partial temporal segregation (cf., Feder et al., Chapter 16, this volume). Within metapopulations, partial spatial separation of adults in different habitat patches may also generate some level of assortative mating. Assortative mating is a prerequisite if separate demes are to establish, associated with different resource types.

It is not always easy to distinguish among these four alternatives, and they are not mutually exclusive in any case. When samples of insects from different plant individuals or species are found to differ genetically, the process of differentiation has often been termed *deme formation* (Boecklen and Mopper, Chapter 4, this volume). However, the same pattern may be generated when there are preference-performance correlations within insect demes (Singer and Thomas 1996). A case in point is the associated variation in preference for, and performance on, different *P. semibarbata* plants among Rabbit Outcrop *E. editha* (Ng 1988), described earlier. Plants belonging to two categories (accepted and rejected by specialist females) grew interspersed in the habitat, and searching female butterflies encountered many plants of both categories. Even though the microdistribution of larvae from the two categories of females differed initially, postdiapause larvae forage together in the same habitat, emerge together, and females are mated on emergence by males that are flying throughout the habitat. So, the insects were not segregated into demes, even though transplantation of newly hatched larvae among plants showed that they performed less well when the plant category was changed, just as in the "deme formation" experiments (Mopper et al. 1995; Boecklen and Mopper, Chapter 4, this volume). We urge researchers to be very cautious when interpreting divergence of insect traits associated with different plants in fairly small geographic areas. Differentiation may be an example of deme formation, which could potentially lead to sympatric speciation, or could simply reflect preference–performance correlations. These alternatives are not easy to distinguish without detailed analyses of mobility and genetics.

## 14.9 A Model for *Euphydryas editha* Diet Evolution

Based on our data gathered from 10 evolutionarily relatively stable populations of *E. editha*, and two rapidly evolving populations (Singer et al. 1993), we present a verbal model for the evolution of diet in *E. editha* (Table 14.4). Most local populations lay eggs on one to three species of Scrophulariaceae and/or *Plantago* (Fig. 14.6) out of three to six potential host-plant species growing in the area. All females in the population agree on the rank order of oviposition preference. Within these populations, there is, however, variation in behavioral preferences along a specialist–generalist axis: Some females will strongly prefer the top-ranked plant (specialists), but others will readily accept second- or even third-ranked plants (generalists). This type of variation is probably maintained by selection within populations and perhaps by population histories. But, for *E. editha*, we have too many hypotheses and too little data to say exactly how. Almost all females are behavioral specialists in populations where only one plant species receives eggs, whereas generalists are common where two or more plant species receive eggs. In all populations we have examined, the top-ranked plant (on the basis of female behavior) confers the highest survival on egg/larval offspring in the field (Fig. 14.1).

Following habitat perturbation in two populations, relative survival changed. In one case, logging changed relative survival on different native hosts, such that one species in the Scrophulariaceae, which would have previously conferred low relative fitness, changed in size and phenology, resulting in very high offspring survival. In the other, introduced European *Plantago lanceolata* experienced higher survival than any of the native Scrophulariaceae. In both cases, rapid evolution took place in response to measured natural selection on genetic variation in oviposition preference. Although both populations we studied had been perturbed *in situ*, similar consequences might be expected from the successful (population dynamic) colonization of a habitat with novel combinations of potential hosts, or where the same plant species were growing but in conditions that conferred different patterns of relative survival.

*Table 14.4* Proposed scheme for evolution of oviposition patterns in plant-feeding insects, modeled on *Euphydryas editha*. The actual spatial scale at which local evolutionary changes take place will vary greatly, depending on the mobility of each insect and the spatial scale at which that insect experiences variation in selection.

*Existing evolutionarily stable population:*
• specialist-generalist axis of variation for behavioral oviposition preference
• concordance between preference and relative survival

*Perturbation:*
• arrival of new plant species, extinction of existing plant species
• change in relative plant qualities through habitat modification
• colonization of new site with different plant species of relative qualities

*Response to perturbation:*
• change in population size, or extinction (inability to adapt)
• mismatch between preference and survival
• natural selection on exiting or new variation
• extensive genetic variation for rank oviposition preference (different specialists present), resulting in some oviposition on (virtually) all potential hosts

*Stabilization:*
• selection against specialists with low fitness leads to decline in variation
• return to specialist-generalist variation
• possible long-term increase in specialization with time, in some populations

*Spread:*
• invasion of habitats where new adaptations are favored (usually <10 to 50 km in *E. editha*).

During periods of rapid evolution, many different types of behavioral specialist are found (Table 14.3) and most potential Scrophulariaceae or *Plantago* hosts in the habitat receive some eggs. Three plant species received eggs at Schneider's Meadow (the fourth potential host present was very rare) and four at Rabbit Outcrop in the Generals' Highway metapopulation. Because different specialists tend to oviposit on different plants (discussed earlier), which confer different fitnesses on their offspring, oviposition preferences are then subject to selection. During the early to mid-1980s, both disturbed populations showed a mismatch between behavior and survival (Fig. 14.2), with most females preferring to lay on resources that conferred relatively low offspring fitness. By the early 1990s, the Schneider population had evolved to the point at which most females now preferred *P. lanceolata*, which conferred highest survival (Fig. 14.3). This simple result stemmed from the fact that natural selection favoring oviposition on *P. lanceolata* was consistent in both space (across habitat patches) and time (across years). In contrast, natural selection for use of the novel host at Generals' Highway (*Collinsia torreyi*) was variable in space and time (Singer and Thomas 1996). At the beginning of our study, use of *C. torreyi* was strongly favored in some habitat patches but opposed in others. The overall rate of evolution was determined by the combination of patch-specific selection and exchange rates between patches, resulting in relatively slow evolution toward preference for the novel host. The direction of both natural selection and evolution was sharply reversed when survival on *C. torreyi* declined again in the late 1980s and early 1990s (Singer and Thomas 1996). In time, we expect both Schneider and Generals' Highway to settle down to a new specialist–generalist pattern of variation, but with a new rank order of preference, at least at Schneider.

In summary, most *E. editha* populations maintain relatively low levels of variation for preference rank. However, unidimensionality of preference in stable populations cannot be used to deduce how populations will respond to changes in selection. If they are perturbed or colonize new environments, where there are new selection pressures for a new diet, diversity of preference rank may be generated rapidly. During this period of rapid diet evolution, most Scrophulariaceae are evolutionarily "sampled" before the diet settles down to a new pattern of local specialization.

Our results emphasize that conclusions based on observed patterns of variation within existing populations (probably under stabilizing selection) give little idea of the scope for evolutionary change in those populations when faced with habitat change. Studies of most *E. editha* populations reveal only specialist–generalist variation within populations and might lead us to conclude erroneously that different types of specialist could not evolve. But we know that they have evolved many times because (1) there is extensive (evolved) geographic variation in oviposition behavior, with completely different preference hierarchies expressed in different populations (Table 14.1); (2) the two disturbed populations we have examined both show extreme within-population variation in behavior (Table 14.3); and (3) host races are apparently polyphyletic, indicating that changes in diet have evolved more times than the number of different diets that exist in the species (Radtkey and Singer 1995). Wasserman (1986) and Courtney et al. (1989) have suggested that preference hierarchies are unidimensional within a species/population, and that this hierarchy constrains diet evolution in plant-feeding insects. Their argument is based on the hypothesis that an evolutionary response to incorporate one low-ranked host into the diet (e.g., because a change in plant quality has recently made it more suitable) would result in a correlated increase in use of other low-ranked plants, on which fitness is still low. In *E. editha*, preference hierarchies are unidimensional within undisturbed populations (Table 14.1) and may act as temporary constraints, but this does not prevent the evolution of new hierarchies (Tables 14.1 and 14.3) when conditions change and selection for an alternative hierarchy is strong. Thompson (1993, 1994) has argued that there may be some basic genetic constraint preventing deviation from specialist–generalist patterns of variation in swallowtails, and there may be, but we note that a small amount of variation in rank order of preference has been seen in swallow-tails, and that local evolutionary responses to recent habitat changes will be much more difficult in these relatively mobile insects. We should not, however, overemphasize the power of evolutionary responses to habitat perturbation. Most modern changes in land use result in butterfly population extinctions (inability to adapt) rather than successful evolutionary responses (New et al. 1995).

With modifications to take account of the specific biologies of other organisms, we think the verbal model we present (Table 14.4) is likely to be widely applicable to plant-feeding insects. However, being free to evolve local adaptations does not mean that they will always evolve narrow diets, for example, if there is density-dependent parasitism independently on each potential host, or if females are time-limited and face a trade-off between fecundity and fitness per offspring.

## 14.10 The Scale of Diet Differentiation and Speciation

Patterns of local adaptation can be generated by the balance of selection and migration in many insects, but at different real scales in different species (Peterson and Denno, Chapter 12, this volume). For example, swallowtail host races usually occur over hundreds to thousands of kilometers (Scriber 1986; Scriber et al. 1991; Scriber and Lederhouse 1992; Thompson 1993, 1994; Shapiro 1995), whereas flightless stick-insect color morphs may evolve in parallel with local changes in the vegetation over tens to hundreds of meters (Sandoval 1994). *Euphydryas editha* is somewhere in the middle, with a typical scale of population differentiation of five to tens of kilometers, when selection pressures differ between environments.

Related to this, speciation is apparently only very rarely associated with host shifts in *Euphydryas* and its relatives. The evidence for this is as follows:

1. Almost all members of the Melitaeinae subfamily or tribe to which *E. editha* belongs show similar patterns of local diet differentiation, with many diets per species (e.g., *Mellicta athalia*, Warren 1987). Most diet shifts do not apparently result in speciation.

2. Closely related species show more within- than between-species variation in diet (Thomas et al. 1990; R. R. White personal communication).

3. We have not found any problems obtaining fertile crosses between different populations of *E. editha* (however, we have not undertaken formal tests of viability; this is a problem in the field, because crosses will usually be at a selective disadvantage in both "home" environments, but they might be at an advantage in habitats that neither parental population occupies).

4. "Ecotypes" that have a single diet are polyphyletic (Radtkey and Singer 1995).

5. The two rapidly evolving, disturbed populations have not resulted in increasing parapatric divergence, but preference for one of the host plants has "won."

6.  *Euphydryas* butterflies show regional variation in adult and larval color patterns, perhaps owing to a combination of thermoregulation and mimicry (Bowers 1981). Adult "color pattern races" cut across "host races" (personal observations), suggesting that regional adaptations may be free to evolve separately for different traits.

Why do we conclude that host shifts are rarely associated with speciation in *Euphydryas*, when Thompson has concluded that fundamental shifts in swallowtail preference hierarchies are usually accompanied by speciation? Evolution of diet differentiation is difficult in relatively mobile insects; selection for a new diet must take place over a large area if it is to establish. In mobile species, only those major shifts in preference hierarchies that are accompanied by strong assortative mating and/or hybrid inviability are likely to result in dietary divergence. Without speciation, local diets will usually either fail to establish ("the old diet wins") or succeed and spread at the expense of the preexisting diet ("the new diet wins"). In either case, what we observe is one species with one preference hierarchy over large areas, unless we happen to catch a species in the act of shifting preference hierarchy. In contrast, if dietary divergence is accompanied by speciation, the new species and diet may spread, giving rise to two swallowtail species with different diets. It is quite possible that local diet shifts are sometimes associated with speciation in *Euphydryas* butterflies, but this does not have to be so because of the lower migration rate. Comparing the two groups of butterflies, the smaller migratory scale in *Euphydryas* permits a higher ratio of diet changes to speciation events, given climate-dependent limits to distributions and finite land surfaces that populations can inhabit.

## Acknowledgments

We thank the many people who have discussed *Euphydryas* evolution with us, and who have collected much of the data summarized here. We also thank David Boughton, Andrew Davis, Susan Mopper, and Sharon Strauss for their valuable comments, and National Science Foundation and the University of Texas for support. We thank many landowners for facilitating this work, but particularly the Schneider family, and the U.S. Forest and National Parks Services.

## 14.11 References

Bowers, M. D. 1981. Unpalatability as a defense strategy of western checkerspot butterflies (*Euphydryas*). *Evolution* 35:367–375.

Courtney, S. P. 1982a. Coevolution of pierid butterflies and their cruciferous foodplants: III. *Anthocharis cardamines* (L.) survival, development and oviposition on different hostplants. *Oecologia* 51:91–96.

Courtney, S. P. 1982b. Coevolution of pierid butterflies and their cruciferous foodplants: IV. Crucifer apparency and *Anthocharis cardamines* (L.) oviposition. *Oecologia* 51:91–96.

Courtney, S. P., G. K. Chen and A. Gardner. 1989. A general model for individual host selection. *Oikos 55*:55–65.

Dempster, J. P. 1994. The ecology and conservation of *Papilio machaon* in Britain. Pp. 137–149 *in* A. S. Pullin (Ed.), *Ecology and Conservation of Butterflies*. Chapman & Hall, London.

Ehrlich, P. R., D. D. Murphy, M. C. Singer, C. B. Sherwood, R. R. White, and I. L. Brown. 1980. Extinction, reduction, stability and increase: The responses of checkerspot butterfly (*Euphydryas*) populations to the California drought. *Oecologia 46*:101–105.

Ehrlich, P. R., R. R. White, M. C. Singer, S. W. McKechnie, and L. E. Gilbert. 1975. Checkerspot butterflies: A historical perspective. *Science 188*:221–228.

Fox, L. R. and P. A. Morrow. 1981. Specialization: Species property or local phenomenon. *Science 211*:887–893.

Futuyma, D. J., R. P. Cort, and I. van Noordwijk. 1984. Adaptation to host plants in the fall cankerworm (*Alsophila pometaria*) and its bearing on the evolution of host affiliation in phytophagous insects. *Am. Nat. 123*:287–296.

Gibbs, G. W. 1962. The New Zealand genus *Metacris* Meyrick (Lepidoptera: Arctiidae) systematics and distribution. *Trans. Roy. Soc. New Zealand, Zool.* 2:154–159.

Gilpin, M. E. and I. Hanski (Eds.) 1991. *Metapopulation Dynamics: Empirical and Theoretical Investigations*. Academic Press, London.

Harrison, S. 1989. Long-distance dispersal and colonization in the bay checkerspot butterfly. *Ecology 70*:1236–1243.

Harrison, S. 1991. Local extinction in a metapopulation context: An empirical evaluation. *Biol. J. Linn. Soc. 42*:73–88.

Harrison, S., D. D. Murphy, and P. R. Ehrlich. 1988. Distribution of the bay checkerspot butterfly, *Euphydryas editha bayensis*: Evidence for a metapopulation model. *Am. Nat. 132*:360–382.

Kareiva, P. 1983. Experimental and mathematical analyses of herbivore movement: Quantifying the influence of plant spacing and quality on foraging discrimination. *Ecol. Monogr. 52*:261–282.

Lima, S. L. and P A. Zollner. 1996. Towards a behavioural ecology of ecological landscapes. *Trends Ecol. Evol. 11*:131–135.

Mackay, D. A. 1985. Pre-alighting search behavior and host plant selection by ovipositing *Euphydryas editha* butterflies. *Ecology 66*:142–151.

Malcolm, S. B. and M. P. Zalucki, (Eds.). 1993. *Biology and Conservation of the Monarch Butterfly*. Natural History Museum of Los Angeles County, Los Angeles, CA.

Moore, R. A. 1987. Patterns and consequences of within-population variation in reproductive strategies. Ph.D. dissertation, University of Texas at Austin.

Moore, S. D. 1989. Patterns of juvenile mortality within an oligophagous insect population. *Ecology 70*:1726–1737.

Mopper, S., M. Beck, D. Simberloff, and P. Stiling. 1995. Local adaptation and agents of selection in a mobile insect. *Evolution 49*:810–815.

New, T.R., R. M. Pyle, J. A. Thomas, C. D. Thomas, and P. C. Hammond. 1995. Butterfly conservation management. *Annu. Rev. Entomol. 40*:57–83.

Ng., D. 1988. A novel level of interaction in plant-insect systems. *Nature 334*:611–612.

Parmesan, C. 1991. Evidence against plant "apparency" as a constraint on evolution of insect search efficiency. *J. Insect Behav. 4*:417–430.

Parmesan, C., M. C. Singer, and I. Harris. 1995. Absence of adaptive learning from the oviposition foraging behavior of a checkerspot butterfly. *Anim. Behav. 50*:161–175.

Radtkey, R. R. and M. C. Singer. 1995. Repeated reversals on host-preference evolution in a specialist insect herbivore. *Evolution 49*:351–359.

Rausher, M. D. 1983. Ecology of host-selection behavior in phytophagous insects. Pp. 223–257 *in* R. F. Denno and M. S. McClure (Eds.), *Variable Plants and Animals in Natural and Managed Systems*. Academic Press, New York.

Rausher, M. D., D. A. Mackay and M. C. Singer. 1981. Pre- and post-alighting host discrimination by *Euphydryas editha* butterflies: The behavioral mechanisms causing clumped distributions of egg clusters. *Anim. Behav. 29*:1220–1228.

Sandoval, C. P. 1994. The effects of the relative geographic scales of gene flow and selection on morph frequencies in the walking-stick *Timema cristinae*. *Evolution 48*:1866–1879.

Scriber, J. M. 1986. Origins of the regional feeding abilities in the tiger swallowtail butterfly: Ecological monophagy and the *Papilio glaucus australis* subspecies in Florida. *Oecologia 71*:94–103.

Scriber, J. M. and R. C. Lederhouse. 1992. The thermal environment as a resource dictating patterns of feeding specialization of insect herbivores. Pp. 429–465 *in* M. D. Hunter, T. Ohgushi and P. W. Price (Eds.), *Effects of Resource Distribution on Animal–Plant Interactions*. Academic Press, New York.

Scriber, J. M., R. C. Lederhouse, and R. H. Hagen. 1991. Food plants and evolution within *Papilio glaucus* and *Papilio troilus* species groups (Lepidoptera: Papilionidae). Pp. 341–373 *in* P. W. Price, T. M. Lewinsohn, G. W. Fernandes and W. W. Benson (Eds.), *Plant–Animal Interactions: Evolutionary Ecology of Tropical and Temperate Regions*. Wiley-Interscience, New York.

Shapiro, A. M. 1995. From the mountains to the prairies to the ocean white foam: *Papilio zelicaon* makes itself at home. Pp. 67–99 *in* A. R. Krukeberg, R. B. Walker and A. E. Leviton (Eds.), *Genecology and Ecogeographic Races*. Pacific Division American Association for the Advancemnt of Science (AAAS), San Francisco, CA.

Shorrocks, B. J., J. Rosewell, and K. Edwards. 1990. Competition on a divided and ephemeral resource: Testing the assumptions. II. Associations. *J. Anim. Ecol. 59*:1003–1017.

Singer, M. C. 1971. Evolution of food-plant preferences in the butterfly *Euphydryas editha*. *Evolution 25*:383–389.

Singer, M. C. 1982. Quantification of host preference by manipulation of oviposition behavior in the butterfly *Euphydryas editha*. *Oecologia 52*:224–229.

Singer, M. C. 1983. Determinants of multiple host use by a phytophagous insect population. *Evolution 37*:389–403.

Singer, M. C. 1995. Behavioral constraints to the evolution of insect diet breadth. Pp. 279–296 *in* L. Real (Ed.), *Behavioral Mechanisms in Evolutionary Ecology*. University of Chicago Press, Chicago, IL.

Singer, M. C. and P. R. Ehrlich. 1979. Population dynamics of the checkerspot butterfly *Euphydryas editha. Forschritte Zoologishe 25*:53–60.

Singer, M. C., D. Ng, and R. A. Moore. 1991. Genetic variation in oviposition preference between butterfly populations. *J. Insect Behav. 4*:531–535.

Singer, M. C., D. Ng, and C. D. Thomas. 1988. Heritability of oviposition preference and its relationship to offspring performance within a single insect population. *Evolution 42*:977–985.

Singer, M. C. and C. Parmesan. 1993. Sources of variation in patterns of plant–insect association. *Nature 361*:251–253.

Singer, M. C. and C. D. Thomas. 1996. Evolutionary responses of a butterfly metapopulation to human and climate-caused environmental variation. *Am. Nat., 148*:S9–S39.

Singer, M. C., C. D. Thomas, H. L. Billington, and C. Parmesan. 1989. Variation among conspecific insect population in the mechanistic basis of diet breadth. *Anim. Behav. 37*:751–759.

Singer, M. C, C. D. Thomas, H. L. Billington, and C. Parmesan. 1994. Correlates of speed of evolution of host preference in a set of twelve populations of the butterfly *Euphydryas editha. Écoscience 1*:107–114.

Singer, M. C., C. D. Thomas, and C. Parmesan. 1993. Rapid human-induced evolution of insect-host associations. *Nature 366*:681–683.

Singer, M. C., R. R. White, D. A. Vasco, C. D. Thomas, and D. A. Boughton. 1995. Multi-character ecotypic variation in Edith's checkerspot butterfly. Pp 101–114 *in* A. R. Krukeberg, R. B. Walker, and A. E. Leviton (Eds.), *Genecology and Ecogeographic Races.* Pacific Division American Association for the Advancement of Science (AAAS), San Francisco, CA.

Thomas, C. D., D. Ng, M. C. Singer, J. L. B. Mallet, C. Parmesan, and H. L. Billington. 1987. Incorporation of a European weed into the diet of a North American herbivore. *Evolution 41*:892–901.

Thomas, C. D. and M. C. Singer. 1987. Variation in host preference affects movement patterns in a butterfly population. *Ecology 68*:1262–1267.

Thomas, C. D., M. C. Singer, and D. A. Boughton. 1996. Catastrophic extinction of population sources in a butterfly metapopulation. *Am. Nat. 148*:957–975.

Thomas, C. D., D. Vasco, M. C. Singer, D. Ng, R. R. White, and D. Hinkley. 1990. Diet divergence in two sympatric congeneric butterflies: Community or species level phenomenon? *Evol. Ecol. 4*:62–74.

Thompson, J. N. 1988a. Variation in preference and specificity in monophagous and oligophagous swallowtail butterflies. *Evolution 42*:118–128.

Thompson, J. N. 1988b. Evolutionary genetics of oviposition preference in swallowtail butterflies. *Evolution 42*:1223–1234.

Thompson J. N. 1993. Preference hierarchies and the origin of geographic specialization in host use in swallowtail butterflies. *Evolution 47*:1585–1594.

Thompson, J. N. 1994. The geographic mosaic of evolving interactions. Pp. 419–431 *in* S. R. Leather, A. D. Watt, N. J. Mills, and K. F. A. Walters (Eds.), *Individuals, Populations and Patterns in Ecology.* Intercept Ltd., Andover, UK.

Via, S. 1986. Genetic covariance between oviposition preference and larval performance in an insect herbivore. *Evolution 40*:778–785.

Warren, M. S. 1987. The ecology and conservation of the heath fritillary butterfly, *Mellicta athalia*: I. Host selection and phenology. *J. Appl. Ecol. 24*:467–482.

Wasserman, S. S. 1986. Genetic variation in adaptation to foodplants among populations of the southern cowpea weevil, *Callosobruchus maculatus*: Evolution of oviposition preference. *Entomol. Exp. Appl. 42*:201–212.

White, R. R. and M. C. Singer. 1974. Geographical distribution of hostplant choice in *Euphydryas editha*. *J. Lepid. Soc. 28*:103–107.

Wiklund, C. 1975. The evolutionary relationship between adult oviposition preferences and larval host plant range in *Papilio machaon*. *Oecologia 18*:185–197.

Wiklund, C. 1981. Generalist vs. specialist oviposition behaviour in *Papilio machaon* (Lepidoptera) and functional aspects of the hierarchy of oviposition preference. *Oikos 36*:163–170.

# 15

# Factors Affecting Gene Flow between the Host Races of *Eurosta solidaginis*

*Joanne K. Itami*
Department of Life Sciences, Arizona State University West, Phoenix, AZ

*Timothy P. Craig*
Department of Life Sciences, Arizona State University West, Phoenix, AZ

*John D. Horner*
Department of Biology, Texas Christian University, Fort Worth, TX

## 15.1 Introduction

### 15.1.1 Speciation Models and the Evolution of Reproductive Isolation

Speciation involves the evolution of reproductive isolation between populations (Mayr 1963; Bush 1994). Despite many controversies about how species are defined, and how they evolve, the study of reproductive isolating mechanisms remains central to understanding speciation. Different speciation models make very different assumptions about how reproductive isolation evolves (Mayr 1963; Bush 1975a, 1994; White 1978). Host races and recently evolved sibling species of phytophagous insects offer unique opportunities to study the speciation process. Because the process of speciation is ongoing or recently completed, these extant populations retain the characteristics that were responsible for the evolution of their reproductive isolation. By examining the ecological, behavioral, and genetic characteristics of these populations, we can test assumptions about the evolution of reproductive isolation made in different models of speciation.

Allopatric and sympatric speciation models make different assumptions about how reproductive isolation evolves in phytophagous insects (Bush 1975a, 1994; Bush and Howard 1986). These models differ not only in their assumptions about the need for a period of geographic isolation, but also about many other processes of speciation. There are different models of allopatric speciation (Bush 1975a; White 1978), but they share key assumptions about the evolution of isolating mechanisms (Bush 1975a). Allopatric speciation models assume that reproductive isolation evolves as the result of drift and/or adaptation that evolve in the absence of the ancestral species. There is no direct selection for reproductive isolation, as there is no opportunity for selection against hybrids, because the two populations do not encounter each other. Prezygotic behavioral isolating mechanisms fortuitously arise during the development of other adaptations in the new geographic area. Reproductive isolating mechanisms are either complete when populations come back into contact, or are refined via reinforcement in a narrow hybrid zone (Bush 1975a).

There has been strong, long-term opposition to the possibility of sympatric speciation by leading authorities on speciation (Mayr 1988; Carson 1989), but both theory and empirical data support its existence (Bush 1994). Sympatric speciation minimally requires (1) alteration of the insect's feeding and oviposition preferences for feeding and/or oviposition, (2) appropriate physiological adaptation to the new host plant, and (3) assortative mating to maintain genetic changes important to the use of the new host. Genetic models indicate that sympatric speciation can occur under restrictive conditions (Tauber and Tauber 1989). If traits for host recognition, physiological adaptation to a new host, and assortative mating are mediated by different gene loci, then there must be genetic linkage of these traits to ensure sympatric speciation (Felsenstein 1981). Linkage is necessary to prevent recombination from breaking down favorable gene combinations as rapidly as they are formed. However, this constraint can be relaxed if mate choice is associated with habitat choice (Bush 1975a, 1975b; Diehl and Bush 1984, 1989). As a result of this association, mate choice and host choice can proceed as correlated characters (Rausher 1984; Rice 1984, 1987). Such coupling is possible in herbivorous insects, because mating frequently takes place on the host plant (Price 1980). Adaptation to a host plant may also lead to a shift in emergence times, and allochronic isolation can also result in reproductive isolation in sympatry (Wood 1980; Wood and Guttman 1983).

Sympatric models for speciation in herbivorous insects make five important predictions that differ from allopatric models.

1. Sympatric speciation models for herbivorous insects assume that the evolution for reproductive isolation accompanies adaptation to a new host plant (Tauber and Tauber 1989; Bush 1975a, 1975b; Wood 1980; Rice 1987). Phytophagous insects are hypothesized to be able to speciate sympatrically via a host shift (Bush 1969, 1975b). In brief, these models hypothesize that mutations for a change in host preference and the ability to survive on a new host plant occur in an ancestral population. A population with the new mutations colonizes a new host plant. The two populations achieve reproductive isolation in sympatry through assortative mating on the host-plant species. This is subsequently followed by the evolution of additional adaptations to the new host. Prezygotic isolating mechanisms that are plant mediated include mating on the host plant and allochronic isolation due to emergence times that are matched to host-plant phenology. Postzygotic isolation can also be host-plant mediated. Poor hybrid survival can occur for two reasons. First, hybrids may have low fitness because the two populations may be genetically incompatible, indicating that the populations have diverged to the point where they can be considered species. Alternatively, hybrids may not suffer from genetic incompatibilities, but instead have an un-

usual genotype that is not well adapted to either habitat (Rice and Hostert 1993). Hybrid survival will depend on the availability of benign habitats (i.e., suitable genotypes) on one or both host species. This would indicate that differentiation was primarily the result of adaptation to the host plant, which would be consistent with a sympatric origin of the differences.

Adaptations to the host plant resulting in reproductive isolation could also occur in allopatric speciation models. However, as White (1978) points out, if the same characters could have evolved without a period of geographic isolation, it is more parsimonious not to assume the extra steps in the speciation process required by allopatric models.

2. Sympatric models assume that changes in relatively few genes are required for reproductive isolation and speciation. In its simplest form, sympatric speciation in herbivorous insects requires changes in only one gene for preference, and one gene for performance.

3. Sympatric models also assume that there is direct selection for reproductive isolation, because the same genes that adapt an organism to its new habitat act to isolate it from the ancestral populations (Bush 1994). Thus, mating on the host plant is adaptive, because it results in offspring that will be able to survive on the host plant that the female prefers.

4. Sympatric models propose that prezygotic isolating mechanisms will arise previous to or simultaneously with adaptation to a new niche (host plants). Therefore, populations that are largely behaviorally isolated may still be capable of interbreeding to some extent.

5. Sympatric speciation models propose that all stages in the speciation process will be observable in sympatric populations. Intermediate stages of reproductive isolation, along the continuum of host races to species, may be available to study. In allopatry, reproductive isolation arises by chance prior to the interaction of the populations, or it is perfected in a narrow hybrid zone.

If populations are in intermediate stage of sympatric speciation, we can study variation in the factors producing reproductive isolation to understand their importance. Variation in reproductive isolation between sympatric herbivorous insect populations may be caused by variation in the host plant, the environment, or in the insect subpopulations.

### 15.1.2 Host-Plant Variation, Environmental Variation, and Gene Flow

A key assumption of sympatric speciation models is that reproductive isolation between populations results from adaptation to different host plants. Host plants vary within a species, and this variation may influence the expression of characters that lead to reproductive isolation between host-associated populations. The potential for gene flow will be influenced by the cascading effects of the plant–genotype–environment interactions between herbivore populations using different host species. Some of the possible interactions that might influence gene flow are illustrated in Figure 15.1. The environment will influence the density and distribution of each host species, and as the degree of sympatry changes, so will the

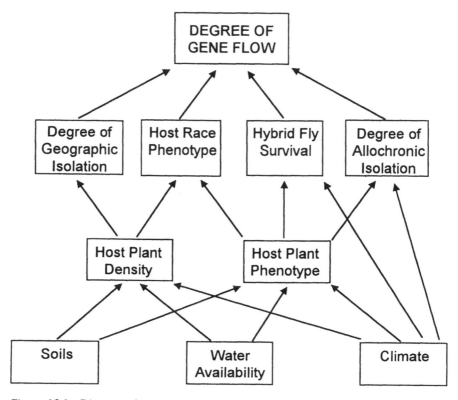

*Figure 15.1* Diagram of the cascading effects that forces from the abiotic environment and lower trophic levels can have on the evolutionary trajectory of the hybridizing populations of *Eurosta solidaginis*.

potential for assortative mating. Allochronic isolation between populations will also be influenced by the host plant and environment. The phenology of herbivore emergence may be influenced both directly by climate and indirectly through changes in plant phenology. In some years and/or locations, the emergence periods of the populations may be broadly overlapping, and there will be opportunity for gene flow, while in other times and locations, there may be total allochronic isolation. Host-plant recognition is crucial for assortative mating on the host plant, and environmental and/or genotypic variation in the cues could influence gene flow. Different sites will favor different genotypes within each host-plant species. If there are intraspecific genotypic and phenotypic differences among plants in the cues that the herbivores use for mating and oviposition site recognition, this could influence gene flow. In some sites, plant cues might be altered so that the rates of mating-site and oviposition-site "mistakes" are increased, thus increasing the potential for gene flow.

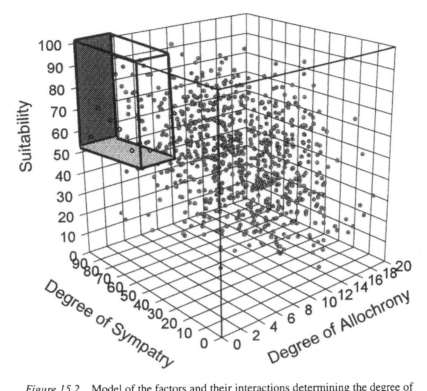

*Figure 15.2* Model of the factors and their interactions determining the degree of gene flow at each site. Only a fraction of the potential combinations of factors will lead to the possibility for gene flow: high suitability, high degree of sympatry, and low degree of allochrony. Under these conditions, the host races overlap physically and temporally with susceptible host plants available.

Reproductive isolation may also be influenced by variation in survival of the host-associated populations and hybrids between the populations. Plant suitability for herbivore development is highly variable. Even if host plants of both species are present, suitable genotypes for survival of both host-associated populations may not be present. This could influence the sympatry of populations. If hybrids require specific plant genotypes for survival, then the distribution of these plant genotypes will influence the distribution of gene flow and survival rates of insect genotypes. Some sites may have genotypes on which hybrids can survive, while other sites may not.

Each of these host-plant-associated factors may interact in a complex manner to determine the degree of gene flow in each site or period in time. We have illustrated this with a simplified model that varies on three axes: degree of sympatry, degree of allochrony, and host-plant suitability (Fig. 15.2). Examination of individual variables may indicate that gene flow is possible, whereas examination of the variable's interactions with other variables may show that gene flow is impossible. For example, hybrid matings may seem inevitable in a site where the host plants are mixed together and there is little allochronic isolation between the host races. However, if suitable host-plant genotypes are not present for offspring survival, then gene flow may not occur. Alternatively, the host species may be well mixed, and plants suitable for the hybrid development may be present, but allochronic isolation may prevent gene flow. Taken together, the interaction of these environmental factors will lead to geographically varying patterns of gene flow. These factors may interact with each other either additively or multiplicatively to determine the degree of gene flow between populations.

### 15.1.3 Insect Variation and Gene Flow

Differences among the subpopulations of each host-associated population may influence reproductive isolation between the host-associated populations in different areas. Populations utilizing a single host species often show considerable genetic variation among sites (McCauley and Eanes 1987; Rank 1992; Guttman and Weigt 1989). These differences may be due to periods of geographic isolation or selection. As host plants and environments vary, the adaptations of herbivorous insects necessary to utilize those the host plants vary (Thompson 1994). If populations of herbivorous insects are utilizing two different host plants, then their characteristics will vary in response to host-plant variation. If a host shift occurs in a single geographic area, the new host-associated population may encounter herbivore populations with a range of genetic differentiation on the ancestral host plant as it spreads to new areas. As a result, the degree of reproductive isolation may differ between the two host-associated populations that encounter each other in different areas.

The degree of gene flow between populations in an intermediate state of divergence may vary because of insect, host-plant, and environmental variation. Thus, gene flow may vary temporally or geographically. We have sought to utilize this variation as a tool for understanding reproductive isolating mechanisms.

We are examining the evolution of reproductive isolating mechanisms in two host-associated populations of *Eurosta solidaginis* that form galls on *S. gigantea* and *S. altissima*. It has been hypothesized that these populations are differentiated at the level of host races (Craig et al. 1993; Abrahamson and Weis 1996). *Host races* are defined by Diehl and Bush (1994) as populations that are partially reproductively isolated due to their association with a specific host plant. We will discuss how well the populations fit the definition of host races later in the chapter. We have taken two approaches toward understanding the importance of various reproductive isolating mechanisms between these closely related populations. First, we will report our efforts to experimentally determine what are the important factors that influence gene flow between host-associated populations. In the next section we will attempt to determine the degree of gene flow in the field. Finally, we will discuss the factors that are likely to influence gene flow in different geographic areas.

### 15.1.4 *The* Eurosta–Solidago *System*

The goldenrod ball gallformer is a narrowly oligophagous herbivore found *on Solidago altissima* and *S. gigantea*. Genetic individuals of these two goldenrod species can form extensive clones through lateral spread of rhizomes. Individual goldenrod ramets can occur side by side with other ramets of the same goldenrod species or the other goldenrod species (Craig et al. 1993). We will refer to the fly population found on *S. gigantea* as gigantea flies and to the population found on *S. altissima* as altissima flies. Uhler (1951) has described the life history of *E. solidaginis,* and an extensive review of the evolution and ecology of this species can be found in Abrahamson and Weis (1996). Adults emerge in mid- to late May in Minnesota. The gigantea flies have a mean emergence time from 2 to 21 days earlier than the altissima flies (Craig et al. 1993; Craig et al. unpublished data). Each host race mates on the bud of its own host plant. After mating, females inject an egg into the unexpanded leaves of the host plant's terminal bud. Insertion of the ovipositor leaves a visible mark that we term an *ovipuncture.* Not all ovipunctures result in oviposition. After hatching, the larva burrows through several millimeters of stem, and settles just below the apical meristem. Galls become apparent 21 days after oviposition. The larvae reach maximum size within the gall during winter, then pupate in spring. Adults live approximately 10 days.

*Eurosta* larvae have a number of natural enemies, among them the parasitoid wasp, *E. gigantea. Eurytoma gigantea* attacks galls after they have reached their maximum size, so gall size can influence successful attack of larvae by this parasitoid (Weis et al. 1985; Weis and Abrahamson 1986). Another parasitoid, *E. obtusiventris,* preferentially attacks galls on *S. altissima* (Brown et al. 1995).

## 15.2 Factors Influencing Gene Flow in *Eurosta*

### 15.2.1 *Prezygotic Isolation*

#### 15.2.1.1 *Mixing of Host Plants and Fly Populations*

Assortative mating of *Eurosta solidaginis* host races is produced by a combination of mating on the host plant and differences in emergence times (Craig et al. 1993). Each host race mates on its own host plant. In the presence of the host plant, there is strong assortative mating, whereas in the absence of the host plant, there is very weak assortative mating (Fig.15.3). Each host race appears to recognize its own host plant, but its ability to recognize members of its own host race in the absence of the host plant is imperfect.

Assortative mating is primarily based on host-plant preference, but any spatial separation of the populations would tend to lower the rate of interpopulation

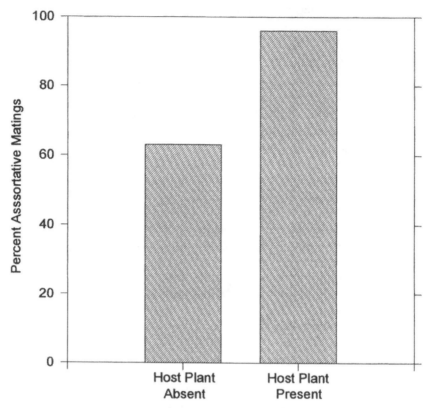

*Figure 15.3* The impact of host-plant cues on assortative mating in *Eurosta solidaginis*. The bars represent the percent of matings that occurred within both fly host races in the presence or absence of their host plant.

matings. The lower the mean distance between host plants, the higher the probability that the host races would encounter each other and mate with each other.

Since each host race mates on its own host plant, an individual mating on the alternative host plant is likely to mate with the other host race. We hypothesized that *Eurosta* are more likely to make "mistakes" in mating-site choice when host plants are mixed together than when they are in blocks. They are also more likely to encounter a mate of the alternate host race. Failure in host fidelity could result from each fly in the population having the same low probability of choosing the alternate host, or because there is variability in host fidelity among individuals in the population. Some individuals may have a strong tendency to mate on the alternate host plant. In either case, the presence of the alternate host is required for this trait to be expressed.

We experimentally altered the pattern of host-species distribution on a very small scale and examined its influence on the frequency of interhost race matings.

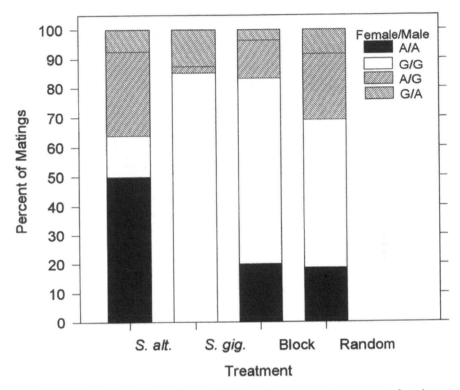

*Figure 15.4* The effects of host-plant availability and arrangement on percent of matings within and between host races of *Eurosta solidaginis*. Abbreviations: A = altissima flies; G = gigantea flies. Treatments consisted of various arrangements of host plants in cage; *S. alt* = 50 *Solidago altissima* plants; *S. gig* = 50 *S. gigantea* plants; Block = 25 *S. altissima* and 25 *S. gigantea* plants arranged in two monospecific blocks; Random = 25 *S. altissima* and 25 *S. gigantea* plants arranged randomly. Values exclude matings on cage.

We placed potted plants of the two host species in 1 × 2 m cages in four different treatments: pure stands of one host, blocks of the two hosts, or a mixture of the two hosts. There was assortative mating in all treatments. However, we found that matings between races were twice as frequent when host plants were mixed randomly as opposed to occurring in blocks (Fig.15.4). We also found that matings between fly races are more than twice as likely to involve gigantea males × altissima females than altissima male × gigantea females. This pattern of mating could be due to low choosiness of altissima females, attractiveness of gigantea males, promiscuity of gigantea males, low fidelity of gigantea males to their natal host, or some combination of these factors. This experiment and others we have conducted indicate that gigantea males have lower fidelity to their host plant than gigantea females or altissima males and females. Therefore, both plant distribution and differences in behavior between host races will affect gene flow between host races. If the degree of spatial separation on this very small scale can influence the frequency of matings between host races, then it is reasonable to assume that larger differences in separation of host species would have an even greater influence on the degree of assortative mating.

### 15.2.1.2 Experience

Gene flow between the host races could occur if either larval or adult experience influences host-species preference. Adult experience could alter host-species preference and assortative mating in two ways. First, if the fly has difficulty locating its own host plant because it is rare, it may switch host preference. Polyphagous herbivores will change their oviposition hierarchy if their preferred host is not available (Singer 1971, 1983; Jaenike 1990). It is possible that the typically monophagous host races (Craig et al. 1993) would change their threshold of what they would accept as a mating or oviposition site if deprived of their normal host. Second, if *Eurosta* finds its own host plant but has low mating success because of the rarity of individuals of its own host plant, it may switch mating-site preference.

We conducted an experiment that measured how the absence of the normal host of the host race could alter host preference and assortative mating. Flies of both host races were placed in cages with either exclusively *S. altissima* or *S. gigantea*. Assortative mating predominated, largely because the host race without access to its host plant was less active (Fig.15.4). There was a strong difference between the host races in their willingness to utilize the alternate host plant. Altissima flies rarely mated on *S. gigantea*. In contrast, gigantea flies frequently mated on the *S. altissima*. Gigantea males would frequently sit on *S. altissima* plants and mate with altissima flies. This resulted in the highest number of nonassortative mating occurring in the pure *S. altissima* cage.

If the environment in which the larva develops influences host preference, then an oviposition mistake can lead to gene flow between the host races. For example, if a gigantea larva developed in an *S. altissima* plant, the larval-conditioning hypothesis would predict that it would prefer to mate and oviposit on *S. altissima*, leading to nonassortative mating. Larval conditioning has long been hypothesized to influence oviposition choice, but there is little evidence that this phenomenon has ever been documented (Mitter and Futuyma 1983).

We have been unsuccessful in testing the larval-conditioning hypothesis in *Eurosta*. However, even if larval conditioning occurs, it is unlikely to be an important source of gene flow. Both host races are unlikely to oviposit in the alternate host plant (Craig et al. 1993); if they do, larvae rarely survive (Fig.15.5). We attempted to induce oviposition on the alternate host by altering adult experience. We placed females on the alternate host plant for two days before offering them a choice of both host plants. Flies very rarely oviposited on the alternate host during either the no-choice or choice segments of the experiment. Even when de-

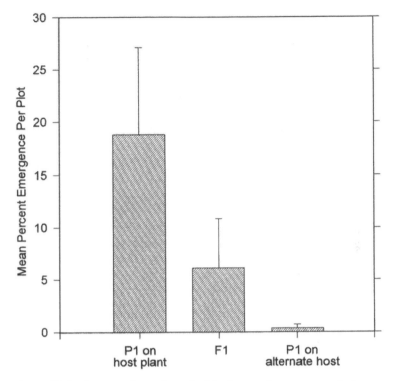

*Figure 15.5* A comparison of survival in the two host races on their own host plant versus survival on the alternative host-plant species and F1 hybrid survival on both host-plant species.

prived of their own host plant throughout their life, most females never oviposited on the alternate host plant (Craig et al. in press). Together, these factors suggest that neither larval conditioning nor oviposition "mistakes" resulting from experience are likely to contribute significantly to gene flow between these populations.

Our studies of the impact of experience on host preference indicate that difference in experience determined by the distribution of host plants could influence gene flow through mating mistakes. Gene flow is most likely when experience causes a deterioration of host fidelity, leading to nonassortative mating. Gene flow between the host races could occur in situations where one host-plant species was a small proportion of the total plants in a field. A fly that dispersed a short distance from its emergence site might be unable to locate either its own host species or mates of its own host race, thus switching host preferences. Long-distance dispersal events could also leave a fly in a situation where it would have no option but to shift host preference to obtain matings. In particular, the relatively low host fidelity of gigantea males is likely to lead to interhost race matings in this situation. Gene flow is not likely to occur due to oviposition mistakes by females.

### 15.2.1.3 Importance of Allochronic Isolation

The incomplete reproductive isolation between the host races is partially due to differences in emergence times (Craig et al. 1993). The gigantea host race emerges before the altissima host race. However, this difference varies among years depending on the weather (Fig. 15.6).

Experiments indicate that fly age influences the probability of matings between the host races. In cage experiments with marked flies, over 93.3% of 60 gigantea females and 91.3% of 23 altissima females mated only once (Craig et al. unpublished data). In these experiments, all of these matings occurred within three days. The females that remated did so when males encountered them when ovipositing. The behavior of the females indicated that these were "forced copulations." Mating times were shorter, and females frequently attempted to groom males off and escape. Some of these matings may have been artifacts of the high fly densities in our cages. We conclude that under normal circumstances, females mate rapidly after emergence and are not likely to remate. As a result, a difference in emergence time of as little as three days could be effective in contributing to reproductive isolation.

As flies aged, they were less likely to mate, which indicates that the interval between emergence times of the host races may be important in reproductive isolation. Since gigantea flies emerge earlier, we conducted an experiment that measured their mating success at different ages. One-half of the gigantea were seven days old, and the other half were newly emerged. Seven days were picked as a typical interval between the emergence of the two host races. These gigantea flies were placed in a cage without the host plant, and with newly emerged altissima flies. All of the flies were unmated at the start of the experiment. Gigantea males again demonstrated their vigor and aggressive mating tendencies. There were equal numbers of old gigantea males, young gigantea males, and young altissima males. Old gigantea males obtained 70% of the matings with altissima females. Gene flow in this direction is possible if gigantea males can survive this long in the field. Flies can live 10–14 days in cage conditions, depending on the temperature and humidity. Mean survival time in the field is difficult to estimate. There were no matings between altissima males and gigantea females, although this mating has occurred in other experiments.

To determine the degree of allochronic isolation, we have measured emergence time over a seven-year period. The differences in emergence times are variable, and the probability of gene flow is higher in some years than in others (Fig.15.6). The differences in emergence times indicate that gigantea flies could survive to mate with altissima in some years. Emergence times tend to be closer together when springs are warm, with a succession of days with above normal

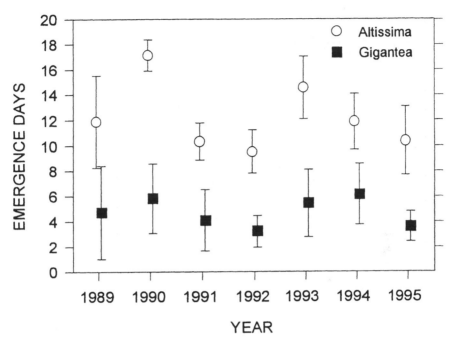

*Figure 15.6* Means and standard deviations of days of emergence are plotted for the two host races for seven years. Emergence for both host races occurred approximately over a total of 20 days. Gigantea flies always emerged before altissima flies, but the difference varied greatly between years.

temperatures. The difference in emergence times of the host races is larger when temperatures are below normal. For example, in 1993, a series of warm days in early in May that initiated gigantea emergence was followed by two weeks with temperatures 10°C below normal that inhibited altissima emergence. The result was a large degree of allochronic isolation in that year. We hypothesize that the host races have different thresholds at which they begin to accumulate degree days. The gigantea host race begins to develop at a lower threshold than altissima. Thus, in cool years, when temperatures are above the threshold for gigantea emergence and below the threshold for altissima emergence, there will be a large difference in emergence times (Fig.15.7). In a year when temperatures rise rapidly above the threshold for development for both host races, there will be little difference in emergence times.

Abrahamson et al. (unpublished, cited in Abrahamson and Weis 1996) conducted a growth-chamber study that supports the hypothesis that the host races respond differentially to temperature variation. Both host races were exposed to a range of temperatures after breaking diapause, and both host races developed

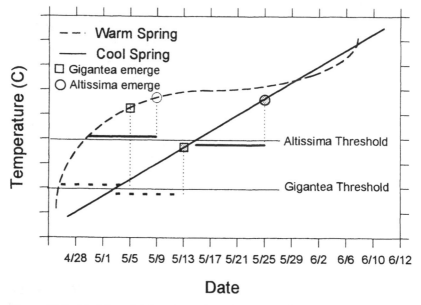

*Figure 15.7* Model explaining the variance in time difference between gigantea and altissima flies' emergence. We hypothesize that the host races have different thresholds at which they begin to accumulate degree days. In this model, metamorphosis from resting-stage larva to adult takes 20 days, once temperatures are above the threshold. In this example, in a warm spring, the host races would emerge on average four days apart, with hybridization between the host races possible, whereas, in a cool spring, the host races would emerge on average 12 days apart, with very little chance for the host races to be active simultaneously.

more rapidly as temperature increased. However, the difference in emergence times of the host races became smaller as temperature increased. At 13°C, the mean emergence of gigantea flies was 14 days earlier than that of altissima, while at 28°C, gigantea emerged only two days earlier.

## 15.2.2 Postzygotic Isolation

Gene flow between host races is influenced by the adaptation of each host race to its own host plant: Neither host race can survive well on the alternate host plant. Hybrids between the host races have low survival rates, impeding gene flow (Fig. 15.8). However, hybrids between the host races survived well in some plots. This led us to hypothesize that poor hybrid survival was not due to genetic incompatibilities of the host races, but due to adaptation of each host race to its own host plant. We hypothesize that hybrids are able to survive in unusual, benign genotypes of each host species. A lack of ability of hybrids to survive on any genotype would indicate that the host-associated populations were species.

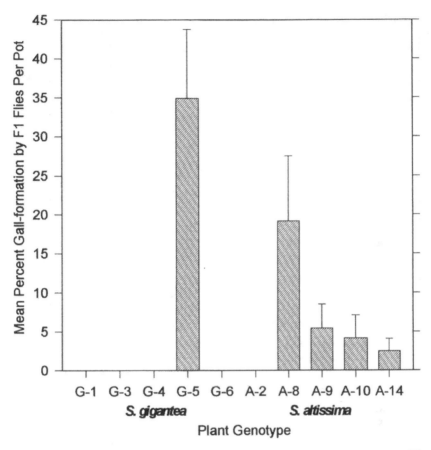

*Figure 15.8* Experimental results showing the impact of host-plant genotype on F1 hybrid flies survival. G = *S. gigantea* genotypes, A = *S. altissima* genotypes.

### 15.2.2.1 Within-Site Variation in Host Plants

The probability of gene flow between the host races is determined by the availability of host plants in which the development of hybrids is possible. Once an interhost race mating has occurred, gene flow will not occur unless plants on which the larva can survive are attacked. The suitability of plant genotypes for development of both *Eurosta* host races and hybrids between the host races is highly variable. We tested the hypothesis that both host plants vary in their suitability for development of different genotypes of each race. We generated 12 full-sib families of each host race. These families were then reared on each of four clonally replicated genotypes of their natal host plant. We found strong interactions between fly genotype (family) and plant genotype on survival (Horner et al. unpublished data).

We tested the hypothesis that the genotypes of both host species varied in their suitability for F1 development. We had F1 hybrid flies oviposit eggs on 20 clonally replicated and potted ramets of five genotypes of *S. gigantea* and five genotypes of *S. altissima*. Gall formation rates differed dramatically among the genotypes in both host plant species (Fig.15.8). On *S. gigantea,* only one of the genotypes produced galls. This genotype produced galls at a rate that is comparable to the rate of gall induction of P1 flies on the most suitable host-plant genotype of their own host species. There were also differences among the rates of gall induction on *S. altissima*. These data suggest that gene flow at a site is dependent on the frequency of susceptible plant genotypes at that site.

The suitability of host plants for hybrid development could be dependent on the natural enemies at a site. Gall size determines the susceptibility of larva to parasitism by *Eurytoma gigantea* (Weis et al. 1985; Weis and Abrahamson 1986). Hybrid galls appear smaller and misshapen. Craig et al. (1994) reported that in 1993, hybrid gall sizes were significantly smaller than the parentals on the same host plant, although Craig et al. (unpublished data) did not find significant differences in size in the 1994 and 1995 generations. Hybrid galls in the experiment shown in Figure 15.8 suffered very high rates of parasitism because of their small gall size. They would have had much higher survival in a site with lower parasitism. Parasitism (Weis et al. 1992), and therefore potential hybrid survival, is highly variable among sites. Parasitism by *Eurytoma obtusiventris* is also highly variable among populations (Brown et al. 1995). Brown et al. found that *E. obtusiventris* had a preference for *S. altissima* galls, the hypothesized ancestral host plant. Its preference for hybrid galls is unknown, but it again has the possibility of influencing gene flow. *Eurytoma obtusiventris* is rare in some areas surveyed by Brown et al., and it is absent from the areas we sampled in Minnesota.

## 15.3 Gene Flow between Populations of *Eurosta solidaginis*

### 15.3.1 Direct and Indirect Methods of Measuring Gene Flow

We can experimentally create combinations where gene flow occurs, but to determine whether these combinations naturally occur, we must measure gene flow in the field. Measuring gene flow among populations is a difficult problem, and the data obtained using any method may be difficult to interpret or contradict information obtained using other methods (Slatkin 1987). Populations can show genetic subdivision for many reasons including genetic drift, selection, and gene flow from other populations. Partitioning the genetic variation among populations into these categories is difficult. Gene flow can be measured using either direct methods or indirect methods. Direct methods involve measuring the dispersal and breeding success of individuals. Indirect methods estimate gene flow by measuring allele frequencies in populations. Direct methods have the advantage of actually being able to measure the movement of individuals among populations at a particular place and time. However, this also means their predictive value is limited in space and time (Slatkin 1987). To directly measure gene flow, it is also critical to measure the breeding success of dispersers, since a disperser has no evolutionary impact unless it leaves offspring in the new habitat. Gene flow can be highly variable geographically and temporally, and rare and unpredictable events can have a large impact on gene flow. Indirect measures cannot quantify the movement of individuals at a particular time and place, but they can provide estimates of gene flow averaged over long periods of time (Slatkin 1987). Another problem with indirect methods of measuring gene flow is that all methods measure only a fraction of genome. The analysis of allozyme variation with electrophoresis measures only a small fraction of the genetic variation in a population. Larger numbers of individuals, loci, geographic sites, and temporal periods should optimally be sampled than are practical (Berlocher 1989). The addition of even a single locus can alter the interpretation of the relationships among populations. Unfortunately, there is no universal rule that can be used to determine when enough loci have been sampled to obtain an accurate picture of gene flow.

We have not directly measured gene flow between populations of *Eurosta* in the field, although we have measured important components of gene flow in experiments. The practical difficulties of mark and recapture experiments with *Eurosta* in the field are immense. The flies are extremely cryptic and inactive for large periods of time. Despite years of searching, we have only rarely seen *Eurosta* ovipositing in the field, and we have never found a mating pair. In experiments where hundreds of *Eurosta* were released in the field to measure dispersal, we were unable to visually find or recapture any individuals. In addition, because we believe that the conditions that facilitate gene flow require a specific combination of conditions, we would have to replicate the study over many years and sites to get an accurate measure of gene flow. Even rare instances of gene flow can be important: It is calculated that even the exchange of one individual per generation will prevent allele fixation in two populations (Slatkin 1987).

Three studies have been conducted to indirectly measure gene flow between populations of *E. solidaginis*. Previously, Waring et al. (1990) used starch-gel electrophoresis to measure populations across a broad geographic range from Minnesota to Maine, and Brown et al. (1996) measured mtDNA across a similar geographic range. We are currently using starch-gel electrophoresis to survey genetic variation at many sites in Minnesota. All three studies have found evidence of population subdivision along host-plant lines and geographic variation within host-associated populations that we will discuss.

### 15.3.2 Variation between Populations in South-Central Minnesota

We are conducting an ongoing study in electrophoretic variation among *E. solidaginis* populations. We found 22 variable loci in an examination of 42 populations. We report here the results from 9 of these loci from 13 populations. The loci reported are those that we can presently score reliably, and for which we have adequate sample sizes (minimum $n = 20$, maximum $n = 43$). We report the results for the following loci: glyceraldehyde-3–phosphate-dehydrogenase, GAPDH; malate dehydrogenase 1 and 2, MDH-1, MDH-2; hydroxy acid dehydrogenase, HADH; isocitrate dehydrogenase, IDH; lactate dehydrogenase, LDH; phosphoglucomutase 1 and 2, PGM-1, PGM-2; and superoxide dimutase, SOD. We anticipate reporting more sites and loci when the full data set is published.

The degree of sympatry may influence gene flow (Fig. 15.2). We collected sympatric populations on the two host plants at six sites and one site with only *S. altissima*. In the sympatric populations, the proportion of altissima flies out of the total fly population varied from 1% to 90%. Because *S. gigantea* and *S. altissima* differ slightly in their habitat requirements, these sites may have different environmental conditions. All of the populations were within 100 km of central Minneapolis, and the greatest distance between any two populations was 155 km. However, climatic and environmental differences may show important variation on this scale.

We analyzed the electrophoretic data with the program BIOSYS-1 (Swofford and Selander 1981). We analyzed the differences among populations using Wright's $F_{ST}$ (Wright 1931, 1978). The $F_{ST}$ value can be used to estimate the degree of genetic subdivision of populations. We used the hierarchical $F_{ST}$, developed by Wright (1978) to partition the differences in $F_{ST}$ caused by geography and host-plant affiliation. To determine if the $F_{ST}$ values were significantly different from zero, we used the method developed by Workman and Niswander (1970).

The results indicated that there is significant differentiation both among and within the host-associated populations. First, there is a large differentiation among altissima populations. The pattern of variation suggests that there may be two species on *S. altissima* from different geographic areas within the region we sampled. Second, there is a significant but smaller amount of variation within the gigantea flies among localities. Third, there are small but significant differences between the host-associated populations, supporting the idea that these are host races. We will discuss each of these patterns in turn.

### 15.3.2.1 Variation between Altissima Populations

There is a large degree of differentiation among the flies from *S. altissima* from different geographic areas. An unweighted pair-group method using an arithmetic average (UMPGA) cluster analysis, using Nei's (1978) unbiased genetic distance measure (Fig.15.9), defined two primary groups, one consisting of the altissima population from Waseca, Minnesota, and the other cluster containing all other populations. The Waseca population contained an allele at the MDH-2 locus that was not found in any other population. No heterozygotes were found between the two alleles at this locus, and so these alleles were not in Hardy–Weinberg equilibrium. There were also strong frequency differences at other loci at Waseca when compared to all other populations. $F_{ST}$ analysis of all altissima populations

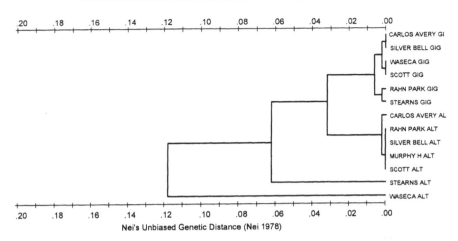

CLUSTER ANALYSIS USING UNWEIGHTED PAIR GROUP METHOD

*Figure 15.9*  A phenogram of 13 *Eurosta solidaginis* populations on *Solidago altissima* and *S. gigantea* from south-central Minnesota. The tree is based upon a hierarchical cluster analysis of Nei's (1978) genetic distance measures for nine variable loci using the unweighted pair-group method with arithmetic averaging of Sneath and Sokal (1973).

showed that there were significant differences among altissima populations at eight of the nine loci (Table 15.1). The mean $F_{ST}$ was 0.102, which is large for a single insect species (McCauley and Eanes 1987; Rank 1992; Costa and Ross 1994). The mean $F_{ST}$ was reduced to half of its former value (Table 15.1) by removing the Waseca population. This value is well within the range typical of within-species variation. Without the Waseca population, the altissima flies from different localities showed significant differences at only three loci (Table 15.1).

When loci show significant uniform differences, it suggests that populations have low gene flow between them, and that the populations have become differentiated by genetic drift. When the differences among loci are highly heterogeneous, then it may indicate that selection is acting on those loci that are highly differentiated (Slatkin 1987; McPheron et al. 1988). The Waseca altissima population shows significant differences at many loci, suggesting that at least some of the Waseca altissima population has been isolated from other populations for a significant period of time. The variation among loci of the other populations suggests that different selection regimes are acting on populations at different sites.

We hypothesize that there may be a new, undescribed species of *Eurosta* on *S. altissima* in the western part of its range. The Waseca population may contain both species *on S. altissima.* Other electrophoretic differences have also been found in populations on *S. altissima* to the west of this site (Itami et al. unpublished data). In addition, Ming (1989) has identified a western subspecies of *E. solidaginis* based

*Table 15.1*  $F_{ST}$ Values for the Two Host Races and the Hierarchical Comparison between the Two Host Races

| HOST | GAPDH | MDH-1 | MDH-2 | HADH | IDH | LDH | PGM-1 | PGM-2 | SOD | Total |
|---|---|---|---|---|---|---|---|---|---|---|
| All *S. altissima* populations | 0.164*** | 0.103*** | 0.327*** | 0.047* | 0.013 | 0.266*** | 0.033*** | 0.019*** | 0.066** | 0.102 |
| All *S. gigantea* populations | 0.075*** | 0.024** | 0.049*** | 0.019 | 0.047*** | 0.015 | 0.015 | 0.024 | — | 0.027 |
| *S. altissima* populations minus Waseca | 0.191*** | 0.058* | 0.015 | 0.020 | 0.010 | 0.040*** | 0.021 | 0.017 | 0.010 | 0.049 |
| Hierarchical | 0.008 | 0.058** | −0.018* | 0.040** | −0.002 | −0.022* | 0.038*** | 0.152*** | 0.003 | 0.055 |

Significance levels are indicated by asterisks: *$p < 0.05$; **$p < 0.05$; ***$p < 0.001$. Gigantea SOD has no value listed as it was fixed at a single allele.

on wing morphology. We have found the western wing pattern predominately on altissima flies from the west of the Waseca site (outside our experimental gall-collection area). The western morphs exist as a small minority to the eastern wing pattern in Minnesota. We have no data to speculate on the origin of the differences at this time, but obviously, host-race formation was not involved.

### 15.3.2.2 Variation between Gigantea Populations

The gigantea populations formed one cluster in the UMPGA cluster analysis and had much smaller genetic distances among populations from different locations than were found among the altissima flies (Table 15.1). There were significant differences among gigantea populations in the $F_{ST}$ at four loci. Again, this heterogeneous response may indicate that selection is acting on these loci differently at different sites.

### 15.3.2.3 Variation between the Host Races

The populations on the two host plants formed distinct clusters in the UMPGA cluster analysis, and they showed significant differences in the $F_{ST}$ at six loci in a hierarchical analysis of variance (Table 15.1). To examine differences between the host-associated populations, we used hierarchical $F_{ST}$ analysis (Wright 1978); this analysis allowed us to partition variance between the host races and among different geographical locations (Table 15.1). The hierarchical analysis indicates that there is significant differentiation between the populations. To determine if comparisons between the host-associated populations were influenced by the inclusion of the unusual Waseca altissima population, we ran the analysis with and without this population. Removing the Waseca population had little influence on the degree of differentiation of the populations: With Waseca included, the mean hierarchical $F_{ST}$ was 0.055, and without Waseca, it was 0.058. The loci that differentiate the host races and Waseca altissima from the remaining altissima, and those that differentiate the host races, are not the same. The Waseca altissima differ from the rest of the altissima at the MDH-2 and LDH loci. The host races were most strongly differentiated at the PGM-2 locus.

We attempted to assess whether gene flow between the host races could vary among sites within a limited geographic area by looking at the $F_{ST}$ between pairs of populations. The sites may have differed in many of the ecological conditions we identified in the first section that could influence gene flow: suitability of host plants for hybrids, the degree of sympatry, the degree of allochronic isolation. In this experiment, we measured only one of these variables: the relative proportions of the two host races in the field. Two observers visually estimated the proportions of each host race in a field at each of our sympatric sites. We then calculated the $F_{ST}$ between the host races at each site. The $F_{ST}$ values covered a wide range from 0.05 to 0.21. There was a significant positive relationship when $F_{ST}$ was regressed on square-root arc-sine transformed percent gigantea ($y = 0.0097 + 0.137x$, $r^2 = 69.1\%$,

$p < 0.05$). This indicates that as the proportion of gigantea in a site increases, the rate of gene flow between populations decreases. A possible mechanism producing this pattern is the low host fidelity of gigantea flies. Our experiments have shown that gigantea flies will move much more readily than altissima flies to the alternative host plant if their own host plant is unavailable. This behavior could produce more inter-host-race matings as gigantea flies became a smaller proportion of the population and had more difficulty locating mates of their host race or mating sites on their own host plant. The result would be higher gene flow as the proportion of gigantea in the population declines. Alternative hypotheses could also explain this pattern. One hypothesis is that the altissima population at Waseca includes a different species and is much more reproductively isolated from all other *Eurosta* populations: By far the largest $F_{ST}$ between pairs of populations at different sites was found at Waseca.

## 15.3.3 Large-Scale Geographic Patterns of Variation

### 15.3.3.1 Origin of the Host Shift

The data provided by the allozyme studies (Waring et al. 1990; Itami et al. unpublished data) and the mtDNA studies (Brown et al. 1996) paint a somewhat contrasting picture of gene flow between the host races and the history of the host shift. Waring et al. (1990) reported significant variation between the host-associated populations. A UMPGA cluster analysis showed that, with one exception, flies from each host plant clustered together regardless of geographic origin. They reported results from six variable loci and found significant variation between the host-associated populations in HBDH (= our HADH) and PGM (= our PGM-1). They found that the altissima population was much more variable than the gigantea population. On this basis, they hypothesized that the altissima population was the ancestral population and gigantea the derived population. In their comparison of populations from New Hampshire, Vermont, Maine, Pennsylvania, Michigan, and Minnesota, they found the populations in the northeast were relatively well differentiated, with the gigantea flies being nearly fixed for one allele at both of these loci. However, in the Midwest, the gigantea flies were more variable at these loci and similar in their frequency to those of altissima flies. Waring et al. suggested that the host shift could have originated in a single location in the Midwest and, as it spread east, lost genetic variation due to the founder effect. Our results pose another alternative hypothesis: The host shift occurred in the East, and the lower degree of differentiation in the Midwest is due to greater gene flow. Our allozyme data indicate that there is limited but significant gene flow between the host races. The $F_{ST}$ for differences between the host races are well within the range found within species and of similar magnitude to the geographical variation within the host races. Additional surveys of molecular variation could reveal markers that would clarify the origin of the host shift and the degree and geographic variation in gene flow between the host races.

Brown et al. (1996), examining populations from Maine to Minnesota, found variation in mtDNA between the host races and among populations on the same host plant. The gigantea flies had lower genetic variation than the altissima flies, supporting the hypothesis that gigantea flies are the derived population. All gigantea, with one exception, had a single haplotype. The altissima shared the same haplotype found in the gigantea in the East, but had different haplotypes in the western part of the range. This pattern led Brown et al. (1996), in contrast to Waring et al. (1990), to hypothesize that the host shift from *S. altissima* to *S. gigantea* occurred in the east and spread west. The fixation of the gigantea haplotype could indicate a lack of gene flow between populations (Feder, personal communication). There are several alternative explanations of this distribution of haplotypes that may be consistent with some gene flow between the host races. First, only three gigantea individuals were haplotyped from Minnesota, where our data suggest a relatively high rate of gene flow. Rare haplotypes indicating gene flow could easily not have been sampled. Second, because of the maternal inheritance of mtDNA, the most likely mode of gene flow would not be detected. In our studies, males are much more likely to mate on the alternative host plant than females. Gene flow from males would not be detected with mtDNA techniques. Finally, there may be selection acting on either the allozymes or mtDNA that would inhibit our ability to detect gene flow (Avise 1994). Selection could be removing altissima haplotypes from the gigantea population, while not altering the allozyme frequencies. The reverse could also be true.

Our results, together with those of Waring et al. (1990), suggest that there is variation in the degree of divergence of the host races in different geographic areas. They reported a mean $F_{ST}$ of 0.438 that is among the highest reported for within-species variation for an insect. However, this value included variation both among localities and between host-plant species, because they did not partition the variation among sites and between plants with a hierarchical $F_{ST}$ as we did. There may be two causes of this larger $F_{ST}$. First, their survey covered a much larger area, and there may have been greater geographic variation within one or both of the host races. Second, the host races may have been more well differentiated in some of the geographic areas they studied. Waring et al. showed that the populations were well differentiated in the East, but less differentiated in the Midwest. They found that many individual loci were fixed or nearly fixed in individual gigantea populations in the East, but that these loci were more variable in the Midwest. In contrast to the eastern pattern, we did not find gigantea to be fixed for most loci in Minnesota. Of the nine variable loci, only one was fixed (frequency of greater than 0.95) at a majority of sites in either host race. Our results confirm this trend for the Midwest: The host races do not show as strong frequency differences in Minnesota as they reported from the Northeast.

Two nonexclusive hypotheses could explain the different degrees of differentiation in the different geographic areas. First, as Waring et al. (1990) suggested, the host shift may have occurred somewhere in the Midwest from *S. altissima* to

*S. gigantea.* As the newly derived gigantea population colonized new areas, it may have lost genetic diversity due to founder effect. A second possibility is that the ecological conditions differ in the different localities, influencing the amount of gene flow between the populations.

## 15.4  Reproductive Isolation of Populations

### 15.4.1  Eurosta solidaginis *in South-Central Minnesota*

What is the current relationship of the host-associated populations of *E. solidaginis* on *S. altissima* and *S. gigantea* in central Minnesota? Answering this question requires a great deal of biological information; we believe that we can answer this for the populations that we have most intensely studied in Minnesota. However, because the relationship between the host-associated populations may vary geographically, we cannot be definitive about the relationship between all populations.

There is no clear-cut means of determining whether host races exist. Host races have been defined in many different ways, and their very existence remains highly controversial (Bush 1994). A determination of whether you have host races depends on how you define host race. We favor the definition by Diehl and Bush (1984), who define host races as being populations that are partially reproductively isolated due to their association with a particular host plant. Host races are an intermediate position on a continuum from undifferentiated populations to well-defined species (Bush 1969, 1975b, 1993b). Host races may occupy very different places on this continuum and we believe that establishing the exact position of populations on this continuum is less important that determining the process by which the populations became differentiated.

Bush (1993a) outlined five criteria for establishing the existence of host races. We believe that these criteria serve as a useful guide to determining the status of the populations, and we tested them in populations in south central Minnesota. Our paraphrases of Bush's (1993a) criteria are indicated in italics below.

1. *Populations are sympatric.* The populations of *Eurosta* on the two species of *Solidago* in central Minnesota are highly sympatric (Craig et al. 1993).

2. *Populations are genetically differentiated.* Results from allozyme electrophoresis (Waring et al. 1990; this chapter), mtDNA (Brown et al. 1996), and behavior (Craig et al. 1993) studies indicate that the populations are differentiated. The question arises: Have they become so differentiated that gene flow is nearly nonexistent and they should be considered species? Our studies support the conclusion that the populations differentiated at the level of host races and not species. First, we have found no fixed differences between the populations in the nine variable loci we have extensively surveyed. In the remaining 12 variable loci on which we have less extensive information, we have also found no indication of a fixation of allelic differences. The migration rate calculated from the hierarchical $F_{ST}$ would correspond to 4.29 migrants between the host-associated popula-

tions per generation, indicating that reproductive isolation is not complete. A single migrant between populations per generation is sufficient to keep populations from diverging significantly (Slatkin 1987). An $F_{ST}$ of 0.055 is well within the range found within species variation (see references cited earlier).

In addition, we have behavioral evidence of hybrids. Field collected flies have a strong preference for that host plant from which they emerged (Craig et al. 1993). Known $F_1$ hybrid flies exhibit a different host preference from presumed pure-host-race flies. In an experiment comparing oviposition preference, 0% of presumed pure-host-race flies and 18% of $F_1$ hybrid flies oviposited on both host plants (Craig et al. unpublished data). Most $F_1$ flies oviposited exclusively on *S. gigantea*. Intermediate-host preference may therefore be an indicator of hybrid status. In other experiments using field-collected flies, a small proportion of presumed pure host race flies demonstrated an intermediate-host preference. For example, Craig et al. (1993) found in two tests of host preference of the presumed host races that 2.5% and 3.0% of these flies oviposited on both host-plant species. This intermediate preference may indicate that there are naturally occurring hybrids in our field collections.

3. *Populations assortatively mate due to host-plant preference.* Craig et al. (1993) found that each host race mates on its own host plant, resulting in assortative mating.

4. *There are trade-offs in fitness, with each population having its highest fitness on its own host plant.* Both host races survive at a higher rate on their own host plant than on the alternate host species (Fig.15.5).

5. There is no evidence of hybrid inviability. Hybrids between the host races from the populations in Minnesota are fully viable and fertile. They mate and oviposit at the same rates as the parental host races (Craig unpublished data). We have mated $F_1$'s to produce fertile and viable $F_2$'s and backcrosses in both directions. However, these flies were from populations that had the smallest genetic distances recorded in a broad geographic sample (Waring et al. 1990). The altissima flies with the largest genetic distances from all other populations, those from Waseca and Stearns, were not used in our experimental crosses. Hybrids do survive at lower rates (Craig et al. unpublished data), but we have shown that lower hybrid survival is due to poor adaptation of the hybrid genotype to most host plants, not to genetic incompatibilities in the hybrid.

The *Eurosta* host races in south-central Minnesota show a higher degree of differentiation in allozymes, the ability to survive on the alternative host, and in host preference, than in the most widely acknowledged example of host race, that *of Rhagoletis pomonella* (see Feder et al., Chapter 16, this volume). The classification of "host race" may cover a wide range of degrees of differentiation from populations that show only a slight degree of genetic subdivision, to those that approach the species level of differentiation. It seems clear that the *Eurosta* populations are farther down the path to speciation than *R. pomonella*, but they have not reached the point of being species yet.

## 15.4.2 Variation in the Status of the Host-Associated Populations in Other Regions and Times

The relationship between the host-associated populations may be at different positions along the host race–species continuum in different regions. We have argued that in our study area in south-central Minnesota, the host-associated populations are host races: There is evidence of limited gene flow between the populations. These populations fall in a region of Figure 15.2 where gene flow can occur. If we studied populations on the two host plants in other areas, would the result be the same? As we have outlined earlier, the degree of differentiation of the *host races* could depend on adaptations to local variation in the host plants, or be due to the patterns of past geographic isolation. The populations that we have termed host races may have diverged to the point that, in some regions, hybrids could suffer from genetic incompatibilities which may deter gene flow. The mechanisms producing assortative mating could be so effective in some areas that reproductive isolation would be nearly complete. In the regions where this is the case, the populations would not fit the criteria for host races that has been proposed by Bush (1993a), and they would be considered separate species. Alternatively, in other regions, assortative mating could be weaker, and the host-associated populations and gene flow may have obliterated differences between the populations.

We do not have the detailed ecological, behavioral, and genetic data to determine the status of the host-associated populations to answer this question in regions other than Minnesota. Preliminary data indicate that the relationship between the host-associated populations may differ geographically. The following data suggest there may be geographic variation in gene flow between the host associated populations:

1. The degree of sympatry of populations varies across the range of *Eurosta solidaginis*. In Pennsylvania, both host-plant species occur, but only the altissima race is found. In the Northeast, both host races occur, but sympatric sites are rare (Brown et al. 1995). In Minnesota, almost all sites have sympatric populations of the host races.

2. The suitability of host plants for *Eurosta* varies geographically. The gigantea population is not found in Pennsylvania, although the host plant is. How et al. (1993) exposed *S. gigantea* from Minnesota and Pennsylvania to attack by gigantea flies from Minnesota. The *S. gigantea* from Minnesota grew more rapidly and were preferred by the gigantea. Larvae developed in both populations of plants, but the galls and larvae were larger on the Minnesota plants.

3. Differences in emergence times have not been measured in different geographic areas. The possibility of differences in the degree of allochronic isolation is strongly suggested by the differential response of the two host races to temperature. If our hypothesized model is valid (Fig.15.7), it would indicate that gene flow may vary geographically as well as annually. Different years have different

weather patterns, leading to different differences in development times. Similarly, different geographic areas with different climatic regimes will have different development times. For example, in some regions, such as Minnesota, spring usually arrives quite suddenly, with temperatures staying cool until temperatures rise rapidly late in May. Under these circumstances, the differences in emergence times would be minimized. There would be only a small interval between the time when the gigantea and altissima thresholds for development were reached. In more moderate climates, the rise in temperatures would be more gradual, and there would be a larger interval between the emergence times between the host races (Fig.15.7). We have shown earlier that development time differed among years with different mean temperatures. Since both host races occur from Minnesota to Virginia, there is likely to be a large variation in the emergence times.

4. Other factors that differ geographically may also have an impact. Natural-enemy attack varies through time and among sites (Weis et al. 1992). Larvae in small galls are more susceptible to parasitoid attack by *E. gigantea* (Weis et al. 1985). If hybrid galls were smaller than those of the pure host races, they would suffer higher rates of parasitism. High parasitism rates could lead to low hybrid survival and low gene flow in some sites. A second parasitoid, *E. obtusiventris*, that might also influence gene flow is absent in Minnesota. It shows a preference for galls on *S. altissima,* and thus may influence the relative fitness of the pure host races and hybrids.

### 15.4.2.1 Genetic Differentiation

1. The degree of genetic variation within populations differs in the Northeast and in the Midwest. Allozyme data indicate that the gigantea populations are more different from altissima populations in the East (Waring et al. 1990). The mtDNA data suggests that the altissima population is more variable in the East (Brown et al. 1996).

2. Allozyme data also suggest that there may be two species of *Eurosta* on *S. altissima,* and that one may have lower gene flow to the gigantea population.

3. The allozyme data also suggest that gene flow may differ between host races among sites in one geographic region.

The study of two host-associated populations that are in a variety of positions along the host race-species continuum in different geographic locations could provide insight into the factors that are important in the evolution of reproductive isolation. Simultaneous studies of genetic and ecological variation could establish correlations between potential reproductive isolating mechanisms and gene flow. The host races of *Eurosta solidaginis* may provide such an opportunity.

## 15.5 Mode of Host-Race Formation

Our studies indicate that the host races could have originated in sympatry, as our data on reproductive isolation meet the assumptions of the sympatric model presented in the introduction. First, reproductive isolation is plant mediated: Assortative mating on the host plant, allochronic isolation, and poor survival hybrids because of adaptation to different host plants are all plant mediated. Second, a scenario requiring few genetic changes would be required for the host shift and speciation. Only a mutation for host preference would evidently be required to initiate the process. The low levels of allozymic divergence also support this contention. Third, there is strong direct selection for reproductive isolation mechanisms: Hybrids have low survival due to their adaptation to different host plants. Fourth, behavioral reproductive isolating mechanisms evidently have originated prior to complete isolation by postzygotic mechanisms. Finally, there is evidence of intermediate stages of divergence of the host races in sympatry.

Our studies suggest the ecological situation that would favor a host shift and divergence of the host races. We agree with the hypothesis (Waring et al. 1990; Brown et al. 1996) that the shift was from populations on *S. altissima* to *S. gigantea*. We have found wide variation among genotypes of both host-plant species in their susceptibility to attack by their own host race, hybrids, and the alternate host race (Craig et al. unpublished data). A host shift would be favored by the presence of *S. gigantea* genotypes on which at least some altissima flies could survive. Because *S. gigantea* genotypes capable of supporting altissima flies are rare, the host shift would be favored by having both host plants within easy dispersal range of each other. This would allow the rare flies with mutations for the ability to survive on *S. gigantea* to have a higher probability of colonizing the new host.

Differentiation after colonization would be favored by the absence of host plants on which hybrids could survive. Differentiation would also be favored by separation of the host species into discrete patches, so that fewer mating mistakes would be made. Equal numbers of both host-plant species would also favor differentiation, because there would be a lower chance of one of the host races failing to find a host plant on which to mate. Climatic conditions that accentuated any genetic differences in emergence time would also favor divergence. We hypothesize that a slow, steady increase in spring temperatures would favor maximum allochronic isolation.

Once populations had differentiated into host races, they would disperse to different sites with different environmental conditions producing different levels of gene flow. Our studies also suggest that there are conditions in which an intermediate stage of convergence can be maintained indefinitely. In the populations we have studied, variation in ecological conditions will create small levels of gene flow. There may not be strong enough selection to eliminate this gene flow. In other sites, ecological conditions could favor further divergence or merging of the populations. The evolutionary fate of the host races, speciation, merging into an undifferentiated species, or maintenance of the current level of differentiation, depends on how ecological conditions change in the future. The present distribution of pure host races and hybrids is dynamic and depends on variation in ecological conditions.

It is impossible to know exactly the conditions that existed when any populations started to diverge into two species, unless we actually observe the divergence. However, as we have argued in this chapter, the recognition that studying the impact of environmental variation on the interaction of recently diverged or diverging populations will give us a window to examine this endlessly controversial and fascinating evolutionary problem.

## Acknowledgments

Special thanks to Dr. James V. Craig for years of meticulous collection of fly emergence data. Warren G. Abrahamson was instrumental in initiating this research project and in helping design the experiments reported here. The electrophoretic analysis and interpretation would not have been possible without the help of Daniel J. Howard, who spent many hours reviewing the gels and data with us. However, any errors are those of the authors alone. The following people helped gather the data used in this chapter: Betty Jo Barton, Cathy Bell-Horner, Dan Buckles, Kathryn Cazares, Jean Craig, Julie Cross, Ruben Marchosky, Heidi Mist, Nigel Mist, and Brian Richardson. We thank the staff and resources support from Cedar Creek Natural History Area in Bethel, Minnesota, especially David Bosanko. Jeffery L. Feder and James T. Costa III provided insightful reviews of an earlier draft that greatly aided in the revision of the chapter. Financial support was provided through grants from the National Science Foundation (BSR-9111433 to TPC and JKI, BSR-9107150 to WGA and JDH, and BSR-8614768 to WGA). Support was also provided through grants from Arizona State University, the Texas Christian University Research Fund, and Bucknell University.

## 15.6 Summary

*Eurosta solidaginis* has formed host races on two species of goldenrod, *Solidago altissima* and *S. gigantea*. We refer to flies from *S. altissima* as altissima flies and those from *S. gigantea* as gigantea flies. Craig et al. (1993) found that assortative mating due to mating on the host plant by each host race and differences in emergence times created partial reproductive isolation between the host plants. We have found that variation in a number of factors has the potential to influence the degree of reproductive isolation between the host races. First, the relative frequency and degree of sympatry of the two host plants determine the opportunities for mating and formation of hybrids. Second, the difference in emergence times between the host races varies, influencing the degree of allochronic isolation and the possibilities for gene flow. Third, host plants of both species vary in their suitability for hybrid development, and the distribution of these host-plant phenotypes will determine the degree of gene flow that is possible. Fourth, the degree of differentiation within host races may influence the amount of gene flow between the host races. All of these factors will interact to determine the degree of gene flow at a particular site, and may vary among sites on different geographical scales. The reproductive isolating mechanisms are consistent with a sympatric origin of the host races. Genetic studies using allozymes and mitochondrial DNA (mtDNA) have supported the classification of the two host-associated populations as host races in at least part of their geographic range. Measures of gene flow suggest that in at least some areas, the populations are not completely reproductively isolated. Altissima flies have a much higher degree of genetic variability than the gigantea fly population, and therefore may be the ancestral population. The clustering together of all gigantea populations suggests that they may all have originated from a host shift in a single location.

## 15.7 References

Abrahamson, W. G., J. M. Brown, S. K. Roth, D. G. Sumerford, J. D. Horner, M. D. Hess, S. Togerson How, T. P. Craig, R. A. Packer, and J. K. Itami. 1986. Gallmaker speciation: An assessment of the roles of host-plant characters, phenology, gallmaker competition, and natural enemies. Pp. 208–221 *in* P. W. Price , W. J. Mattson, and Y. N. Baranchikov (Eds.), *The Ecology and Evolution of Gall-Forming Insects.* USDA Forest Service, North-Central Experiment Station, General Technical Report NC-174.

Abrahamson, W. G. and A. E. Weis. 1996. *The Evolutionary Ecology of a Tritrophic Interaction: Goldenrod, the Stemgaller and Its Natural Enemies.* Princeton University Press, Princeton, NJ.

Avise, J. C. 1994. *Molecular Markers, Natural History, and Evolution.* Chapman & Hall, New York.

Berlocher, S. H. 1989. The complexities of host races and some suggestions for their identification by enzyme electrophoresis. Pp. 51–68 *in* H. D. Loxdale and J. den Hollander (Eds.), *Electrophoretic Studies on Agricultural Pests, Systematics Association Special,* Vol. 39. Clarendon Press, Oxford, UK.

Brown, J. M., W. G. Abrahamson, R. A. Packer, and P. A. Way. 1995. The role of natural-enemy escape in a gallmaker host-plant shift. *Oceologia 104:* 52–60.

Brown, J. M., W. G. Abrahamson, and P. A. Way. 1996. Mitochondrial DNA phylogeography of host races of the goldenrod ball gallmaker, *Eurosta solidaginis* (Diptera: Tephritidae). *Evolution 50:*777–786.

Bush, G. L. 1969. Sympatric host race formation and speciation in frugivorous flies of the genus *Rhagoletis* (Diptera, Tephritidae). *Evolution 23:*237–251.

Bush, G. L. 1975a. Modes of animal speciation. *Annu. Rev. Ecol. Syst. 6:*339–369.

Bush, G. L. 1975b. Sympatric speciation in phytophagous parasitic insects. Pp. 187–206 *in* P. W. Price (Ed.), *Evolutionary Strategies of Parasitic Insects and Mites.* Plenum Press, New York.

Bush, G. L. 1993a. Host race formation and sympatric speciation in *Rhagoletis* fruit flies (Diptera: Tephritidae). *Psyche 99:*335–355.

Bush, G. L. 1993b. A reaffirmation of Santa Rosalia, or why are there so many kinds of small animals. Pp. 229–249 *in* D. R. Lees and D. Edwards (Eds.), *Evolutionary Patterns and Processes.* The Linnean Society of London, London, UK.

Bush, G. L. 1994. Sympatric speciation in animals: New wine in old bottles. *Trends Ecol. Evol.* 9:285–288.

Bush, G. L. and D. J. Howard. 1986. Allopatric and non-allopatric speciation: Assumptions and evidence. Pp. 411–437 *in* S. Karlin and E. Nevo (Eds.), *Evolutionary Processes and Theory.* Academic Press, New York.

Carson, H. L. 1989. Genetic imbalance, realigned selection and the origin of species. Pp. 345–362 *in* L. V. Giddings, K. Y. Kaneshiro, and W. W. Anderson (Eds.), *Genetics, Speciation and the Founder Principle.* Oxford University Press, Oxford, UK.

Costa, J. T. III and K. G. Ross. 1994. Hierarchical structure and gene flow in macrogeographic populations of the eastern tent caterpillar (*Malacosoma americanum*). *Evolution* 48:1158–1167.

Craig, T. P., J. K. Itami, W. G. Abrahamson, and J. D. Horner. 1993. Behavioral evidence for host-race formation in *Eurosta solidaginis.* Evolution 47:1696–1710.

Craig, T. P., J. K. Itami, J. D. Horner, and W. G. Abrahamson. 1994. Host shifts and speciation in gall-forming insects. Pp. 194–207 *in* P. W. Price, W. J. Mattson, and Y. N. Baranchikov (Eds.), *The Ecology and Evolution of Gall-Forming Insects.* USDA Forest Service, North Central Experiment Station, General Technical Report NC-174.

Diehl, S. R., and G. L. Bush. 1984. An evolutionary and applied perspective of insect biotypes. *Annu. Rev. Entomol. 29*:471–504.

Diehl, S. R., and G. L. Bush. 1989. The role of habitat preference in adaptation and speciation. Pp. 345–365 *in* D. Otte and J. A. Endler, (Eds.), *Speciation and its Consequences.* Sinauer, Sunderland, MA.

Felsenstein, J. 1981. Skepticism toward Santa Rosalia, or why are there so few kinds of animals. *Evolution 35*:124–138.

Guttman, S. I. and L. A. Weigt. 1989. Macrogeographic genetic variation in the *Enchenopa binotata* complex (Homoptera: Membracidae). *Ann. Entomol. Soc. Am. 82*:156–165.

How, S. T., W. G. Abrahamson, and T. P. Craig. 1993. Role of host plant phenology in host use by *Eurosta solidaginis* (Diptera: Tephritidae) on *Solidago* (Compositae). *Environ. Entomol. 22*:388–396.

Jaenike, J. 1990. Host specialization in phytophagous insects. *Annu. Rev. Ecol. Syst. 21*:243–273.

Mayr, E. 1963. *Animal Species and Evolution.* Harvard University Press, Cambridge, MA.

Mayr, E. 1988. *Toward a New Philosophy of Biology.* Harvard University Press, Cambridge, MA.

McCauley, D. E. and W. F. Eanes. 1987. Hierarchical population structure analysis of the milkweed beetle, *Tetraopes tetraophthalmus* (Forster). *Heredity 58*:193–201.

McPheron, B. A., D. Courtney Smith, and S. H. Berlocher. 1988. Microgeographic genetic variation in the apple maggot *Rhagoletis pomonella. Genetics 119*:445–451.

Ming, Y. 1989. A revision of the genus *Eurosta* Loew with scanning electron microscopic study of taxonomic characters (Diptera: Tephritidae). M.S. thesis, Washington State University, Pullman, WA.

Mitter, C. and D. J. Futuyma. 1983. An evolutionary–genetic view of host plant utilization by insects. Pp. 427–459 *in* R. F. Denno and M. S. McClure (Eds.), *Variable Plants and Herbivores in Natural and Managed Systems.* Academic Press, New York.

Nei, M. 1978. Estimation of average heterozygosity and genetic distance from a small number of individuals. *Genetics 89*:583–590.

Price, P.W. 1980. Evolutionary Biology of Parasites. Princeton University Press, Princeton, NJ.

Rank, N. E. 1992. A hierarchical analysis of genetic differentiation in a montane leaf beetle *Chrysomela aeneicollis* (Coleoptera: Chrysomelidae). *Evolution 46*:1097–1111.

Rausher, M. D. 1984. The evolution of habitat preference in subdivided populations. *Evolution 38*:596–688.

Rice, W. R. 1984. Disruptive selection on habitat preference and the evolution of reproductive isolation: A simulation study. *Evolution. 38*:1251–1260.

Rice, W. R. 1987. Speciation via habitat specialization: The evolution of reproductive isolation as a correlated character. *Evol. Ecol. 1*:301–314.

Rice, W. R. and E. E. Hostert. 1993. Laboratory experiments on speciation: What have we learned in 40 years? *Evolution. 47*:1637–1653.

Singer, M. C. 1971. Evolution of food-plant preference in the butterfly *Euphydryas editha.* *Evolution 25*:383–389.

Singer, M. C. 1983. Determinants of multiple host use by a phytophagous insect population. *Evolution 37*:389–403.

Slatkin, M. 1987. Gene flow and the geographic structure of natural populations. *Science 236*:787–792.

Sneath, P. H. A. and R. R. Sokal. 1973. *Numerical Taxonomy.* W. H. Freeman, San Francisco, CA.

Swofford, D. L. and R. B. Selander. 1981. BIOSYS-1: A FORTRAN program for the comprehensive analysis of electrophoretic data in population genetics and systematics. *J. Hered. 72*:281–283.

Tauber, C. A. and M. J. Tauber. 1989. Sympatric speciation in insects: Perception and perspective. Pp. 307–344 *in* D. Otte and J. A. Endler (Eds.), *Speciation and Its Consequences.* Sinauer Associates, Sunderland, MA.

Thompson, J. N. 1994. *The Coevolutionary Process.* University of Chicago Press, Chicago, IL.

Uhler, L. D. 1951. Biology and ecology of the goldenrod gall fly, Eurosta solidaginis (Fitch). Memoir 300, Cornell University Agricultural Experiment Station, Ithaca, NY.

Waring, G. L., W. G. Abrahamson, and D. J. Howard. 1990. Genetic differentiation among host-associated populations of the gallmaker *Eurosta solidaginis. Evolution 44*:1648–1655.

Weis, A. E. and W. G. Abrahamson. 1986. Evolution of host-plant manipulation by gall makers: Ecological and genetic factors in the *Solidago–Eurosta* system. *Am. Nat. 127*:681–695.

Weis, A. E., W. G. Abrahamson, and M. C. Andersen. 1992. Variable selection on *Eurosta's* gall size: I. The extent and nature of variation in phenotypic selection. *Evolution* 46:1674–1697.

Weis, A. E., W. G. Abrahamson, and K. D. McCrea. 1985. Host gall size and oviposition success by the parasitoid *Eurytoma gigantea*. *Ecol. Entomol. 10*:341–348.

White, M. J. D. 1978. *Modes of Speciation*. W. H. Freeman, San Francisco, CA.

Wood, T. K. 1980. Divergence in the *Enchenopa binotata* Say complex (Homoptera: Membracidae) effected by host plant adaptation. *Evolution 34*:147–160.

Wood, T. K. and S. I. Guttman. 1983. *Enchenopa binotata* complex: Sympatric speciation? *Science 220*:310–312.

Workman, P. L. and J. D. Niswander. 1970. Population studies in western Indian tribes: II. Local genetic differentiation in the Papago. *Am. J. Hum. Genet. 22*:24–49.

Wright, S. 1931. Evolution in Mendelian populations. *Genetics 16*:97–159.

Wright, S. 1978. *Evolution and the Genetics of Populations: Vol. 4. Variability within and among Natural Populations*. University of Chicago Press, Chicago, IL.

# 16

## Sympatric Host-Race Formation and Speciation in *Rhagoletis* (Diptera: Tephritidae): A Tale of Two Species for Charles D.

*Jeffrey L. Feder*
Department of Biological Sciences, University of Notre Dame, Notre Dame, IN

*Stewart H. Berlocher*
Department of Entomology, University of Illinois, Urbana, IL

*Susan B. Opp*
Department of Biology, California State University at Hayward, Hayward, CA

Sympatric speciation is like the Lernaean Hydra which grew two new heads whenever one of its old heads was cut off. There is only one way in which final agreement can be reached and that is to clarify the whole relevant complex of questions to such an extent that disagreement is no longer possible.

—Ernst Mayr (*Animal Species and Evolution* 1963, p. 451)

### 16.1 It Was the Best of Hosts, It Was the Worst of Hosts?

The question of why there are so many host plant–specific phytophagous insects has long perplexed entomologists, ecologists, and evolutionary biologists alike. In this chapter, we will argue that part of the answer resides in the relationship between host-plant specialization and reproductive isolation. Plants (either different parts, varieties, or species) represent different niches to phytophagous insects. Traits adapting an insect to one species or variety of plant may prevent an insect from efficiently utilizing alternative hosts. Do such host-associated traits also result in reproductive isolation (Walsh 1864; Thorpe 1930; Bush 1966; Futuyma and Keese 1992)? Can isolation evolve as an inadvertent, pleiotropic by-product of a phytophagous insect adapting to a new host plant (Rice 1987; Berlocher 1989; Rice and Hostert 1993)? If so, then the plethora of host specialists is due, at least in part, to numerous plant niches that have imposed divergent selection pressures on phytophagous insects (Hutchinson 1968; Rosenzweig 1978).

The idea that host specialization and speciation are interrelated has a long, contentious history. As early as 1864, Benjamin Walsh proposed that many phytophagous insect varieties arise in the absence of geographic isolation when they attack and adapt to new plants. In particular, Walsh (1867) cited the shift of the apple maggot fly, *Rhagoletis pomonella* (Walsh), from its native host hawthorn (*Crataegus* L. spp.) to domestic apple (*Malus pumila* L.) as an example of an incipient speciation event. Subsequently, the term *host race* has been used to describe this initial stage in the sympatric speciation process, host races being partially isolated, conspecific populations that owe their isolation to host-associated adaptations (Diehl and Bush 1984). Walsh, therefore, not only presented a mechanism to account for the large number of insect specialists but also framed the process in the geographic context of sympatry.

Since Walsh's time, the topics of host specialization and sympatric speciation have remained intertwined. This need not be the case. Populations may also commonly adapt to different host plants in allopatry. There is no reason that host-associated traits that evolve in allopatry should be any less effective in isolating populations than traits that evolve in sympatry, and, of course, there are many who feel that geographic isolation is necessary for divergence (Mayr 1963). The question of whether host specialization can arise in sympatry is therefore somewhat separate from the role it plays in reproductively isolating insect populations. Both questions are important. Establishing that gene pools can be split by ecological factors in the face of gene flow runs counter to the view of many concerning speciation. But the implications go beyond this to whether speciation is often the direct outcome of niche shifts or just a correlated event (i.e., niche shifts increase the likelihood of speciation by increasing the persistence and rate of establishment of isolated populations; Schluter in press).

Our objectives in this chapter are twofold. First, we want to establish that host-plant specialization can act as a strong reproductive barrier between phytophagous insect populations. Second, we want to see whether host-associated traits can evolve in sympatry. We will focus on the apple maggot fly (*R. pomonella*) that prompted Walsh to propose his hypothesis of sympatric speciation, and its closely related sister species the blueberry maggot, *R. mendax* (Curran). We refer the reader to other chapters in this book and several recent reviews (Futuyma and Meyer, 1980; Futuyma and Peterson 1985; Diehl and Bush 1984; Tauber and Tauber 1989; Bush 1992, 1994), as well as work on other insect systems (Wood and Keese 1990; Wood et al. 1990; Waring et al. 1990; Carroll and Boyd 1992; Menken et al. 1992; Craig et al. 1993) for additional information on sympatric speciation.

## 16.2 The Principles: The Apple Fly and the Blueberry Maggot

We begin by introducing our two main characters, the apple and blueberry maggots. Both are members of the *R. pomonella* sibling-species group, which currently contains four described species, the others being *R. zephyria*, which attacks snowberries (*Symphoricarpos*), and *R. cornivora*, which primarily infests the silky dogwood *Cornus amomum* (Bush 1966; Berlocher 1984). Genetic and life-history studies suggest that populations of "*pomonella*-like" flies infesting flowering dogwood (*Cornus florida*) and sparkleberry (*Vaccinium arboreum*) should also be considered separate species (Berlocher et al. 1993; Payne and Berlocher 1995a, 1995b).

The *R. pomonella* group has a number of interesting attributes that have been used as circumstantial evidence for sympatric divergence via host shifts (Bush 1966). These flies are morphologically very similar, yet each species infests a unique, mutually exclusive set of host plants (Bush 1966). All *R. pomonella* group species are indigenous to North America, with taxa being either partially or broadly sympatric in their geographic distributions (Bush 1966). Although several pairs of *R. pomonella* group species can be crossed in the laboratory (Reissig and Smith 1978; Smith 1988a, 1988b; Feder and Bush 1989a; Smith et al. 1993), these species are genetically diverged in nature and do not appear to hybridize often (Berlocher 1995; Berlocher and Bush 1982; Berlocher et al. 1993; Feder et al., 1989a; Feder and Bush 1989a).

The apple maggot is a well-known pest of domestic apples. But apples are not the fly's native host. This distinction belongs to *Crataegus* or hawthorn (Bush 1966; Berlocher and Enquist 1993). Agricultural records compiled from farmers document the shift from hawthorns to apples ~150 years ago in the Hudson Valley region of New York (Bush 1969a). Since that time, apple-infesting populations have spread across much of eastern North America and have recently moved into parts of the West (McPheron 1987, 1989).

It has been suggested by some that apple flies did not originate from sympatric hawthorn-fly populations. For instance, apple flies could have been introduced from a source outside of the known range of *R. pomonella* or shifted to apples from a host other than hawthorns (Carson 1989). Perhaps different races of *R. pomonella* exist on different allopatrically distributed species of hawthorns, and the apple race was derived from one of these races (Carson 1989). As discussed elsewhere (Bush et al. 1989; Feder and Bush 1989b), all available evidence argues against an allopatric origin or nonhawthorn source for apple flies. Genetic studies of fly populations infesting different species of hawthorns from across the United States (*including C. mollis, C. viridis, C. punctata, C. brachyacantha, C. douglasii* and *C. monogyna*) have not revealed the existence of cryptic species or races of hawthorn flies (McPheron, 1987, 1989; McPheron et al. 1988a; Feder et al. 1990a; Berlocher 1976; Berlocher and McPheron 1996).

Another myth to dispel is that the apple race is an "unnatural" system, representing a pest that has adapted to an extensive monoculture crop (Mayr 1992). Commercial apple orchards do not support large populations of flies. *R. pomonella* is too sensitive to insecticide spraying for this to happen. Rather, the primary source of apple flies in the Northeast is feral apples, which often grow in proximity to *Crataegus* in old fields (Feder personal observation). Apple flies, therefore, did not evolve in seclusion from hawthorn flies. It is more accurate to view the genesis of apple flies, as one in which a species was suddenly exposed to a new, open niche within its range.

The mendacious blueberry maggot was aptly named by Curran (1932). For years *R. mendax* was considered a "variety" of the apple maggot (Woods 1915; Patch and Woods 1922; Lathrop and Nickels 1932; Diehl and Prokopy 1986). But allozyme studies have confirmed that *R. mendax* and *R. pomonella* are distinct siblings species, as these flies possess "species-specific" alleles at 11 loci (Feder et al. 1989a; Berlocher and Bush 1982; Berlocher 1995). High- and low-bush blueberries were thought to be the native hosts for *R. mendax*. However, it has recently been found that the fly specializes on another species of endemic blueberry (deerberries, *Vaccinium stamineum*) through much of the Southeast (Payne and Berlocher 1995a; Berlocher 1995). This discovery has broadened the known geographic range of *R. mendax* to the extent that the fly is almost totally sympatric with *R. pomonella* across eastern North America.

## 16.3 The Life Cycle of Apple and Blueberry Maggot Flies

*Rhagoletis pomonella* and *R. mendax* have similar life cycles (see Fig. 16.1 for a diagram for apple and hawthorn flies). Both species are univoltine (Dean and Chapman 1973; Boller and Prokopy 1976). Females lay their eggs into the ripening fruit of the appropriate host, which they identify by specific visual, olfactory,

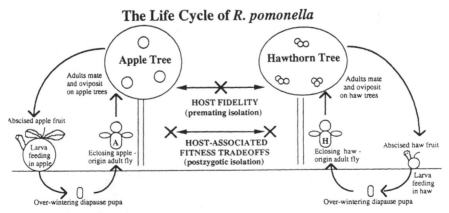

**The Life Cycle of *R. pomonella***

**Q: To what extent do host fidelity and negative genetic fitness tradeoffs reduce gene flow between sympatric apple and hawthorn host races?**

*Figure 16.1* Summary of the life cycle of *R. pomonella*, emphasizing the roles that host fidelity and fitness trade-offs play in isolating apple and hawthorn races of the fly. *R. mendax* has a similar life history, except for having an obligate, rather than facultative, pupal diapause.

and tactile cues (Prokopy 1968a, 1977; Prokopy et al. 1973, 1987; Bush 1969a, 1969b; Moericke et al. 1975; Fein et al. 1982; Owens and Prokopy 1986). Males recognize the same cues, and mating occurs exclusively on or near the fruit of the host (Prokopy et al. 1971, 1972). Studies suggest that *R. pomonella* adults are highly vagile and can travel at least 1.6 km in search of host plants (Maxwell and Parsons 1968). Larval feeding and development, however, is confined to the fruit chosen by a larva's mother for oviposition. When a mature fruit ripens and falls to the ground, larvae leave the fruit and burrow into the soil where, they pupate. Flies overwinter as pupae and usually eclose the following summer in synchrony with the fruiting phenologies of their host plants (Smith 1988b; Feder et al. 1993).

The biologies of *R. mendax* and apple and hawthorn populations of *R. pomonella* differ in important ways that mirror variation in host-plant phenology. The blueberry hosts of *R. mendax* fruit ~ 1–2 weeks earlier than do apple varieties favored by *R. pomonella,* and about a month earlier than hawthorns (Lathrop and Nichols 1932; Payne and Berlocher 1995a, 1995b; Feder personal observation). These phenological differences are reflected in

1.  Blueberry flies eclosing as adults earlier in the season than apple flies, and apple flies eclosing earlier than hawthorn flies (Lathrop and Nickels 1932; Smith 1988b; Feder et al. 1993; Feder personal observation).

2.  Larvae emerging from blueberry fruits and pupating from 1–2 weeks earlier than from apples, and about a month earlier than from hawthorns.

3.  *Rhagoletis mendax* having an obligate diapause, while the apple and hawthorn races of *R. pomonella* are facultative diapausers.

This then is our golden thread: that the interaction between host-plant phenology and fly development is largely responsible for patterns of genetic differentiation and reproductive isolation among *Rhagoletis* flies.

## 16.4 A Model for Sympatric Host-Race Formation in *R. pomonella:* The Wine Shop

Guy Bush (1966, 1969a, 1969b, 1975a, 1975b, 1992, 1994) integrated several aspects of *R. pomonella*'s biology into a model for sympatric speciation. First, Bush argued that because *R. pomonella* mate on their host plants, variation in host preference could establish a system of positive assortative mating between populations infesting alternative plants. We shall refer to host-specific mating under the broader term of *host fidelity,* or the tendency of an insect to mate on and oviposit into the same species of host plant that it utilized as a larva. Host fidelity acts as a premating reproductive barrier. Several factors may contribute to host fidelity, including genetically based differences in preference, allochronic isolation due to asynchrony between fly eclosion and host phenology (Smith 1988b; Feder et al.,

1993, 1994), limited adult movement, and learning or conditioning in adults (Prokopy et al. 1982a, 1986; Papaj and Prokopy 1986).

The second major component of Bush's sympatric speciation model centers on larval survivorship; in particular, that negative genetic-fitness trade-offs associated with larval feeding in host fruits serve as postmating barriers to gene flow. We define *trade-offs* as traits increasing the performance of a fly on one host plant that have detrimental fitness effects on alternative hosts. Such trade-offs are necessary to counterbalance any "leakiness" in host fidelity.

There is currently little empirical evidence for host-related fitness trade-offs in phytophagous insects (Futuyma and Moreno 1988; Jaenike 1990; but see Gould 1979; Mitter et al. 1979; Fry 1990; Karowe 1990; Via 1991; see Mackenzie 1996 for possible exceptions). On the surface, *Rhagoletis* appears to be no exception. One may infer from Bush's emphasis on larval survivorship that *Rhagoletis* flies should be intimately adapted to the chemical and nutritional composition of their host fruits. But this does not appear to be true, as reciprocal transplant experiments performed by Prokopy et al. (1988) gave no indication of any feeding specialization in *R. pomonella*. Larval-to-pupal survivorship was much higher for both apple- and hawthorn-origin flies in hawthorn fruits than in apple fruits, consistent with hawthorns being the ancestral host of *R. pomonella*. These results present a paradox. If apples are such a poor fruit for larval survivorship, then why do *R. pomonella* females not avoid ovipositing into apples altogether?

Part of the solution to this conundrum concerns the concept of enemy-free space (Hairston et al. 1960; Gilbert and Singer 1975; Price et al. 1980; Jeffries and Lawton 1984; Bernays and Graham 1988; Jaenike 1990). Levels of braconid parasitism, interspecific competition (from a number of different moth species and plum curculio weevils), and intraspecific competition are much lower for fly larvae infesting apples than hawthorns (Feder 1995; Feder et al. 1995). These factors were excluded from Prokopy et al. (1988) survivorship estimates. The inferiority of apples as a food resource is therefore counterbalanced by the protection apples afford flies from parasitoids and competitors. A lack of genetic variation for feeding performance related to host-plant chemistry–nutrition may therefore constrain the diet breadth of many phytophagous insects to those host plants, where biotic–ecological factors adequately balance the survivorship equation (Jaenike 1990; Futuyma et al. 1995; Mopper et al. 1995).

An escape from parasitoids or competitors does not necessarily constitute the type of fitness trade-off needed for sympatric host-race formation and speciation. If a hawthorn-origin female were to oviposit into apples, her offspring would receive the same beneficial escape from parasitoids that apple-origin larvae enjoy. So, while enemy-free space can act in a density-dependent manner and lead to a stable polymorphism (e.g., in maintaining host choice for an inferior but underutilized plant), such a polymorphism does not guarantee the existence or evolution of reproductive isolation. Mating may still be random and not host-specific, such that host choice does not result in any isolation (Wilson and Turelli 1986; Wilson 1989). Biotic factors alone are therefore unlikely to affect the frequency with which apple and hawthorn flies mate or to select against "hybrid" offspring. An escape from enemies can help explain why and how *R. pomonella* successfully expanded its diet to include apples but still leaves open the question of negative genetic trade-offs between apple and hawthorn resources.

## 16.5 Evidence Required to Support Sympatric Race Formation and Speciation in *Rhagoletis*

We are now in a position to clarify what information is needed to support sympatric race formation and speciation in the *R. pomonella* group. In general, three criteria must be met:

1.  It must be shown that genetically differentiated and partially reproductively isolated host races of *R. pomonella* actually exist in nature and are formed in sympatry.

2.  It must be established that these races owe their partial reproductive isolation to the very traits that adapt them to their respective host plants.

3.  It must be documented that the same host-associated traits that partially isolate races can continue to evolve in sympatry, resulting in fully isolated sibling species.

The logic behind this approach is that genetic and ecological field experiments can be done on extant races and species to test whether reproductive isolation stems from host-associated adaptation. While it is true that these populations are snapshots in time, they depict stages in a historical process, stages in the ontogeny of divergence. These stages can be actively studied to see whether they fit predictions of the sympatric hypothesis. The known history of *R. pomonella*'s shift to apples is important here, because it makes it probable that any host specialization and isolation that is detected evolved in sympatry. Satisfying the third criterion, that an essentially complete closure of a host-race system not only occurred but happened in sympatry, is the most difficult. Through the analysis of sibling species in the *R. pomonella* group, it is possible to investigate whether the same traits partially isolating apple and hawthorn races of *R. pomonella* are also involved in completely isolating sibling species. This is what motivated our studies of the blueberry maggot. However, evidence that the common ancestor of the clade leading to *R. pomonella* and *R. mendax* split in sympatry and not in allopatry must be inferential. For instance, the ranges of *R. mendax* and *R. pomonella*, as well as their respective native host plants, are currently sympatric in North America. It is unlikely that either these flies or their hosts were ever fully allopatric in the past (microallopatric isolation notwithstanding). However, alternative allopatric scenarios are always possible, although less parsimonious.

## 16.6  Patterns of Genetic Differentiation in *R. pomonella*

The first issue to address is whether apple and hawthorn populations of *R. pomonella* are actually genetically differentiated and partially reproductively isolated host races. To answer this question, we and co-workers conducted a series of allozyme surveys of pairs of sympatric apple and hawthorn populations collected from across the eastern United States (Feder et al. 1988, 1990a, 1990b; McPheron et al. 1988a; Feder and Bush 1989b). The highlights of these studies are as follows:

1. Genetic differences were found between apple and hawthorn populations. Six loci ([*Me*] Malic enzyme, [*Acon-2*] Aconitase-2, [*Mpi*] Mannose phosphate isomerase, [*Dia-2*] NADH-Diaphorase-2, [*Aat-2*] Aspartate amino transferase-2, and [*Had*] Hydroxyacid dehydrogenase) showed consistently significant allele frequency differences between paired apple and hawthorn populations (Feder et al. 1988, 1990a, 1990b; Feder and Bush 1989b; McPheron et al. 1988a). Seven other polymorphic loci displayed little differentiation. We must emphasize that the host races differed only in allele frequencies; no electromorph diagnostically distinguished apple from hawthorn flies, as the races shared even rare alleles in common (Feder et al. 1990a).

2. Only a few regions of the genome differentiated the host races. The six allozyme loci displaying interhost differentiation map to only three different regions on the six chromosomes constituting the *R. pomonella* genome (Berlocher and Smith 1983; Feder et al. 1989b). *Aat-2* and *Dia-2* map together on linkage group I, *Me, Acon-2,* and *Mpi* are tightly linked on group II, and *Had* is on group III (Berlocher and Smith 1983; Feder et al. 1989b). Linkage disequilibrium was found between nonallelic genes within each of these three regions (Feder et al. 1988, 1990a), with linkage disequilibrium referring to the nonrandom association of genes in gametes or haplotypes. No disequilibrium was observed among the three regions displaying host-associated differentiation, suggesting that strong epistatic interactions do not exist across regions. The seven polymorphic allozyme loci not displaying host-associated variation were generally found to be in linkage equilibrium with other loci.

3. Temporal and microgeographic patterns of genetic variation existed within an old field. To determine the temporal (within- and between-years) and microgeographic (among-trees) stability of interhost variation, we analyzed flies collected from an old field near the town of Grant, Michigan over an 11–year period beginning in 1984. The Grant site contains $> 30$ trees each of apples and hawthorns, evenly distributed over an $\sim 0.2$ km$^2$ area that has remained undisturbed since at least 1922 (see Feder et al. 1990b, 1993 for map of site). Four main points emerged from these studies: First, the apple and hawthorn fly populations at the Grant site displayed consistently significant allele frequency differences for *Me 100, Acon-2 95, Mpi 37, Dia-2 100*, and *Aat-2 100* from 1984 until 1994, whereas *Had 100* differed significantly in some years but not others (see Fig. 16.3 for data for *Acon-2*; Feder et al. 1990b, 1993, for data for other loci).

Second, interrace differences persisted across different life-history stages of the fly, including newly eclosing adults, adults collected directly from host trees, and larvae dissected from host fruits (Fig. 16.2; Feder et al. 1993). This is important, because it indicates that apple and hawthorn populations are not completely panmictic, and that disruptive selection is not the sole factor differentiating the races; some form of host fidelity must also be at play.

Third, allele frequency differences between the host races were consistent across individual apple and hawthorn trees (Feder et al. 1990b, 1993). In addition, samples of adults and larvae taken from individual trees on a weekly basis over the course of the 1987 season did not differ greatly in allele frequencies (Feder et al. 1993). Intrahost variation was not completely absent, however. Significant allele frequency variation was observed for several loci among apple trees in a number of different years at the Grant site (Feder et al. 1990b, 1993). McPheron et al. (1988b) have also found significant intrahost genetic variation among hawthorn trees at a site in Urbana, Illinois. But intrahost variation was, at the very least, almost an order of magnitude less than interhost differentiation (Feder et al. 1990b, 1993).

Fourth, significant allele frequency shifts were observed across years at the Grant site, especially for the hawthorn race (see Fig. 16.3 for data for *Acon-2*; Feder et al. 1990b, 1993) .

4. Gene flow is likely between the host races. The most likely cause for the observed pattern of linkage disequilibrium in apple and hawthorn races is disruptive selection coupled with interhost gene flow. Population genetics theory can be used to estimate the level of gene flow ($m$) between apple and hawthorn populations according to the formula (Barton et al. 1988)

$$m \approx \frac{Rr\sqrt{p1q1p2q2}}{\Delta p1 \Delta p2}$$

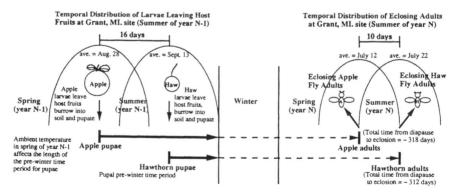

*Figure 16.2* Key life-history differences between apple and hawthorn host races of *R. pomonella*. Data are from the 1987–1992 field seasons at the Grant, Michigan study site. *R. mendax*'s life history is shifted ~ 1–2 weeks earlier than that of the apple race of *R. pomonella*, depending upon the particular study site.

$m$ = % interhost gene flow per generation (% of flies in a race coming from the other race)

$R$ = standardized disequilibrium between nonallelic genes $p^1$ and $p^2$ at loci 1 and 2 within the hawthorn (or apple) race

$r$ = recombination distance between loci 1 and 2

$\Delta p1$ = difference in the frequency of allele $p$ at locus 1 between the host races

$p1$ = frequency of allele $p$ at locus 1 in the hawthorn (or apple) race; $q1 = 1-p1$

Data for the *Aat-2/Dia-2* region of the *R. pomonella* genome from the Grant site in 1985 produced an estimate of $m$ of 4.3 % (*Note:* Barton et al. [1988] originally

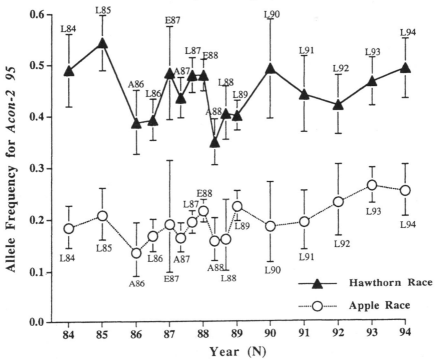

*Figure 16.3* Allele frequencies for *Acon-2 95* over an 11 year period (1984–1994) for different life-history stages of apple and hawthorn flies at the Grant, Michigan site. L = larvae dissected from host fruits. E = newly eclosing adults captured in field traps (nets) constructed beneath host trees. A = adults captured off of host trees. Upper and lower 95% confidence intervals for allele frequencies are indicated by bars. (For data for other loci see Feder et al. 1988, 1990a, 1990b, 1993, submitted).

estimated gene flow to be 20%, but we used more extensive cross-data in our estimate of *r* that were not available at the time of their original calculation). This suggests that a fairly high level of host fidelity exists for *R. pomonella*, but that premating isolation is not complete between apple and hawthorn races.

5. Latitudinal clines exist in both host races. *Me, Acon-2, Mpi, Dia-2, Aat-2,* and *Had* display latitudinal frequency clines among both apple and hawthorn populations (Fig. 16.4; Feder and Bush 1989b; Feder et al. 1990a; Berlocher and McPheron 1996). These clines show several perturbations that coincide with differences in local ambient temperature conditions (Feder and Bush 1989b, 1991; Feder et al. 1990a). The slopes of the frequency clines were also steeper for the hawthorn than the apple race (Fig. 16.4). The geographic pattern is therefore complex, with latitudinal genetic variation within the races superimposed on interhost differences.

Paradoxically, the clines provide additional evidence that the apple race was not formed from a sister taxon to the hawthorn race. If this were true, then we might expect the apple race to form a distinct genetic cluster from the hawthorn race. But it does not. For example, the difference in the slopes of the clines between the races results in hawthorn populations from northern Wisconsin being genetically more similar to apple than hawthorn populations from Illinois.

In contrast to the clinal pattern for *Me, Acon-2, Mpi, Dia-2, Aat-2,* and *Had*, the seven polymorphic loci not displaying host-associated differentiation ([*Pep-2*] Peptidase-2, [*Idh*] Isocitrate dehydrogenase, [*Ak*] Adenylate kinase, [*Pgm*] Phosphoglucomutase, [*Pgi*] Phosphoglucose isomerase, [*Aat-1*] Aspartate amino transferase-1, and [*Acy*] Aminoacylase) showed little geographic variation among apple or hawthorn populations (Feder et al. 1990a; Berlocher and McPheron 1996). This suggests that migration (gene flow) between the host races and among local populations is sufficient to homogenize frequencies for genes not directly experiencing selection or linked to loci under selection, like the aforementioned three regions.

Our allozyme studies also suggest that the frequency clines are primary in origin and due to selection, for if hawthorn flies were geographically separate into different demes for any appreciable time, then we might expect at least some neutral alleles to have drifted to different frequencies in the demes. Gene flow following secondary contact produced the clines in the hawthorn race: clines that would be mirrored in the apple race due to gene flow between local hawthorn and apple populations. But this is not the case. True, the apple and hawthorn races show clines for the same loci, but the slopes of the clines differ for apple and hawthorn flies. Furthermore, the clines display several shifts that correspond to local environmental conditions. The data, therefore, suggest that selection, not history, is primarily responsible for the pattern of allozyme variation.

6. Environmental and genetic correlates exist at the Grant site. Allele frequencies for *Me, Acon-2, Mpi, Dia-2,* and *Aat-2* in the hawthorn race were significantly related to ambient temperature conditions at the Grant site from 1984 to 1994

# Growing Degree Days versus Average Allele Frequency

## Michigan / Indiana transects

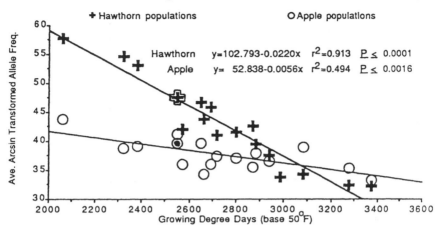

## Wisconsin / Illinois transects

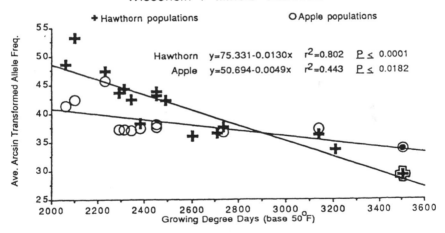

*Figure 16.4* Average arcsine transformed allele frequency for the six loci, *Me 100, Acon-2 95, Mpi 37, Dia-2 100, Aat-2 100,* and *Had 100,* in apple- and hawthorn-fly populations plotted against growing degree days base 50°F (GDD = Average of the daily high and low temperatures above 50°F) for collecting sites along latitudinal transects sampled through Michigan/Indiana and Wisconsin/Illinois. A total of 34 different sites were analyzed in the survey. Outlined symbols in the Michigan/Indiana and Wisconsin/Illinois graphs are the Grant, Michigan and Urbana Illinois sites, respectively. GDD values are 30- or 50-year averages, as compiled by the National Weather Service. (For data for individual loci, see Feder and Bush 1989b).

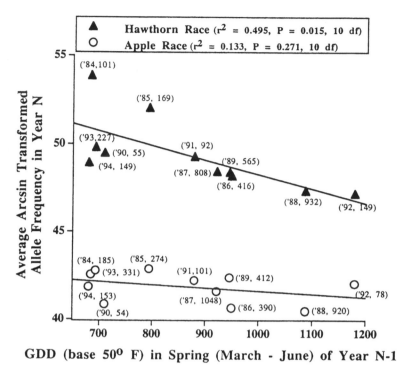

*Figure 16.5* Average arcsin transformed allele frequencies for the six loci, *Me 100, Acon-2 95, Mpi 37, Dia-2 100, Aat-2 100,* and *Had 100,* in apple and hawthorn races at Grant, Michigan from 1984 to 1994 (designated year *N*) plotted against growing degree days (GDD base 50°F) in the spring months (March–June) of the preceding year (designated year *N* − 1). Linear regressions ($r^2$ and *p* values) are given in legend. Year numbers (*N*) and sample sizes (# of individuals genetically scored) appear in parentheses. (For data for individual loci see Feder et al., submitted).

(Fig. 16.5; Feder et al. 1993). The apple race showed similar trends, but the regressions were not significant.

7. Developmental/genetic correlates exist at the Grant site. Allele frequencies for the six loci showing interhost differentiation correlate with the timing of adult eclosion for both apple and hawthorn flies at the Grant site (see Fig. 16.6 for *Me 100* results; Feder et al. 1993).

The allozyme studies show that genetically differentiated and partially reproductively isolated host races of *R. pomonella* exist and can be maintained in sympatry. The results suggest that ambient temperature and, by inference, host phenology is involved in differentiating the races—our first inkling of the axis along which disruptive selection may act on *Rhagoletis* flies.

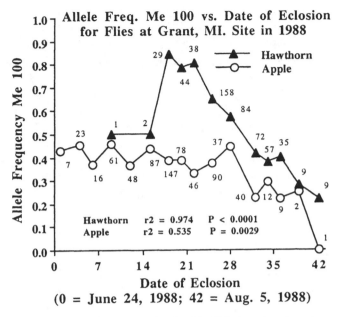

*Figure 16.6* Allele frequencies for *Me 100* plotted against dates that samples of apple and hawthorn flies eclosed at the Grant, Michigan site in 1988. Newly eclosing adults were collected every other day from field traps (nets) constructed beneath host trees. Results are therefore indicative of the natural eclosion patterns of flies in the field. Linear regressions ($r^2$ and $p$ values) and sample sizes (# of individuals genetically scored) are also given. Zero = June 24, 1988; 42 = August 5, 1988. (For data for other loci, see Feder et al. 1993.)

## 16.7 Host Fidelity in *R. pomonella* Races: Evidence from Mark and Recapture Studies

Much of the foundation for behavioral studies in *Rhagoletis* was established by Ron Prokopy's group, who showed that *R. pomonella* flies court and mate on or near host fruits. They also found that naive apple- and hawthorn-origin adults differ in their host-acceptance behaviors (Prokopy et al. 1988). Interestingly, both apple and hawthorn females will readily accept hawthorn fruits for oviposition, consistent with hawthorns being the ancestral host of *R. pomonella*. But the two races display a clear difference with respect to apples. Apple flies will accept apples. Hawthorn flies, on the other hand, have an aversion for apples. It therefore appears that apple flies have evolved the ability to recognize apples as a new host fruit, while retaining their predilection for hawthorns. While these results point to a genetically based difference in host preference between the races, they also reveal a possible difficulty with Bush's host fidelity model. If apple flies readily accept hawthorns, then what stops them from moving to hawthorn trees?

In order to resolve the issue of host fidelity, we conducted three mark and re-capture studies at the Grant field site in 1991 and 1992 (see Feder et al. 1994 for complete details). We shall refer to these experiments as the *field release* study, the *host switch* study and the *net release* study (Figs. 16.7a–c). The rationale be-hind the *field release* study (Fig. 16.7a) was to simultaneously release "naive" apple and hawthorn flies in the middle of the study site near neither host species and monitor their subsequent distribution on apple and hawthorn trees. Multiple release days spaced through the season were used to minimize the effects of phe-nology, so the fly distribution on hosts would primarily depend on genetically based differences in preference. In the *host switch* study, we released adults of both races under their own and the other race's host trees (Fig. 16.7b). Data from the *host switch* experiment therefore provided a second test for genetic differ-ences in host preference, as well as a means for estimating host fidelity based on the combined effects of preference and eclosion under the "correct" host species. In this context, adult eclosion under the correct host species encompasses both limited dispersal and adult conditioning. The goal of the *net release* study was to measure overall levels of gene flow, taking into account all potential factors af-fecting host fidelity, including allochronic isolation (Figs. 16.7c).

### 16.7.1 The Field Release Experiment

Results from the *field release* study implied that genetically based differences in preference contribute to host fidelity. We calculated that the relative preference of hawthorn flies for hawthorn over apple trees in the field experiment was 91.2%, whereas for apple flies toward apple trees it was 55.2%. It is interesting that these relative preference values coincide well with Prokopy et al. (1988) behavioral as-says of host choice; hawthorn flies display an aversion to apples, whereas apple flies are more willing to accept either host.

The *field release* experiment suggested that genetically based differences in host preference reduce gene flow to ~ 33% per generation from the apple into the hawthorn race and ~ 14% in the reverse direction (Feder et al. 1994). The only possible nongenetic explanation for the data is preimaginal conditioning. But such imprinting has not been convincingly documented for any phytophagous in-sect, including *R. pomonella* (Papaj and Rausher 1983; Futuyma and Peterson 1985; Prokopy et al. 1982b, 1988; Courtney and Kibota 1990; Jaenike 1990).

### 16.7.2 Host Switch Experiment

Data from the *host switch* experiment suggested that eclosing under apple trees significantly reduced the movement of apple-origin flies to hawthorn trees. When this "correct host effect" was taken into account, gene flow from the apple to the hawthorn race was only ~ 9% per generation. The same was not true for hawthorn flies, however, as the estimated level of gene flow from the hawthorn to the apple

# Mark and Recapture Experiments at the Grant, MI. site

**a.) Field Release Experiment**

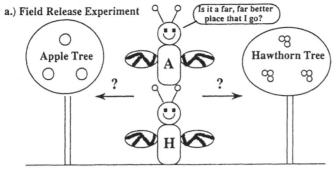

Concept:  - Release naive adults in middle of study site under neither host tree .
          - Where they end up tests for innate host preference differences between the races.

**b.) Host Switch Experiment**

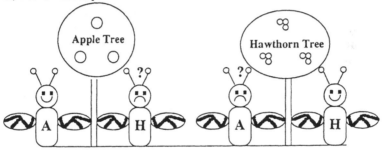

Concept:  - Release naive apple - and hawthorn - origin adults under both host trees .
          - Represents a second test for genetically based host preference differences between races.
          - Allows for eclosion under the "correct" host tree to be factored into host fidelity estimates.

**c.) Net Release Experiment**

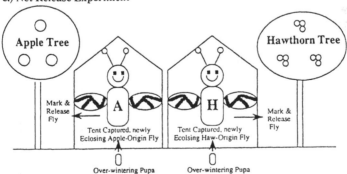

Concept:  - Capture naive adults in field tents (nets) as they eclose, mark and release them.
          - Release schedule for flies therefore mirrors natural eclosion patterns of host races.
          - This allows allochronic isolation to be factored into host fidelity estimates.

*Figure 16.7*  A synopsis of the mark and recapture studies conducted at the Grant, Michigan study site during the 1991 and 1992 field seasons.

race from the *host switch* study (15%) was almost identical to the value from the *field release* experiment (14%).

What is responsible for the correct host effect for apple flies? Both limited adult dispersal and adult conditioning could be involved. Overall, 18.6% (82/440) of apple flies released under apple trees and recaptured in the *host switch* study were migrants (i.e., flies that left their release trees). Many of these migrants were caught on trees hundreds of meters away. Apple flies, therefore, have a fair dispersal capacity and are always within the "cruising range" of hawthorn flies (Mayr 1963). Data from the *host switch* study also implied that adult conditioning was not involved. If we consider the movement patterns of apple flies released under apple trees in the *host switch* study, 60.8% of emigrating flies moved to a different apple tree, whereas 39.2% moved to a hawthorn tree ($n = 82$ total migrants). These values are in close agreement with the relative preference estimate of 61.1% for apple flies toward apple over hawthorn trees derived from nonmigrant flies in the *host switch* study. Eclosing beneath an apple tree, therefore, had little effect on whether a migrating apple fly subsequently chose to alight on an apple or hawthorn tree. Similarly, 76.7% of apple flies released under hawthorn trees returned to apple trees, whereas 23.3% immigrated to a different hawthorn tree ($n = 162$ total migrants). This is also inconsistent with naive adults learning to reject novel host species. Instead, these data imply an overriding influence of genetically based host preference. The "correct host effect" observed in the *host switch* study therefore appears to be due to apple flies remaining on their apple-release (eclosion) trees when suitable fruit is available, and not to adult conditioning, or to adults being inherently sedentary.

## 16.7.3 The Net Release Experiment

The *net release* study was designed to take into account all potential factors affecting host fidelity, including allochronic isolation. Results from 1991 and 1992 confirmed that overall levels of host fidelity are high, but not absolute, as inter-host movement averaged 5.5% per generation (1991 = 5.05 %, 1992 = 5.95%, mean = 5.5% ± 0.45% S.E.).

## 16.7.4 Take-Home Messages from the Mark and Recapture Studies at the Grant Site

First, host fidelity acts as a fairly effective premating barrier between apple and hawthorn flies, confirming a critical part of Bush's sympatric model. Second, we were able to partition host fidelity into its seminal elements. With respect to apple flies, the most important factor turned out to be eclosing from beneath apple trees, followed by a genetically based preference for apples and then allochronic isolation. In contrast, allochronic isolation and a genetically based preference for hawthorns were the two main contributors to host fidelity for the hawthorn race.

Surprisingly, eclosing under hawthorn trees had little effect on host fidelity for hawthorn flies. This is probably due to the fact that ripe hawthorn fruits are not abundant when a majority of hawthorn flies eclose from under hawthorn trees. Hawthorn flies are therefore likely to disperse from their eclosion tree, returning to hawthorn trees after the flies have reached sexual maturity and hawthorn fruits are suitable for oviposition. This is reflected in the observation that 32.5% (135/416) of all hawthorn-origin flies released under hawthorn trees in the *host switch* experiment were recaptured on a different tree, compared to only 18.6% of apple-origin flies released under apple trees (*G*-contingency test = 21.7, $p <$ 0.0001, 1 *df*).

Third, the mark–recapture studies revealed how reproductive isolation can arise in sympatry as a pleiotropic by-product of host-associated adaptation. In particular, traits related to host preference and the timing of adult eclosion were shown to contribute to premating isolation.

A key unresolved question is whether the mark and recapture studies we performed at the Grant site are pertinent to other *R. pomonella* populations. Results from a field study by Luna and Prokopy (1995) suggest that host fidelity is a general feature of *R. pomonella*. These authors present data on the host-acceptance behaviors of apple and hawthorn flies at a field site near Amherst, Massachusetts, that coincide very well with our findings from the *host switch* experiment. Host fidelity therefore appears to be similar for *R. pomonella* flies from two widely separated sites.

But is host fidelity the only premating barrier? Could apple and hawthorn flies be ethologically isolated? Given that a fly alights on the "wrong" host, will it mate with other flies on that tree? To answer this, we compared frequencies that marked apple and hawthorn flies were observed mating on host trees. Although apple flies mated slightly more often than hawthorn flies on apple trees (3.99% compared to 3.58%), this difference was not significant (*G*-contingency test = 0.10, $p = 0.67$, 1 *df*). Similarly, hawthorn flies mated more often than apple flies on hawthorn trees (6.76% compared to 7.89%), but again this difference was not significant (*G*-contingency test = 0.50, $p = 0.48$, 1 *df*). Consequently, there is no clear evidence for ethological isolation between the races.

Further reductions in interhost gene flow could also result from the reluctance of females to oviposit into the fruit of the alternative host species. Marked apple-origin females were observed ovipositing into apples more often than hawthorn-origin females (16.4% [52/317] compared to 11.3% [15/133]), but the difference was not significant (*G*-contingency test = 2.03, $p = 0.16$, 1 *df*). Marked females of both races were observed ovipositing into hawthorn fruits at nearly identical frequencies (10.6% [18/170] for apple-origin females compared to 10.9% [29/267] for hawthorn-origin females; *G*-contingency test = 0.008, $p = 0.93$, 1 *df*).

Finally, one could argue that sterility or fertility barriers exist that further curtail gene flow. However, experimental crosses have given no indication for such barriers (Reissig and Smith 1978; Smith 1988b). Consequently, our estimate of 5.5% gene flow derived from the mark–recapture study would seem to accurately reflect levels in nature, and it is in good agreement with our earlier estimate of 4.3% based on the allozymes.

## 16.8 Negative Genetic Trade-Off in the *R. pomonella* Host Races: Still Knitting

An important implication of the allozyme and mark–recapture studies is that although host fidelity greatly reduces gene flow between apple and hawthorn races, enough gene flow still occurs to genetically homogenize the races in the absence of some form of host-related selection. We have built a *prima facie* case that it is the interaction of host phenology, local ambient temperature conditions, and fly development that is responsible for this postzygotic selection. However, we have been rather vague as to the details of how this may work.

One imperative for a *Rhagoletis* fly is to coordinate its eclosion with the phenology of its host plant, such that when the fly reaches sexual maturity, host fruits are in prime condition for oviposition. As we discussed earlier, the eclosion patterns of apple and hawthorn flies, as well as *R. mendax*, mirror the phenologies of their hosts. Smith (1988b) has shown that eclosion time differences between the host races are genetically based. Reciprocal F1 "hybrids" between apple and hawthorn flies had intermediate eclosion times compared to their respective parents, with apple flies eclosing the earliest. These results were taken to show that seasonal asynchrony is sufficient both to initiate and maintain restricted gene flow between the races (Smith 1988b). The observation that allozymes correlate with the timing of eclosion lends some credence to this claim (Fig. 16.6; Feder et al. 1993). But as we found in the mark–recapture studies, host fidelity reduces genetic exchange between apple and hawthorn races to ~ 5.5% per generation: allochronic isolation stemming from eclosion time differences was included in this estimate. Therefore, seasonal mating asynchrony alone cannot account for the continued differentiation of the races.

Another important consequence of the early phenology of apples is that fly larvae emerge from apples and pupate at the Grant site an average of 16 days earlier in the season than they do from hawthorns (Fig. 16.2). Apple flies are therefore exposed to more growing degree days and longer photoperiods before they overwinter than are hawthorn flies. Because *R. pomonella* are facultative diapausers, apple flies that develop too rapidly in the summer run the risk of bypassing diapause and developing into adults. Almost all *R. pomonella* larvae do fail to diapause when held at a temperature above 28°C (Prokopy 1968b), and small second generations of apple flies have been reported in the field (Caesar and Ross 1919; Porter 1928; Phipps and Dirks 1933). Nondiapausing flies are inevitably doomed: either they eclose at times when suitable host fruit is no longer available or they are committed to, but do not complete, adult development before the onset of winter and subsequently freeze to death. Selective pressures are different for hawthorn flies. The relatively late phenology of hawthorns means that slow-developing hawthorn flies run the risk of not entering diapause quickly enough before the first frost.

The temporal pattern of genetic variation at the Grant site supports the diapause hypothesis. The significant regressions for the hawthorn race that we eluded to earlier were between growing degree days in the spring of year $N - 1$ and allele frequencies the following year ($N$; Fig. 16.5). Because *R. pomonella* are univoltine, hawthorn flies collected in year $N$ represent larvae and pupae that survived the preceding summer and winter (year $N - 1$). Spring temperature is a major determinant of the length and quality of the growing season. High spring temperatures generally mean an early and long field season, whereas the converse is true following cold springs. The selective effects of the growing season were reflected in genotype-specific mortality between larvae in year $N - 1$ and the adult hawthorn flies in year $N$. In warm seasons in year $N - 1$, allele frequencies in year $N$ resembled those found in southern populations (Fig. 16.5). Conversely, following cold seasons, allele frequencies at the Grant site resembled those in more northern populations. The diapause hypothesis can therefore account for both genetic differentiation between the races and allele frequency clines within the races, since the length of the growing season and host-plant phenology vary latitudinally. Or can it? The diapause hypothesis is based on the premise that hawthorn flies, on average, develop faster than apple flies. But a study by Smith (1988b) indicated that apple flies from Illinois are genetically programmed to eclose earlier than hawthorn flies. Also, data from field eclosion traps at the Grant site showed that apple flies eclosed an average of 10 days earlier than hawthorn flies (Feder et al. 1993; Feder 1995). How could this be when the apple flies should eclose later than hawthorn flies based on their lower frequencies at *Me 100, Acon-2 95, Mpi 37, Dia-2 100, Aat-2 100,* and *Had 100* loci?

Part of the answer to the eclosion time paradox is that the flies in this study (Smith 1988b) study were from south-central Illinois. These sites are located south of where allele frequency clines cross for the races (Fig. 16.4). Allozyme frequencies for hawthorn flies in this part of Illinois are more similar to the apple than the hawthorn population at the Grant site (Fig. 16.4). Apple flies from Illinois should develop faster and eclose earlier than hawthorn flies from Illinois based on their higher *Me 100, Acon-2 95, Mpi 37, Dia-2 100, Aat-2 100,* and *Had 100* frequencies, and they do. By the same line of reasoning, however, apple flies should eclose later than hawthorn flies in Michigan, since they have lower *Me 100, Acon-2 95, Mpi 37, Dia-2 100, Aat-2 100,* and *Had 100* frequencies than the hawthorn race here. However, the earlier chronology of apple-fly eclosion at the Grant site suggests that this is not the case. Recall that larvae emerge from apple fruits and pupate ~ 16 days earlier than they do from hawthorns in year $N - 1$

(Fig. 16.2). Yet the mean eclosion date for apple adults the following summer (year $N$) is only 10 days earlier than that for hawthorn flies (Fig. 16.2). Consequently, the developmental period from pupation to eclosion is actually almost a week longer for apple flies at the Grant site (16 days earlier to pupate minus only 10 days earlier to eclose; Fig. 16.2), consistent with the diapause hypothesis.

We have extensive circumstantial support for the diapause trade-off hypothesis, but such evidence does not constitute proof. Selection experiments testing for genetic responses to varied environmental conditions would greatly bolster our case. These selection experiments are currently underway in which we systematically vary the time period between pupation and overwintering for flies, as well as the length of the overwintering period. Preliminary results match the predictions of the developmental trade-off hypothesis. Longer pre-and over-wintering treatments (i.e., conditions simulating early fruiting host plants) selected for genotypes more common in the apple than the hawthorn race at the Grant site.

An important message from studies of host races is that it may be unwise to focus exclusively on larval feeding if one is interested in detecting trade-offs. It can be critical for an insect to grow quickly on its host plant, but the life history of the insect must also be in synchrony with the phenology of its host plant. Developing too rapidly in one stage of the life cycle can upset the timing of other stages. In *R. pomonella*, for example, rapid pupal development can disrupt the match between adult eclosion and host availability. Failure to properly consider the interplay of development, host phenology, and microclimatic conditions could help explain why there are so few examples of host-associated trade-offs for phytophagous insects (Rausher 1988, 1992).

## 16.9 Patterns of Genetic Differentiation in *R. mendax*

*Rhagoletis mendax* has a different pattern of geographic variation than *R. pomonella*. For a 17-locus allozyme set resolved for both species, the composite $F_{ST}$ value among *R. mendax* populations across the eastern United States was 0.015 compared to 0.148 for *R. pomonella* (Berlocher 1995; Berlocher and McPheron 1996). There is no evidence for latitudinal clines in *R. mendax* (Berlocher 1995). In addition, there appears to be no host-associated variation among *R. mendax* populations infesting different *Vaccinium* host species (Berlocher 1995).

The paucity of geographic variation in *R. mendax* came as a surprise. It cannot be explained by a lack of allozyme polymorphism in the blueberry maggot, as *R. mendax* and *R. pomonella* have comparable average heterozygosities (*R. men.* = 0.176 ± 0.038; *R. pom* = 0.221 ± 0.044) and numbers of alleles per locus (*R. men.* = 2.3 ± 0.2; *R. pom.* = 2.7 ± 0.3; Berlocher, 1995). Blueberries require acidic soils. Consequently, *R. mendax* should have a patchy distribution that should promote differentiation through genetic drift. But this is apparently not the case.

## 16.10  Field Studies of Host Fidelity Involving *R. mendax*

An important criterion of Bush's sympatric speciation model is that the same host-associated traits partially isolating host races can evolve completely to isolate species. We have seen that host fidelity causes premating isolation between apple and hawthorn host races of *R. pomonella*. Does host fidelity play a role in reproductively isolating *R. mendax* and *R. pomonella*?

The unique allozyme alleles possessed by *R. pomonella* and *R. mendax* permit a direct test of host fidelity in these flies. Using the genotype of a fly for the loci possessing species-specific alleles, it is possible to type a majority of adults as either *R. mendax* or *R. pomonella,* and thereby infer whether an individual infested either an apple, hawthorn, or blueberry as a larva. Genetic analysis of field-captured adults from sympatric sites would therefore provide a convenient measure of host fidelity. If reproductively active *R. pomonella* and *R. mendax* adults frequently come into contact on host plants, then strong ethological premating or postmating isolation must also be involved in maintaining the genetic integrity of these species.

Field experiments involving *R. mendax* and *R. pomonella* were conducted at a study site near the town of Chickaming (a.k.a. Sawyer), Michigan (Feder and Bush 1989a). This site is ideal because not only are apple trees and cultivated high bush blueberries found together, but an apple tree is actually in physical contact with a row of blueberry bushes. We therefore collected adult flies from the "microsympatric" apple tree and blueberry bushes at the Chickaming site on July 7, 16, and 23, 1987. Later in the summer, on August 10, larvae were dissected from infested apple and blueberry fruits within the same area from which the adults were sampled.

Genetic analysis of the field-captured adults ($n = 114$ flies from the apple tree, $n = 130$ flies from blueberries) gave no evidence of any movement between apples and blueberries (Feder and Bush 1989a). Not a single adult *R. mendax* or *R. pomonella* was captured on the wrong host plant. In addition, all 85 larvae dissected from blueberries were genetically *R. mendax,* and all 120 larvae from apples were *R. pomonella.*

The allozyme study suggests that host fidelity is strong in *R. mendax* and *R. pomonella.* Blueberry and apple maggots from the Chickaming site, as well as laboratory-created F1 hybrids, also respond differently to host-fruit volatiles in oviposition and electroantennal studies (Bierbaum and Bush 1988, 1990a; Frey and Bush 1990, 1996; Frey et al. 1992). It is therefore likely that genetically based differences in host preference exist for these species. However, the results are just from one site and may not be indicative of the entire range of these flies. Furthermore, host fidelity was assessed only between apples and blueberries; hawthorns were not considered. While these criticisms are valid, allozyme studies of larvae collected from blueberries, apples, and hawthorns from across the eastern

United States give no indication that host fidelity is lax (Feder et al. 1989a; Berlocher 1995).

Nevertheless, just because host fidelity exists in extant populations does not mean that it played a role in speciation, for it may have evolved after blueberry and apple flies split. Ethological or postmating sterility barriers could have originally isolated these flies. Perhaps nutritional and/or chemical differences between blueberries and apples also select against larvae infesting the wrong fruit, as well as F1 hybrids. Our study shows only that genetic divergence is currently being maintained between *R. mendax* and *R. pomonella*, because host-preference differences ensure that these flies rarely come into contact in nature. The study tells us nothing about what would happen if they did happen to meet on a fruit. If, however, we could show that *R. mendax* and *R. pomonella* flies readily hybridize and produce viable and fertile offspring when they meet, then this would argue for an important role for host fidelity in speciation. Field and laboratory hybridization experiments were therefore also conducted on Chickaming flies to investigate the possibilities of ethological isolation and postmating sterility (Feder and Bush 1989a). In summary:

1. Mating experiments performed in the field using wild-captured adults gave no sign of any ethological isolation (Feder and Bush 1989a).

2. Hybrid viability and survivorship (Feder and Bush 1989a) are sufficient, such that if blueberry and apple maggot flies were ever to mate, then their offspring would be easily detected by genetic analysis. We have yet to score a definitive hybrid individual in surveys of North America (Feder et al. 1989a; Berlocher 1995; Berlocher and McPheron 1996).

3. Interspecific sperm competition cannot explain the lack of hybrids in nature. Field-captured females used in laboratory hybridization experiments produced only genetically homospecific offspring prior to the addition of heterospecific males to mating cages (Feder and Bush 1989a). These females therefore carried homospecific sperm with them into the lab from prior matings in the field. Hybrid offspring were produced immediately after heterospecific males were added to the cages. This indicates that homospecific sperm cannot completely exclude heterospecific sperm from fertilizing eggs. Females also did not simply run out of homospecific sperm in their spermathecae, because genetically homospecific offspring were recovered from crosses up until the last day of the experiment.

All available evidence therefore suggests that host fidelity was involved when the ancestor of *R. mendax* and *R. pomonella* split into new species. Of course, we cannot be certain of the plants involved in this split or its geographic context, but other isolating mechanisms cannot adequately account for the lack of gene flow between the species. Juerg Frey (personal communication) found that although F1 hybrids are fertile and produce viable F2 and backcross offspring, the fecundity of F1 hybrids can be reduced by up to 50% compared to pure crosses. Postmating isolation therefore further reduces the likelihood of successful introgression between *R. pomonella* and *R. mendax*. F1 hybrids are far from sterile, however, and reduced hybrid fecundity cannot explain the lack of F1-hybrid larvae in nature. So, although further studies of interspecific fertility and hybrid breakdown are needed, these studies will not detract from the importance of host fidelity in the divergence of blueberry and apple flies.

## 16.11  Fitness Trade-Offs in the Blueberry Maggot

We have seen that host fidelity can evolve to completely reproductively isolate *Rhagoletis* sibling species. Host fidelity is also involved in partially isolating the apple and hawthorn host races of *R. pomonella*. We saw that developmental trade-offs stemming from differences in host phenology also appear to maintain the genetic integrity of the host races. Is there any compelling evidence for host specialization in *R. mendax*?

We begin by asking whether *R. mendax* and *R. pomonella* larvae are differentially adapted to any chemical or nutritional differences in their host plants. Unfortunately, the only available data are from larval transplant experiments between high bush blueberries (*V. corymbosum*) and apples using *R. mendax* and the apple race of *R. pomonella* (Bierbaum and Bush 1990b). This is unfortunate because apples are the derived host of *R. pomonella,* and Prokopy et al. (1988) showed that apples are an inferior resource for the fly. In addition, it is not certain that high bush blueberry is the ancestral host of *R. mendax* (Payne and Berlocher 1995a). The ideal comparison would therefore have been between deerberry, which is the more widely distributed host of *R. mendax*, and hawthorns. Still, *R. mendax* survivorship was higher in blueberries than in apples, but *R. mendax* larvae did only slightly, albeit significantly, worse than *R. pomonella* larvae in apples (Bierbaum and Bush 1990b). Furthermore, *R. pomonella* larvae did equally well in blueberries and apples. F1 hybrids between *R. pomonella* and *R. mendax* either had intermediate or lower survivorships compared to pure parental types when reared in apples or blueberries. The results are once again equivocal for larval feeding trade-offs related to fruit chemistry. While it is true that blueberry larvae did best in blueberries, this did not preclude them from a fair showing in apples relative to apple flies. Furthermore, there appear to be no barriers to the apple maggot fly in utilizing blueberries as a host. It would be interesting to see how parasitoids and competition balance the survivorship equation in blueberries, as *R. mendax* is heavily parasitized (Lathrop and Nichols 1932; Berlocher unpublished data).

Do other aspects of the ecology of blueberries hold the key to *R. mendax* specialization? As we discussed earlier, the *Vaccinium* species that *R. mendax* infests have earlier fruiting phenologies than apples. *Vaccinium* fruits are also often small and rot quickly once they abscise from plants. These factors may put conflicting constraints on *R. mendax* development. Fly larvae must develop rapidly to leave fruits before blueberries decay in quality. But if flies develop too rapidly and enter a pupal stage too early in the field season, then they run the risk of bypassing diapause and developing directly into adults. *Rhagoletis mendax* appears to have circumvented this problem by decoupling development rates in larval and pupal life-history stages. Bierbaum and Bush (1990b, unpublished data) have shown that *R. mendax* develop faster as larvae and pupate at smaller body masses than *R. pomonella*. We hypothesize that to combat the resulting nondiapause problem, *R. mendax* has evolved an obligate diapause.

Why has the apple race not evolved an obligate diapause? A likely reason is the balance between pupal diapause and the timing of adult eclosion. Flies in very deep diapause also eclose significantly later than other flies (Feder et al. unpublished data). Consequently, we suspect that obligate-diapausing apple flies would eclose very late in the field season, well after apples have peaked in abundance.

Preliminary data indicate that life-history differences between *R. mendax* and *R. pomonella* are genetically based (facultative diapause appears to be dominant to obligate diapause, whereas larval development rates and pupal size are partially dominant traits; Bierbaum and Bush 1990b; Berlocher unpublished data). These results again underscore the importance of host phenology as an axis for host specialization in *Rhagoletis*. They also reveal a possible role for fruit decay as a selective force on *Rhagoletis* larvae. Further work on the genetics of developmental differences between *R. pomonella* and its related siblings could therefore prove quite fruitful.

We can now return to the question of why allozyme variation is more pronounced among *R. pomonella* than *R. mendax* populations. The six loci displaying host-related differences in *R. pomonella* either code for or are linked to genes affecting development rates, and these allozymes are attuned to local temperature conditions. In contrast, *R. mendax*'s need for rapid larval development and the early phenology of blueberries has resulted in an obligate pupal diapause. Such a life style apparently obviates *R. mendax* from genetically tracking local climatic conditions, at least with respect to the allozymes. In *R. pomonella*, it is primarily only the allozymes related to race formation that show marked geographic differentiation. If these loci are discounted, then *R. pomonella* also displays limited geographic variation. In retrospect, the paucity of variation for *R. mendax* is not that surprising. The only conundrum is the perceived patchy distribution of its host plants. It would be interesting to see whether other DNA markers, which, *a priori*, we would assume to be neutral, display much higher levels of site-to-site variation in *R. mendax* than the allozymes. If they do, then this would suggest that the constancy of allozyme frequencies among *R. mendax* populations may be due to balancing selection.

## 16.12 A Synopsis and Synthesis of the Data: The Knitting Done or Just Begun?

We see from *R. mendax* and *R. pomonella* that host-plant specialization can act as a strong reproductive barrier between phytophagous insects. We see that host fidelity essentially eliminates gene flow between sympatric blueberry and apple-fly populations. We see that other host-associated differences between these two sibling species related to larval development and diapause also appear to act as effective postmating barriers to introgression, although more work is needed in this area. There is no reason to believe that what we see in *R. mendax* and *R. pomonella* should be limited to just these two taxa or to the genus *Rhagoletis*. A single suite of multipurpose life-history traits and behaviors may often be an unstable (inflexible) compromise to the divergent selective pressures exerted by alternative host plants, especially for an univoltine insect. The result is host-plant specialization and the potential for speciation. Why do we see so many phytophagous insect specialists? One reason is because there are so many different host plant niches that pull insect populations in different evolutionary directions.

We also see from apple and hawthorn flies that host-associated traits can maintain the genetic integrity of races in the face of gene flow, and that these traits and races are likely to have arisen in sympatry. Whether such traits will continue to evolve in sympatry to cause the "closure" of a host-race system is open to debate. Clearly, host specialization can evolve to the point that it completely isolates populations. Also, the biogeographic data for *R. pomonella, R. mendax,* and their siblings suggest that closure can occur in sympatry. But since we cannot see into the distant past, our argument must be one of probability, not certainty, as with all speciation work. The sympatric hypothesis is currently the most parsimonious explanation for divergence in the *R. pomonella* group. But this would change if we were to find that the most closely related sibling to *R. pomonella* is allopatric. At the present time, this is not the case; the flowering dogwood and sparkleberry flies are sympatric with *R. pomonella*. Additional surveys of North America are still needed until we can definitively say that we have identified all members of the group.

Are apple and hawthorn races of *R. pomonella* moving toward permanent separation? Certainly the partial reduction in gene flow already present will facilitate the fixation of more subtle, host-related adaptations. Such a runaway process has been postulated by Rice and Hostert (1993) as a model for rapid, nonallopatric speciation via resource specialization. The apple and hawthorn races could therefore be a speciation event waiting to happen, contingent upon new, host-specific mutations (Berlocher 1989). Perhaps, however, the races have run their course and no further differentiation is possible, given the current ecological status quo. For while it is true that apples and hawthorns differ in their phenologies, apple and hawthorn fruits still overlap for a fair period of time at most field sites. As the field season progresses, apples become a comparatively poorer resource for *R. pomonella*. It does not seem logical to us that apple-origin flies should completely ignore hawthorns late in the field season, considering that apple flies are not behaviorally excluded from recognizing hawthorns as potential hosts, and larval survivorship is high for apple flies in hawthorns. The same holds true for early eclosing hawthorn flies that are active before hawthorns ripen. Perhaps what is needed for sympatric speciation in the *R. pomonella* group is a slightly larger seasonal gulf between host-plant "islands" than is provided by apples and hawthorns, such as that between blueberries and hawthorns. Or alternatively, host plants such as apples could serve as a temporal atoll in a plant archipelago, opening up the possibility of future movement to new hosts with earlier phenologies. Only time will tell.

In conclusion, we see that Darwin (1859) appears justified in his view that speciation is often simply part of the process of evolution by natural selection. In the case of the *R. pomonella* group, divergence following a host shift may just happen faster than Darwin imagined. Benjamin Walsh also seems to have known what the dickens he was talking about when he said that certain phytophagous insects speciate in sympatry when they shift and adapt to new host plants. But aren't two heads, after all, better than one?

## Acknowledgments

Thanks to all those who have assisted in *Rhagoletis* projects over the years, and whose names would fill this whole volume. Special thanks to Guy and Dorie Bush, and Marty Kreitman, however, for nurturing and supporting much of the research described in this chapter.

## 16.13 References

Barton, N. H., J. S. Jones, and J. Mallet. 1988. No barriers to speciation. *Nature 336*:13–14.

Berlocher, S. H. 1976. The genetics of speciation in *Rhagoletis* (Diptera: Tephritidae). Ph.D. dissertation, University of Texas at Austin, TX.

Berlocher, S. H. 1984. A new North American species of *Rhagoletis* (Diptera: Tephritidae), with records of host plants of *Cornus*-infesting *Rhagoletis. J. Kan. Ent. Soc. 57*:237–242.

Berlocher, S. H. 1989. The complexities of host races and some suggestions for their identification by enzyme electrophoresis. Pp. 51–68 *in* H. D. Loxdale and J. den Hollander (Eds.), *Electrophoretic Studies of Agricultural Pests. Syst. Assoc. Spec.,* Vol. 39. Oxford University Press, Oxford, UK.

Berlocher, S. H. 1995. Population structure of *Rhagoletis mendax*, the blueberry maggot. *Heredity 74*:542–555.

Berlocher, S. H. and G. L. Bush. 1982. An electrophoretic analysis of *Rhagoletis* (Diptera: Tephritidae) phylogeny. *Syst. Zool. 31*:136–155.

Berlocher, S. H. and M. Enquist. 1993. Distribution and host plants of the apple maggot fly, *Rhagoletis pomonella* (Diptera: Tephritidae) in Texas. *J. Kan. Entomol. Soc. 66* (1):51–59.

Berlocher, S. H. and B. A. McPheron. 1996. Population structure of *Rhagoletis pomonella*, the apple maggot fly. *Heredity, 77*:83–99.

Berlocher, S. H., B. A. McPheron, J. L. Feder, and G. L. Bush. 1993. Genetic differentiation at allozyme loci in the *Rhagoletis pomonella* (Diptera: Tephritidae) species complex. *Ann. Entomol. Soc. Am. 86* (6):716–727.

Berlocher, S. H. and D. C. Smith. 1983. Segregation and mapping of allozymes of the apple maggot fly. *J. Hered. 74*:337–340.

Bernays, E. and M. Graham. 1988. On the evolution of host specificity in phytophagous arthropods. *Ecology 69* (4):886–892.

Bierbaum, T. J. and G. L. Bush. 1988. Divergence in key host examining and acceptance behaviors of the sibling species *Rhagoletis mendax* and *R. pomonella* (Diptera: Tephritidae). Pp. 26–55 *in* M.T. AliNiazee (Ed.), *Fruit Flies of Economic Importance: Bionomics, Ecology and Management.* Oregon State University Press, Corvallis, OR.

Bierbaum, T. J. and G. L. Bush. 1990a. Host fruit chemical stimuli eliciting distinct ovipositional responses from sibling species of *Rhagoletis* fruit flies. *Entomol. Exp. Appl. 56*:165–177.

Bierbaum, T. J. and G. L. Bush. 1990b. Genetic differentiation in the viability of sibling species of *Rhagoletis* fruit flies on host plants, and the influence of reduced hybrid viability on reproductive isolation. *Entomol. Exp. Appl. 55*:105–118.

Boller, E. F. and R. J. Prokopy. 1976. Bionomics and management of *Rhagoletis. Annu. Rev. Entomol. 21*:223–246.

Bush, G. L. 1966. *The Taxonomy, Cytology and Evolution of the Genus* Rhagoletis *in North America (Diptera: Tephritidae).* Museum of Comparative Zoology, Cambridge, MA.

Bush, G. L. 1969a. Sympatric host race formation and speciation in frugivorous flies of the genus *Rhagoletis* (Diptera: Tephritidae). *Evolution 23*:237–251.

Bush, G. L. 1969b. Mating behavior, host specificity, and the ecological significance of sibling species in frugivorous flies of the genus *Rhagoletis* (Diptera: Tephritidae). *Am. Nat. 103*:669–672.

Bush, G. L. 1975a. Sympatric speciation in phytophagous parasitic insects. Pp. 187–206 *in* P. W. Price (Ed.), *Evolutionary Strategies of Parasitic Insects and Mites.* Plenum Press, New York.

Bush, G. L. 1975b. Modes of animal speciation. *Annu. Rev. Ecol. Syst. 6*:339–364.

Bush, G. L. 1992. Host race formation and sympatric speciation in Rhagoletis fruit flies (Diptera: Tephritidae). *Psyche 99* (4):335–358.

Bush, G. L. 1994. Sympatric speciation in animals: New wine in old bottles. *Trends Ecol. Evol. 9*:285–288.

Bush, G. L., J. L. Feder, S. H. Berlocher, B. A. McPheron, D. C. Smith, and C. A. Chilcote. 1989. Sympatric origins of *R. pomonella. Nature 339*:346.

Caesar, L. and W. A. Ross. 1919. The apple maggot. Ontario Department of Agriculture Bulletin. No. 275.

Carroll, S. P. and C. Boyd. 1992. Host race radiation in the soapberry bug: Natural history with the history. *Evolution 46*:1052–1069.

Carson, H. 1989. Sympatric pest. *Nature 338*:304.

Courtney, S. P. and T. T. Kibota. 1990. Mother doesn't know best: Selection of hosts by ovipositing insects. Pp. 161–188 *in* E. A. Bernays (Ed.), *Insect/Plant Interactions,* Vol. 2. CRC Press, Boca Raton, FL.

Craig, T. P., J. K. Itami, W. G. Abrahamson, and J. D. Horner. 1993. Behavioral evidence for host-race formation in *Eurosta solidaginsis. Evolution 47*:1696–1710.

Curran, C. H. 1932. New North American diptera with notes on others. *Amer. Mus. Nov. 526*:1–13.

Darwin, C. 1859. *On the Origin of Species by Means of Natural Selection.* John Murray, London.

Dean, R. W. and P. J. Chapman. 1973. Bionomics of the apple maggot in eastern New York. Search Agric. Entomol. Geneva No. 3. Geneva, New York.

Diehl, S. R. and G. L. Bush. 1984. An evolutionary and applied perspective of insect biotypes. *Annu. Rev. Entomol. 29*:471–504.

Diehl, S. R. and R. J. Prokopy. 1986. Host-selection behavior differences between the fruit fly sibling species *Rhagoletis pomonella* and *R. mendax* (Diptera: Tephritidae). *Ann. Entomol. Soc. Am. 79*:266–271.

Feder, J. L. 1995. The effects of parasitoids on sympatric host races of the apple maggot fly, *Rhagoletis pomonella* (Diptera: Tephritidae). *Ecology 76*:801–813.

Feder, J. L. and G. L. Bush. 1989a. A field test of differential host plant usage between two sibling species *Rhagoletis pomonella* fruit flies (Diptera: Tephritidae) and its consequences for sympatric models of speciation. *Evolution 43*:1813–1819.

Feder, J. L. and G. L. Bush. 1989b. Gene frequency clines for host races of *Rhagoletis pomonella* (Diptera: Tephritidae) in the Midwestern Unites States. *Heredity 63*:245–266.

Feder, J. L. and G. L. Bush. 1991. Genetic variation among apple and hawthorn host races of *Rhagoletis pomonella* (Diptera: Tephritidae) across an ecological transition zone in the Mid-Western United States. *Entomol. Exp. Appl. 59*:249–265.

Feder, J. L., C. A. Chilcote, and G. L. Bush. 1988. Genetic differentiation between sympatric host races of *Rhagoletis pomonella. Nature 336*:61–64.

Feder, J. L., C. A. Chilcote, and G. L. Bush. 1989a. Are the apple maggot, *Rhagoletis pomonella*, and the blueberry maggot, *R. mendax*, distinct species? Implications for sympatric speciation. *Entomol. Exp. Appl. 51*:113–123.

Feder, J. L., C. A. Chilcote, and G. L. Bush. 1989b. Inheritance and linkage relationships of allozymes in the apple maggot fly. *J. Hered. 80*:277–283.

Feder, J. L., C. A. Chilcote, and G. L. Bush. 1990a. The geographic pattern of genetic differentiation between host associated populations *of Rhagoletis pomonella* (Diptera: Tephritidae) in the eastern United States and Canada. *Evolution 44*:570–594.

Feder, J. L., C. A. Chilcote, and G. L. Bush. 1990b. Regional, local and micro-geographic allele frequency variation between apple and hawthorn populations of *Rhagoletis pomonella* in western Michigan. *Evolution 44*:595–608.

Feder, J. L., T. A. Hunt, and G. L. Bush. 1993. The effects of climate, host plant phenology and host fidelity on the genetics of apple and hawthorn infesting races of *Rhagoletis pomonella. Entomol. Exp. Appl. 69*:117–135.

Feder, J. L., S. Opp, B. Wlazlo, K. Reynolds, W. Go, and S. Spisak. 1994. Host fidelity is an effective pre-mating barrier between sympatric races of the apple maggot fly. *Proc. Natl. Acad. Sci. USA 91* (17):7990–7994.

Feder, J. L., K. Reynolds, W. Go, and E. C. Wang. 1995. Intra- and interspecific competition and host race formation in the apple maggot fly, *Rhagoletis pomonella* (Diptera: Tephritidae). *Oecologia 101*:416–425.

Feder, J. L., J. B. Roethele, B. Wlazlo, and S. H. Berlocher. The selective maintenance of allozyme differences between sympatric host races of the apple maggot fly, *submitted.*

Fein, B. L., W. H. Reissig, and W. L. Roelofs. 1982. Identification of apple volatiles attractive to the apple maggot, *Rhagoletis pomonella. J. Chem. Ecol. 8*:1473–1487.

Frey, J. E., T. J. Bierbaum, and G. L. Bush. 1992. Differences among sibling species *Rhagoletis mendax* and *R. pomonella* (Diptera: Tephritidae) in their antennal sensitivity to host fruit compounds. *J.Chem. Ecol. 18*(11):2011–2024.

Frey, J. E., and G. L. Bush. 1990. Rhagoletis sibling species and host races differ in host odor recognition. *Entomol. Exp. Appl. 7*:123–131.

Frey, J. E. and G. L. Bush. 1996. Impaired host odor perception in hybrids between the sibling species *Rhagoletis pomonella* and *R. mendax. Entomol. Exp. Appl., 80*:163–165.

Fry, J. D. 1990. Trade-offs in fitness on different hosts: Evidence from a selection experiment with a phytophagous mite. *Am. Nat. 136*:569–580.

Futuyma, D. J. and M. C. Keese. 1992. Evolution and coevolution of plants and phytophagous arthropods. Pp. 439–475 *in* G. A. Rosenthal and M. R. Berenbaum (Eds.), *Herbivores: Their Interaction with Secondary Plant Metabolites,* Vol. 2. Academic Press, San Diego, CA.

Futuyma, D. J., M. C. Keese, and D. J. Funk. 1995. Genetic constraints on macroevolution: The evolution of host affiliation in the leaf beetle genus *Ophraella.* Evolution 49:797–809.

Futuyma, D. J. and G. C. Meyer. 1980. Non-allopatric speciation in animals. *Syst. Zool.* 29:254–271.

Futuyma, D. J. and G. Moreno. 1988. The evolution of ecological specialization. *Annu. Rev. Ecol. Syst. 19*:207–233.

Futuyma, D. J. and S. C. Peterson. 1985. Genetic variation in the use of resources by insects. *Annu. Rev. Entomol. 30*:217–238.

Gilbert, L. E. and M. C. Singer. 1975. Butterfly ecology. *Annu. Rev. Ecol. Syst. 6*:365–397.

Gould, F. 1979. Rapid host range evolution in a population of the phytophagous mite *Tetranycchus urticae* Koch. *Evolution 33*:791–802.

Hairston, N. G., F. E. Smith, and L. B. Slobodkin. 1960. Community structure, population control and competition. *Am. Nat. 94*:421–425.

Hutchinson, G. E. 1968. When are species necessary? Pp. 177–186 *in* R. C. Lewontin (Ed.), *Population Biology and Evolution.* Syracuse University Press, Syracuse, NY.

Jaenike, J. 1990. Host specialization in phytophagous insects. *Annu. Rev. Ecol. Syst. 21*:243–273.

Jeffries, M. J. and J. H. Lawton. 1984. Enemy-free space and the structure of biological communities. *Biol. J. Linn. Soc. 23*:269–286.

Karowe, D. N. 1990. Predicting host range evolution: Colonization of *Coronnilla varia* by *Colios philodice* (Lepidoptera: Pieridae). *Evolution 44*:1637–1647.

Lathrop, F. H. and C. B. Nickels. 1932. The biology and control of the blueberry maggot in Washington County Maine. Tech. Bull. No. 275, United States Department of Agriculture, Washington, DC.

Luna, I. and R. J. Prokopy. 1995. Behavioral differences between hawthorn-origin and apple-origin *Rhagoletis pomonella* flies in patches of host trees. *Entomol. Exp. Appl. 74*:277–282.

Mackenzie, A. 1996. A trade-off for host plant utilization in the Black Bean Aphid, *Aphis fabae. Evolution 50*:155–162.

Maxwell, C. W. and E. C. Parsons. 1968. The recapture of marked apple maggot adults in several orchards from one release point. *J. Econ. Entomol. 61*:1157–1159.

Mayr, E. 1963. *Animal Species and Evolution.* Harvard University Press, Cambridge, MA.

Mayr, E. 1992. Controversies in retrospective. Pp. 1–34 *in* D. J. Futuyma and J. Antonovics (Eds.), *Oxford Surveys in Evolutionary Biology,* Vol. 8. Oxford University Press, New York.

McPheron, B. A. 1987. The population genetics of the colonization of the western United States by the apple maggot, *Rhagoletis pomonella* (Walsh) (Diptera: Tephritidae). Ph.D. dissertation, University of Illinois, Urbana, IL.

McPheron, B. A. 1989. Patterns of genetic variation in western apple maggot populations and its implications for understanding population biology. *In* R. V. Dowell (Ed.), *Apple Maggot in the Western United States: History, Biology and Control.* University of California Press, Berkeley, CA.

McPheron, B. A., D. C. Smith, and S. H. Berlocher. 1988a. Genetic differences between *Rhagoletis pomonella* host races. *Nature 336*:64–66.

McPheron, B. A., D. C. Smith, and S. H. Berlocher. 1988b. Microgeographic genetic variation in the apple maggot, *Rhagoletis pomonella. Genetics 119*:445–451.

Menken, S. B. J., W. M. Herrebout, and J. T. Wiebes. 1992. Small ermine moths (*Yponomeuta*): Their host relations and evolution. *Annu. Rev. Entomol. 37*:41–66.

Mitter, C., D. J. Futuyma, J. C. Schneider, and J. D. Hare. 1979. Genetic variation and host plant relations in a parthenogenic moth. *Evolution 33*:770–790.

Moericke V., R. J. Prokopy, S. H. Berlocher, and G. L. Bush. 1975. Visual stimuli eliciting attraction of *Rhagoletis pomonella* (Diptera: Tephritidae) flies to trees. *Entomol. Exp. Appl. 18*:497–507.

Mopper, S., M. Beck, D. Simberloff, and P. Stiling. 1995. Local adaptation and agents of selection in a mobile insect. *Evolution 49*:810–815.

Owens, E. D. and R. J. Prokopy. 1986. Relationship between reflectance spectra of host plant surfaces and visual detection of host fruit by *Rhagoletis pomonella* flies. *Physiol. Entomol. 11*:297–307.

Papaj, D. R. and R. J. Prokopy. 1986. Phytochemical basis of learning in *Rhagoletis pomonella* and other herbivorous insects. *J. Chem. Ecol. 12*:1125–1143.

Papaj, D. R. and M. D. Rausher. 1983. Individual variation in host location by phytophagous insects. Pp. 77–124 *in* S. Ahmad (Ed.), *Herbivorous Insects: Host-Seeking Behavior and Mechanisms.* Academic Press, New York.

Patch, E. M. and W. C. Woods. 1922. The blueberry maggot in Washington County. Pp. 77–93 *in Bull. No. 308 of Maine Agricultural Exp. Station.* University of Maine, Orono, Maine.

Payne, J. A. and S. H. Berlocher. 1995a. Distribution and host plants of the blueberry maggot, *Rhagoletis mendax* (Diptera: Tephritidae) in Southeastern North America. *J. Kan. Entomol. Soc. 68*:133–142.

Payne, J. A. and S. H. Berlocher. 1995b. Phenological and electrophoretic evidence for a new blueberry-infesting species in the *Rhagoletis pomonella* sibling species complex. *Entomol. Exp. Appl. 75*:183–187.

Phipps, C. R. and C. O. Dirks. 1933. Notes on the biology of the apple maggot fly. *J. Econ. Entomol. 26*:349–358.

Porter, B. A. 1928. The apple maggot. U.S. Department of Agriculture Technical Bulletin. 66.

Price, P. W., C. E. Bouton, P. Gross, B. A. McPheron, J. N. Thompson, and A. E. Weis. 1980. Interactions among three trophic levels: Influence of plants on interactions between insect herbivores and natural enemies. *Annu. Rev. Ecol. Syst. 11*:41–65.

Prokopy, R. J. 1968a. Visual responses of apple maggot flies, *Rhagoletis pomonella*: Orchard studies. *Entomol. Exp. Appl. 11*:403–422.

Prokopy, R. J. 1968b. The influence of photoperiod, temperature and food on the initiation of diapause in the apple maggot. *Can. Entomol. 100*:318–329.

Prokopy, R. J. 1977. Attraction of *Rhagoletis* flies (Diptera: Tephritidae) to red spheres of different sizes. *Can. Entomol. 109*:593–596.

Prokopy, R. J., M. Aluja, and T. A. Green. 1987. Dynamics of host odor and visual stimulus interaction in host finding behavior of apple maggot flies. Pp. 161–166 *in* V. Labeyrie, G. Fabres, and D. Lachaise (Eds.), *Insects–Plants.* Junk, The Netherlands.

Prokopy, R. J., A. L. Averill, S. S. Cooley, and C. A. Roitberg. 1982a. Associative learning in egglaying site selection by apple maggot flies. *Science 218*:76–77.

Prokopy, R. J., A. L. Averill, S. S. Cooley, C. A. Roitberg, and C. Kallet. 1982b. Variation in host acceptance pattern in apple maggot flies. Pp. 123–129 *in* Proc. 5th Int. Symp. Insect–Plant Relationships. Pudoc, Wageningen.

Prokopy, R. J., E. W. Bennett, and G. L. Bush. 1971. Mating behavior in *Rhagoletis pomonella* (Diptera: Tephritidae): I. Site of assembly. *Can. Entomol. 103*:1405–1409.

Prokopy, R. J., E. W. Bennett, and G. L. Bush. 1972. Mating behavior in *Rhagoletis pomonella* (Diptera: Tephritidae): II. Temporal organization. *Can. Entomol. 104*:97–104.

Prokopy, R. J., S. R. Diehl, and S. S. Cooley. 1988. Behavioral evidence for host races in *Rhagoletis pomonella* flies. *Oecologia 76*:138–147.

Prokopy, R. J., V. Moericke, and G. L. Bush. 1973. Attraction of apple maggot flies to odor of apples. *Environ. Entomol. 2*:743–749.

Prokopy, R. J., D. R. Papaj, S. S. Cooley, and C. Kallet. 1986. On the nature of learning in oviposition site acceptance by apple maggot flies. *Anim. Behav. 34*:98–107.

Rausher, M. D. 1988. Is coevolution dead? *Ecology 69*:898–901.

Rausher, M. D. 1992. Natural selection and the evolution of plant–insect interactions. *In* B. D. Roitberg and M. B. Isman (Eds.), *Insect Chemical Ecology: An Evolutionary Approach.* Chapman & Hall, New York.

Reissig, W. H. and D. C. Smith. 1978. Bionomics of *Rhagoletis pomonella* in *Crataegus. Ann. Entomol. Soc. Am. 71*:155–159.

Rice, W. R. 1987. Speciation via habitat specialization. *Evol. Ecol. 1*:301–314.

Rice, W. R. and E. E. Hostert. 1993. Laboratory experiments on speciation: What have we learned in 40 years? *Evolution 47*:1637–1653.

Rosenzweig, M. L. 1978. Competitive speciation. *Biol. J. Linn. Soc. 10*:275–289.

Schluter, D. Ecological causes of speciation. *In* D. Howard and S. H. Berlocher (Eds.), *Endless Forms: Species and Speciation.* Oxford University Press, Oxford, UK., *in press.*

Smith, D. C. 1988a. Reproductive differences between *Rhagoletis* (Diptera: Tephritidae) fruit parasites of *Cornus amomum* and *C. florida* (Cornaceae). *J. NY Entomol. Soc. 96* (3):327–331.

Smith, D. C. 1988b. Heritable divergence of *Rhagoletis pomonella* host races by seasonal asynchrony. *Nature 336*:66–67.

Smith, D. C., S. A. Lyons, and S. Berlocher. 1993. Production and electrophoretic verification of F1 hybrids between the sibling species *Rhagoletis pomonella* and *Rhagoletis cornivora. Entomol. Exp. Appl. 69*:209–213.

Tauber, C. A. and J. J. Tauber. 1989. Sympatric speciation in insects: Perception and perspective. Pp. 307–344. *in* D. Otte and J. Endler (Eds.), *Speciation and Its Consequences.* Sinauer Associates, Sunderland, MA.

Thorpe, W. H. 1930. Biological races in insects and allied groups. *Biol. Rev. 5*:177–212.

Via, S. 1991. The genetic structure of host plant adaptation in a spatial patchwork: Demographic variability among reciprocally transplanted pea aphid clones. *Evolution 45*: 827–852.

Walsh, B. J. 1864. On phytophagous varieties and phytophagous species. *Proc. Entomol. Soc. Philadelphia 3*:403–430.

Walsh, B. J. 1867. The apple-worm and the apple maggot. *J. Hortic. 2*:338–343.

Waring, G. L., W. G. Abrahamson, and D. J. Howard. 1990. Genetic differentiation in the gallformer *Eurosta solidaginis* (Diptera: Tephritidae) along host plant lines. *Evolution 44*:1648–1655.

Wilson, D. S. 1989. The diversification of single gene pools by density- and frequency-dependent selection. Pp. 366–385 *in* D. Otte and J. A. Endler (Eds.), *Speciation and Its Consequences.* Sinauer Associates, Sunderland, MA.

Wilson, D. S. and M. Turelli. 1986. Stable underdominance and the evolutionary invasion of empty niches. *Am. Nat. 127*:835–850.

Wood, T. K. and M. C. Keese. 1990. Host-plant-induced assortative mating in *Enchenopa* treehoppers. *Evolution 44*:619–628.

Wood, T. K., K. C. Olmstead, and S. I. Guttman. 1990. Insect phenology mediated by host-plant water relations. *Evolution 44*:629–636.

Woods, W. C. 1915. Blueberry insects in Maine. Pp. 249–292. *In Bull. 244 of Maine Agricultural Exp. Station.* University of Maine, Orono, Maine.

# Index